Maurizio Pandolfi
Renzo Piva (Eds.)

Proceedings of the Fifth GAMM-Conference on Numerical Methods in Fluid Mechanics

D1749477

Notes on Numerical Fluid Mechanics
Volume 7

Series Editors: Ernst Heinrich Hirschel, München
Maurizio Pandolfi, Torino
Arthur Rizzi, Stockholm
Bernard Roux, Marseille

Volume 1 Boundary Algorithms for Multidimensional Inviscid Hyperbolic Flows (Karl Förster, Ed.)

Volume 2 Proceedings of the Third GAMM-Conference on Numerical Methods in Fluid Mechanics (Ernst Heinrich Hirschel, Ed.) (out of print)

Volume 3 Numerical Methods for the Computation of Inviscid Transonic Flows with Shock Waves (Arthur Rizzi/Henri Viviand, Eds.)

Volume 4 Shear Flow in Surface-Oriented Coordinates (Ernst Heinrich Hirschel/Wilhelm Kordulla)

Volume 5 Proceedings of the Fourth GAMM-Conference on Numerical Methods in Fluid Mechanics (Henri Viviand, Ed.) (out of print)

Volume 6 Numerical Methods in Laminar Flame Propagation (Norbert Peters/Jürgen Warnatz, Eds.)

Volume 7 Proceedings of the Fifth GAMM-Conference on Numerical Methods in Fluid Mechanics (Maurizio Pandolfi/Renzo Piva, Eds.)

Volume 8 Vectorization of Computer Programs with Application to Computational Fluid Dynamics (Wolfgang Gentzsch) (in preparation)

Manuscripts should have well over 100 pages. As they will be reproduced fotomechanically they should be typed with utmost care on special stationary which will be supplied on request. In print, the size will be reduced linearly to approximately 75 %. Figures and diagrams should be lettered accordingly so as to produce letters not smaller than 2 mm in print. The same is valid for handwritten formulae. Manuscripts (in English) or proposals should be sent to the general editor Prof. Dr. E. H. Hirschel, MBB-UFE 122, Postfach 80 11 60, D-8000 München 80.

Maurizio Pandolfi/Renzo Piva (Eds.)

Proceedings of the Fifth GAMM-Conference on Numerical Methods in Fluid Mechanics

Rome, October 5 to 7, 1983

With 263 Figures

Friedr. Vieweg & Sohn Braunschweig/Wiesbaden

CIP-Kurztitelaufnahme der Deutschen Bibliothek

Conference on Numerical Methods in Fluid Mechanics:
Proceedings of the ... GAMM Conference on Numerical Methods in Fluid Mechanics. – Braunschweig; Wiesbaden: Vieweg

NE: Gesellschaft für Angewandte Mathematik und Mechanik; HST

5. Rome, October 5 to 7, 1983. – 1984.
(Notes on numerical fluid mechanics; Vol. 7)
ISBN 3-528-08081-7

NE: GT

Prof. Karl Förster, Stuttgart, laid down the editorship for private reasons. The Vieweg Verlag and the new editors would like to thank for his vast and untiring efforts in establishing the series.

All rights reserved
© Friedr. Vieweg & Sohn Verlagsgesellschaft mbH, Braunschweig 1984

No part of this publication may be reproduced, stored in a retrieval system or transmitted mechanically, by photocopies, recordings or otherwise, without prior permission of the copyright holder.

Produced by IVD, Industrie- und Verlagsdruck, Walluf
Printed in Germany

ISBN 3-528-08081-7

FOREWORD

The GAMM-Committee for Numerical Methods in Fluid Mechanics (GAMM-Fachausschuss für Numerische Methoden in der Strömungsmechanik) organizes the GAMM-CONFERENCE ON NUMERICAL METHODS IN FLUID MECHANICS every two years.
The previous four Conferences were held at the DFVLR in Köln (1975-77-79) and at ENSTA in Paris (1981). The fifth Conference was held at the University of Rome, October 5-7, 1983.
The GAMM-Conference is intended to bring together scientists who are working on numerical methods in fluid mechanics. The main objective is to foster exchanges between the various fields of development of computational fluid mechanics such as Aerodynamics, Hydrodynamics, Propulsion, Fluidmachinery, Nuclear Reactor Technology, Meteorology, Biofluidmechanics etc.
The subjects covered in the Conference are mainly related to theoretical aspects of numerical methods in fluid mechanics (finite difference methods, finite element methods, spectral methods, etc.) or to particular applications to fluid problems which may enhance the novelties of the methods themselves.
Moreover reports are presented on the GAMM-WORKSHOPS promoted by the Committee where very definite subjects have been investigated by scientists working in those particular fields.
The 1983 Conference was attended by more than 100 scientists from 16 different countries. There were 48 contributed papers and the activity on 4 GAMM-Workshops have been reported. The contributions are here presented in alphabetical order according to the first author.
The editors, who have also been the chairmen of this Conference, would like to acknowledge the support from the Faculty of Engineering of the University of Rome and the Italian National Research Council (C.N.R.) and to express their gratitude to all colleagues and personnel of the University of Rome and the Politechnic Institute of Turin for the cooperation in organizing the Conference.

December 5, 1983

Maurizio PANDOLFI

Renzo PIVA

CONTENTS

SHORT REPORTS ON GAMM WORKSHOPS

ON THE NUMERICAL SOLUTION OF THE NAVIER STOKES EQUATIONS FOR INTERNAL INCOMPRESSIBLE FLOWS IN THE PRESENCE OF FILTRATING WALLS

R. Albanese, F. Grasso, and C. Meola
Istituto di Gasdinamica
P.le Tecchio 80, Napoli, Italy, 80125

SUMMARY

A numerical algorithm has been developed to model a non standard boundary value Navier Stokes problem. The method is a variation of a scheme developed by the authors, and successfully applied to the steady state problem of two-dimensional incompressible laminar flow confined by permeable walls. Such a method saves the implicit character of the pressure/velocity correlation on the permeable boundary, thus yielding an accurate description of the transient evolution of the phenomena. Moreover it reduces the stiffness of the pressure matrix. The latter property suggests that the model can be applied as a regularization process for non permeable walls (provided that the permeability constant approaches zero), leading to the concept of "artificial permeability".

INTRODUCTION

The numerical solution of viscous incompressible laminar flows, confined by permeable walls, was recently studied in a primitive variable formulation by the present authors [1]. The particular boundary conditions imposed along such walls (normal suction/injection velocity assumed to be proportional to the pressure jump across the permeable boundary) introduced a strong coupling between velocity and pressure fields. The implicit character of the problem was effectively bypassed by assuming a sort of delay time between pressure jump and velocity without affecting the steady state solution. However the above approach with the assumed explicit pressure/velocity correlation is not adequate to study the transient of the flow evolution and is not suitable for an implicit numerical solution of the equations.

In the present work a modified algorithm has been developed by implicitly treating the coupling between pressure and velocity along the permeable walls, so as to satisfy the implicit character of the particular boundary value problem. The proposed algorithm yields meaningful detailed informations during the transient of the phenomena.

A careful analysis of the physical and mathematical correlations between boundary conditions, continuity properties of the solution, and the proposed numerical discretization, has shown that the present treatment of this non standard boundary value Navier Stokes (BVNS) problem can also be exploited for non permeable walls. This seems to lead to the concept of "artificial permeability", in analogy with other regularization and/or optimization techniques as the "artificial compressibility" of Chorin, the "artificial viscosity" of V.Neuman etc.

The method has been tested comparing the results with the ones obtained by the approach of Ref. [1] . In the present work the effects of the gravity forces and the exit velocity and pressure boundary conditions on the flow field have also been studied. Finally the applicability of the "artificial permeability" model has been tested for a driven cavity flow where the velocity boundary conditions are exactly known.

THE MODEL

The model equations are:

$$\left.\begin{aligned} \underline{v}_t &= \underline{R} - \underline{\nabla} p + \underline{f} \\ \underline{\nabla}\cdot\underline{v} &= 0 \end{aligned}\right\} \quad \text{in } \Omega \qquad (1)$$

where $\underline{f}$ represents the mass forces, and $\underline{R} = - \underline{v}\cdot \underline{\nabla}\,\underline{v} + \nabla^2\underline{v}\,/Re$.

Boundary conditions are:

$$\underline{v} = \underline{v}_1 \qquad \text{on } B\Omega_1 \qquad (2)$$

$$\left.\begin{aligned} \underline{s}\cdot\underline{v} &= v_{s2} \\ p &= p_2 \end{aligned}\right\} \quad \text{on } B\Omega_2 \qquad (3)$$

$$\left.\begin{aligned} \underline{n}\cdot\underline{v} &= kp \\ \underline{s}\cdot\underline{v} &= 0 \end{aligned}\right\} \quad \text{on } B\Omega_3 \qquad (4)$$

The strong coupling between injection/suction velocity and pressure, introduced by Eqn. (4) shows that the pressure field cannot be determined only by a constant. Moreover if the effects of conservative external forces are included in the pressure potential, then Eqns. (3)-(4) must be consequently modified.

NUMERICAL SOLUTION

According to the conclusions of Ref. [1] , the discretized governing equations have been obtained following a finite volume approach. For rectangular geometries the grid points have been evenly spaced in x and y with mesh size $\Delta x = \Delta y$. The velocity components have been defined at the grid nodes, the pressure at the center of the geometric cell.

For internal momentum cell (i,j) (centered around a velocity node) the equations are:

$$u^{np}_{ij} = u^{n}_{ij} -\alpha((u^{n}_{ipj}+u^{n}_{ij})^{2}-(u^{n}_{ij}+u^{n}_{imj})^{2}+(u^{n}_{ijp}+u^{n}_{ij})(v^{n}_{ijp}+v^{n}_{ij})-(u^{n}_{ij}+u^{n}_{ijm})(v^{n}_{ij}+v^{n}_{ijm}))+$$

$$+\beta(u^{n}_{ipj}+u^{n}_{imj}-4u^{n}_{ij}+u^{n}_{ijp}+u^{n}_{ijm})-2\alpha(p^{np}_{ij}+p^{np}_{ijm}-p^{np}_{imj}-p^{np}_{imjm}) \qquad (5)$$

$$v^{np}_{ij} = v^{n}_{ij} -\alpha((u^{n}_{ipj}+u^{n}_{ij})(v^{n}_{ipj}+v^{n}_{ij})-(u^{n}_{ij}+u^{n}_{imj})(v^{n}_{ij}+v^{n}_{imj})+(v^{n}_{ijp}+v^{n}_{ij})^{2}-(v^{n}_{ij}+v^{n}_{ijm})^{2})+$$

$$+\beta(v^{n}_{ipj}+v^{n}_{imj}-4v^{n}_{ij}+v^{n}_{ijp}+v^{n}_{ijm})-2\alpha(p^{np}_{ij}+p^{np}_{imj}-p^{np}_{ijm}-p^{np}_{imjm}) \qquad (6)$$

For every mass cell but the ones along the permeable walls, the discretized conservation equation is:

$$u^{np}_{ipj}+u^{np}_{ipjp}-u^{np}_{ij}-u^{np}_{ijp}+v^{np}_{ijp}+v^{np}_{ipjp}-v^{np}_{ij}-v^{np}_{ipj} = 0 \qquad (7)$$

On cells adjacent to the permeable walls Eqn. (7) becomes:

$$u^{np}_{ipj}+u^{np}_{ipjp}-u^{np}_{ij}-u^{np}_{ijp}+2k_{i}\,p^{np}_{ij}-v^{np}_{ipj}-v^{np}_{ij} = 0 \qquad (8)$$

Boundary conditions are:

$$\underline{v}^{np}_{ij} = \underline{v}^{np}_{1ij} \qquad \text{on } B\Omega_1 \qquad (9)$$

$$\underline{s}\cdot\underline{v}^{np}_{ij} = vs^{np}_{2ij}, \quad p^{np}_{ij} = p^{np}_{2ij} \qquad \text{on } B\Omega_2 \qquad (10)$$

$$u^{np}_{ij} = 0, \quad v^{np}_{ij} = (k_{im}\,p^{np}_{imj}+k_{i}\,p^{np}_{ij})/2 \qquad \text{on } B\Omega_3 \qquad (11)$$

From Eqns. (7)-(8) observe the different discretization of the divergence operator along the permeable walls (consistent with a different definition of the mass flux through such boundaries). Furthermore note the implicit treatment of the filtrating boundary conditions.

In a quasi matrix form the governing equations are:

$$\begin{pmatrix} I & -\Delta t B^{T} & 0 \\ B & B_{pm} & 0 \\ 0 & M_{p} & I \end{pmatrix}\cdot\begin{pmatrix} \underline{v} \\ p \\ \underline{v}_{m} \end{pmatrix}^{np}=\begin{pmatrix} I+\Delta t A & 0 & A \\ 0 & 0 & 0 \\ 0 & 0 & 0 \end{pmatrix}\cdot\begin{pmatrix} \underline{v} \\ p \\ \underline{v}_{m} \end{pmatrix}^{n}+\begin{pmatrix} \underline{c}_{1} \\ c_{2} \\ 0 \end{pmatrix} \qquad (12)$$

where $\underline{v}_m$ is the injection/suction velocity, and $\underline{c}_1$, c_2 account for boundary conditions and external forces. A, B and $-B^T$ represent respectively the discretized $\underline{R}$, divergence and gradient operators (note that the adjoint character of B and B^T is maintained for the assumed discretization). The definitions of B_{pm} and M_p ("membrane flux" and "membrane permeability" matrices) follow from Eqns. (8),(11).

The solution of the system (12) requires the simultaneous solution of pressure and velocity. Premultiplying Eqn. (12) by the non singular matrix T, defined as:

$$T = \begin{pmatrix} I & 0 & 0 \\ -B & I & 0 \\ 0 & 0 & I \end{pmatrix}$$

the following equation for p is obtained:

$$(\Delta t B \cdot B^T + B_{pm}) \cdot p^{np} \equiv M \cdot p^{np} = -B \cdot (I + \Delta t A) \cdot \underline{v}^n + c_2' \equiv Q \qquad (13)$$

From the definitions of B, B^T and B_{pm}, M is shown to satisfy the following properties [2] : i) it is symmetric and positive definite; ii) with an appropriate reordering (i+j even/odd) it can be reduced to a two block diagonal matrix; iii) property A; iv) weak diagonal dominance (with strong character for the rows along the permeable boundary).

Each of the two block matrices (M', M'') satisfies properties i),iii) and iv), hence is 2-cyclic in the sense of Varga [2] .

From definition of the pressure matrix M, Eqn. (13) is shown to be formally consistent with the usual elliptic equation for p

$$\nabla^2 p = \underline{\nabla} \cdot (\underline{R} + \underline{f}) - \underline{\nabla} \cdot \underline{v}_t \qquad (14)$$

generally obtained by taking the divergence of the momentum equation [3] - [4] .

The closure of Eqn. (14) can be obtained by assuming that $\underline{v}$ satisfies some smoothness properties so that Eqn. (1) can be extended to the boundary by a limit process, thus yielding boundary conditions in terms of pressure gradient [1] , [5] , [6]. However such a naïve procedure may lead to paradoxes especially if pressure tangential derivatives are deduced [1] . Eqns. (9)-(11) seemingly allow the closure of the elliptic pressure equation yielding Neumann conditions on $B\Omega_1$, Dirichlet conditions on $B\Omega_2$, and Robin b.c. on $B\Omega_3$. Moreover the strong solution of Eqn. (14) implies $p \in C^2$, while Eqn. (1) only requires $p \in C^1$.

The solution of Eqn. (13) does not require such smoothness assumptions and it bypasses the whole closure problem for the differential BVP formulation by simply imposing:

$$\int_{\partial\Sigma} \underline{n} \cdot \underline{v} \, dS = 0 \qquad \forall \; \Sigma \subseteq \Omega$$

in a discretized form.

For an internal cell (i,j), Eqn. (13) yields:

$$p^{np}_{imjm} + p^{np}_{ipjp} - 4p^{np}_{ij} + p^{np}_{imjp} + p^{np}_{ipjm} = 2\Delta x^2 q^n_{ij} \qquad (15)$$

For a cell adjacent to a non permeable boundary, Eqn. (13) gives:

$$p^{np}_{ipjm} - 2p^{np}_{ij} + p^{np}_{imjm} = \Delta x\ q^n_{ij} \qquad (16)$$

For a cell along $B\Omega_3$ one has:

$$p^{np}_{ipjm} - (1 + k_i\ \Delta x/\Delta t)p^{np}_{ij} + p^{np}_{imjm} = \Delta x\ q^n_{ij} \qquad (17)$$

Eqn. (17) shows the effect on the structure of $\mathcal{M}$ due to the implicit treatment of the b.c. and the particular definition of the numerical divergence operator (Eqns. (8),(11)). Such an equation is consistent with a Robin type b.c. for the differential equation for p; i.e. :

$$k\ p_t + p_n = \underline{n} \cdot (\underline{R} + \underline{f})$$

Such a boundary condition reduces the computational effort to obtain the pressure field with respect to the standard BVNS problem. The advantage of employing the above formulation for non permeable walls thus follows. In this case the "differential" b.c. for p would be:

$$(kp)_t + p_n = \underline{n} \cdot (\underline{R} + \underline{f})$$

with k approaching zero as t increases.

In such a case the algorithm can be interpreted as an iterative method yielding the correct non permeable steady state solution. In other words the concept of the "artificial permeability" (AP) can be viewed as an artificial compressibility limited to the cells adjacent to the solid boundary and vanishing at steady state.

RESULTS AND DISCUSSIONS

The incompressible laminar NS equations with non standard b.c. have been solved by using a finite difference algorithm that saves the implicit character of the problem (due to the incompressibility and the particular pressure/velocity correlation), guaranteeing the mass conservation for every computational cell.

For the chosen staggering (with velocity components at the geometrical nodes, and pressure unknowns at the center of the mass cells) both velocity components can be assigned at the boundaries. It can be shown that such a grid configuration corresponds to the overlapping of two grids with the classical staggering [3] , having mesh sizes $\sqrt{2}\Delta x$, and grid lines rotated of 45° with respect to x and y. Consequently the method yields two uncoupled systems of equations for the pressure by separating the p unknowns in even and odd ones. The coupling of the fluiddynamic field is obtained by an appropriate discretization of the momentum flux $\underline{R}$. Moreover proper care must be taken for the assignement of the b.c., since a well posed problem must be imposed on each of the two "overlapping" grids.

The model has been tested to describe the flow motion in the presence of permeable walls for a variety of geometries and operating conditions. Figs. 1.a-e show the effects of different values of the exit pressure for given inlet mass flow rate, Reynolds number (Re=10), and filtrating constant. Observe that the membrane flux does vary linearly with the exit pressure. Moreover for the selected Re (Figs. 1.a-c) by simply imposing $\underline{s}\cdot\underline{v} = 0$ and p = constant at the exit, the Poiseuille flow is recovered in the outlet region. Figs. 2-3 show the computed results in forced percolators. The effects of gravity are illustrated in Figs. 3.b-c; the differences on the velocity field are due to the particular boundary conditions (Eqn. (4)). The results in a "shear filtrating pump" configuration are plotted in Fig . 4 . Finally Fig. 5 shows the application of the method to investigate the validity of the AP concept in a driven cavity flow configuration. The AP method is equivalent to a regularization one, yielding an accurate description of the steady state when k approaches zero.

In conclusions the applicability of the proposed method to calculate a variety of transient and steady flow configurations has been shown, even in the presence of external forces. The concept of artificial permeability has been introduced (with some analogies with the artificial compressibility); however the advantages of using it depend on the filtrating law k(t) and on the number of grid points.

REFERENCES

[1] Albanese, R., Grasso, F., Meola, C., "On the critical problem of f.d. pressure treatment for laminar flows confined by permeable walls", submitted for publication to Int. Jour. of Numerical Methods in Engineering, and to be presented at Third International Conference on Numerical Methods in Laminar and Turbulent Flows, Seattle, Washington, August 8-11, 1983.

[2] Young, D., "Iterative solution of large linear systems", Acad. Press, 1971.

[3] Harlow, F.H., Welch, J.A., "Numerical calculation of time dependent viscous incompressible flow of fluid with a free surface", Phys. of Fluids, 8 (1965), pp 2182-2189.

[4] Chorin, A.J., "A numerical method for solving incompressible viscous flow problems", Jour. of Comp. Phys., 2 (1967), pp 12-26.

[5] Temam, R., "Navier Stokes equations", North Holland, 1979.

[6] Moin, P., Kim, J., "On the numerical solution of time dependent viscous incompressible fluid flows involving solid boundaries", Jour. of Comp. Phys., 35 (1980), pp 381-392.

(a) $p_{ex} = 50$

(b) $p_{ex} = 45$

(c) $p_{ex} = 40$

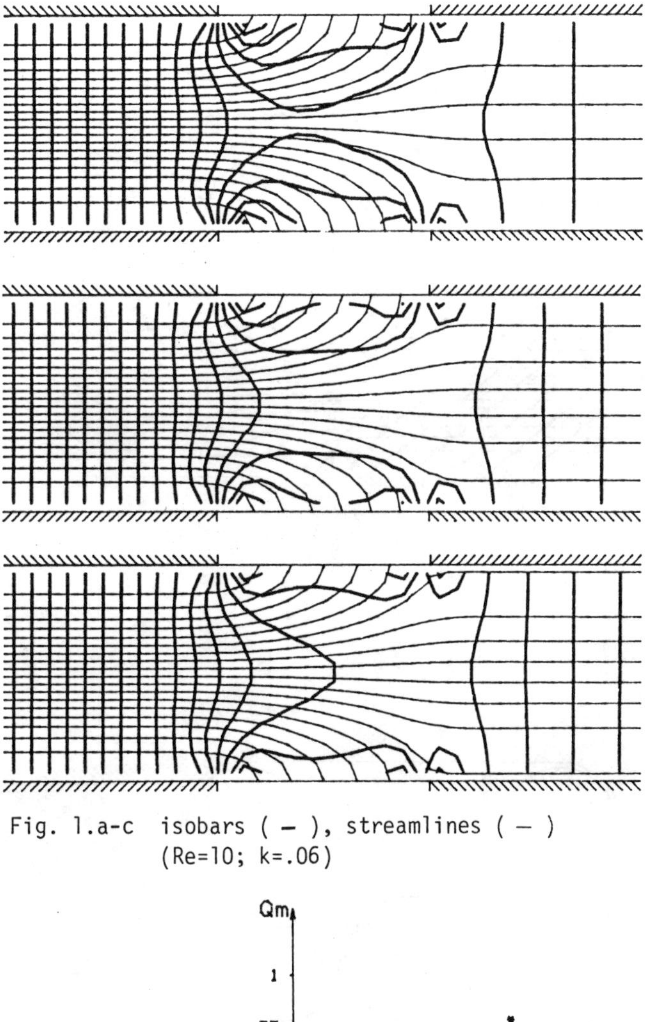

Fig. 1.a-c isobars (—), streamlines (—)
(Re=10; k=.06)

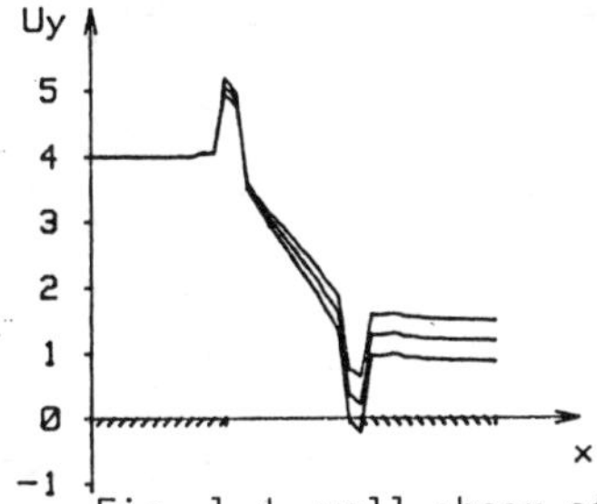

Fig. 1.d wall shear stress

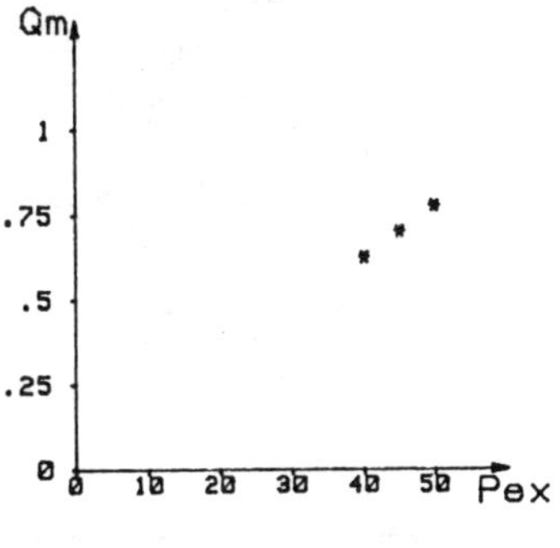

Fig. 1.e membrane mass flow rate

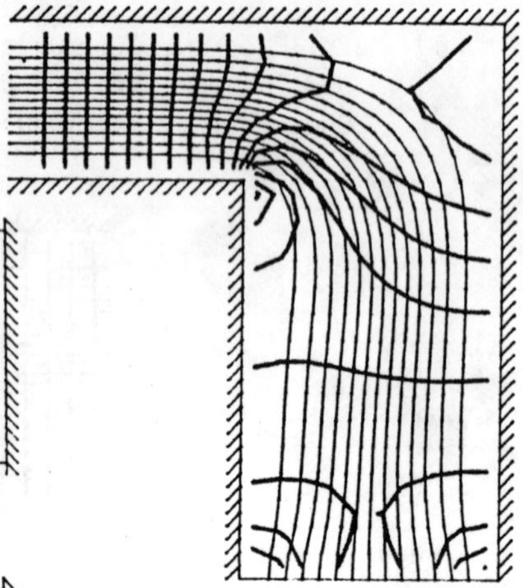

Fig. 2 (Re=10; k=.02)

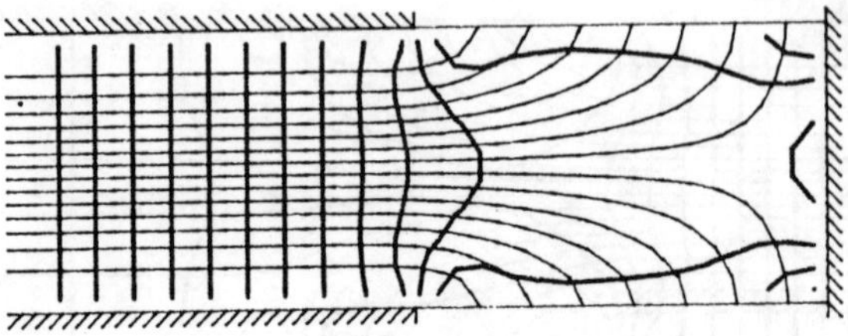

Fig. 3.a Fr=∞ (Re=10; k=.02)

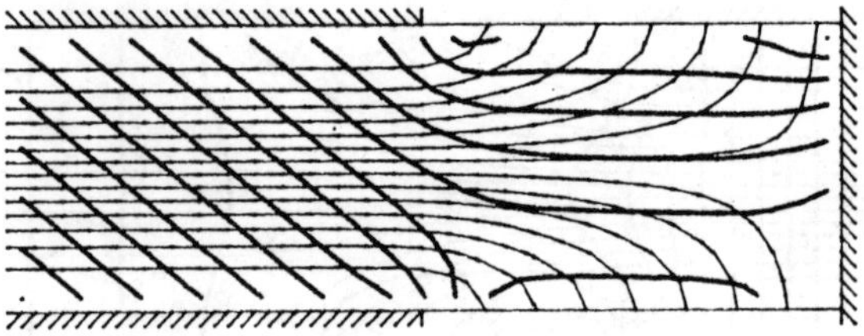

Fig. 3.b Fr=.1

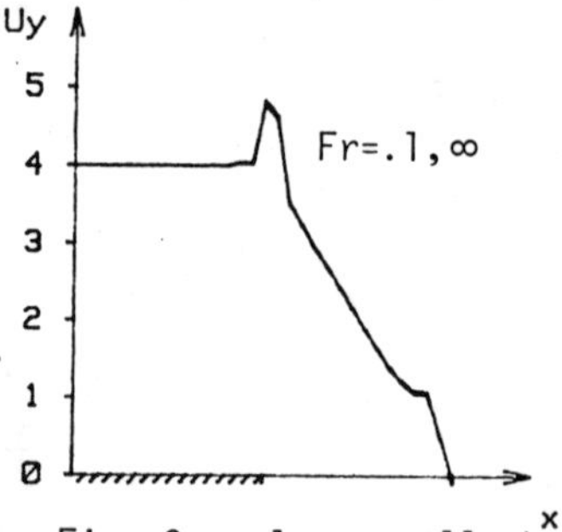

Fig. 3.c lower wall shear stress

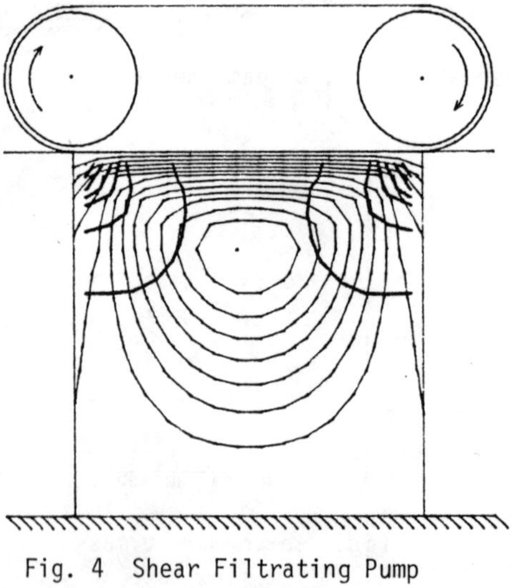

Fig. 4 Shear Filtrating Pump (Re=10; k=.02)

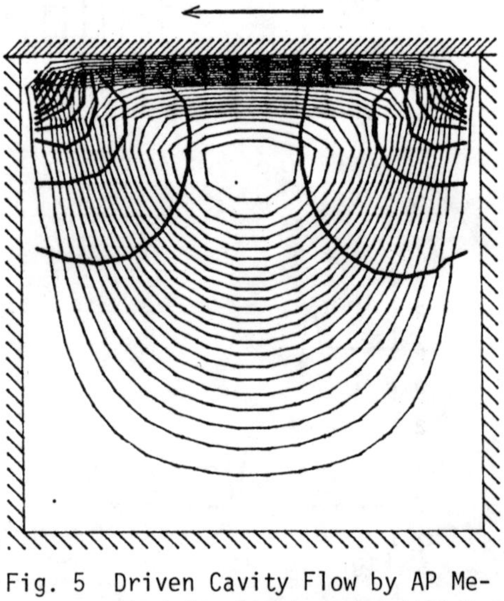

Fig. 5 Driven Cavity Flow by AP Method (Re=10; k=.02exp(-3t/1-3t))

NUMERICAL SOLUTIONS OF A TSL-MODEL FOR FREE-SURFACE FLOWS

Helge I. Andersson
Institutt for mekanikk
Norwegian Institute of Technology
N-7034 Trondheim - NTH, Norway

SUMMARY

A finite-difference procedure is presented for calculating numerical solutions to a Thin-Shear-Layer (TSL) model for steady two-dimensional free-surface flows. The parabolic system of equations is solved by an implicit two-point central-difference method employed in an iterative scheme, and the normalstress condition on the free-surface is satisfied by successive approximations to the unknown free-surface position. Solutions are presented for both gravity-driven laminar film flow along a sloping wall and for highly supercritical hydraulic (turbulent) flow on an adverse incline. Comparisons are made with experimental results and simple integral method solutions.

INTRODUCTION

Free surface flows in open channels or along sloping walls represent interesting cases in viscous fluid flow, with various technological applications. Liquid films flowing on steep slopes are encountered in many types of heat and mass transfer equipment, such as in distillation columns, wetted wall towers and falling film evaporators. The supercritical flow under a sluice gate or over an overflow spillway are typical subjects in open-channel hydraulics. In all these problems the flow is governed by a set of partial differential equations together with boundary conditions specified on the known part of the boundary and on the free surface of the flow, which moves in a way that depends on the solution of the differential equations. Thus, a crucial element in any numerical solution procedure lies in the determination of the free surface position. This difficulty can be surmounted by the use of a transformation of the independent variables (e.g. the von Mises transformation) which maps the region between the wall and the free surface into a fixed rectangular domain. This type of approach was applied to laminar film flow problems by Cerro and Whitaker [1], Yilmaz and Brauer [2] and recently by Bertschy et al. [3].

A different approach to free surface flow problems is the "rigid lid" approximation recently used by Rastogi and Rodi [4] and Leschziner and Rodi [5]. They treated the pressure as an unknown variable, and replaced the free surface by a fictious plane boundary parallel to the bottom. As a result of this constraint on the flow, a nonzero pressure gradient was predicted at the surface, and this pressure gradient accounted for and was interpreted as the surface slope.

The purpose of the present paper is to present a numerical procedure for obtaining solutions to a steady, two-dimensional thin-shear-layer model for free surface flows. In this approach the pressure is treated as an unknown variable, and the predicted surface pressure is used to obtain successive approximations to the free surface position at each streamwise station, until a prescribed normal-stress condition is satisfied. Other important features of the present solution method are: 1) Solutions are obtained at discrete points on a fixed grid and no coordinate transformation is required, 2) the differential equations are solved using the efficient Keller Box scheme [6] for parabolic equations, and 3) an algebraic eddy viscosity model can be included to account for turbulent shear stresses.

BASIC EQUATIONS

We consider the gravity-driven two-dimensional free-surface flow of liquid along an inclined wall, as depicted in Fig.1. It is assumed that the liquid surface is smooth and waveless, and that the fluid motion can be properly described by a Thin-Shear-Layer approximation [7] to the steady Navier-Stokes equations:

$$\frac{\partial u}{\partial x} + \frac{\partial v}{\partial y} = 0 \tag{1}$$

$$u\frac{\partial u}{\partial x} + v\frac{\partial u}{\partial y} = -\frac{1}{\rho}\frac{\partial p}{\partial x} + g\cdot\sin\alpha + \frac{1}{\rho}\frac{\partial \tau}{\partial y} \tag{2}$$

$$0 = -\frac{1}{\rho}\frac{\partial p}{\partial y} - g\cdot\cos\alpha \tag{3}$$

$$u(x,y) = v(x,y) = 0 \qquad \text{at} \quad y = 0 \tag{4}$$

$$p(x,y) = p_o \quad , \quad \frac{\partial u}{\partial y} = 0 \qquad \text{at} \quad y = h(x) \tag{5}$$

$$\int_o^{h(x)} u(x,y)\,dy = Q \tag{6}$$

Here, equations (1-3) represent the conservation of mass and momentum components, respectively. The boundary conditions (4) and (5) are the usual impermeability and no-slip conditions imposed at the wall, and the approximated normal and tangential stress conditions on the unknown free surface. Furthermore, conservation of the flow rate Q gives the additional condition (6).

A distinct feature of this TSL-model is that the x-momentum equation (2) includes the streamwise pressure gradient term, which has been neglected in most of the previous papers on laminar film flow, e.g. [1,2,8]. This term couples the cross-stream pressure-gravity balance (3) with the streamwise momentum balance (2) through the free surface boundary conditions, thus allowing for an important viscous-gravity interaction.

In the case of laminar flow of a Newtonian liquid film, the shear stress is related to the streamwise velocity gradient by the relation $\tau = \mu(\partial u/\partial y)$, and the present model becomes essentially identical to that of Bertschy et al. [3]. In dimensionless form the model equations then become:

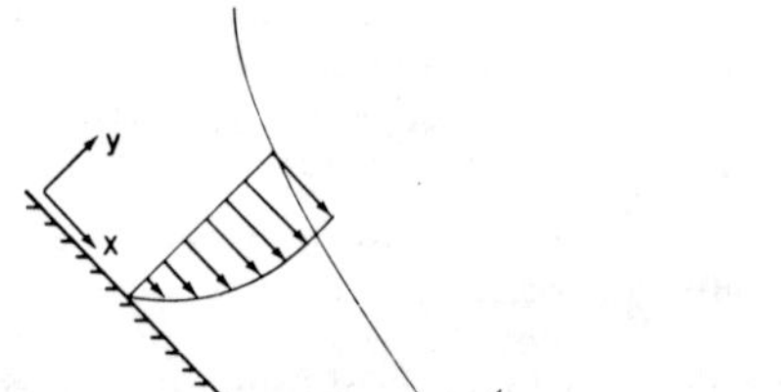

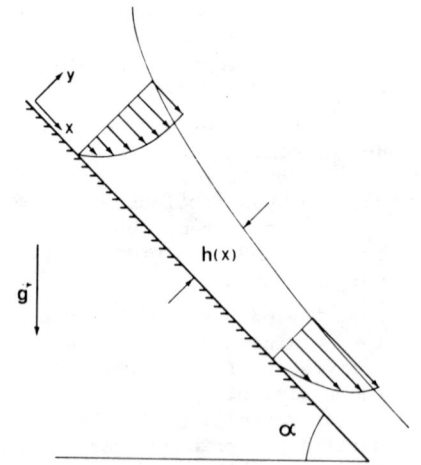

Fig. 1 Flow configuration

$$f''' = f' \frac{\partial f'}{\partial \xi} - f'' \frac{\partial f}{\partial \xi} + \frac{\partial P}{\partial \xi} - 3 \tag{7}$$

$$P' = - F_\infty^{-2} \tag{8}$$

$$f(\xi,0) = f'(\xi,0) = f''(\xi,\delta) = P(\xi,\delta) = 0 \quad ; \quad f(\xi,\delta) = 1 \tag{9}$$

where

$$\xi = \frac{x}{h_\infty Re} \quad , \quad \eta = \frac{y}{h_\infty} \tag{10}$$

$$\bar{u} = \frac{u}{u_\infty} = \frac{\partial f}{\partial \eta} \quad , \quad \bar{v} = \frac{v}{u_\infty} . Re = - \frac{\partial f}{\partial \xi} \quad , \quad P = \frac{p - p_o}{\rho u_\infty^2} \tag{11}$$

$$u_\infty = g \sin\alpha \cdot h_\infty^2/3\nu \quad , \quad h_\infty = (3\nu Q/g\sin\alpha)^{1/3} \tag{12}$$

$$Re = Q/\nu \quad , \quad F_\infty^2 = u_\infty^2/(g\cos\alpha h_\infty) = \frac{1}{3} Re \cdot \tan\alpha \tag{13}$$

and u_∞ and h_∞ are the downstream asymptotic mean velocity and flow depth, respectively. f is a dimensionless stream function, Re is the Reynolds number and F_∞ is the asymptotic Froude number of the flow. δ denotes the unknown value of the dimensionless cross-stream coordinate η at the free surface, i.e. $\delta(x) = h(x)/h_\infty$, and the primes denote differentiation with respect to η.

SOLUTION METHOD AND FINITE-DIFFERENCE EQUATIONS

The present TSL-model is similar to a steady two-dimensional boundary layer model, but the streamwise pressure gradient $\partial P/\partial \xi$ in eq. (7) cannot be prescribed as in classical boundary layer theory. We therefore treat $P(\xi,\eta)$ as an unknown function, in accordance with the Mechul function approach [9,10] introduced by Cebeci and Keller. Traditionally, the parabolic system of equations (7-8) can be solved when initial values are given at the upstream cross-section bounding the calculation domain, and four conditions are prescribed on known boundaries. However, in the present problem, as in a moving boundary problem, an additional boundary condition is required in order to find the position of the free surface. The solution of the free surface problem can therefore be obtained by an iterative solution procedure, in which the following sequence of calculation steps is performed at each streamwise position ξ^n:

1. Let $\delta^{(0)}$ be an initial guess or estimate for the apriori unknown free surface position $\delta(\xi^n)$.

2. Use any finite-difference scheme for parabolic systems to solve equations (7-8) with four of the boundary conditions (9) imposed on the wall $\eta = 0$ and on $\eta = \delta^{(0)}$, while the remaining normal-stress condition $P(\xi,\delta) = 0$ is not involved at this stage of the solution procedure.

3. The predicted surface pressure $P(\xi^n,\delta^{(i)})$ is then interpreted as a surface elevation, from which an improved estimate for $\delta(\xi^n)$ can be calculated:

$$\delta^{(i+1)} = \delta^{(i)} - \Phi \cdot F_\infty^2 \cdot P(\xi^n,\delta^{(i)}) \tag{14}$$

where the dimensionless parameter Φ is adjusted by linear interpolation to improve the convergence.

4. Steps 2 and 3 are repeated until the predicted surface pressure has become smaller than a prescribed value, i.e.

$$|P(\xi^n,\delta^{(i)})| \leqslant \varepsilon\, P(\xi^n,0) \tag{15}$$

where ε is taken as 10^{-3}.

In order to obtain numerical solutions of the nonlinear differential equations (7-8), i.e. step 2 above, we employ the Keller Box scheme [6]. First, we write eqs. (7-8) as a system of first-order equations, introducing new dependent variables $g(\xi,\eta) = \partial f/\partial\eta$ and $q(\xi,\eta) = \partial g/\partial\eta = \partial^2 f/\partial\eta^2$. Let f_j^n, g_j^n, q_j^n and P_j^n denote the values of the unknowns at the discrete net points

$$\xi^o = 0 \quad ; \quad \xi^n = \xi^{n-1} + k^n \qquad n = 1,2,\ldots N \tag{16}$$

$$\eta_o = 0 \quad ; \quad \eta_j = \eta_{j-1} + d_j \qquad j = 1,2,\ldots H \tag{17}$$

For laminar flow calculations equal cross-stream grid intervals $d_j = d$ is used. However, an unequal grid spacing is usually required next to the free surface, i.e. $0 < d_H \leqslant d$, so that the surface is located on the H'th grid point. Using central-difference approximations to the derivatives, the finite-difference approximations to the model equations (7-9) become

$$(f_j^n - f_{j-1}^n)/d_j = \frac{1}{2}(g_j^n + g_{j-1}^n) \tag{18}$$

$$(g_j^n - g_{j-1}^n)/d_j = \frac{1}{2}(q_j^n + q_{j-1}^n) \tag{19}$$

$$\begin{aligned}(q_j^n - q_{j-1}^n)/d_j &= -(q_j^{n-1} - q_{j-1}^{n-1})/d_j - 6 \\ &+ \gamma\left[-(g^2)_{j-\frac{1}{2}}^{n-1} + (qf)_{j-\frac{1}{2}}^{n-1} - 2P_{j-\frac{1}{2}}^{n-1}\right] \\ &+ \gamma\left[f_{j-\frac{1}{2}}^{n-1}\, q_{j-\frac{1}{2}}^{n} - q_{j-\frac{1}{2}}^{n-1}\, f_{j-\frac{1}{2}}^{n} + 2P_{j-\frac{1}{2}}^{n} + (g^2)_{j-\frac{1}{2}}^{n} + (fq)_{j-\frac{1}{2}}^{n}\right]\end{aligned} \tag{20}$$

$$(P_j^n - P_{j-1}^n)/d_j = -F_\infty^{-2} \tag{21}$$

$$f_o^n = g_o^n = q_H^n = P_H^n = 0 \quad ; \quad f_H^n = 1 \tag{22}$$

where $\gamma = 1/k^n$ and the notation

$$f_{j-\frac{1}{2}} = \frac{1}{2}(f_j + f_{j-1}) \tag{23}$$

is employed for quantities midway between netpoints. According to the Box scheme, we apply Newton's method to obtain solutions to the nonlinear system of difference equations (18-22). The resulting linearized system of 4H+4 algebraic equations exhibits a block tridiagonal structure, the blocks consisting of 4x4 matrices. This system can be solved by the block tridiagonal factorization scheme [11] due to Keller.

COMPUTATION OF TURBULENT FLOWS

The present solution procedure has recently been applied to the study of highly supercritical turbulent flows on an adverse incline, as reported in [7,12]. In the case of a turbulent flow field the governing equations (1) - (6) represent the TSL-approximation to the Reynolds-averaged Navier-

Stokes equations, with u and v denoting mean velocity components. The shear stress τ then includes the Reynolds stress term, i.e.

$$\tau = \mu\frac{\partial u}{\partial y} - \rho\overline{u'v'} = \rho(\nu+\varepsilon)\frac{\partial u}{\partial y} \tag{24}$$

Here, the eddy-viscosity ε is conveniently modelled by a modified version of the Cebeci and Smith [13] two-layer expression:

$$\varepsilon = \min\left\{\varepsilon_i, \varepsilon_o\right\} \qquad \begin{aligned} \varepsilon_i &= 0.40\ u_\tau y[1 - \exp(-\ yu_\tau/26\nu)]^2 \\ \varepsilon_o &= 0.07\ u_\tau h(x) \end{aligned} \tag{25}$$

where $u_\tau = (\tau_w/\rho)^{\frac{1}{2}}$ is the friction velocity. Further details concerning the turbulent computations can be found in references [7,12].

RESULTS AND DISCUSSION

The first case considered is the laminar film flow along a vertical wall. Figure 2 shows the computed surface velocity u(x,h) for values of the flow depth ratio h_o/h_∞ equal to 0.5 and 1.2. Here, $h_o \equiv h(0)$ denotes the initial value of the flow depth. The TSL-predictions compare favourably with the finite-difference results of Cerro and Whitaker [1] given in ref. [14]. This is not remarkable, since for $\alpha = 90^o$ the effect of the streamwise pressure gradient vanishes, and the present TSL-formulation becomes identical to the model considered in refs. [1,2,8].

The hydrodynamic entrance length x_e is an interesting quantity in many applications of film flow in chemical engineering. Here, x_e is defined as the streamwise position at which the film thickness h(x) has reached its asymptotic value h_∞ (eq. 12) to within 2 percent. Figure 3 shows that the present predictions of x_e are close to the analytical solution proposed by Stücheli and Özisik [8] and the finite-difference results of Yilmaz and

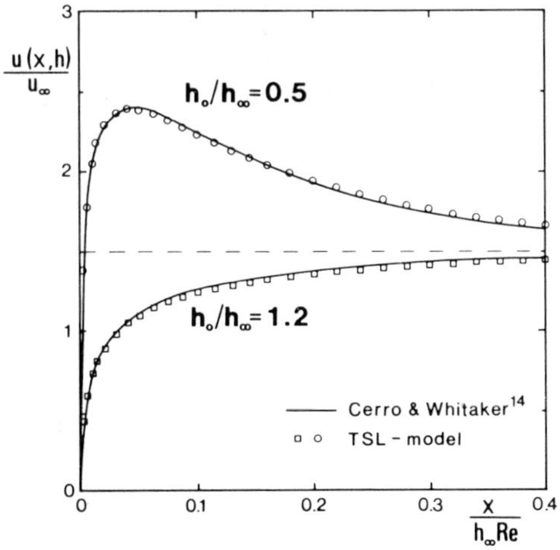

Fig. 2 Predicted surface velocity. Vertical wall; parabolic initial velocity profile.

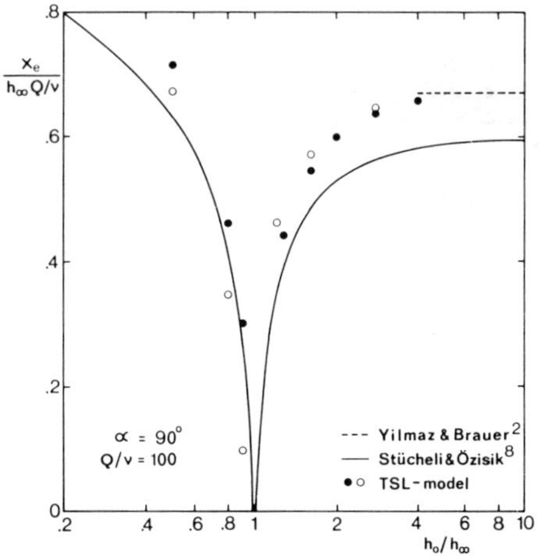

Fig. 3 Entrance length predictions. Vertical wall.
○ parabolic initial vel. profile
● semi-parabolic initial profile

Brauer [2]. The latter results, which were based on a uniform initial velocity profile, have also been found in reasonably good agreement with experimental data [2]. It is furthermore apparent from Fig.3 that the predicted entrance length is essentially independent of the choice of initial profile for values of h_o/h_∞ above 3.

In most of the previous analyses on laminar film flow, e.g. [8], the solutions depend only on the prescribed initial conditions and implicitly on the Reynolds number. Using the present TSL-formulation, however, the pressure couples the streamwise and cross-stream momentum equations (2-3) so that the model explicitly depends also on the angle of inclination α. The entrance length predictions in Fig.4 show the same α-dependence as the results obtained by an extended integral analysis [15] based on the TSL-model equations (1-6). Fig. 5 shows that the predicted surface velocity for flow along a slightly inclined wall is consistent with the experimental data of Lynn [16].

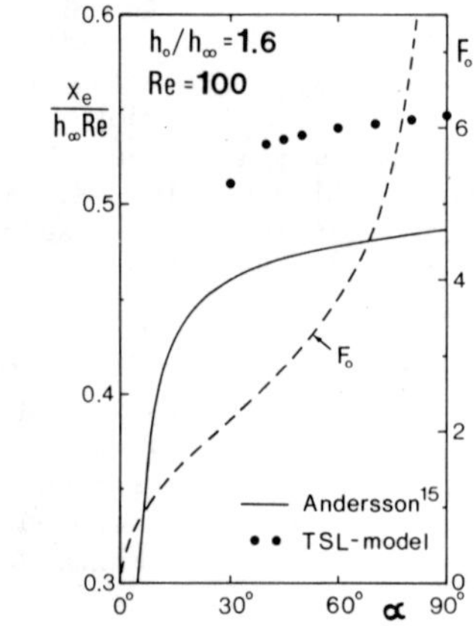

Fig. 4 Entrance length.
● u(0,y) semi-parabolic
—— u(x,y) sinusoidal

Note that an improved accuracy in the initial guess $\delta^{(o)}$ of the surface position at the first streamwise position ξ^1 is required as the angle of inclination is reduced below $\alpha = \pi/2$. This difficulty is probably related to the corresponding abrupt decrease in the initial Froude number

$$F_o = (Q/h_o)/(g\cos\alpha h_o)^{\frac{1}{2}} \qquad (26)$$

shown in Fig.4. As an example, consider the case $\alpha = 40^o$ for which $F_o = 2.61$. The predicted surface pressure $P(\xi^1,\delta^{(i)})$ at $\xi = \xi^1$ for various estimates $\delta^{(i)}$ of the surface position is shown in Fig.6. It is evident from this figure that two distinct solutions exist, both satisfying the normal-stress condition $P(\xi^1,\delta) = 0$. The solution $\delta(\xi^1) > \delta(\xi^o)$ represents a retarded flow, while $\delta(\xi^1) < \delta(\xi^o)$ corresponds to the accelerated solution which approaches the asymptotic solution for large ξ-values.

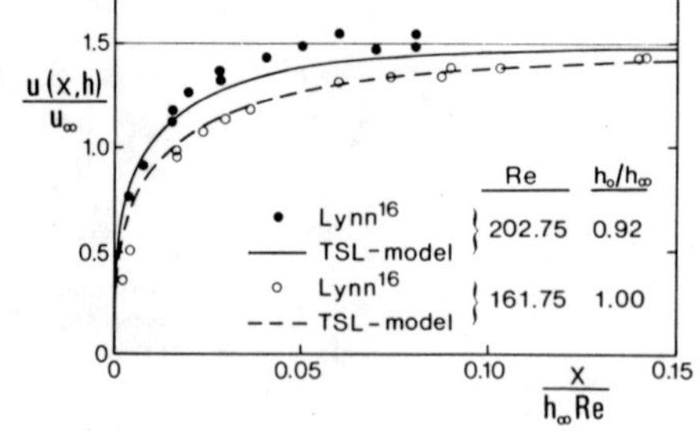

Fig. 5 Surface velocity.
Sloping wall; $\alpha = 6^o$.
Parabolic initial velocity profile.
Experiments[16] and predictions.

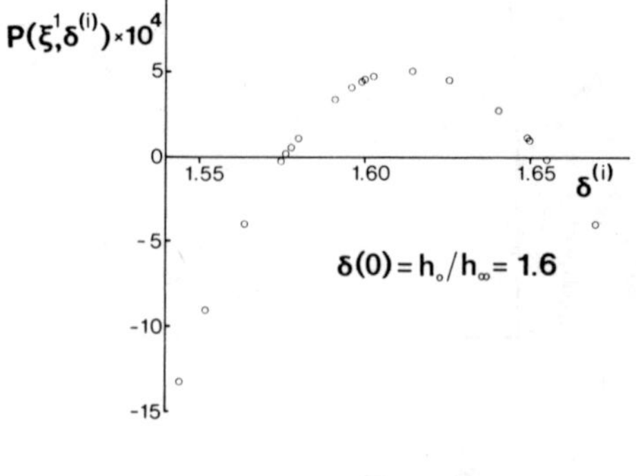

Fig. 6 Successive surface iterations.
$\alpha = 40^o$; Re = 100 ; $\xi^1 = 0.0002$.

It should furthermore be mentioned that, according to the integral analysis of Andersson [15], the integrated version of the momentum equation (2) exhibits a singularity for the initial Froude number close to 1. Assuming a semi-parabolic velocity profile, for instance, the singular behaviour occurs for $F_o = (5/6)^{\frac{1}{2}}$. This fact offers a possible explanation for the difficulties observed in starting the computations for only slightly supercritical inflow conditions.

As an example on turbulent free-surface flow calculations, some selected results for a highly supercritical hydraulic flow along an adverse incline are presented. In Fig.7 the experimental data points [17] fall close to the predicted flow depth variation along the incline. Here, the dimensionless streamwise coordinate $\tilde{x}$ is defined as $x|\sin\alpha|/(u(0,h)^2/2g)$. It is observed that the surface level $h(x)$ increases in the downstream direction. The increase in the surface slope dh/dx is accompanied by a reduction in the bottom shear, and a breakdown position x_{br} is defined as the streamwise station at which the bottom shear vanishes, i.e. as the separation point. Predicted values of x_{br} for various values of F_o in Fig.8 are consistent with experimentally observed breakdown positions. In the experimental flume [17], however, the appearance of an unsteady hydraulic jump or bore defined the critical position x_{br}, the jump completely quenching the originally stable flow along the adverse incline. Thus, the results displayed in Fig.8 indicate an energy-recovery of about 60 percent in the transformation from kinetic energy at $x = 0$ to potential energy at $x = x_{br}$.

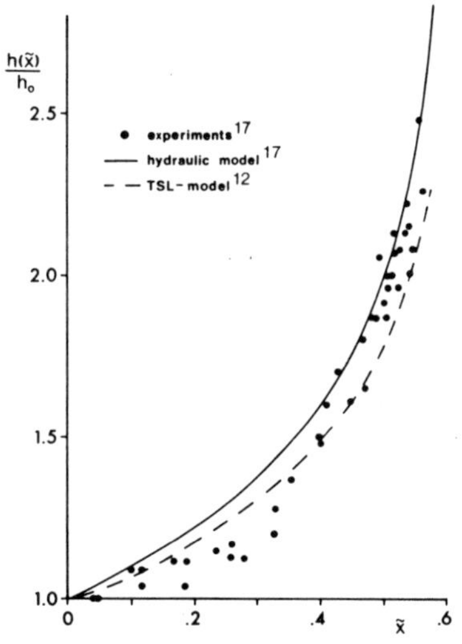

Fig. 7 Flow depth variation. Turbulent flow along an adverse slope; $\alpha = -9.59^o$; $F_o = 6.13$; $Re = 6.6 \cdot 10^4$.

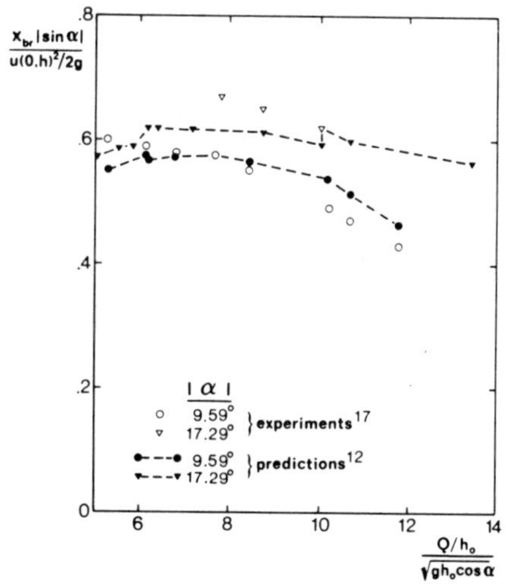

Fig. 8 Breakdown predictions compared with experimental results. Turbulent flow along an adverse slope.

CONCLUSIONS

A finite-difference procedure has been developed for calculating solutions to a TSL-model for steady, two-dimensional free-surface flows. The present technique, which is based on the Keller Box method employed in an iterative scheme, has been applied to laminar film flows and turbulent hydraulic flows. The presented predictions, which are generally in accord with other results and experimental data, demonstrate the capabilities of the solution procedure in treating both laminar and turbulent free surface flows.

REFERENCES

[1] Cerro, R.L., Whitaker, S., "Entrance region flows with a free surface: the falling liquid film", Chem. Engng. Sci., 26, pp. 785-798, 1971.

[2] Yilmaz, T., Brauer, H., "Beschleunigte Strömung von Flüssigkeitsfilmen an ebenen Wänden und in Füllkörperschichten", Chemie-Ing. -Techn.,45, pp. 928-934, 1973.

[3] Bertschy, J.R., Chin, R.W., Abernathy, F.H., "High-strain-rate free-surface boundary-layer flows", J. Fluid Mech., 126, pp. 443-461, 1983.

[4] Rastogi, A.K., Rodi, W., "Predictions of heat and mass transfer in open channels", ASCE J. Hydraul. Div., 104, pp. 397-420, 1978.

[5] Leschziner, M.A., Rodi, W., "Calculation of strongly curved open channel flow", ASCE J. Hydraul. Div., 105, pp. 1297-1314, 1979.

[6] Keller, H.B., "A new difference scheme for parabolic problems", Numerical Solutions of Partial Differential Equations, Vol. II (J. Bramble, ed.), Academic Press, New York, 1970.

[7] Andersson, H.I., "Thin-shear-layer models for two-dimensional free-surface flows in sloping channels", Dr.ing. thesis, Norwegian Institute of Technology, 1982.

[8] Stücheli, A.,Özisik, M.N., "Hydrodynamic entrance lengths of laminar falling films", Chem. Engng. Sci.,31, pp. 369-372, 1976.

[9] Cebeci, T., Keller, H.B., "Laminar boundary layers with assigned wall shear", Proc. 3rd Int. Conf. on Numerical Methods in Fluid Dynamics, Lecture Notes in Physics, 19, pp. 79-85, Springer Verlag, Berlin, 1973.

[10] Cebeci, T., "Calculation of momentum and heat transfer in internal flows with small regions of separation", Turbulent Forced Convection in Channels and Bundles, Vol. I (S. Kakac, D.B. Spalding, eds.), McGraw-Hill-Hemisphere, Washington, 1979.

[11] Keller, H.B., "Accurate difference methods for nonlinear two-point boundary-value problems", SIAM J. Num. Anal., 11, pp. 305-320, 1974.

[12] Andersson, H.I., Ytrehus, T., "Thin-shear-layer model in supercritical hydraulic flow", to appear in ASME J. Appl. Mech.

[13] Cebeci, T., Smith, A.M.O., "A finite-difference method for calculating compressible laminar and turbulent boundary layers", J. Basic. Eng., 92, pp. 523-535, 1970.

[14] Cerro, R.L., Whitaker, S., "Stability of falling liquid films", Chem. Engng. Sci., 26, pp. 742-745, 1971.

[15] Andersson, H.I., "On integral method predictions of laminar film flow", to appear in Chem. Engng. Sci.

[16] Lynn,S., "The acceleration of the surface of a falling film", A.I.Ch. E. J., 6, pp. 703-705, 1960.

[17] Andersson, H.I., Madsen, T., Ytrehus, T., "Hydraulic flow on adverse slopes", Rep. no. 81:2, Institutt for mekanikk, Norwegian Institute of Technology, 1981.

A SECOND-ORDER LAGRANGIAN-EULERIAN METHOD FOR COMPUTATION OF TWO-DIMENSIONAL UNSTEADY TRANSONIC FLOWS

by M. Borrel and Ph. Morice
Office National d'Etudes et de Recherches Aérospatiales (ONERA)
92320 Châtillon (France)

SUMMARY

An improved numerical method, based on a Lagrangian-Eulerian approach, is presented for computing two-dimensional unsteady flows governed by Euler equations. This method uses both a finite volume technique and a finite element approximation. Numerical results are shown for steady and unsteady transonic flows in a channel.

INTRODUCTION

Numerical methods in current use for computing unsteady transonic flows can be roughly classified in two categories. To the first one belong spatially centered schemes of second order accuracy, which necessitate the adjunction of artificial viscosity near the discontinuities. The second category consists of non centered schemes which behave well in shocks but are often less accurate in smooth parts of the flow. Currently, in most of schemes of this class, the upwinding is more or less based on use of characteristic relations or of flux splitting.

In the present paper, a somewhat intermediate way is adopted. Proper upstream influence is taken in account through a Lagrangian-Eulerian approach. Some preliminary results on 2-D flows have been presented in [1] . The corresponding schemes were only first order accurate. A new second order accurate scheme is presented herein which retains the robustness of the previous method while the dissipative errors are decreased. The basic ideas of the formulation used have been investigated first by Van Leer [2 ,3] following Godunov for the one-dimensional problem. However, we decided to make the extension to the two-dimensional case without resorting to some splitting algorithms. The presented scheme is closely related to the Fluid-in-Cell method [4] , in which pressure fluxes and convective terms are also treated separately. Though our method can be formulated over arbitrary grids, with triangular elements for example, for computational considerations, quadrilateral meshes have been used over a grid of finite difference type.

1. GOVERNING EQUATIONS

The Euler equations for an inviscid compressible ideal gas can be written in the divergence form :

$$W_t + F_x + G_y = 0 \qquad \text{(I.1)}$$

with $W = \begin{bmatrix} \rho \\ \rho u \\ \rho v \\ \rho E \end{bmatrix}$, $F = \begin{bmatrix} \rho u \\ \rho u^2 + p \\ \rho u v \\ (\rho E + p)u \end{bmatrix}$, $G = \begin{bmatrix} \rho v \\ \rho u v \\ \rho v^2 + p \\ (\rho E + p)v \end{bmatrix}$, where ρ, p, u, v and E are respectively the density, the pressure, the components of the fluid velocity and the total energy ($E = p/(\gamma-1)\rho + (u^2 + v^2)/2$ for polytropic ideal gas). The flux F and G can be splitted up in convective fluxes and pressure fluxes :

$$\left.\begin{aligned} F &= uW + F_p \\ G &= vW + G_p \end{aligned}\right\} \qquad (I.2)$$

with $F_p = (o, p, o, pu)^t$
$G_p = (o, o, p, pv)^t$.

This leads to the Lagrangian formulation (cf [1]) :

$$\left.\begin{aligned} \widetilde{W}_\tau + \widetilde{F}_a + \widetilde{G}_b &= 0 \\ x_\tau &= u \\ y_\tau &= v \end{aligned}\right\} \qquad (I.3)$$

with : $\widetilde{W} = (J\rho,\ J\rho u,\ J\rho v,\ J\rho E,\ J)^t$
$\widetilde{F} = (0,\ y_b p,\ -x_b p,\ (uy_b - vx_b)p, -(uy_b - vx_b))^t$
$\widetilde{G} = (0, -y_a p,\ x_a p,\ -(uy_a - vx_a)p,\ (uy_a - vx_a))^t$

where a, b are the Lagrangian coordinates and $J = x_a y_b - x_b y_a$ is the Jacobian of the transformation from the reference coordinate system to the coordinate system which moves with the fluid.

2. NUMERICAL SCHEME

The explicit scheme used consists for each time step, in alternating a Lagrangian phase and an Eulerian remapping which takes into account the convective terms.

The approximation chosen is based on piecewise constant values for the conservative variables, though, temporary, a higher approximation is introduced, for example, in calculating slopes. These slopes must not be understood as new degrees of freedom but as functions which makes the solution smoother in the regular zones and more stiff in high gradient zones.

The Lagrangian phase uses a second order accurate predictor-corrector scheme, built, for the predictor, with a finite element weak formulation and, for the corrector, with a finite volume formulation of the conservation laws. During this phase, the nodes of the grid are moved with the fluid velocity, so that a remapping consists in projecting the Lagrangian solution just calculated onto the Eulerian grid in a conservative donor cell manner.

In [1], we have investigated two possibilities : the first one consists in making periodically an exact but time consuming projection, the second possibility is to make an approximate projection at each time step. This last possibility has been selected in the presented scheme.

a - Predictor in the Lagrangian phase

The mesh displacement and the pressure needed for the computation of Lagrangian fluxes are calculated at this time. Previously, this was done by solving Riemann problems at the middle of each interface. In the present method the equations (I.3) are discretized over a half time step with a piecewise constant approximation for $J\rho$ and a continuous piecewise bilinear approximation (Q_1 approximation) for u, v, E and J.

The weak formulation for these equations can be written as follows:

$$-\int_{t^n}^{t^{n+1/2}} dt \int_{\Omega} (\widetilde{W}\varphi_t + \widetilde{F}\varphi_a + \widetilde{G}\varphi_b)\, da\, db + \int_{\Omega} \widetilde{W}\varphi\, da\, db \Big|_{t^n}^{t^{n+1/2}} - \int_{t^n}^{t^{n+1/2}} dt \int_{\partial\Omega} \varphi(\widetilde{F}n_a + \widetilde{G}n_b)\, d\sigma = 0 \quad \text{(II.1)}$$

where Ω is the computation domain and φ belongs to the test-functions set.

Making use of the approximation functions previously described, if we set $\varphi_t = 0$ and evaluate the integrals in time according to :

$$\int_{t^n}^{t^{n+1/2}} \Phi(t)\, dt \simeq \frac{\Delta t}{2}\, \Phi(t^n) \quad ,$$

then, the weak formulation leads to linear systems of the type :

$$MX^{n+1/2} - MX^n = \frac{\Delta t}{2} \quad P \qquad \text{(II.2)}$$

where X^n, $X^{n+1/2}$ represent the vector of the nodal values of each one of u, v, E and J at $t = t^n$ and $t = t^{n+1/2}$, P the right hand-side which depends on the linear system to be solved, and M the mass matrix, the coefficients of which being (with the systems for u, v and E) :

$$m_{ij} = \sum_{K \in \mathcal{T}_h} \int_K (J\rho)\, \varphi_i \varphi_j\, da\, db$$

Owing to the small size of the support of the test-functions φ_i, these coefficients vanish outside of a band ; however in order to avoid solving linear systems, and following [5] , the matrix M becomes a diagonal matrix when use is made of an appropriate numerical integration formula :

$$\int_K (J\rho)\, \Phi\, da\, db \simeq \sum_{k=1}^{4} \Phi(M_k)\, \alpha_k \quad , \quad \alpha_k = \int_K \varphi_k (J\rho)\, da\, db \qquad \text{(II.3)}$$

where M_k, $k = 1, \ldots, 4$ are the four vertices of the cell K.

The equations (II.2) give then $X^{n+1/2}$ in function of X^n through explicit formulae. The nodal values X^n are calculated by averaging the mean cell values of the conservative quantities W over a staggered grid. On a Cartesian grid, this leads to the same predictor as in the Burstein's variant of the Richtmyer 2-D scheme. In fact, precision has been improved if, from the piecewise constant values of W^n, piecewise linear values are restored by calculating slopes in two directions. We will see later how these slopes are calculated (formulae II.10). From the nodal values of $u^{n+1/2}$ and $v^{n+1/2}$, the nodes are moved with the law :

$$\left.\begin{aligned} x^{n+1} &= x^n + \Delta t\ u^{n+1/2} \\ y^{n+1} &= y^n + \Delta t\ v^{n+1/2} \end{aligned}\right\} \qquad \text{(II.4)}$$

At the beginning of each time step, we suppose that the Eulerian and the Lagrangian grids coïncide ; so, the values of J^n are set to 1. From the nodal values of ρ^n , therefore the nodal values of $(J\rho)^{n+1/2}$, and the nodal values of $u^{n+1/2}$, $v^{n+1/2}$, $E^{n+1/2}$ and $J^{n+1/2}$, it is possible to calculate the nodal values of the pressure $p^{n+1/2}$ with :

$$p^{n+1/2} = (\gamma-1)\frac{(J\rho)^{n+1/2}}{J^{n+1/2}}\left(E^{n+1/2} - .5\left[(u^{n+1/2})^2 + (v^{n+1/2})^2\right]\right) \tag{II.5}$$

The value of the pressure at the middle of each side is then obtained by an arithmetic average.

The weak formulation (II.1) includes, in a natural way, the boundary conditions such as a zero normal speed on rigid walls. In some test calculations, slight improvements on results, specially on curved walls, were reached by imposing strongly this boundary condition by means of compatibility relations (cf [7]).

b - Corrector in the Lagrangian phase :

The discretization of the equations (I.3) with the finite volume technique is well-known. It ensures conservation of mass, momentum and energy and can be written as follows :

$$[\tilde{W}]_K^{n+1} = [\tilde{W}]_K^n - \frac{\Delta t}{|K|}\sum_{\partial K}\int_{\partial K}\left[\tilde{F}(x,y,p)^{n+1/2} n_a + \tilde{G}(x,y,p)^{n+1/2} n_b\right] d\sigma \tag{II.6}$$

where $[\tilde{W}]_K^{n+1}$ represents the mean value of the Lagrangian conservative variables over the cell K, and (n_a, n_b) the exterior normal. Let us recall that we have supposed $J^n = 1$, therefore :

$[\tilde{W}]_K^n = ([W]_K^n, 1)$

C. Eulerian remapping :

The aim of this phase is to calculate the Eulerian quantities $[W]_K^{n+1}$ from the Lagrangian ones $[\tilde{W}]_K^n$. We have already presented in [1] a projection which was strongly related to the procedure used in the Fluid-in-Cell method [4]. This projection can be roughly expressed by the formula :

$$[W]_K^{n+1} = [\tilde{W}]_K^{n+1} - \frac{1}{|K|}\sum_{\ell=1}^{4}[\hat{W}]_{K_\ell}^{n+1}\,\delta(\partial K_\ell) \tag{II.7}$$

where $\delta(\partial K_\ell) = \Delta t\left[u_\ell(y_{i+1}^{n+1/2} - y_i^{n+1/2}) - v_\ell(x_{i+1}^{n+1/2} - x_i^{n+1/2})\right]$

represents the algebraic area swept by each side ∂K_ℓ of the mesh K and where :

$$[\hat{W}]_{K_\ell}^{n+1} = [\tilde{W}]_{K_\ell}^{n+1} / [J]_{K_\ell}^{n+1} \tag{II.8}$$

represents the value of the Eulerian quantities over K_ℓ: $K_\ell = K$ if $\delta(\partial K_\ell) \geq 0$ and K is the adjacent mesh if $\delta(\partial K_\ell) < 0$. In order to reach the second order accuracy, we have modified the (II.8) formula by introducing slopes $\delta\hat{W}/\delta x$ in the Lagrangian solution at $t = t^{n+1}$. This technique has been studied by Van Leer. Over a grid of finite difference type, the simplest way to calculate these slopes consists in setting :

$$\left.\begin{aligned} \delta_i\hat{W} &= \text{ave}\left([\hat{W}]_{i+1,j} - [\hat{W}]_{i,j},\ [\hat{W}]_{i,j} - [\hat{W}]_{i-1,j}\right) \\ \delta_j\hat{W} &= \text{ave}\left([\hat{W}]_{i,j+1} - [\hat{W}]_{i,j},\ [\hat{W}]_{i,j} - [\hat{W}]_{i,j-1}\right) \end{aligned}\right\} \tag{II.9}$$

with the non linear averaging function [3] :

$$ave(a, b) = \left[(b^2 + \varepsilon^2) a + (a^2 + \varepsilon^2) b\right] / \left[a^2 + b^2 + 2\varepsilon^2\right] \qquad \text{(II.10)}$$

(ε = small parameter)

In the formula (II.7), the quantity $[\hat{W}]^{n+1}_{K_\ell} \, \mathcal{S}(\partial K_\ell)$ represents the integral of constant quantities over the swepted area. In order to evaluate the correction provided by the slopes, a one point rule for integrating linear quantities gives the correction to be made. In a regular grid, this correction boils down to replace in (II.7), by the expression :

$$[\hat{\hat{W}}]^{n+1}_{K_\ell} = [\hat{W}]^{n+1}_{K_\ell} + \frac{1}{2}\,\mathrm{sgn}(\mathcal{S}(\partial K_\ell))\left[(\delta_\ell \hat{W})^{n+1}_{K_\ell}\left(1 - |\mathcal{S}(\partial K_\ell)| / |K_\ell|\right)\right] \qquad \text{(II.11)}$$

where $\delta_\ell \hat{W}$ means $\delta_i \hat{W}$ or $\delta_j \hat{W}$ according to the side ∂K_ℓ . With an irregular grid, or with a discretization by triangles, the formulae (II.10), (II.11) and (II.12) are not well adapted. It is possible to find a truly 2-D version of these formulae, but our first numerical tests have shown much larger CPU time.

d - Boundary conditions

On solid walls, the boundary conditions are prescribed in the Lagrangian phase by the weak formulation, and in the remapping phase by setting the boundary convective fluxes to zero. On fluid boundaries, the boundary conditions are dealt with fictitious cells and compatibility relations (cf [1] , [7]).

e - Time step limitation

Owing to the fact that the present scheme is explicit, the classical C.F.L. condition has to be imposed for the Lagrangian phase. On the other hand, the time step restriction concerning the remapping phase is that nodes must not be moved over more than one mesh.

3. NUMERICAL RESULTS

In this section, some numerical results are presented for the flow of an ideal gas in a channel. Three examples are given herein.

The first example concerns the stationary flow past a 4.2 % circular bump in a channel at Mach 0.85. A 71 x 50 grid with a uniform distribution of nodes in the y-direction was used. On Fig. 1 are shown the results obtained after convergence within 10^{-6} tolerance. We have already presented results for this test problem in [1] , over a 71 x 20 grid. The improvements between the present results and those obtained with the first order scheme are, in particular, a better resolution of the shock and a lower level of entropy errors. For one time step, the calculation takes 30% more CPU time than a calculation done with the first order scheme and about twice as much as a calculation done with the MacCormack scheme.

The two other test problems concern unsteady flows in a channel containing either a forward facing step, or a wedge with a slope of 20%. The initial solution is a uniform flow at Mach 3 for the step problem and at Mach 1.6 for the wedge problem.

The step problem, known as the Emery test, has been used by Woodward and Colella in [6] in order to compare their high order accuracy scheme (Piecewise Parabolic Method) with others. The results at time 6.11T* with T* = L/C* (L = unit length and C* = critical sound velocity) are plotted on Fig. 2. The corner of the step is a singular point of the flow. In [6], a local treatment sets the entropy and the total enthalpy at a constant value in a neighbourhood of this point. In our calculation nothing special is done. Consequently the numerical errors generated near this point produce a boundary layer which interacts with the reflected shock (see the entropy contours). Except some wiggles behind oblique shocks, the results are satisfactory and compare well with those presented in [6].

In [1] we have presented some results from the first order scheme or the wedge problem. The time evolution of isomach lines for this problem is shown on Fig. 3. The reference time is as above T* = L/C*. For this calculation, a 150 x 50 grid has been used. The formation of the Mach stem and the evolution of the lambda shock and the contact discontinuity is shown as an illustration of the ability of the method to treat truly unsteady flow with strong shocks.

4. CONCLUSIONS

The second order technique used in this paper greatly improve the accuracy of calculations of transonic flows, compare to those given by first order Lagrangian-Eulerian schemes. It would be relatively easy to implement moving grids and multi-zone techniques. However, progress have to be made in order to handle arbitrary grids and to suppress some non physical oscillations.

REFERENCES

[1] Borrel, M. and Morice, Ph. "A Lagrangian-Eulerian Approach to the computation of unsteady transonic flows", Proc. fourth GAMM-Conference on Numerical Methods in Fluid Mechanics (H.Viviand, ed.), Braunschweig, Vieweg Verlag, 1982, Vol.5 of Notes on Numerical Methods in Fluid Mechanics.

[2] Van Leer, B. "Towards the ultimate conservative scheme ; V-A Second Order sequel to Godunov's method", J.Comp.Phys.32,101-136, 1979.

[3] Van Leer, B. "Report ICASE N° 81-11, March 1981. On the relation between the upwind differencing scheme of Godunov, Engquist-Osher and Roe.

[4] Gentry, R.A., Martin, R.E., Daly, B.J. "An Eulerian differencing method for unsteady compressible flow problems. J. Comp. Phys. 1, 87-118 (1966).

[5] Lascaux, P. "Numerical Methods for time-dependent equation applications to fluid flow problems. Lecture notes in Tata Institute of Fundamental Research - Bombay (1976).

[6] Woodward, P., Colella, P. "The Numerical simulation of two-Dimensional Fluid Flow with Strong Shocks. To be published.

[7] Cambier, L., Ghazzi, W., Veuillot, J.P., Viviand, H. in 3rd Vol. of series on Recent Advances in Numerical Math. in Fluids, Ed. W.G. HABASHI, To be published.

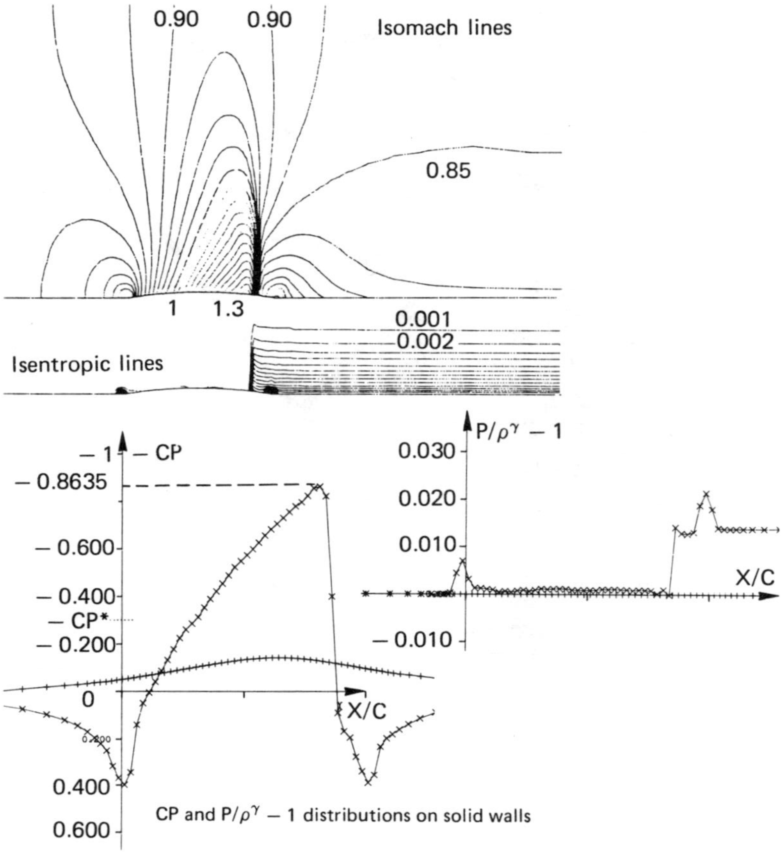

Fig. 1 – Channel flow past a circular bump.

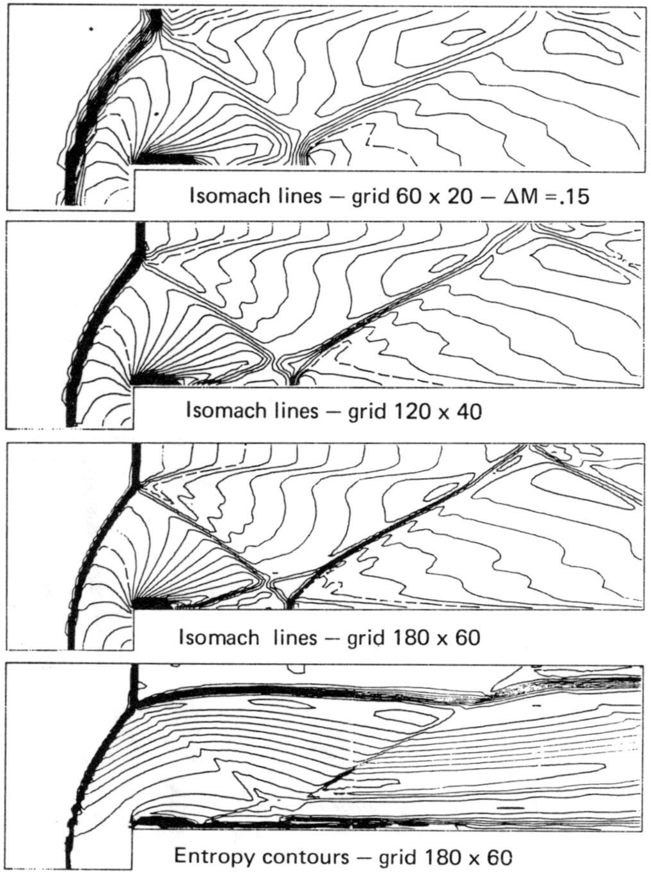

Fig. 2 – Results at time 6.11 T^* for the Mach 3 step problem.

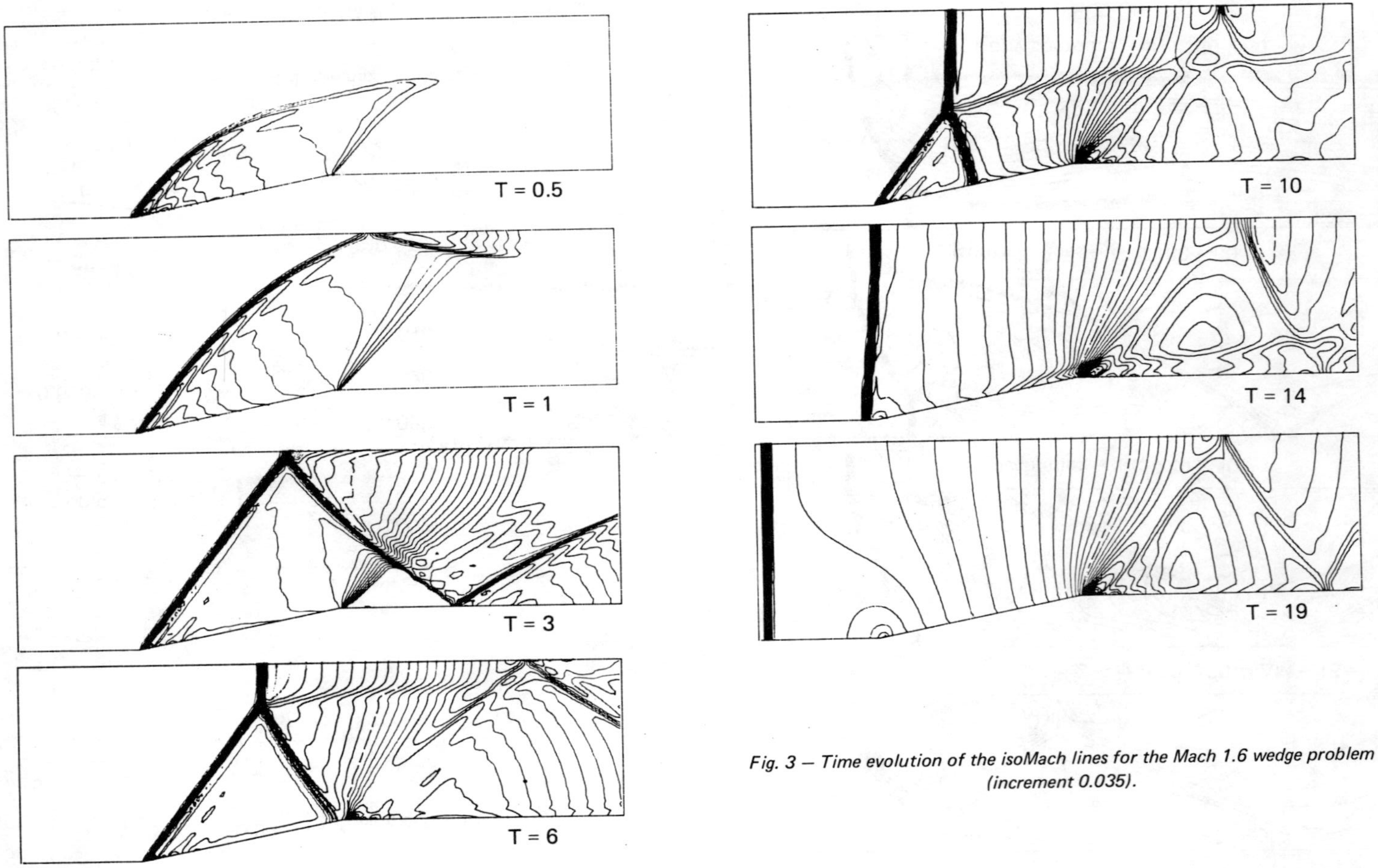

Fig. 3 – Time evolution of the isoMach lines for the Mach 1.6 wedge problem (increment 0.035).

NUMERICAL SOLUTION OF THE THREE DIMENSIONAL, TIME DEPENDENT NAVIER-STOKES EQUATIONS

U. Bulgarelli, V. Casulli
Istituto per le Applicazioni del Calcolo
Viale del Policlinico 137, Roma, Italy
D. Greenspan
Department of Mathematics
The University of Texas at Arlington, Arlington Texas 78019

INTRODUCTION

In this paper we will develop and apply a family of powerful finite difference methods for the numerical solution of the time-dependent Navier-Stokes equations for incompressible fluids in three dimensions. The power is derived by combining advantageous aspects of the Marker-And-Cell (M.A.C.) method and several special techniques for hyperbolic and parabolic equations. The Navier-Stokes equations, which are considered in the primitive variables $(\vec{q},p)$ can be written as:

$$\frac{\partial \vec{q}}{\partial t} + (\vec{q}\cdot\nabla)\vec{q} = -\nabla p + \nu\Delta\vec{q}$$
$$\nabla\cdot\vec{q} = 0, \qquad (1)$$

where $\vec{q} = (u,v,w)^T$ is the velocity vector, t is the time, p is the pressure and ν is the kinematic viscosity coefficient.

If $\Omega \subset \mathbb{R}^3$ denotes the spatial domain for equations (1), and Γ is the boundary of Ω, the following initial and boundary conditions are associated with equations (1):

$$\vec{q}(x,y,z,o) = \vec{q}_o(x,y,z), \quad (x,y,z) \in \Omega, \qquad (2)$$

$$\vec{q}(x,y,z,t) = \vec{q}_b(x,y,z,t), \quad (x,y,z) \in \Gamma, \quad t > 0. \qquad (3)$$

SEMIDISCRETE FORMULATION

Equations (1) are approximated at the mesh points of a finite difference grid using centered differences for the pressure, and appropriate differences for the convective and viscous terms. Depending in the choice for the latter, there results a system of ordinary differential equations (see[3]):

$$\begin{cases} \dfrac{d\vec{Q}}{dt} = F(\vec{Q})\vec{Q} - A^T\vec{P} \\ A\vec{Q} = 0 \end{cases} \tag{4}$$

where $\vec{Q}$ is a global vector containing all nodal velocity components, $\vec{P}$ is a global vector of nodal pressures, A is a rectangular divergence matrix, and $F(\vec{Q})$ is a matrix which results from the discretization of the convective and viscous terms.

Once a discretization for the convective and for the viscous terms has been chosen, $F(\vec{Q})$ is determined and the differential equations in (4) are readily solvable numerically with any one of the several finite difference methods for ordinary differential equations. The selected method can be one step or multistep, explicit or implicit.

EXPLICIT METHODS

The simplest way to solve system (4) is to discretize it with the explicit Euler scheme:

$$\begin{cases} \dfrac{1}{\Delta t}(\vec{Q}^{n+1} - \vec{Q}^{n}) = F(\vec{Q}^{n})\,\vec{Q}^{n} - A^T\,\vec{P}^{n+1} \\ A\vec{Q}^{n+1} = 0 \end{cases} \tag{5}$$

Equations (5), at each time step, constitute a linear system of equations with unknowns $\vec{Q}^{n+1}$ and $\vec{P}^{n+1}$.

Since the velocity, not the pressure, is given as a boundary condition on Γ, the matrix of coefficients for the linear system of equations (5) is singular and hence the pressure field $\vec{P}^{n+1}$ is determined up to an arbitrary additive constant which cancels in first equation of system (5), because A^T is the gradient matrix. Thus, the velocity field $\vec{Q}^{n+1}$ is determined uniquely. In practice, linear system (5) can be solved iteratively by using the successive overrelaxation iterative method in a very efficient way to yield, simultaneously, and $\vec{Q}^{n+1}$ $\vec{P}^{n+1}$ (see, e.g. [2]).

Thus for the structure of the matrix $F(\vec{Q}^n)$ has not been considered. The structure of this term plays a very important role for the consistency, accuracy and stability of (5). Note that, since $F(\vec{Q}^n)$ contains the finite differences corresponding to the spatial discretization of the convective and viscous terms of the Navier-Stokes equations (1), an appropriate choice of $F(Q^n)$ can be determined by considering the parabolic (hyperbolic if $\nu = 0$) equations that one obtains from the momentum equations in (1) when the pressure terms are neglected. Particular forms for $F(\vec{Q}^n)$ are provided and discussed in reference [2].

STABILITY CONSIDERATIONS

Once a space mesh has been chosen, the choice of the time increment and, at times, even the space steps have to satisfy special conditions for stability. Interestingly enough, these can be derived without assigning a specific structure to $F(\vec{Q}_n)$.

By combyning the two equations of system (5) one has

$$AA^T \vec{P}^{n+1} = \frac{1}{\Delta t} A\vec{Q}^n - A\, F(\vec{Q}^n)\vec{Q}^n \qquad (6)$$

Assuming the pressure to be given at, at least, one boundary point, the matrix AA^T is nonsingular and a solution for $\vec{P}^{n+1}$ can be expressed as follows:

$$\vec{P}^{n+1} = (AA^T)^{-1} A \left[\frac{1}{\Delta t} I - F(\vec{Q}^n)\right] \vec{Q}^n. \qquad (7)$$

Substitution of (7) into the first equation of system (5) yields

$$\vec{Q}^{n+1} = [I-A^T(AA^T)^{-1}A]\ [I-(\Delta t)F(\vec{Q}^n)]\ \vec{Q}^n. \qquad (8)$$

To determine the condition under which (8) is stable, consider the discrete L_2 norm of $\vec{Q}^{n+1}$. Then,

$$\|\vec{Q}^{n+1}\|_2 \leq \|I-A^T(AA^T)^{-1}A\|_2 \quad \|I-(\Delta t)F(\vec{Q}^n)\|_2 \quad \|\vec{Q}^n\|. \qquad (9)$$

Note now that since the matrix $[I-A^T(AA^T)^{-1}A]$ is a projector, its L_2 norm is unity and hence (9) implies that (5) is stable provided the following inequality is satisfied at each time step:

$$\|I - (\Delta t)F(\vec{Q}^n)\|_2 \leq 1. \qquad (10)$$

The stability condition (10) is independent of the pressure. It depends only on the structure of the matrix $F(\vec{Q}^n)$, which depends, in turn on the discretization of the convective and diffusive terms.

IMPLICIT METHODS

The above considerations apply also when one uses an implicit discretization for the ordinary differential system [4]. As an example we consider the following modified implicit Euler scheme:

$$\begin{cases} \frac{1}{\Delta t}(\vec{Q}^{n+1} - \vec{Q}^{n}) = F(\vec{Q}^{n})\vec{Q}^{n+1} - A^T \vec{P}^{n+1} \\ A\vec{Q}^{n+1} = 0. \end{cases} \tag{11}$$

Note that the finite difference scheme one obtains from (11), for any particular choice of $F(\vec{Q}^n)$, leads to an algebraic system which is linear, even though the convective terms are included, and this will simplify the solution algorithm for (11).

With regard to the stability of the implicit formulas (11) assume that $F(\vec{Q}^n)$ is chosen in such a fashion that $[I + (\Delta t)F(\vec{Q}^n)]$ is nonsingular. Thus (11) implies

$$\vec{Q}^{n+1} = [I + (\Delta t)F(\vec{Q}^n)]^{-1} [\vec{Q}^n - \Delta t\, A^T \vec{P}^{n+1}]. \tag{12}$$

Substitution of (12) into the second equation of system (11) yields

$$(\Delta t)\, A\, [I+(\Delta t)F(\vec{Q}^n)]^{-1} A^T \vec{P}^{n+1} = A[I+(\Delta t)F(\vec{Q}^n)]^{-1}\vec{Q}^n \tag{13}$$

Assuming the pressure to be known at, at least, one boundary point, system (13) has a unique solution $\vec{P}^{n+1}$ provided Δt is sufficiently small. (In practice, this sufficient condition appears to be unnecessarily restrictive [1]). The solution $\vec{P}^{n+1}$ of (13) is given by

$$\vec{P}^{n+1} = \frac{1}{\Delta t}(A\, R^n A^T)^{-1} A\, R^n\, \vec{Q}^n \tag{14}$$

where R^n denotes the matrix $[I+(\Delta t)F(\vec{Q}^n)]^{-1}$. Substitution of (14) into (12) yields

$$\vec{Q}^{n+1} = R^n\, [I-A^T(A\, R^n A^T)^{-1} A\, R^n]\, \vec{Q}^n. \tag{15}$$

Note now that since $[I-A^T(A\, R^n A^T)^{-1} A\, R^n]$ is a projector, (15) is stable if the following inequality is satisfied

$$\|R^n\|_2 = \|I + (\Delta t)F(\vec{Q}^n)\|_2 \leq 1. \tag{16}$$

Thus, one can observe that the stability condition (16) is independent of the pressure. It depends only on the structure of the matrix $F(\vec{Q}^n)$, that is, on the discretization chosen for the convective and viscous terms. Moreover, since (16) is also the stability condition for the implicit scheme that one obtains from (11) by neglecting the pressure term, we can state that any stable implicit discretization for the parabolic (hyperbolic if $\nu = 0$) equations that one obtains from the momentum equations in (1) by neglecting the pressure gradient terms can be adapted to obtain a stable method of the form (11) for the complete Navier-Stokes equations (1).

The family of methods (5) and (11) produce velocities that are exactly discrete divergence free and can be implemented most efficiently.

A DRIVEN CAVITY CALCULATION

As an application of the methods discussed above, consider the familiar driven cavity problem. Let us follow the flow which develops in a cubic cavity whose sides have length 2 (see Fig. 1).

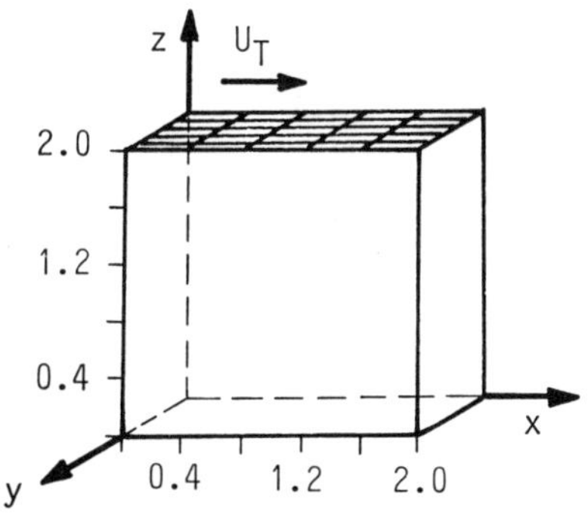

Fig. 1

Assume that the fluid has viscosity $\nu = 0.4$ and that at the initial time $t = 0$ the velocities are

$$\vec{q} = 0 \tag{17}$$

In addition, the bottom and side boundaries are fixed, while the top boundary is assumed to be moving with constant velocity $u_T = 0.5$. The discretization parameters are taken to be $\Delta x = \Delta y = \Delta z = 0.4$ and $\Delta t = 0.02$. The form for the matrix $F(\vec{Q}^n)$ used in this example is the one obtained when centered finite differences are used for the viscous terms and upwind differences for the convective terms (see [2] for details).

Clearly, the flow which must and does develop is a three dimensional lid driven flow which goes to the steady state.

Figures 2 and 3 show typical computed velocities at the steady state (t=1). More precisely, Fig. 2 represents u vs x at z=1.8, y=1; while Fig. 3 represents z vs u at x=1, y=1. The results compare well with those of Gresho and others [4].

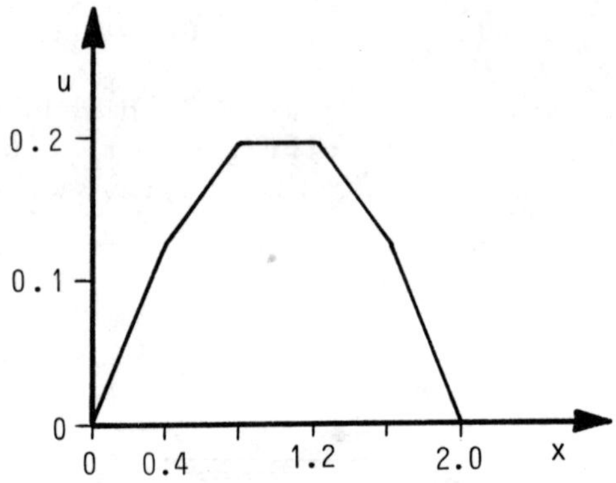

Fig. 2

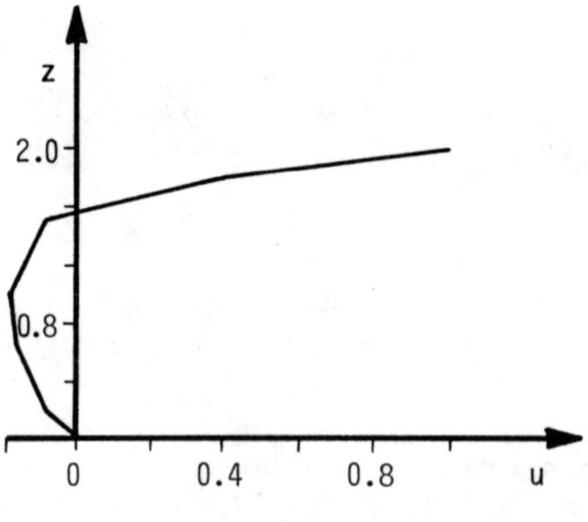

Fig. 3

ACKNOWLEDGMENT

The authors would like to express their appreciation to M. Prosperi who supplied the graphic routines and performed the calculation.

REFERENCES

[1] Amit, R., Hall, C.A., Porsching, T.A., "An Application of Network Theory to the Solution of Implicit Navier-Stokes Difference Equations", Journal of Computational Physics, Vol. 40, No. 1, pp. 183-201, 1981.

[2] Bulgarelli, U., Casulli, V., Greenspan, D., "Pressure Methods for the Numerical Solution of Free Surface Fluid Flows", Pineridge Press, Swansea U.K., to appear.

[3] Bulgarelli, U., Casulli, V., Greenspan, D., "Pressure Methods for the Approximate solution of the Navier-Stokes Equations. In Proceedings of the Third International Conference on Numerical Methods in Laminar and Turbulent Flow", Pineridge Press, Swansea, U.K., 1983.

[4] Gresho, P.M., Chan, S.T.K., Lee, R.L., Upson, C.D., "Solution of the Time-Dependent, Three-Dimensional, Incompressible Navier-Stokes Equations Via FEM. In Proceedings of the Second International Conference on Numerical Methods in Laminar and Turbulent Flow". Pineridge Press, Swansea, U.K., pp. 27-35, 1981.

[5] HARLOW, F.H., WELCH, F.E., "Numerical Calculation of Time-Dependent Viscous Incompressible Flow", Physics of Fluids, Vol. 8, No. 12, pp. 2182-2189, 1965.

[6] Pan, F., Acrivos, A., "Steady Flows in Rectangular Cavities", Gam-Fluid Thech., 28, 643-655, 1967.

Computations in External Transonic Flow

by D.M. Causon and P.J. Ford
Department of Aeronautical and Mechanical Engineering
University of Salford
Salford M5 4WT
U.K.

Summary

A pseudo-time dependent, split, finite-volume method is described for the numerical solution of the Euler equations governing transonic flow. A particular feature is the use of operator-switching between the MacCormack scheme and an explicit upwind scheme according to whether the flow locally is subsonic or supersonic in the direction of the split. It is found that the developed algorithm is more robust and the resolution of captured shock waves and rate of convergence are both improved. Some sample computed results are presented in support of these observations.

Introduction

Our work concerns developments of MacCormack's finite-volume method. This popular method is easily programmed and has found wide use. But critics dislike the fact that the scheme is not generally compatible with local characteristic directions. The excellent flux-vector and flux-difference splitting methods [1,2] *are compatible* and appear to be promising contenders for future development as design methods. However, notwithstanding such criticisms, MacCormack's method is quite robust and capable of useful predictions in a fairly wide range of applications, the limit of applicability perhaps being marked by cases involving particularly strong shock waves. Therefore, one may argue that an exercise in algorithm development, using this method as a basis, may produce a useful design method. Other workers, perhaps most notably Lerat and Sides [3] and Co-workers in France, and Jameson [3] in the U.S.A. are pursuing similar approaches. Our developments of MacCormack's method consist chiefly of the construction of a hybrid algorithm using this scheme and the explicit upwind scheme of Beam and Warming [4]. Operator-splitting is employed and *switching* takes place between the two schemes according to whether the flow locally is subsonic or supersonic in the direction of the split. The upwind scheme, applicable only in supersonic regions, is rather more compatible with the local characteristic directions and is stable for Courant numbers up to two. Also, shockwave resolution is improved, a matter we will return to later. For steady-flow problems local time stepping further improves the rate of convergence. In the following sections the finite-volume method will be outlined and the implementation of operator-switching will be described. Results will be presented for some of the Ref [3] test problems; for a simulated aircraft forebody treated, first axisymmetrically (i.e. as a nose/canopy body of revolution) and then three-dimensionally; and finally for a representative forebody taken from Ref [5].

ANALYSIS

Finite volume method

The unsteady Euler equations in vector-integral conservation law form are

$$\frac{\partial}{\partial t}\int_{vol} U dvol + \iint_S \underline{H}.\hat{n} ds = 0 \qquad (1)$$

Where $U = (\rho, \rho U, \rho V, \rho W)^T$, has components of density and Cartesian components of momentum, $\underline{H}$ is a flux-vector of these quantities across the cell surface whose normal is $\tilde{n}$ and it is assumed that a steady-flow solution is required and we can delete the energy equation, deriving pressure from a relation based on total enthalpy.

The paper by Rizzi [3] and references therein give fuller details of the finite-volume method, boundary conditions and operator-splitting methodology.

Operator - Switching

Each split operator in a sequence of operators represents merely the application of a one-dimensional scheme (like MacCormack's) to a previous plane of data, and we need not use the same scheme in every cell. This is the basis of operator-switching. The full range of operators used comprise the MacCormack scheme (M), the explicit upwind scheme (U) of Beam and Warming [4] and two 'transition' operators (MU) and (UM) connecting schemes (M) and (U) such that strict conservation form is maintained. All four schemes share the same predictor step. For example, an L_1 (Δt) split operator would be defined in terms of a predictor step,

Predictor

$$(volU)_{ij}^{\widetilde{n+1}} = (volU)_{ij}^{n} - \Delta t\,(H_{ij}^{n}.S_{i+\frac{1}{2}} + H_{i-1j}^{n}.S_{i-\frac{1}{2}}) \qquad (2)$$

and one of the four following corrector steps.

M Corrector

$$(volU)_{ij}^{n+1} = \tfrac{1}{2}\,((volU)_{ij}^{\widetilde{n+1}} + (volU)_{ij}^{n} - \Delta t\,(H_{i+1j}^{\widetilde{n+1}}.S_{i+\frac{1}{2}} + H_{ij}^{\widetilde{n+1}}.S_{i-\frac{1}{2}})) \qquad (3)$$

U Corrector

$$(volU)_{ij}^{n+1} = \tfrac{1}{2}\,((volU)_{ij}^{\widetilde{n+1}} + (volU)_{ij}^{n} - \Delta t\,(H_{ij}^{\widetilde{n+1}}S_{i+\frac{1}{2}} + H_{i-1\,j}^{\widetilde{n+1}}.S_{i-\frac{1}{2}})$$
$$- \Delta t\,(H_{ij}^{n}.S_{i+\frac{1}{2}} + 2H_{i-1j}^{n}.S_{i-\frac{1}{2}} + H_{i-2j}^{n}.S_{i+\frac{1}{2}})) \qquad (4)$$

MU Corrector

$$(volU)_{ij}^{n+1} = \tfrac{1}{2}\,((volU)_{ij}^{\widetilde{n+1}} + (volU)_{ij}^{n} - \Delta t\,(H_{ij}^{n}.S_{i+\frac{1}{2}} + H_{i-1j}^{n}S_{i-\frac{1}{2}}$$
$$+ H_{ij}^{\widetilde{n+1}}.S_{i+\frac{1}{2}} + H_{ij}^{\widetilde{n+1}}.S_{i-\frac{1}{2}}) \qquad (5)$$

UM Corrector

$$(\text{volU})_{ij}^{n+1} = \tfrac{1}{2}\,((\text{volU})_{ij}^{\widetilde{n+1}} + (\text{volU})_{ij}^{n} - \Delta t\,(H_{i-2j}^{n}\cdot S_{i+\frac{1}{2}} + H_{i-1j}^{n}\cdot S_{i-\frac{1}{2}}$$

$$+ H_{i+1j}^{\widetilde{n+1}}\cdot S_{i+\frac{1}{2}} + H_{i-1j}^{\widetilde{n+1}}\cdot S_{i-\frac{1}{2}})). \qquad (6)$$

The transition operators are derived as follows. Write down the two schemes to be connected at several points either side of the switch point (i.e. at i = IS - 1, IS - 2 etc) and choose the transition operator appropriately to maintain telescopic cancellation of fluxes through the switch-point. Strict conservation form will thus be maintained.

We have found that stability is assured if both transition operators are applied on the side of the switch-point corresponding to supersonic flow. A criterion for switching which works quite well is a test on the local Mach number, i.e. when the Mach number in any cell becomes supersonic switch from scheme M to U via UM. Of course this simple criterion may fail if, for example, a supersonic pocket is very small and two transition operators appear adjacent to one another. This can be avoided by suitable logic but is in any case not fatal to the running of the program, merely introducing minor localised errors. Also, this criterion is only suitable for a flow-conforming mesh rather than a body-conforming (C-grid) mesh which one might use in a blunt-body computation. In such cases one could adopt a criterion based on, say, the sign change of an eigenvalue, and switching would take place in more than one split operator.

In practice shock waves generally are captured with little or no oscillations, owing to the dispersive errors of schemes M and U being of opposite sign. But sometimes, when a supersonic-supersonic shock wave occurs oscillations will still occur on the high pressure side. A significant advantage of operator-switching, especially when using local time-stepping, is in the fact that the scheme U can advance with twice the time step of scheme M. This can be a very worthwhile saving in cases where the flow is almost entirely supersonic, with one or more subsonic bubbles.

RESULTS

Computational Specifics

The solutions to be presented have been converged to residuals (representing the differential operator multiplied by the cell volume) of less than 10^{-5}, and confirmed by inspection. As an example, calculations on a 64 x 15 mesh utilised for those results depicted in Fig. 2 require less than twenty five seconds on a CDC7600. As to storage, the three-dimensional code requires the equivalent of 8 3D arrays.

Selected Results

A sample of our computed results are shown in Figs. 1-3. Those in Fig. 1 relate to the 1979 GAMM workshop [3] test problems. Fig. 1a depicts the C_p distribution for problem B, a M = 0.85 internal flow through a channel with a bi-convex bump in the lower wall. An essential feature is the crisply captured embedded shock wave. The entropy plot $((P/\rho^{\gamma})/(P/\rho^{\gamma})_{\infty}-1.0)$, shown in Fig. 1b, displays the expected step increase but also a 'spike' appearing at the shock location and a small positive value upstream of it which is probably due to the dissipation implicit in the method. Figs. 1c and d correspond to problem E, a supersonic flow at M = 1.2 around an axisymmetric tangent-ogive defined by $r = -2.4383 + (2.6354^2 - (x-1)^2)^{\frac{1}{2}}$ for $x \leq 1$ and $r = 0.1971$ for $x \geq 1$. Figs. 1e and f depict C_p distributions and contour plots for a NACA 0012 aerofoil at M = 0.85 and $\alpha = 0$ and -1. The results shown in Fig. 1 are in good general agreement with those of other Euler methods reported at the workshop.

The results presented in Fig. 2 relate to a body of revolution formed by rotating a curve representative of an aircraft nose and canopy about the 'fuselage' axis. The contour is defined by $r = 0.365x$, $0 \leq x \leq 1$; $r = 0.365\,(x - \frac{(x-1)^2}{12})$, $1 \leq x \leq 7$; $r = 1.46 + 0.577\,(x-7)$, $7 \leq x \leq 8$; $r = 1.46 + 0.577\,(1.3x - 9.4)^{\frac{1}{2}}\,(1.8 - 0.1\,x)^3$, $8 \leq x \leq 18$. The chosen Mach number is 1.40. Expected features are an attached bow shock and detached canopy shock. The grid chosen was a simple flow-conforming type (with 64 x 15 cells), as shown in Fig. 2a, and with extra definition near the nose and near the canopy. The sudden change in mesh spacing in these areas may account for the spikes in the entropy plot Fig. 2d.

Finally, the results of our initial excursions into three-dimensions are shown in Fig. 3. The first of the two examples, shown in Fig. 3a, is simply a 3D version of the Fig. 2 axisymmetric body in which the canopy is faired out linearly as $\theta \to 90^\circ$. As expected, the canopy shock vanishes in axial distributions for $\theta \geq 90^\circ$. Fig. 3b depicts results for the NASA forebody No. 4 from [5], at Mach 1.70 and $\alpha = 0^\circ$. Agreement with the experimental data is satisfactory in view of the particularly coarse mesh used in our computations (40 x 8 x 8, in half-body representation).

Acknowledgements

This work was supported in part by the D.T.I.(U.K.).

References

1. ROE, P.L. Numerical modelling of shock waves and other discontinuities, Numerical methods in aeronautical fluid dynamics, p. 211, Academic Press, 1982.

2. YANG, J.Y. and LOMBARD, L.K. A characteristic flux difference splitting for the hyperbolic conservation laws of inviscid gas dynamics, AIAA Paper No. 83-0040, AIAA 21st Aerospace Sciences Meeting, Reno, Nevada, 1983.

3. Numerical methods for the computation of inviscid transonic flows with shock waves, A GAMM workshop, Notes on Num. Fluid Mech Vol. 3, Vieweg 1981.

4. WARMING, R.F. and BEAM, R.M. Upwind second-order difference schemes and applications in aerodynamic flows, J.A.I.A.A. No. 114, pp 1241-1249, Sep. 1976.

5. TOWNSEND, J.C. and HOWELL, D.T., COLLINS, I.K. and HAYES, C. Surface pressure data on a series of analytic forebodies at Mach numbers from 1.70 to 4.50 and combined angles of attack and sideslip. NASA TM 80062, 1979.

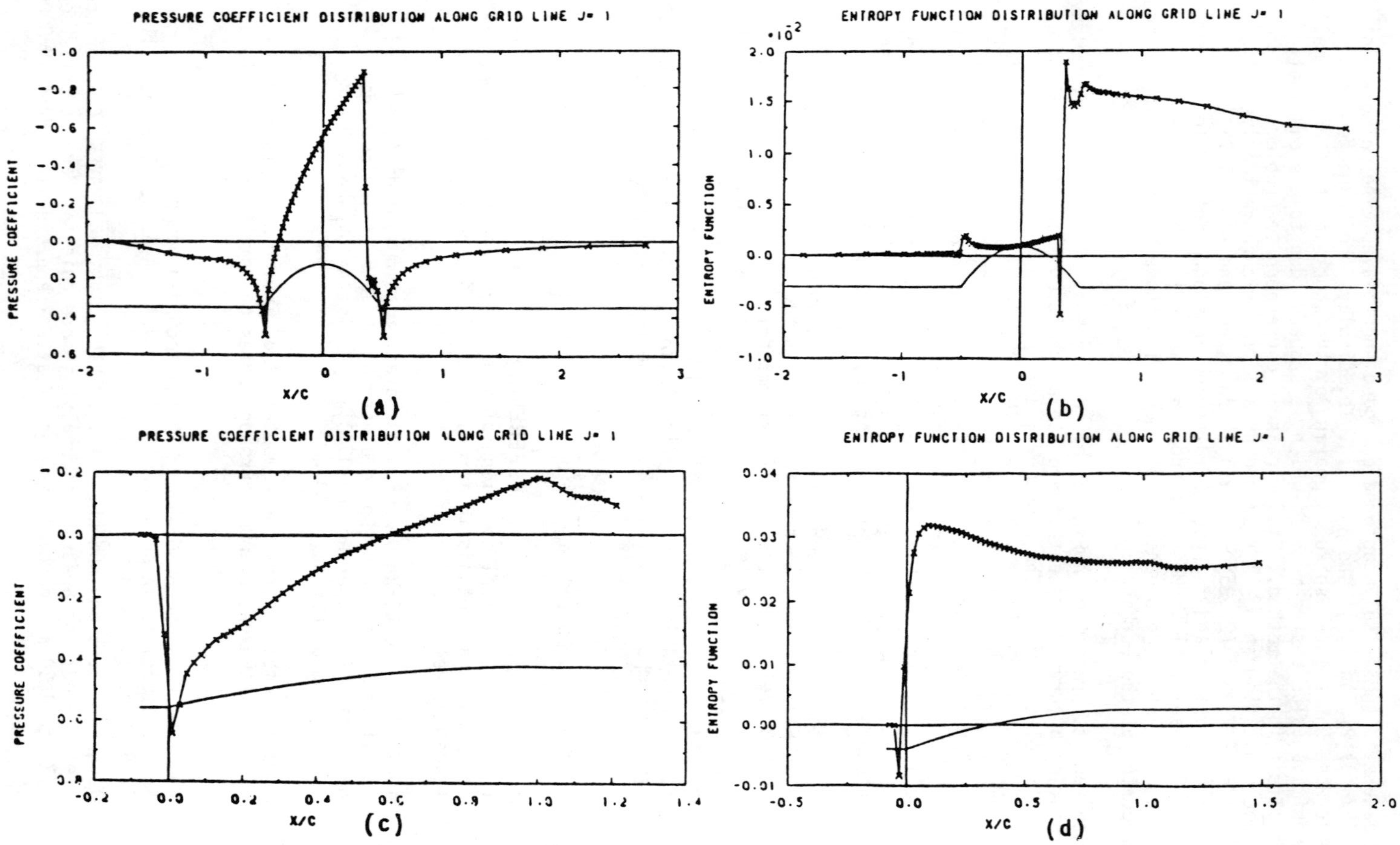

FIG. 1 - 1979 GAMM WORKSHOP TEST PROBLEMS B AND E

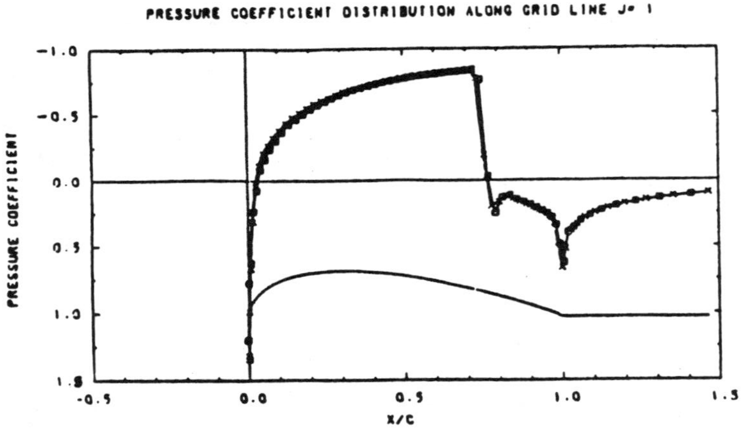

(e) $\alpha = 0^{\circ}$

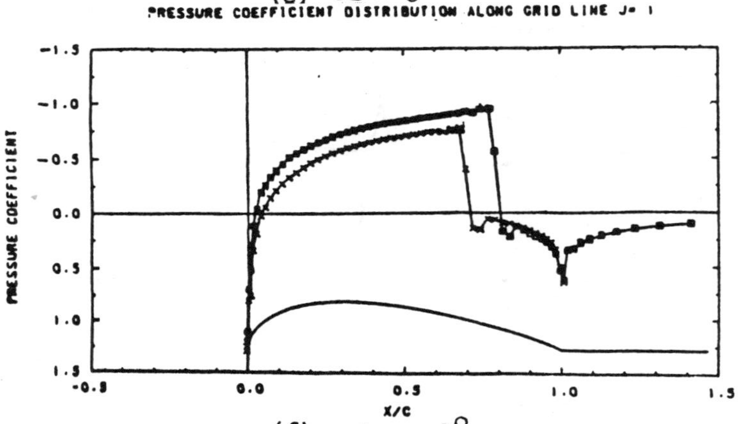

(f) $\alpha = -1^{\circ}$

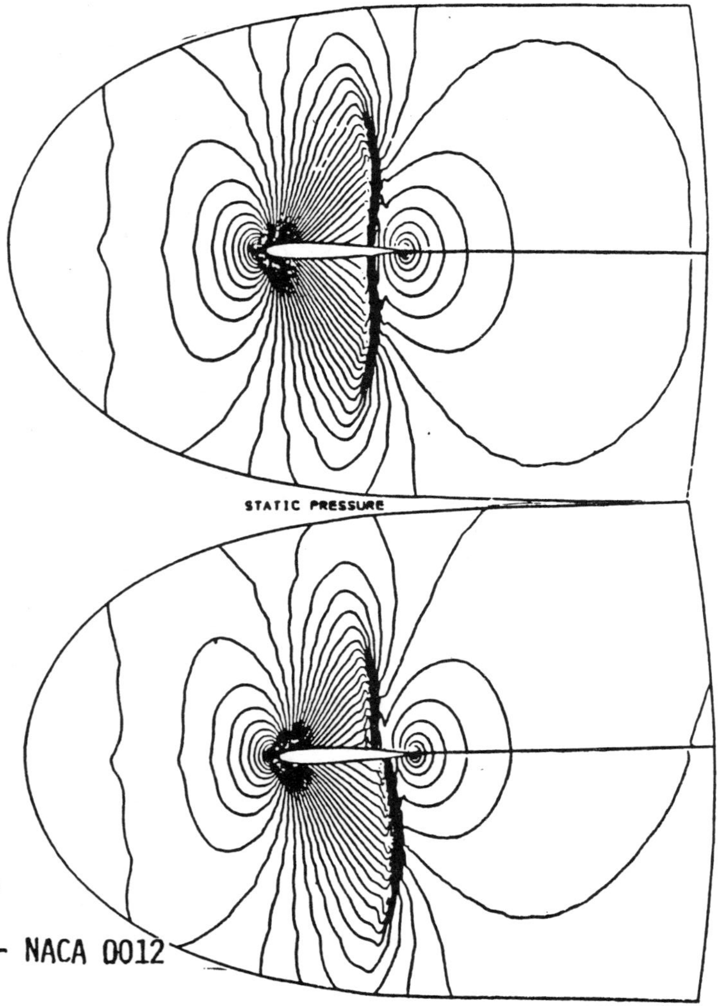

FIG. 1 CONCLUDED - NACA 0012

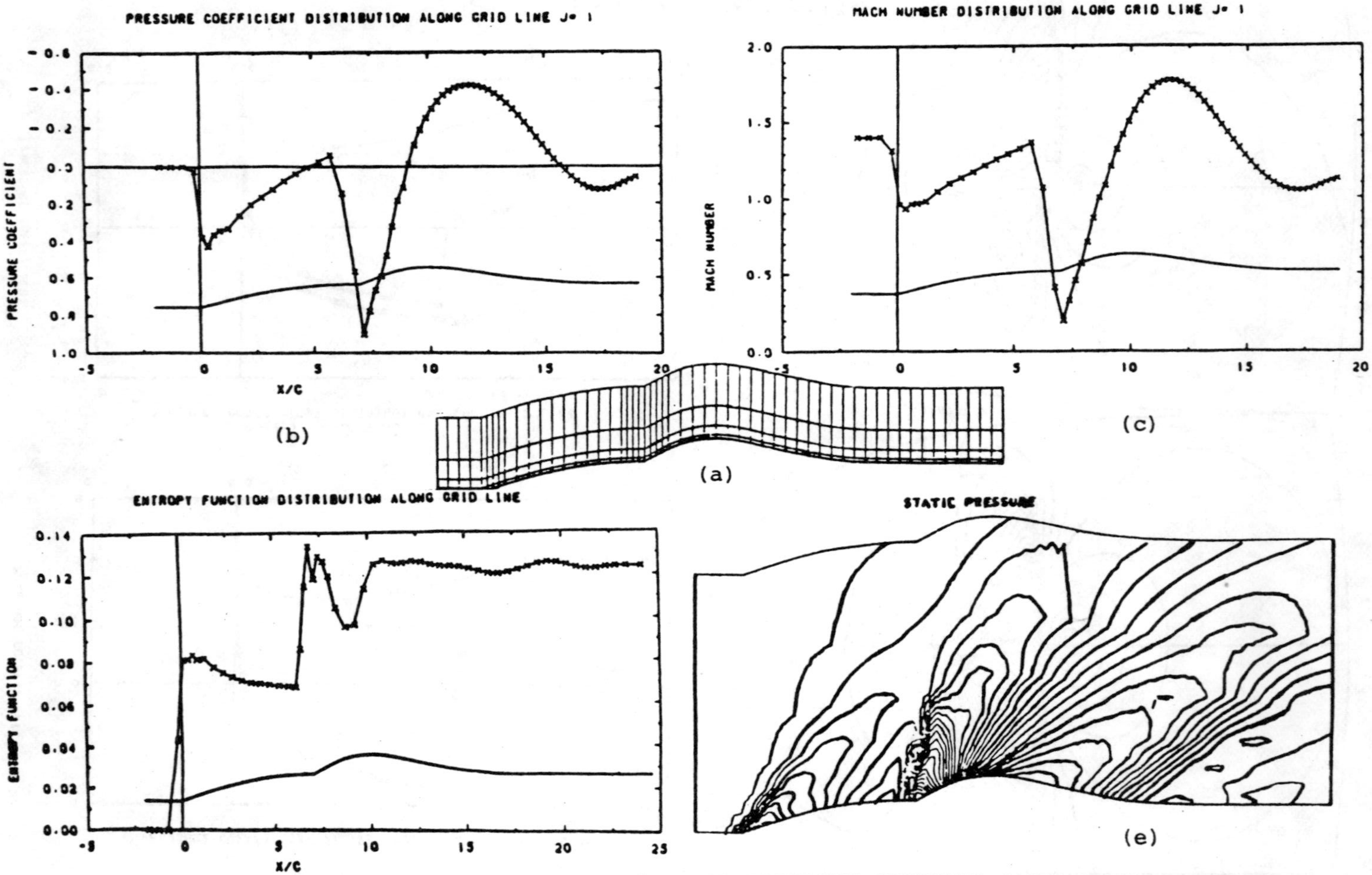

(d) FIG. 2 - SIMULATED AIRCRAFT FOREBODY (TREATED AXISYMMETRICALLY)

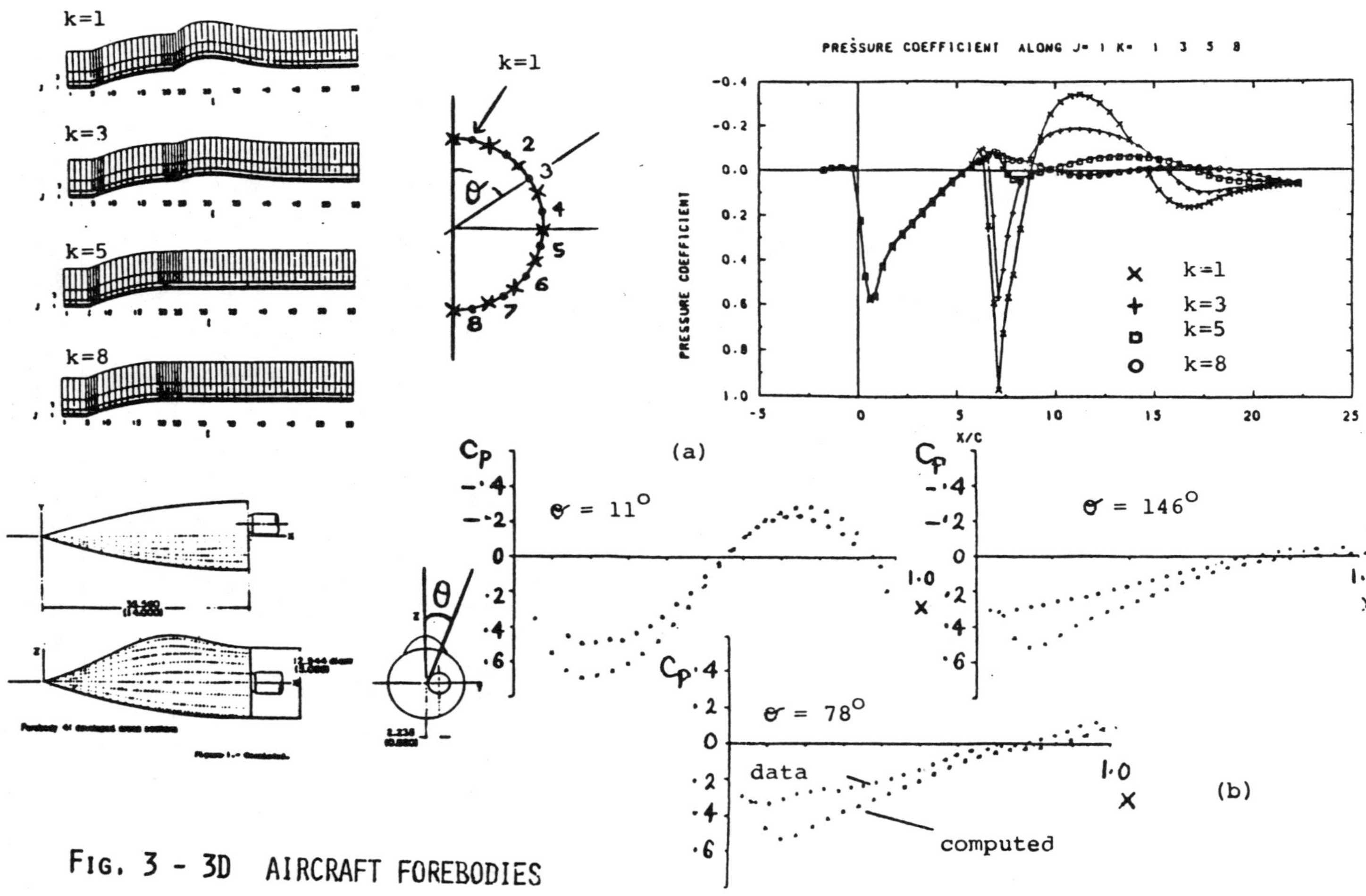

FIG. 3 - 3D AIRCRAFT FOREBODIES

THREE-DIMENSIONAL VORTEX SEPARATION PHENOMENA - A CHALLENGE TO NUMERICAL METHODS -

U. Dallmann
DFVLR, Institute for Theoretical Fluid Mechanics
Bunsenstrasse 10, D-3400 Göttingen, West Germany

1. INTRODUCTION

The calculation of three-dimensional steady separated flows at high Reynolds numbers is certainly one of the most challenging fields of computational fluid dynamics. The physicist's interest in numerical results will be in answering questions like: *How does vorticity develop vortices? How is separation related to the genesis of vortical flow?*
The engineer's interest is to perform such numerical calculations because of the enormous impact of separated flow phenomena on engineering fields like aerodynamic, combustion, pollution etc.. Despite the increasing number of publications no definition of three-dimensional separation has been given yet, none which allows to call a certain limiting streamline, or wall-shear stress trajectory a separation line is known. Hence, no information of the regular or singular behaviour of three-dimensional equations of motion could be used up to date to answer the question:
What are the simplifications to the complete equations ot motion that allow proper numerical methods to get reliable results at least of the gross behaviour of separated vortex flows?
Most of the numerical calculations are still performed, despite of missing definitions and of insufficient knowledge about the qualitative properties of separated vortex flows, i.e. about its topological structure and its structural stability.

The following discussion points out some consequences of our study: *Topological Structures of Three-Dimensional Flow Separations* [1] to the development of numerical methods that deal with separated vortex flows.

2. TOPOLOGICAL STRUCTURES OF THREE-DIMENSIONAL SEPARATED FLOWS

Why do we have to be concerned about the topological structures of flow fields where three-dimensional boundary layer separation takes place?
Because of the sensitive structural dependence of three-dimensional separated vortex flow fields on changes in the parameter space of Reynolds number, Mach number, angle of attack and equally important also on laminar, transitional or turbulent flow conditions.

Many experimental observations (see for instance [2],[3]) have shown that topologically different structures may cause the same streamline pattern on the surface of a body submerged into a flow. Dallmann [1], Hornung and Perry [4], Bippes [2] have pointed out that there is no sense in talking about separating flow patterns of flow separation in terms of wall streamline patterns only. In [1] we have shown that *there is no unique relationship between the wall streamline patterns and the vortex flow field above the surface of a body*. Topologically different flow fields may produce qualitatively the same wall streamline pattern!

Only *the topological structure of steady two-dimensional separation is invariant*, i.e. it is always characterized by a combination of two saddle points connected by one singular streamline, trapping a center point. In some cases this centre together with the saddle point of reattachment, can move downstream into the wake. We have shown in Dallmann [1] that *a two-dimensional separating and reattaching flow is a structurally unstable, transient state of three-dimensional separating and attaching vortcal flows.* We have to be aware of this in order to understand that different small deviations from perfect two-dimensional symmetry allow completely different, i.e. topological different flows to develop.

As a consequence for numerical calculations of flow fields around axially symmetric bodies one has to ensure in the limiting case of axially symmetric flow that there is perfect spatial symmetry in the calculation procedure applied.

If one looks at the limiting streamlines, the degenerate nodes of two-dimensional separation and reattachment disappear when the three-dimensional separation takes place. At separation, strong convergence of wall streamlines is observed, divergence happens near a line of flow attachment. The saddle-centre-saddle character together with uniform convergence of wall streamlines is restricted to pseudo-planar flow behaviour, for instance to the flow around swept cylinders (infinite swept wings). As soon as - for instance - an axially symmetric body is inclined relatively to the oncoming flow a fully three-dimensional flow pattern will be observed, the saddle-centre-saddle character of a separation bubble in general being destroyed. As a consequence *there is no closed region of recirculating fluid, no separation bubble in general three-dimensional flow fields.* This can be seen easily if one considers a flow having a local plane of flow symmetry which contains streamlines of a separating flow. In the purely two-dimensional case it is always the same streamline which separates and reattaches and thus forms a separation bubble. Whenever an arbitrarily small three-dimensional deformation of the flow field takes place the degenerate centre changes into a focus; a certain streamline may leave the point of separation and will start a spiralling motion around that focus. The reattaching streamline is necessarily different from the separating one. As a result the separation bubble will either break up and give rise to an unsteady flow or a steady three-dimensional separated vortex flow may be formed as shown by Fig. (1),(2),(3).

In [1] we theoretically analysed a sequence of structurally stable and unstable separating and attaching flow patterns. Elementary topological structures of associated three-dimensional flow fields have been obtained thereby (Fig. (2), Fig. (3)). We have pointed out: *There is a limited number of three-dimensional elementary topological structures.* These can be used to give a geometrical understanding of the complicated vortical flow fields found in real flow situations.

It is important to realize that *in three dimensions different structurally stable separated vortex flow fields appear and several structural flow bifurcations take place.* For the three-dimensional flow field around a given body, changes in Reynolds number, Mach number, angle of attack etc. will cause changes in the wall shear stress distribution. As soon as new critical points (saddles, foci, nodes) appear in the wall streamline pattern a structural change of the whole three-dimensional flow field takes place. A certain structurally stable pattern will be found only in a certain range of Reynolds number, Mach number, angle of attack etc. and therefore in a range where certain relationships between derivatives of the wall shear stress and the wall pressure will hold at a critical point under consideration.

3. STRUCTURAL STABILITY AND BIFURCATION - CONSEQUENCES TO NUMERICAL FLOW CALCULATIONS ?

What are the proper equations to be chosen for the calculation of steady, three-dimensional separated vortex flows ?

In recent years the great interest in three-dimensional separated flow fields has led to the development of a number of different numerical codes. These procedures have been applied to so called Parabolized Navier-Stokes equations (NS equ.), Parabolic NS equ., Thin-Layer-Approximated NS equ.. A rigorous mathematical treatment of those sets of equations is still lacking to answer the following question:
Do the simplifications applied really give approximations to the correct solutions of the complete equations ?
So far only the classical boundary layer equations have been proven to be a correct approximation to the full NS equ. for wall shear layers at sufficiently high Reynolds numbers.

Turbulent flows as well as laminar flows provide us with stationary flow patterns on and above the surface of a body. Often such a turbulent flow may be considered as a steady flow of a fictitious fluid, which is a mean representation of the turbulent flow. Since there are several topologically different flow patterns possible also in the turbulent regime we may ask:
Is there a sensitive dependence of separated flow structures on models of laminar turbulent transition incorporated in numerical codes ?
Experimental observations of Bippes [2] have shown the sensitive dependence of the structures of separated vortex flows on laminar, transitional or turbulent boundary layer development for fixed Reynolds number, Mach number and angle of attack. Qualitatively different wall streamline patterns have been observed after different methods of boundary layer tripping had been applied.

Whenever one tries to calculate a certain flow pattern observed in experiments, we believe that one has to be aware of these problems of structural stability and bifurcation, regular or chaotic development of flow fields (see also Legendre [5], Tobak and Peake [6]). Therefore we raise the question:
What are the prerequisites in order to numerically resolve these structural changes of three-dimensional separated flows that can be infered from changes of wall streamline patterns ?

4. EQUATIONS GOVERNING SURFACE FLOW PATTERNS

Let us analyse the equations of motion that govern the three-dimensional surface flow itself. Let us consider incompressible flows where the velocity field $\vec{Q}$ and the pressure P are solutions of the Navier-Stokes equations and the continuity equation

$$NS\{\vec{Q},P\} := \partial_t\vec{Q} + \vec{Q}\cdot\text{grad}\,\vec{Q} + \frac{1}{\rho}\,\text{grad}\,P + \nu\,\text{curl}^2\vec{Q} = 0 \tag{1}$$

$$\text{div}\,\vec{Q} = 0\,, \tag{2}$$

where ρ is the density, ν is the kinematic viscosity, ∂_t is the time differential operator; $\text{curl}^m\vec{Q}$ means applying the curl operator m-times.

We expect the velocity field to fulfil the no-slip boundary condition on the surface S of the body submerged into the flow. Let us choose a Cartesian coordinate system (x,y,z) according to Fig. (1) such that τ and σ are shear stress components in x- and y-direction, respectively:

$$\tau = \mu U_z\,,\quad \sigma = \mu V_z \tag{3}$$

where subscripts mean partial derivatives and $\mu = \nu\rho$ is the dynamic viscosity.

By evaluation of the transport equations $\text{curl}^m[NS\{\vec{Q},P\}] = 0$ and by use of $div[NS\{\vec{Q},P\}] = 0$ at the surface S we obtain:

$$\begin{aligned} &\mu(\partial_t P_x + \nu 2P_{xzz}) + \tau(\tau_x - \sigma_y) + \sigma(2\tau_y) = \mu\nu\tau_{zzz} \\ &\mu(\partial_t P_y + \nu 2P_{yzz}) + \tau(2\sigma_x) + \sigma(-\tau_x + \sigma_y) = \mu\nu\sigma_{zzz} \end{aligned} \qquad \text{valid on } S. \tag{4}$$

From this relationship we shall now derive substantial information about the regular or singular behaviour of simplifications of the equations of motion at a point or a line of flow separation. Equ. (4) also draws consequences on numerical resolution necessary for the calculation of three-dimensional separated flows.

5. ON THE CALCULATION OF CRITICAL POINTS IN FLOW PATTERNS

Let us consider steady surface flow patterns which have been observed in many experiments using oil flow visualization techniques or wall shear stress measurements. Oswatitsch [7] was able to show that neither compressibility nor the variation of temperature has an effect on the local behaviour of the flow in the vicinity of a point of separation. Such a point of separation, a critical point in the wall streamline pattern is a point where a streamline may leave the surface of the body. A necessary but not sufficient criterion for a streamline to leave the surface turned out to be vanishing wall shear stress components τ,σ at a point of flow separation. In three dimensions the flow separation phenomenon is always accompanied by strong local convergence and divergence of neighbouring wall streamlines and in most cases also by a structural change in the wall streamline pattern, i.e. by the creation of foci, nodes and saddle points.

Local linear analysis of the flow field near a single critical point C shows (see for instance Oswatitsch [7], Lighthill [8], Perry and Fairly [9]) that foci, nodes and saddles exist if the following relationships are fulfilled on S :

$$\begin{aligned} &\text{Focus: } D<0,\ (J>0) \\ &\text{Node: } D>0,\ J>0 \quad \text{with} \quad \begin{aligned} D &:= (\tau_x - \sigma_y)^2 + 4\tau_y\sigma_x = (\mu W_{zz})^2 - 4J \\ J &:= \tau_x\sigma_y - \tau_y\sigma_x \,. \end{aligned} \\ &\text{Saddle: } (D>0),\ J<0 \end{aligned} \tag{5}$$

We have in general $P_z = \mu W_{zz} \neq 0$ at a nondegenerate critical point. $(D>0)$ for a saddle point and $(J>0)$ for a focus are results of the other conditions.

Let us now return to equ. (4) in its steady version, i.e. set $\partial_t P_x = \partial_t P_y = 0$. Since we expect the Navier-Stokes equations to remain valid also at a point of separation where $\tau = \sigma = 0$ we may differentiate equ.(4) along S to obtain

$$(\tau_x - \sigma_y)^2 + 4\tau_y\sigma_x = -\mu^2\nu\Delta\Delta W_z \,, \qquad \text{valid on S where } \tau=\sigma=0. \tag{6}$$

This relationship holds at a critical point on S where $\tau = \sigma = 0$ under steady flow conditions. $\Delta\Delta W_z$ is twice the three-dimensional Laplacean operator acting on the wall-normal velocity gradient W_z (see Fig. (1)). Comparing equ. (6) with the characterization of critical points given by (5) we realize that the signs of the quantities $\Delta\Delta W_z$ and $\nu\Delta\Delta W_z + (W_{zz})^2$ govern the distinction between saddles, nodes and foci. The spatial distribution of these critical points can be viewed in Fig. (3) and hence we state:
In order to resolve structural changes of wall streamline patterns one has to resolve spatial changes of sign of those characterizing quantities.

6. SINGULAR BEHAVIOUR OF THREE-DIMENSIONAL EQUATIONS OF MOTION AT SEPARATION

The development of numerical schemes for the calculation of flows with separation has led us to study the different equations used today: three-dimensional steady and unsteady boundary layer equations, parabolized, thin-shear-layer-approximated and complete Navier-Stokes equations. The regular or singular behaviour of these equations at separation points and lines is explained in the following.

Structural changes appear in the surface flow pattern as soon as new critical points appear where the wall shear stress vanishes; saddles, nodes and foci may form and streamlines may leave or reach the surface where $\tau = \sigma = 0$. In order to understand the regular or singular behaviour of (simplified) equations of motion, at a point of separation in three-dimensional flow, let us have a look at the equ. (4), governing the surface flow.

Since many patterns of surface flow appear to be stationary in laminar (as well as in turbulent) flow one might look for solutions of the steady equations of motion. This results in setting $\partial_t P_x = 0$, $\partial_t P_y = 0$ in equ. (4). Since we can easily obtain from the equations of motion:

$$\begin{aligned} -2P_{xzz} &= 2\,(P_{xx} + P_{yy})_x = 2\,(\tau_{xx} + \tau_{yy})_z \\ -2P_{yzz} &= 2\,(P_{xx} + P_{yy})_y = 2\,(\sigma_{xx} + \sigma_{yy})_z \end{aligned} \qquad \text{valid on } S\,. \qquad (7)$$

we might also omit these terms in equ. (4) due to thin-shear-layer arguments. But by looking at the steady version of equ. (4) we immediately realize that this assumption is not valid in general at a critical point where the wall shear stress vanishes i.e. where $\tau = \sigma = 0$. The consequences can be derived easily using equ. (4) again.

A. Two-dimensional thin shear layers:

Let us consider two-dimensional flow separation in the y,z-plane. We set $\tau \equiv 0$ and allow no changes of the flow quantities in x-direction. Classical boundary layer theory and thin-layer-approximation provide us (in the steady flow case) with the information given by the simplification of equ. (4), i.e. by

$$\sigma\sigma_y = \mu\nu\sigma_{zzz} \qquad \text{on } S\,.$$

Since the right hand side doesn't vanish at separation it follows immediately that σ_y has to become infinite: A simple way to understand Goldstein's singularity of the steady boundary layer equations at separation. But in addition we realize that the same singular behaviour necessarily appears for any steady thin-shear-layer-model at a point of vanishing wall shear stress numerical marching procedures.

B. Pseudo-planar thin shear layers:

Under the infinite-swept-wing-condition $\partial_x(\cdot) = 0$, indicating no changes of the flow field along the span, the above discussion, concerning thin-shear-layer-assumptions may be repeated for the chordwise shear stress. We get from equ. (4) for steady flows:

$$\begin{aligned} -\tau\sigma_y + \sigma 2\tau_y &= \mu\nu\tau_{zzz} \\ \sigma\sigma_y &= \mu\nu\sigma_{zzz}\,. \end{aligned} \qquad \text{valid on } S\,. \qquad (8)$$

For the spanwise wall shear stress component τ, which doesn't vanish at the separation line, we get from equ. (8)

$$\tau = \mu\nu\left[\frac{2\tau_y}{\sigma_y^2}\sigma_{zzz} - \frac{\tau_{zzz}}{\sigma_y}\right] \neq 0 \qquad \text{where } \sigma = 0 \text{ on } S\,. \tag{9}$$

Since by equ. (8) σ_y becomes infinite at a line of separation under steady thin-shear-layer assumptions, several kinds of singularities may appear for the spanwise wall shear stress component τ as well as can be inferred from equ. (8) and equ. (9).

C. Three-dimensional thin-shear layers:

Let us return to the fully three-dimensional flows where separation is mostly accompanied by the appearance of critical points on the surface. Restricting the calculation of such flows to the steady case and applying thin-shear-layer arguments one would again cause singularities for the wall shear stress gradients at a point of separation where $\tau = \sigma = 0$ unless the pressure terms of equ. (4) become actually zero there, i.e.

$$\bar{\Delta}P = P_{xx} + P_{yy} = \text{extremum!} \qquad \text{where } \tau = \sigma = 0 \text{ on } S\,. \tag{10}$$

This is a necessary condition in order to avoid singular behaviour of the simplified equations of motion. This exceptional situation will not be met in general. On the contrary, any element of the matrix D_{ij} in equ. (4) , i.e.

$$D_{ij} = \begin{pmatrix} \tau_x - \sigma_y & 2\tau_y \\ 2\sigma_x & -\tau_x + \sigma_y \end{pmatrix} \tag{11}$$

may become infinite in a marching procedure based on steady thin-shear-layer-equations.

These results complement the discussion of the elementary topological structures of steady three-dimensional separated flows and show that marching procedures will not be capable of calculating such flows. The only exceptions, where marching procedures might be useful, seem to be those flows where no structural changes (in a topological sense) occur, i.e. flows, where only a gradual formation of (separated) vortices takes place. In such cases (the flow around an inclined prolate spheroid seems to be such a case) the so-called separation lines are formed by strong convergence of neighbouring wall streamlines only , no critical point appears indicating the location from where on one could call a certain wall streamline (wall shear stress trajectory) a separation line. In [1] we have discussed problems associated with such a gradual formation of vortices; stream surface folds, cusps and rips have been shown to be capable of explaining such types of vortex flow separation. In order to avoid such problems of singular behaviour of steady, thin-shear-layer-approximated equations of motion unsteady procedures seem to be a way out.

7. TIME-DEPENDENT CALCULATION OF THREE-DIMENSIONAL STEADY VORTEX FLOWS

First: We realize again from equ. (4) the following necessary condition in order to get unique results. Any unsteady solution that fulfils the simplified equations

$$\begin{aligned} -\mu\partial_t P_x &= \tau(\tau_x - \sigma_y) + \sigma(2\tau_y) - \mu\nu\tau_{zzz} \\ -\mu\partial_t P_y &= \tau(2\sigma_x) + \sigma(-\tau_x + \sigma_y) - \mu\nu\sigma_{zzz} \end{aligned} \qquad \text{on } S \tag{12}$$

has to approach asymptotically in time that solution which fulfils the complete, steady equations of motion

$$\left.\begin{aligned} -2\mu\nu P_{xzz} &= \tau(\tau_x - \sigma_y) + \sigma(2\tau_y) - \mu\nu\tau_{zzz} \\ -2\mu\nu P_{yzz} &= \tau(2\sigma_x) + \sigma(-\tau_x + \sigma_y) - \mu\nu\sigma_{zzz} \end{aligned}\right\} \text{ on } S \ . \qquad (13)$$

Second: As a consequence of those structural changes of the flow fields and the existence of bifurcating flow patterns one might also think about using unsteady calculation procedures instead of steady ones in order to approach asymptotically a steady state solution of the full Navier-Stokes equation. This has to be done using a certain starting solution. For the time being we do not know anything about the dependence of the topological structure to be approached on the initial flow field chosen. Hence the following question arises to those who want to minimize trial and error: *How should one choose proper initial data in order to numerically approach the structurally different steady flow patterns observed in experiment ?*

As a consequence of our qualitative understanding of changing topological structures of three-dimensional separated flows we simply asked questions to stimulate proper developments of numerical means for vortex flow calculations.

LITERATURE

[1] Dallmann, U.: Topological Structures of Three-Dimensional Flow Separations. DFVLR-IB 221-82 A 07, 1982.

[2] Bippes, H.; Turk, M.: Oil Flow Patterns of Separated Flow on a Hemisphere Cylinder at Incidence. DFVLR-IB 222-83 A 07, 1983.

[3] Peake, D.J.; Tobak, M: Three-Dimensional Flows About Simple Components at Angle of Attack. AGARD Lecture Series No.121, Paper No.2, 1982.

[4] Hornung, H.; Perry, A.E.: Streamsurface Bifurcation, Vortex Skeletons and Separation. DFVLR-IB 222-82 A 25, 1982.

[5] Legendre, R.: Regular or catastrophic evolution of steady flows depending on parameters. La Recherche Aérospatiale 1982 - 4, pp.41-49.

[6] Tobak, M.; Peake, D.J.: Topology of Three-Dimensional Separated Flows. Ann. Rev. Fluid Mech., 1982, 14, pp.61-85.

[7] Oswatitsch, K.: Die Ablösungsbedingung von Grenzschichten. IUTAM Symposium on Boundary Layer Research, Freiburg 1957.

[8] Lighthill, M.J.: Attachment and separation in three-dimensional flow. Laminar Boundary Layers, ed. L. Rosenhead, II, 2.6, pp.72-82, Oxford Univ. Press, 1963.

[9] Perry, A.E.; Fairlie, B.D.: Critical Points in Flow Patterns. Advances in Geophysics, 18 B , 1974, pp.299, Academic Press, New York - San Francisco - London.

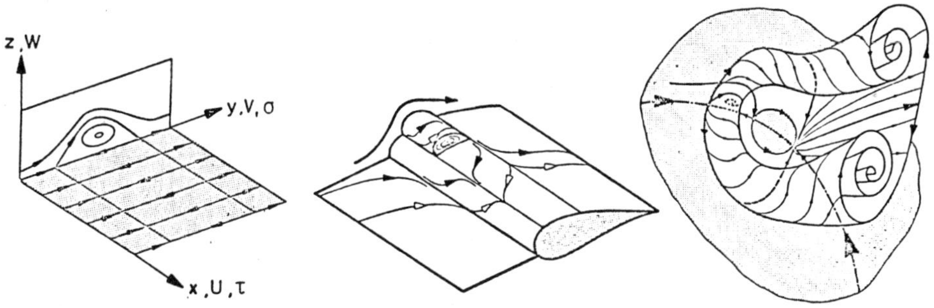

Fig. 1 Sketch of a break-up of a 2D-separation bubble into a 3D-vortex flow. (A: plane separation; B: pseudo-planar separation; C: 3D-separation).

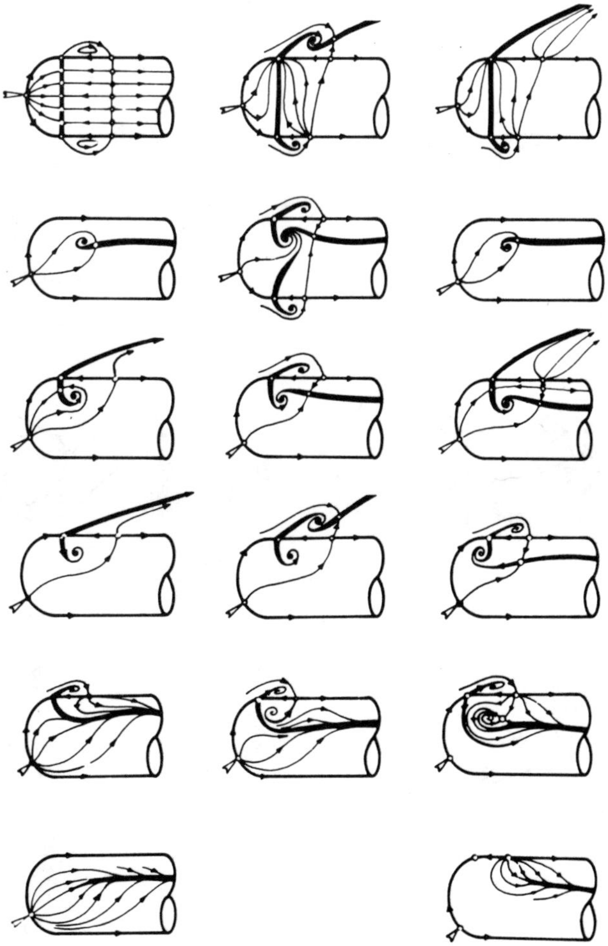

Fig. 2 Development of wall and symmetry plane streamlines of 3D-separated flows around blunt bodies. (Topologically changing patterns).

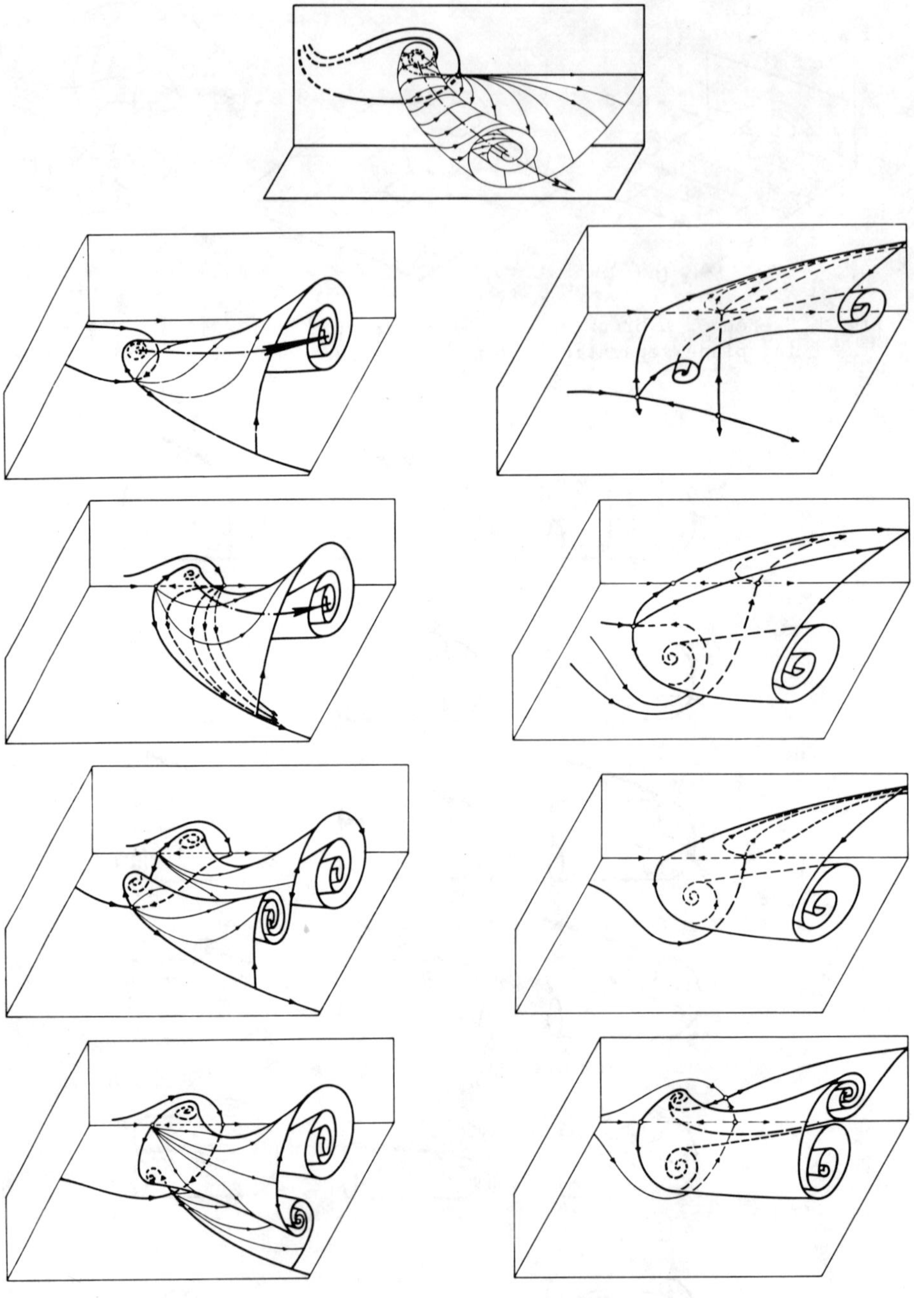

Fig. 3 Streamsurface representations of elementary topological structures of three-dimensional vortex flows. (Separating vortical flows shown only where streamlines spiral into the foci).

THREE-DIMENSIONAL CALCULATION OF FILM COOLING BY A ROW OF JETS

A.O. Demuren and W. Rodi
Institute for Hydromechanics
University of Karlsruhe
D-7500 Karlsruhe 1

SUMMARY

This paper presents three dimensional calculations of the flow and film cooling effectiveness resulting from the injection of coolant from an inclined row of jets into a turbulent mainstream. The governing partial-differential-equations are solved by a finite difference procedure which is capable of accommodating local regions with reverse flow in an efficient manner requiring only a moderate increase in computer storage. The turbulent stresses and heat fluxes are obtained from a modified, (non-isotropic), $k-\varepsilon$ turbulence model. The calculations are compared to measurements at two mainstream turbulence levels and two jet to mainstream velocity ratios. The velocity predictions are satisfactory, but the predicted turbulence intensity distributions show only qualitative agreement with the measurements. The temperature field is also well predicted in one of the cases, but there are some uncertainties in the other. The same applies for the film cooling effectiveness. There is a need for an improved turbulence model.

INTRODUCTION

In many applications of film cooling, design considerations prevent the use of continous slots for the introduction of the coolant. Discrete holes or a slot with discontinuities (due to structural supports) are used for injection. The jet stream, blowing through discrete openings, thus interacts with the essentially two-dimensional flow, which is considerably different from that resulting from a continous slot injection. The film cooling effectiveness is now not only a function of the streamwise distance, but of the cross-stream location as well, and is often considerably less than that for slot injection. It has also been observed [1] that, as the blowing rate is increased beyond a relatively low value (usually about 0.5 for a 30° injection), the film cooling effectiveness for injection through discrete holes falls off rapidly. For design purposes, there is therefore a need for a three-dimensional calculation methods which enable the analysis of the flow field and the computation of the film cooling effectiveness for geometries and blowing rates of practical interest. However, most of the previous calculation methods applied to this problem, have been two-dimensional ones, in which the variations in the lateral direction were either neglected or partially accounted for by empirical coefficients [2].

Calculation methods which adequately account for three-dimensional effects have been presented by Patankar, Rastogi, and Whitelaw [3] for injection through discrete tangential slots. Patankar, Basu, and Alpay [4] have analysed the injection at high blowing rates from a single hole, normal to the mainstream, and similar cases for injection through a row of holes have been considered by Khan et al. [5] and Demuren [6]. Bergeles, Gosman, and Launder [7] presented a calculation method which was restricted to injections at very low rates from a normal hole, or to medium rates from holes inclined an acute angle to a turbulent mainstream.

They showed that in the predominantly near wall flow, downstream of the injection hole, it was necessary to take special account of the anisotropy of the turbulent diffusion coefficients. The standard k - ε turbulence model, (which assumes an isotropic eddy viscosity) employed in [4,5,6] was found to be inadequate.

In the present paper, a three-dimensional numerical procedure is presented for the calculation of the flow and film cooling effectiveness of a row of inclined jets issuing from round holes into a turbulent main stream. The partial-differential equations governing the flow and heat transport are solved by a finite difference technique, incorporating the locally-elliptic procedure of Rodi and Srivatsa [8]. The latter enables local regions of reverse flow to be accounted for in an economical manner. Thus, one does not need to employ such a computer-storage-intensive procedure as in [4, 5], and there is no restriction to low blowing rates as in [7]. The turbulent stresses and heat fluxes are evaluated with the modified k - ε turbulence model, proposed by Bergeles et al. [7]. Predictions with the present numerical procedure are compared with measurements of Kadotani and Goldstein [9, 10].

MATHEMATICAL MODEL

Governing differential equations

The time-averaged partial-differential equations governing the steady uniform-density three-dimensional flow and heat transfer may be written in Cartesian coordinate, for a property Ø as

$$\frac{\partial U\phi}{\partial x} + \frac{\partial V\phi}{\partial y} + \frac{\partial W\phi}{\partial z} = -\frac{\partial \overline{u\varphi}}{\partial x} - \frac{\partial \overline{v\varphi}}{\partial y} - \frac{\partial \overline{w\varphi}}{\partial z} + \frac{S_\phi}{\rho} \tag{1}$$

and the continuity equation as

$$\frac{\partial U}{\partial x} + \frac{\partial V}{\partial y} + \frac{\partial W}{\partial z} = 0 \tag{2}$$

Ø may stand for any of the velocities, temperature or turbulence quantities k or ε. $S_Ø$, then stands for the respective source terms, given in [6].

Turbulence model

According to the standard k - ε turbulence model the turbulent stresses may be expressed in tensor notation as

$$-\overline{u_i u_j} = \nu_t \left(\frac{\partial U_i}{\partial x_j} + \frac{\partial U_j}{\partial x_i} \right) - \frac{2}{3} \rho k \delta_{ij} \tag{3}$$

where i, j = 1, 2, 3 and $(x_1, x_2, x_3) = (x, y, z)$

The turbulent heat fluxes are similarly expressed as

$$-\overline{u_i T} = \frac{\nu_t}{\sigma_T} \frac{\partial T}{\partial x_i} \tag{4}$$

where i = 1, 2, 3. ν_t is the isotropic eddy viscosity which is evaluated from k and ε as

$$\nu_t = c_\mu k^2 / \varepsilon \tag{5}$$

The turbulent kinetic energy k and its rate of dissipation ε are obtained by solving transport equations of type (1) for k and ε.

Bergeles et al. [7], derived a non-isotropic eddy viscosity by simplifying the second of the two Reynolds stress models proposed by Launder, Reece, and Rodi [11], and the turbulent heat flux model of Gibson and Launder [12]. By neglecting the transport terms, and all the gradients of the lateral and vertical velocity components they obtained

$$-\overline{uw} = \nu_{t,x}\frac{\partial W}{\partial x}\ ; \ -\overline{vw} = \nu_{t,y}\frac{\partial W}{\partial y} \tag{6}$$

$$-\overline{uT} = \frac{\nu_{t,x}}{\sigma_T}\frac{\partial T}{\partial x}\ ; \ -\overline{vT} = \frac{\nu_{t,y}}{\sigma_T}\frac{\partial T}{\partial y} \tag{7}$$

where

$$\begin{aligned} \nu_{t,y} &= \nu_t = c_\mu \frac{k^2}{\varepsilon} \\ \nu_{t,x} &= \nu_t\,(1{\cdot}0 + 3{\cdot}5\,(1 - {}^{y}/_{\Delta}));\ \text{for } y \leq \Delta \\ \nu_{t,x} &= \nu_t \qquad ;\ \text{for } y > \Delta \end{aligned} \tag{8}$$

$$\sigma_T = 0 \cdot 9\ ;\ c_\mu = 0 \cdot 09,$$

and, Δ is the boundary layer thickness.

The form of the expression for $\nu_{t,x}$ was obtained from empirical data and indicates a linear decay of the anisotropy from the wall to the edge of the boundary layer. (The streamwise (z) gradients have been neglected, so are the respective stresses and heat flux).

Solution procedure and boundary conditions

The solution procedure is based on the well known SIMPLE algorithm which was also employed in [3 - 8], so it will not be described here. More attention needs to be paid to the boundary conditions. The computational domain has six boundaries, along each of which boundary conditions need to be prescribed for the dependent variables. Along the top and bottom walls, turbulence is assumed to be in a state of local equilibrium so that the wall function method is employed in prescribing the boundary conditions for the velocities and turbulence quantities. The adiabatic condition is prescribed for the temperature. The two side boundaries are symmetry planes along which the normal velocity and the normal gradients for all other variables are prescribed as zero. Measured mainstream conditions are prescribed at the inlet plane, and the exit plane of the computational domain is positioned far enough downstream of the injection port, that it is an outflow boundary where only pressure conditions are required. A constant pressure gradient is prescribed at this plane.

The boundary conditions for the injected jet are more difficult to prescribe. Measurements by Andreopoulos [13] inside the injection pipe for a jet issuing at right angles to the mainstream has shown significant influence of the mainstream on the jet exit profiles. The importance of correctly modelling this exit flow was shown in [6]. In the absence of data on exit velocity profiles, for inclined jets injections the following conditions have been prescribed:

$$\begin{aligned} V_j &= Q_j \sin\alpha \\ W_j &= Q_j \cos\alpha + 0{\cdot}30 * W_{IN} * {}^{z}/D \end{aligned} \tag{9}$$

This implies a uniform profile for the vertical component of the jet exit velocity, but a linearly increasing longitudinal profile. The former will probably not be constant, but previous calculations [6] have shown the latter to be of more significence, and the measurements of [13] did show the longitudinal component to be proportional to the free stream velocity, in the cases studied, where one might have expected it to be zero.

The boundary conditions for k and ε are prescribed from jet data measured in the absence of the cross flow as

$$k_i = 0{\cdot}005\, V_i^2 \;;\; \varepsilon_i = k^{3/2}/0{\cdot}5\, D. \qquad (10)$$

RESULTS AND DISCUSSION

Test case and computational details

The present model predictions are compared with the experimental measurements of Kadotani and Goldstein [9, 10] for a row of inclined jets issuing at an angle of 35° into a turbulent mainstream. The flow situation is illustrated in Fig. 1. Measurements were made at various freestream turbulence levels and various jet to mainstream velocity ratios. In the present paper comparisons are made for two turbulence levels; case 1A with an intensity of 8•2 % and an average length scale of 0•33 D, and case 4A with an intensity of 4•8 % and a length scale of 0.12 D. Two velocity ratios, 0•35 and 1•5 are considered, in each case.

The calculations were carried out with a grid of 14x41x66 in the (x,y,z) directions. This is a relatively fine grid, which experience in [6] indicates should give "grid-independent" results. A typical calculation required 200 iterations for convergence, with a core storage of 2048 kbytes and takes 60 minutes of CPU time on the Siemens 7880 computer.

Comparison of predictions with experiments

The computed velocity fields are compared with measurements for the lower velocity ratio, at the crossstream plane one diameter downstream of the jet hole, in Fig. 2. The agreement is seen to be good, except very close to the wall at $^x/D = 0{\cdot}25$. The observed kink in the velocity profile is not reproduced. The wall function method used to prescribe the boundary conditions near walls, require that the nearest grid point be placed in the fully turbulent region, and it is therefore not possible to resolve some near wall effects in detail. The corresponding fluctuating velocity profiles are compared in Fig. 3. Since the individual components of the Reynolds stresses are not computed in the present model, it is assumed that $\tilde{w} = k^{1/2}$, which is approximately true in wall boundary layers. The comparisons show only a qualitative agreement. The observed complex turbulent structure, presumably resulting from the jet-mainstream interaction is not well predicted. This may partly be due to the crudity of the above assumption for $\tilde{w}$, in this region, and lack of detailed information on jet exit velocity and turbulent quantities. The temperature predictions are compared with the measurements in Fig. 4. At the higher turbulence level (case 1A), there is fairly good agreement between the predicted contours and the measured ones. The correct trends are also predicted at the lower level (case 4A), although the observed bifurcation of the jet is not reproduced as strongly. The reduced mixing, and the corresponding smaller penetrations of the injected fluid, as compared with case

1A, are adequately predicted.

The predicted film cooling effectiveness is compared with the measurements, for all cases, in Fig. 5. There is a striking difference between the measured effectivenesses for cases 1A and 4A, which is a consequence of the bifurcation of the jet, in the latter case, previously discussed. This phenomenon does not appear to result mainly from the free stream turbulence level and length scales alone, but there may be history effects at play, originating at the turbulence-generating grid. Measurements for case 3A, (included in Fig. 5) with the same mainstream turbulence intensity as case 4A but a 50% higher length scale do not show any tendency towards bifurcation. They show similar trends as those for case 1A, but with somewhat higher effectiveness near the jet centre plane and lower values near the mid plane between jets. Such a trend is also predicted by the present model. However, the effectiveness appears to be slightly overpredicted near the jet centre plane and underpredicted near the mid plane between jets. This is partly due to the adiabatic wall temperatures used in computing the measured effectiveness having been corrected for wall conduction. Deficiencies in the turbulence model would also contribute to these deviations, since turbulent diffusion has a stronger effect on the temperature field than on the velocity field. Calculations are now being carried out with an improved turbulence model. This employs the Reynolds stress and heat flux models of Launder, Reece, and Rodi [11], and Gibson and Launder [12] respectively, but relates the transport terms to the transport of the turbulence kinetic energy. The latter enables the Reynold stresses and heat fluxes to be calculated from algebraic expressions. Unlike in [7], no velocity and temperature gradients are neglected; especially in the region of jet-mainstream interaction. The present computations appear to be little affected by numerical diffusion. Test calculations with a higher-order-method did not produce significantly different results.

CONCLUDING REMARKS

A three-dimensional numerical procedure for calculating the flow and film cooling effectiveness resulting from the injection of coolant from a row of jets into a turbulent mainstream was presented. The velocity field was predicted fairly well, but there are some uncertainties in the predictions of the turbulence field and the film cooling effectiveness. There may be a need for employing an improved turbulence model. On the experimental side, more attention needs to be paid to determining the jet exit conditions in the presence of the mainstream.

ACKNOWLEDGEMENT

The calculations were carried out with a highly modified FLAIR computer code of CHAM Ltd. on the Siemens 7880 computer of the University of Karlsruhe. The project was sponsored by the German Research Foundation.

NOMENCLATURE

c_μ	turbulence model constant	α	jet angle
D	jet diameter	Δ	boundary layer thickness
k	turbulence kinetic energy	ν_t	turbulent viscosity

Q_j	jet velocity	σ_ϕ	Prandtl/Schmidt no
$S\phi$	source of dependent variables ϕ	ε	dissipation rate of turbulence kinetic energy
T	Temperature		
U,V,W,	velocity components	ϕ, φ	dependent variable and its fluctuation
x,y,z	coordinates		

Subscripts

IN	mainstream
i	jet
W	wall

REFERENCES

[1] Goldstein, R.J., "Film cooling", Advance in Heat Transfer, 7, (1971) pp. 321-379.

[2] Platten, J.L. and Keffer, J.F., "Entrainment in deflected axisymmetric jets at various angles to the stream", University of Toronto, Mech. Engg. Dept. Rept. TP6808 (1968).

[3] Patankar, S.V., Rastogi, A.K., and Whitelaw, J.H., "The effectiveness of three-dimensional film-cooling slots-II Predictions", Int. J. Heat Mass Transfer, 16, (1973) pp. 1665-1681.

[4] Patankar, S.V., Basu, D.K., and Alpay, S.A., "Prediction of three-dimensional velocity field of a deflected turbulent jet", ASME J. Fluids Engg., 15, (1977) pp. 758-762.

[5] Khan, Z.A., McGuirk, J.J., and Whitelaw, J.H., "A row of jets in cross flow", AGARD CP 308 (1982).

[6] Demuren, A.O., "Numerical calculations of steady three-dimensional turbulent jets in cross flow", Comp. Meth. Appl. Mech. Engg., 37, (1983) pp. 309-328.

[7] Bergeles, G., Gosman, A.D., and Launder, B.E., "The turbulent jet in a cross-stream at low injection rates", Mech. Eng. Dept., University of California, Davis, Rept. TF/78/3 (1978)

[8] Rodi, W. and Srivatsa, S.K., "A locally elliptic calculation procedure for three-dimensional flows and its application to a jet in cross flow", Comp.Meth.Appl.Mech.Engg., 23, (1980) pp. 67-83.

[9] Kadotani, K. and Goldstein, R.J., "On the nature of jets entering a turbulent flow, Part A - Jet-mainstream interactions", Proc. 1977 Tokyo Joint Gas Turbine Congress, pp. 46-54.

[10] Kadotani, K. and Goldstein, R.J., "On the nature of jets entering a turbulent flow, Part B - Film cooling performance", Proc. 1977 Tokyo Joint Gas Turbine Congress, pp. 55-59.

[11] Launder, B.E., Reece, G.J., and Rodi, W., "Progress in the development of a Reynolds-stress turbulence closure", J. Fluid Mech., 68, (1975) pp. 537-566.

[12] Gibson, M.M. and Launder, B.E., "Ground effects on pressure fluctuations in the atmospheric boundary layer", J. Fluid Mech., 86, (1978) pp. 491-511.

[13] Andreopoulos, J., "Measurements in a pipe flow issuing perpendicular into a cross stream", ASME J. Fluids Engg., 104, (1983) pp. 493-499.

FIGURES

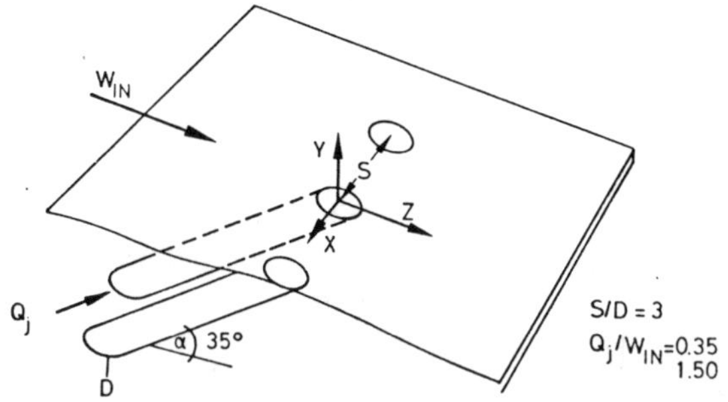

Fig. 1: Flow configuration

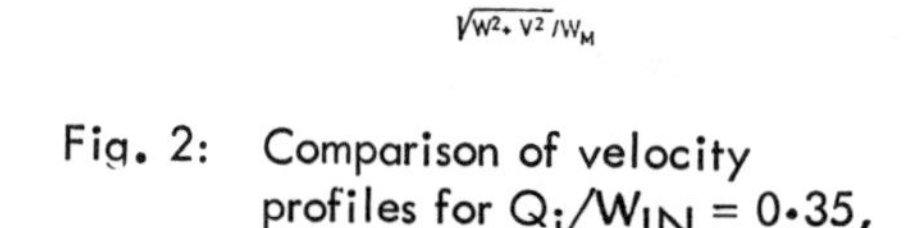

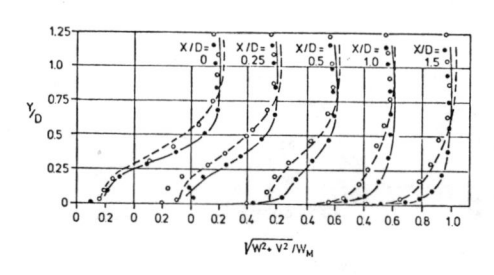

Fig. 2: Comparison of velocity profiles for $Q_j/W_{IN} = 0{\cdot}35$, at $^z/D = 1.87$: Experiments: - • - case 1A; o - case 4A, present predictions: -,—— case 1A; --- case 4A.

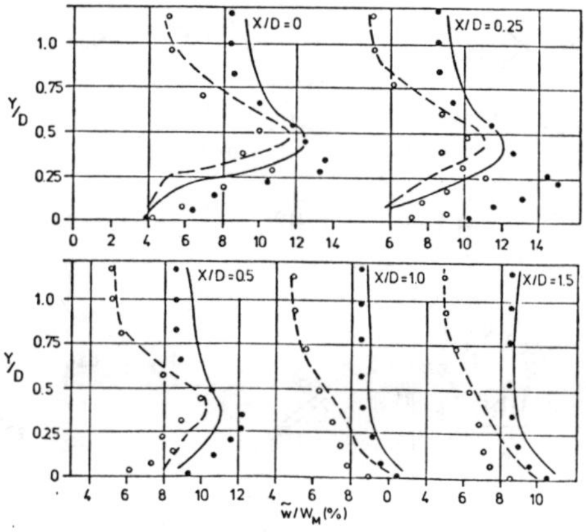

Fig. 3: Comparison of turbulence intensity profiles (Key as in Fig. 2)

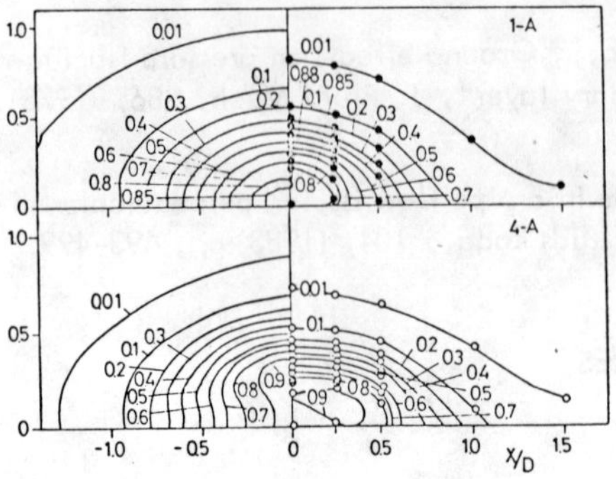

(a) $^{Z}/D = 1{\cdot}87$

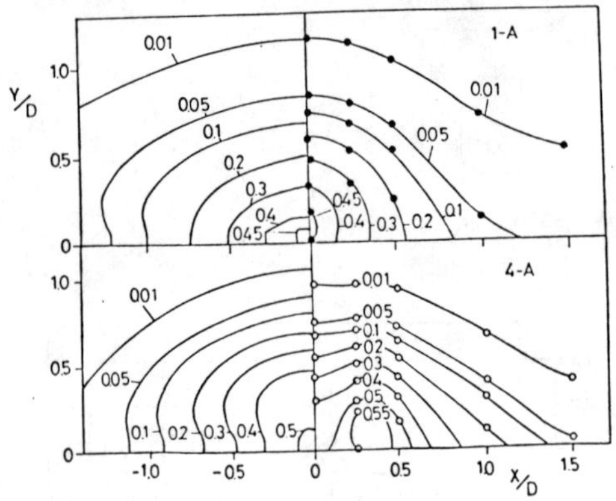

(b) $^{Z}/D = 4{\cdot}82$

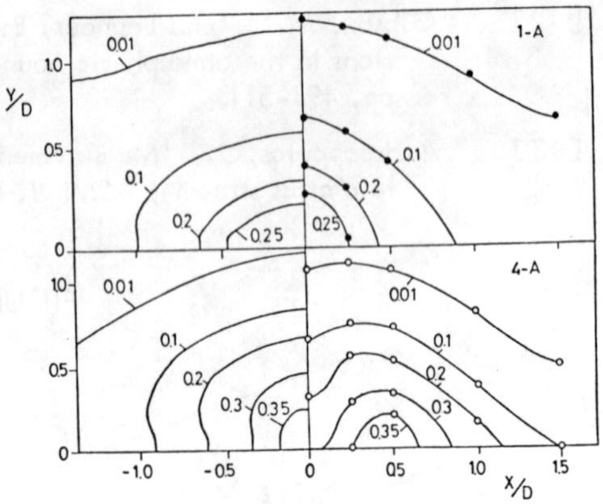

(c) $^{Z}/D = 7{\cdot}52$

Fig. 4: Comparison of normalised Temperature contours;
—●—, —o— measurements
——— present predictions.

$$\frac{T_w - T_{IN}}{T_j - T_{IN}}$$

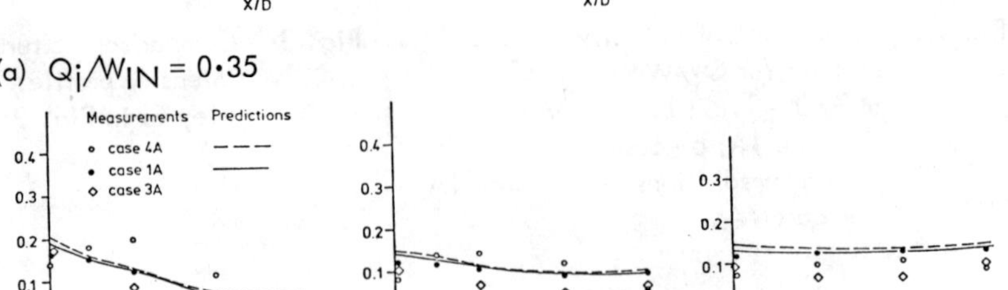

(a) $Q_j/W_{IN} = 0{\cdot}35$

$$\frac{T_w - T_{IN}}{T_j - T_{IN}}$$

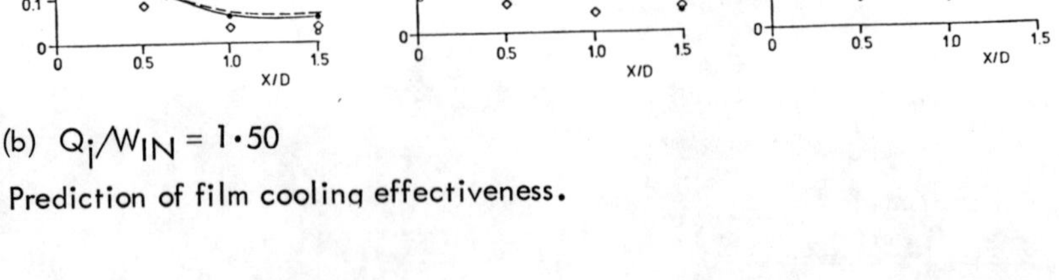

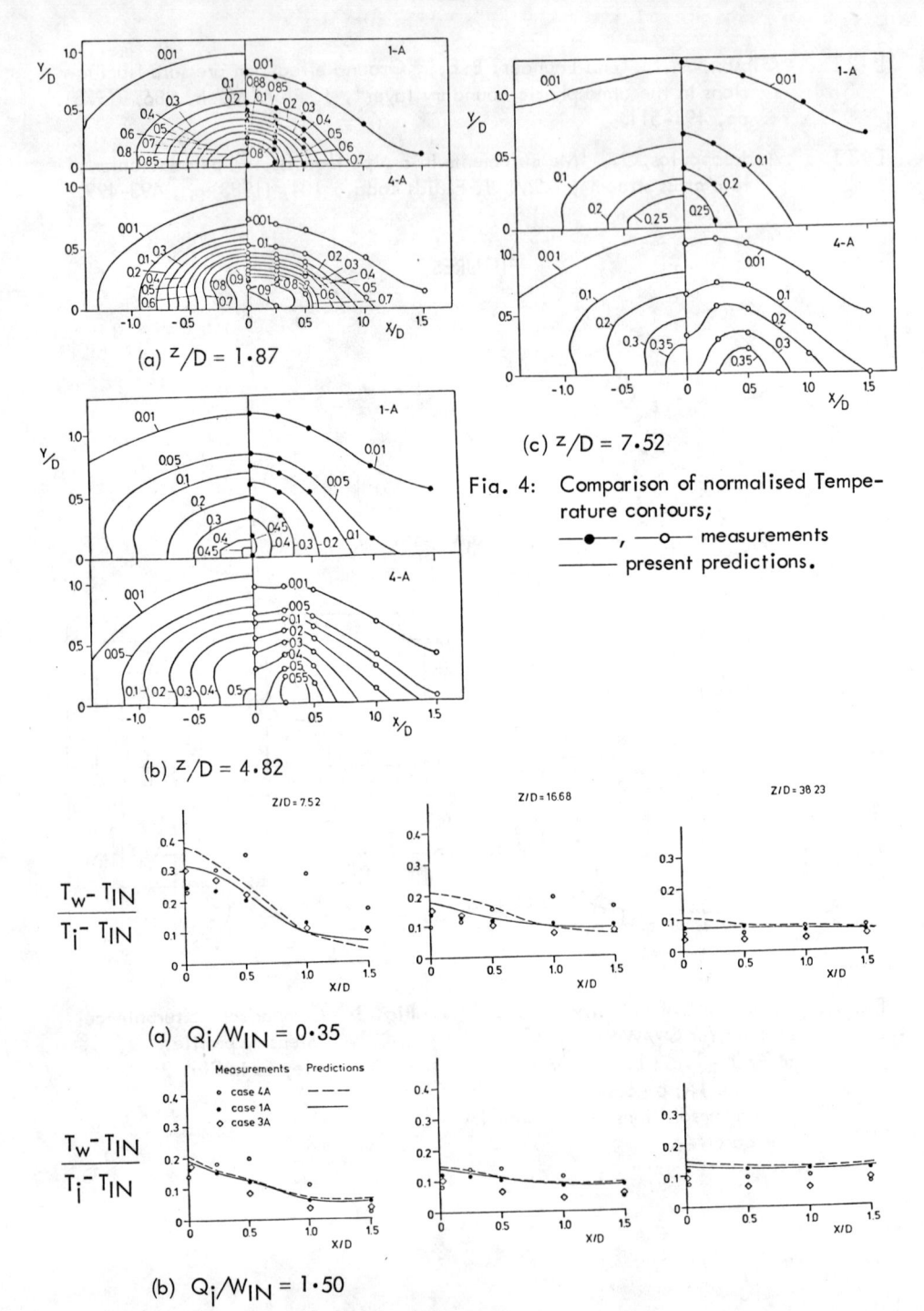

(b) $Q_j/W_{IN} = 1{\cdot}50$

Fig. 5: Prediction of film cooling effectiveness.

SPECTRAL APPROXIMATION FOR BOUSSINESQ DOUBLE DIFFUSION

ELIE F., CHIKHAOUI, A., RANDRIAMAMPIANINA, A., BONTOUX, P., and ROUX, B.
Institut de Mécanique des Fluides
1, rue Honnorat, 13003 Marseille, France

SUMMARY

A Tau-Chebyshev method is proposed for the solution of 2D-Boussinesq double diffusion equations. The method is based on fast Poisson-Helmoltz Solvers (matrix diagonalization technique) and FFT (Singleton, Lhomme algorithms). The numerical stability of AB and AB-CN schemes in time is discussed when using the vorticity-stream function formulation. The optimization of the method is sought for vector computers. Applications are made for buoyancy driven flow problems.

1. INTRODUCTION

The spectral methods were developed as efficient approximation methods at the M.I.T. since 1971 (Orszag and co-workers, 1971,72,76,77, 79,80,82 ...) because of the construction of very fast algorithms, as the FFT and the Poisson Solvers, adapted to special trial functions as the Fourier series and the Chebyshev polynomials. When the calculation are made in the spectral space, the boundary conditions are easily taken into account with the Tau method. The basic rate of efficiency of the Chebyshev polynomials is more than 2 in terms of degrees of freedom, compared to second order and hermitian finite difference schemes (Bondet de la Bernardie, 1980, Bontoux et al, 1980,81, Hirsh et al,1981,82).

The Tau-Chebyshev method is applied to the Navier-Stokes equations with the vorticity and the stream function as dependent variables (Morchoisne,1979, Bondet de la Bernardie, 1980). The initial method was optimized in terms of computing time for its use on vector computers as CRAY01. Applications were made for the simulation of flows driven by buoyancy in 2D differentially heated enclosures. The results concern : (i) the onset of instabilities on the basic convection regimes for mono-component fluid flows, (ii) the simulation of crystal growth by vapor process with the double diffusion equations and interfacial mass flux models.

2. PHYSICAL MODELS AND GOVERNING EQUATIONS

2.1. A rectangular cavity of aspect ratio $\ell = L/H$, with vertical hot ($\bar{T}_2$) and cold ($\bar{T}_1$) active walls is considered as basic physical model (Fig.1). For binary mixture, flat isothermal interfaces are assumed at the hot (source) and cold (sink) walls. They are impermeable to the inert component, noted B while for the vapor component, noted A, the mass fractions are given as $\bar{W}_{A2}$ and $\bar{W}_{A1}$. The passive side walls are supposed perfectly conducting and impermeable to both species.

2.2. With the Boussinesq approximation, the density is linearly related to the variation of the temperature, $\bar{T}$, and the mass fraction, $\bar{W}_A$, as follows :

$$\bar{\rho} = \bar{\rho}_o \; (1 - \beta(\bar{T} - \bar{T}_o) - \alpha_A \; (\bar{W}_A - \bar{W}_{Ao})) \qquad (1)$$

where β is the thermal expansion factor (= $1/\bar{T}_o$ for ideal gas), and

$\alpha_A = \dfrac{\bar{M}_o\ (\bar{M}_B - \bar{M}_A)}{\bar{M}_A\ \bar{M}_B}$ is the solutal expansion factor expressed in terms of the molar masses, the subscript o refering to average quantities.

With the dimensionless vorticity, ζ, streamfunction, ψ, temperature, T, and mass fraction, W_A, the governing equations introduce the physical parameters : Prandtl number, $Pr = \nu/\kappa$, Lewis number, $Le = \kappa/D_{AB}$, thermal Rayleigh number, $Ra_T = \beta g\ (\bar{T}_2 - \bar{T}_1)(2H)^3 /\nu_o \kappa_o$, and solutal Rayleigh number, $Ra_M = \alpha_A g(\bar{W}_{A2} - \bar{W}_{A1})\ (2H)^3 /\nu_o\ D_{AB}$.

$$\frac{\partial \zeta}{\partial t} + \frac{u}{\ell}\frac{\partial \zeta}{\partial x} + v\frac{\partial \zeta}{\partial y} = Pr\ \left(\frac{1}{\ell^2}\frac{\partial^2 \zeta}{\partial x^2} + \frac{\partial^2 \zeta}{\partial y^2}\right) + \frac{Ra_T\ Pr}{16}\ \frac{\partial T}{\partial y} + \frac{Ra_M\ Pr}{16\ Le}\ \frac{\partial W_A}{\partial y} \tag{2}$$

$$\frac{1}{\ell^2}\ \frac{\partial^2 \psi}{\partial x^2} + \frac{\partial^2 \psi}{\partial y^2} = \zeta \tag{3}$$

$$\frac{\partial T}{\partial t} + \frac{u}{\ell}\frac{\partial T}{\partial x} + v\ \frac{\partial T}{\partial y} = \frac{1}{\ell^2}\ \frac{\partial^2 T}{\partial x^2} + \frac{\partial^2 T}{\partial y^2} \tag{4}$$

$$\frac{\partial W_A}{\partial t} + \frac{u}{\ell}\ \frac{\partial W_A}{\partial x} + v\ \frac{\partial W_A}{\partial y} = \frac{1}{Le}\ \left(\frac{1}{\ell^2}\ \frac{\partial^2 W_A}{\partial x^2} + \frac{\partial^2 W_A}{\partial y^2}\right) \tag{5}$$

where $u = \dfrac{\partial \psi}{\partial y}$ and $v = -\ \dfrac{1}{\ell}\dfrac{\partial \psi}{\partial x}$.

2.3. For the two components fluid, the dynamical variables, ζ and ψ, are defined as the mass average quantities. A model for the interfacial mass flux of component A at the active wall is derived from Fick's law (Rosenberger and co-workers, 1979, 80,81). It results Robbin's type boundary conditions :

$$v\ (x, \pm 1) = -\ \frac{1}{Le}\ \frac{1}{E - W_A\ (x, \pm 1)}\ \frac{\partial W_A}{\partial y}\ (x, \pm 1) \tag{6}$$

where E is a dimensionless number characterizing the mass fractions at the source and the sink and, then, related to the molar masses of A and B and the partial pressure conditions corresponding to the sublimation and the cristallization (Elie, 1983)

$$E = \frac{2\ (1 - \bar{W}_{Ao})}{\bar{W}_{A2} - \bar{W}_{A1}} \tag{7}$$

The second model for the interfacial flux is an analytical model constructed on the geometry and the maximum of convective velocity at the conducting regime when $\ell \ll 1$ (Hart, 1972, Klosse and Ullerma, 1973, Bejan and Tien, 1978).

$$v\ (x, \pm 1) = -\ \frac{5}{2048}\ v_d\ Ra_{eq}\ \ell^3\ (x^2 - 1)^2 \tag{8}$$

where the equivalent Rayleigh number is $Ra_{eq} = Ra_T\ \left(\frac{\partial T}{\partial y}\right)_m + \frac{Ra_M}{Le}\ \left(\frac{\partial W_A}{\partial y}\right)_m$, (m refering to the gradients in the middle of the cavity) and v_d an arbitrary interfacial mass flux rate.

The other boundary conditions are the usual no-slip and no - permeability conditions.

3. NUMERICAL APPROXIMATION

3.1. Tau-Chebyshev method

3.1.1. The variables ζ, T, ψ and W_A are approximated with (N+1) and (M+1) Chebyshev polynomials in x and y :

$$\begin{bmatrix} \zeta \\ T \\ \psi \\ W_A \end{bmatrix} = \sum_{n=o}^{N} \sum_{m=o}^{M} \begin{bmatrix} a_{nm} \\ b_{nm} \\ c_{nm} \\ d_{nm} \end{bmatrix} T_n(x)\; T_m(y)$$

The ith and j^{th} derivative respectively in x and y of a variable, ζ for example, is denoted by $a_{nm}^{(i,j)}$ in the spectral plane (Gottlieb and Orszag, 1977).

3.1.2. When applied to the governing equations (2)-(5), the Tau - Chebyshev method gives the following differential system :

$$\frac{da_{nm}}{dt} + \frac{1}{4}\left(\frac{1}{\ell} e_{nm}^{(1,0)} + e_{nm}^{(0,1)}\right) - Pr\left(\frac{1}{\ell^2} a_{nm}^{(2,0)} + a_{nm}^{(0,2)}\right) - \frac{Ra_T Pr}{16} b_{nm}^{(0,1)} - \frac{Ra_M Pr}{16\ Le} d_{nm}^{(0,1)} = 0 \tag{9}$$

$$\frac{1}{\ell^2} c_{nm}^{(2,0)} + c_{nm}^{(0,2)} = a_{nm} \tag{10}$$

$$\frac{db_{nm}}{dt} + \frac{1}{4}\left(\frac{1}{\ell} g_{nm}^{(1,0)} + g_{nm}^{(0,1)}\right) - \left(\frac{1}{\ell^2} b_{nm}^{(2,0)} + b_{nm}^{(0,2)}\right) = 0 \tag{11}$$

$$\frac{dd_{nm}}{dt} + \frac{1}{4}\left(\frac{1}{\ell} h_{nm}^{(1,0)} + h_{nm}^{(0,1)}\right) - \frac{1}{Le}\left(\frac{1}{\ell^2} d_{nm}^{(2,0)} + d_{nm}^{(0,2)}\right) = 0 \tag{12}$$

$$0 \leqslant n \leqslant N-2, \quad 0 \leqslant m \leqslant M-2$$

where $e_{nm}^{(i,j)}$, $g_{nm}^{(i,j)}$ and $h_{nm}^{(i,j)}$ are the components of the convective terms.

3.1.3. The closure of the system with the conventional boundary conditions was already detailed elsewhere (Bondet de la Bernardie and co-authors, 1980,81). When using condition (6), the mass flux profile may be very stiff at low Ra while the mass fraction gradient is nearly one. Then, to avoid oscillations in the spectral approximation, the model is adapted with a function $\varepsilon(x)$ matching continuously zero at $x = \pm 1$, as :

$$\varepsilon(x) = \tanh\ [\alpha(1-x^2)^2] \tag{13}$$

3.2. Time differencing schemes and numerical stability

The optimal evaluation of the convective terms with fast convolution algorithms implies the use of explicit time differencing schemes in the solution of system (9)-(12). The $O(\Delta t^2)$ accurate Adams - Bashforth scheme, noted AB was used to the solution of the complete governing system including the diffusive terms for free convection problems (Bontoux et al, 1981).

$$a_{nm}^{k+1} = a_{nm}^{k} + \frac{\Delta t_a}{2} \left[3 \left(\frac{da_{nm}^{k}}{dt} \right) - \left(\frac{da_{nm}^{k-1}}{dt} \right) \right] \qquad (14)$$

where k refers to the time level ; the time step Δt_a is related to variable a. Severe numerical stability conditions are associated as $\Delta t_a^c \sim 1.3/N^4$, when N = M, $\ell = 1$ and with coefficients of the diffusive terms close to unity. With the false transient solution the critical time step Δt_b^c remains limited as $\sim 2\ \Delta t_a^c$. More restrictive conditions are obtained when $\ell < 1$:

$$\Delta t^c \sim Pr\ \ell^2 / [Min\ (N,M)\ ]^4$$

The alternative consists in using a semi-implicit scheme, noted AB-CN, and based on the AB scheme for the convective terms and the Crank - Nicolson scheme for the diffusive terms (Gottlieb and Orszag, 1977). With $Ra_M = 0$, equation (9) is written as :

$$- \frac{2}{\Delta t} a_{nm}^{k+1} + \frac{1}{\ell^2} a_{nm}^{(2,0)k+1} + a_{nm}^{(0,2)k+1} = G_{nm} \qquad (15)$$

$$\text{where } G_{nm} = - \frac{2}{\Delta t} a_{nm}^{k} + \frac{1}{4\ell} \left[3\ (e_{nm}^{(1,0)} + e_{nm}^{(0,1)})^k - (e_{nm}^{(1,0)} + e_{nm}^{(0,1)})^{k-1} \right]$$

$$- \left(\frac{1}{\ell^2} a_{nm}^{(2,0)} + a_{nm}^{(0,2)} \right)^k + \frac{Ra_T Pr}{16}\ b_{nm}^{(0,1)k}$$

Poisson Solvers are easily extended to the solution of Helmoltz equation. The boundary conditions on the vorticity are imposed similarly to Bontoux et al (1981).

The various methods described in Table I have been studied from the numerical stability point of view, for the Navier-Stokes and energy equations at $Pr \to \infty$ (Catton et al, 1974).

Method	Vorticity Eq. (9)	Energy Eq. (11)
I	AB	AB
II	AB	AB - CN
III	AB - CN	AB - CN

Table I : Definition of the methods

The critical time steps, Δt_a^c and Δt_b^c, are plotted on Fig. 2, in terms of N^2 (N=M) for $\ell = 1$ and $Ra_T = 10^4$. The most severe condition correspond to the fully explicit method I, $\Delta t_a^{cI} \sim 2/N^4$, $\Delta t_b^{cI} \sim 3.4/N^4$. When using the mixt method II, Δt_a^c is unchanged but Δt_b^c is very strongly relaxed. Numerical results give $\Delta t_b^{cII} \sim 1/N^{0.6}$ that is sensibly the criteria proposed by Gottlieb and Orszag (1977) for a semi-implicit two steps scheme ($\Delta t^c \sim 1/N$) and by Hirsh et al (1983) for the finite difference predictor-spectral corrector iterations scheme ($\Delta t^c \sim 4/N$). With method III, Δt_b^c increased again of about 25 % while $\Delta t_a^{cIII} \sim 6\ \Delta t_a^{cI} \sim 10/N^4$. As for the finite difference ADI scheme, the evaluation of the vorticity at the boundaries should be the cause of the this still restrictive condition (Bontoux et al, 1980).

3.3. Basic algorithms and optimization

The basic algorithms are the Fast Poisson (Helmoltz) Solver for equation (10) and semi-implicit scheme (14), and the Fast Convolution Algorithms for the optimal evaluation of the non linear terms in the physical plane and based on the FFT.

3.3.1. The matrix diagonalization method (MD) proposed by Haidvogel and Zang (1979) was used for the solution of Poisson equation (Bondet de la Bernardie, 1980). Apart from an important preprocessing, the method results in the solution of serial quasi-tridiagonal systems. For Poisson equation and Helmoltz equation, $\Delta f + \lambda f = G$, with unvarying $\lambda = 2/\Delta t$ (constant time step), the solution simplifies to matrix products. The improvement in computing time with the IMSL package is about 5 for scalar computer (IBM 3033) and 11 for vector computer (CRAY 01-S). With an efficient vectorization (Farge, 1981) it was possible to raise the gain by a factor of 3 for the matrix products and 15 x 15 Chebyshev polynomials. For Helmoltz type equation with varying Δt the vectorization brings a gain-factor of 10 for the quasi-tridiagonal algorithms. The efficiency of the method was moreover confirmed recently, with respect to other pseudo - spectral and spectral ADI methods when $N < 10^3$ (Abboudi, 1982, Labrosse et al, 1982).

3.3.2. For the evaluation of the non-linear terms different FFT algorithms were tested. The Singleton (1969) algorithm avalaible in the IMSL package is very convenient because it is not restricted to 2^L Fourier series. However there is already of factor 2 in time between $M = 2^L$ and $M \neq 2^L$ cases when computing Chebyshev Transform. Moreover, with $M = 2^L$, the 2D vectorized FFT of Lhomme, Morgenstern and Quandalle (1982) is four times faster than the Singleton FFT on CRAY 01-S. For Chebyshev Transform the factor is only 2.3 with 17 x 17 polynomials, but the factor may be increased to 4.6 with recent improvement of the vectorization and for large arrays.

3.4. Initial conditions and convergence

3.4.1. When the steady solution is sought, the convergence is strongly improved by using adapted time steps. With method III the results are shown on Fig. 3 for 7 x 7 Chebyshev polynomials ($Ra = 10^4$, $Pr \to \infty$). The basic critical $\Delta t_a^{cIII} \sim 2.8 \cdot 10^{-3}$. When using $\Delta t_a = \Delta_b^{cIII}$, the convergence on ψ is reached after about 200 iterations from initial zero conditions. The steady state may be obtained much early after 60 iterations when using $10^{-2} \lesssim \Delta t_b \lesssim 4.5 \cdot 10^{-2}$. With the false transient solution applied to method I the gain would have been much smaller, around 30 % in terms of number of iterations for $Ra = 10^4$ and 10^5.

3.4.2. The improvement of the initial conditions is also determinant for an optimal convergence toward the steady state and for the convergence itself at high Ra.

3.4.2.1. When $\ell \gg 1$, the Navier-Stokes and energy equations simplify with the assumptions of steady and similar solution with thermal stratification (Elder, 1865, Brenier, 1982). The asymptotic solution writes as :

$$\tilde{\psi}_E(y) = \frac{k}{2m} [(1-\rho) \sinh my \sin my - (1+\rho) \cosh my \cos my - (1-\rho) \sinh m \sin m + (1+\rho) \cosh m \cos m] \quad (16)$$

where $m = \left(\frac{Ra\ \beta}{64}\right)^{1/4}$, $k = \frac{2m^2}{k}$ (cos m sinh m + ρ sin m cosh m)$^{-1}$, $\rho = \frac{tg\ m}{tanh\ m}$

and $\beta\ell = \frac{\partial T}{\partial x}$. A model for the variation of the stratification parameter was derived from Thomas and De Vahl Davis (1970) and Roux et al (1980) as : $\beta\ell = 2.10^{-5}\ (Ra\ \ell^{-1.25})^{1.75}$ for $Ra\ \ell^{-1.25} < 400$ and $\beta\ell = 0.7$ for $Ra\ \ell^{-1.25} > 400$. The solution (16) is extended to the entire cavity with the matching function (13) and $\alpha = C_E\ \ell^4/16\ (\ell-1)^2$. Similar solution is also derived for the temperature.

The convergences obtained with the initial conditions derived from (16) are shown on Fig. 4 for $Ra = 10^4$, $\ell = 5$, $Pr \to \infty$. When compared to the case of conducting initial conditions, the transient state may be divided by more than 3 with (16) and adapted $C_E \sim 3$.

3.4.2.2. When $\ell << 1$ (shallow cavities), the asymptotic solution derived from the parallel flow and similarity assumptions in the core writes as follows (Hart, 1972, Klosse and Ullersma, 1973, Bejan and Tien, 1978).

$$\tilde{\psi}_H\ (x) = \frac{Ra_{eq}\ \ell^4}{384}\ (x^2 - 1)^2 \qquad (17)$$

As for long cavities the initial conditions may be derived from (16) with a matching function, $\varepsilon(y)$, similar to (13) and $\alpha = \frac{C_H}{16\ \ell^2\ (1-\ell^2)}$. Figure 5 shows the effect on the convergence of adapted initial conditions ($Ra_T \simeq 10^3$, $Ra_M = 1.3\ .10^3$, $\ell = 0.2$). The results exhibits a reduction by about 10 of the transient state when $C_H \sim 3.3$.

4. RESULTS

4.1. Natural convection in pure gas ($Ra_M = 0$) for long cavities ($\ell >> 1$)

The onset of instabilities on basic unicellular flow was studied by linear analysis (Bergholz, 1978, Brenier, 1982). The transitional Grashof number, Gr = Ra/Pr, expresses in terms of aspect ratio, ℓ (Fig. 6). For large aspect ratios at Pr = 0.71, the first instability is monotonic at low Ra. Results for $\ell = 15$ (Fig. 7) are in good agreement with previous finite difference solutions (Roux et al, 1980). At moderate aspect ratios ($\ell \lesssim 11$) the first instability is predicted as oscillatory for large Ra. For $\ell = 5$ the critical Ra is about 2.10^5. Previous studies (Mallinson and De Vahl Davis, 1973, Lauriat, 1981) have reported about oscillations and difficulties in the convergence, associated with the appearance of secondary cells in the end regions at $Ra \sim 1.8\ .\ 10^5$ and $Pr \sim 1$. The present computations at $Ra = 2.5\ .\ 10^5$ and Pr = 0.71 also exhibited oscillations in the flow. They are slowly damping in the core. In the end region they are modulated within 5 % of the average extrema over $\Delta t \simeq 0.1$. Three transient streamline patterns are shown on Fig. 8 with to the associated thermal pattern which was quickly stabilized.

4.2. Double diffusion in shallow cavities ($\ell << 1$)

For moderate Ra ($Ra < 10^6$) the temperature gradient in the core drives the flow in the enclosure, Hermitian finite difference solutions (Manouelian, 1982) were shown to fit very well with asymptotic relation (18) for 21 x 41 discretizing points (Fig. 9). With the spectral approximation the convergence is obtained with only 12 x 16 Chebyshev polynomials as shown on Fig. 9. With increasing Ra, the boundary layer in the end region becomes the driving mechanism. Some typical features are shown on Fig. 10. The development of the boundary layer is exhibited by the modification of the flow pattern near the active walls. For such regimes the validity of the Boussinesq approximation is to be controlled with analogous spectral approximation but including compressibility effect (Ouazzani and Peyret,1983).

References
Bejan and Tien, J. Heat Transf., 100, 1978.
Brenier, Th. Univ. Aix-Marseille II, 1982.
Bergholz, J.F.M., 84, 1978.
Bondet de la Bernardie, Th. Univ. Aix-Marseille III, 1980.
Bontoux, Gilly, Roux, J. Comp. Physics, 36, 3, 1980.
Bontoux, Bondet de la Bernardie, Roux, GAMM-Workshop, Univ. LLN, Belgique, 1980/Num. Meth. for Coupled Problems, Pineridge, 1981.
Catton, Ayyaswamy, Clever, Int. J. Heat Mass Transf., 17, 1974.
Elder, J.F.M., 23, 1965.
Elie, Th. Univ. Aix-Marseille II, 1983.
Farge, EDF-DER Rap. SIMA n° HI/3912-00, 1981.
Gottlieb, Orszag, CBMS-NSF, Conf. S. Applied Math., 1977.
Hart, J. Atmos. Science, 29, 1972.
Hirsh, Taylor, Nadworny, Euromech 159, Nice, 1982/Comp.Fluids,11,3,1983.
Haidvogel, Zang, J. Comp. Physics, 30, 1979.
Jhaveri, Markham, Rosenberger, Chem. Eng. Commun., 13, 1981.
Klosse, Ullersma, J. Crystal Growth, 18, 1973.
Labrosse, Abboudi, Deville, Haldenwang, Euromech 159,Nice,1982.
Lauriat, Th. Doct. Es-Sciences, CNAM, Paris, 1982.
Lhomme, Morgenstern, Quandalle, Euromech 159, Nice,1982.
Mallinson, De Vahl Davis, J. Comp. Physics, 12,1973.
Manouélian, Mémoire CNAM, Aix-en-Provence, 1982.
Mc Laughlin, Orszag, J.F.M., 122, 1982.
Markham, Rosenberger, Chem. Eng. Commun., 5, 1980.
Morchoisne, ONERA-TP-138, n° 1979-S, 1979.
Orszag, J.F.M., 149, 1971/Studies Applied Math.,L,1971 ; L1, 1972.
Ouazzani, Peyret, GAMM-Conf. Num. Math. Fluid Mech.,Rome, 1983.
Roux, Grondin, Bontoux, De Vahl Davis, Phys. Chem.Hydr.Conf.,3,PCH80,1980.
Acknowledgment : CNES and CNRS (GC_2VR)
P.Extremet,J.M.Lacroix,B.Lhomme and R. Peyret.

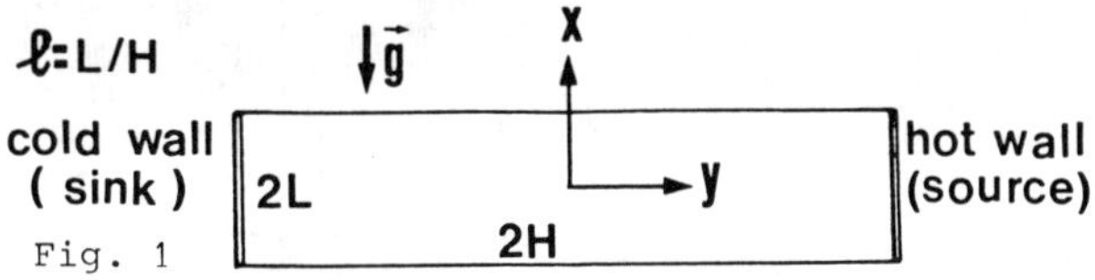

Fig. 1

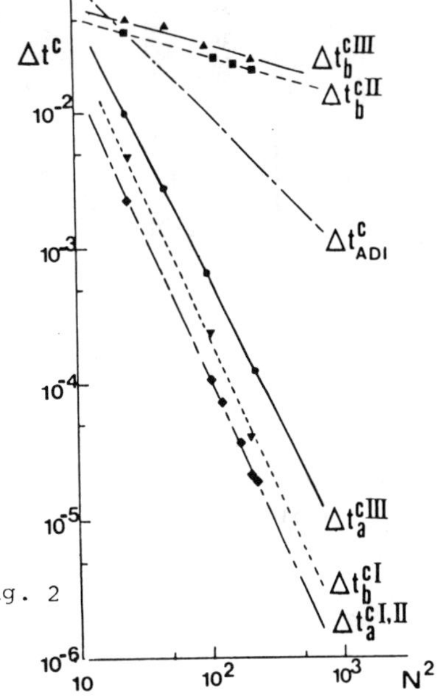

Fig. 2

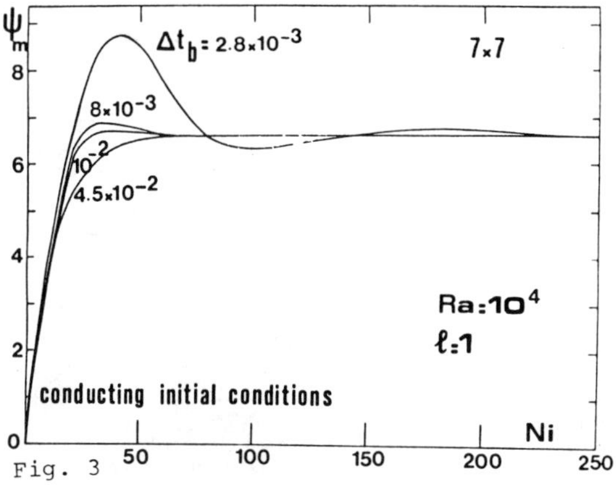

Fig. 3

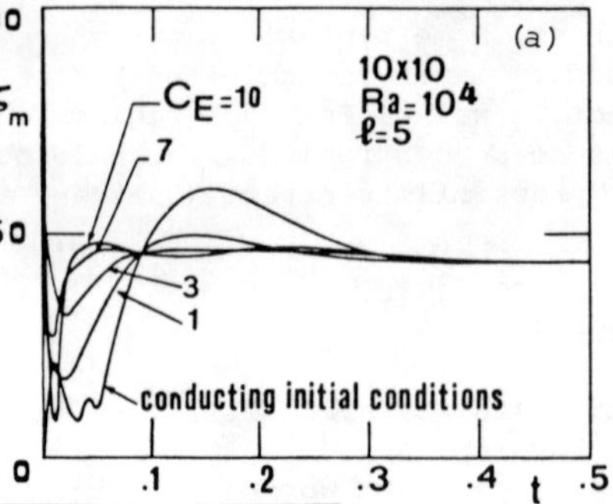

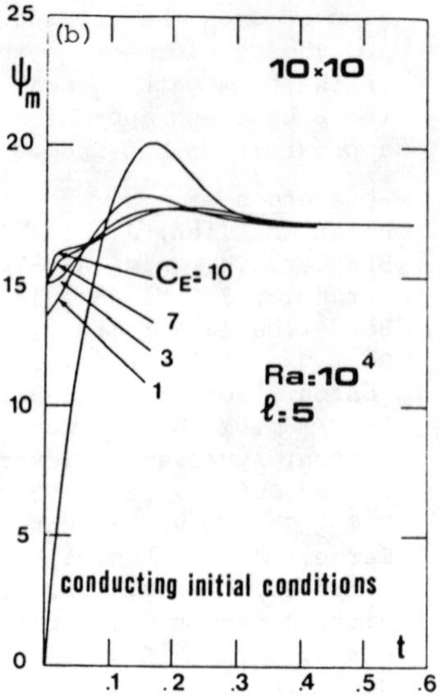

Fig. 4

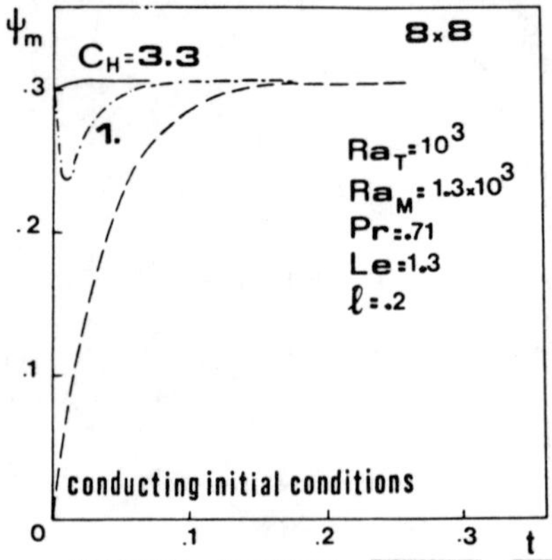

Fig. 5

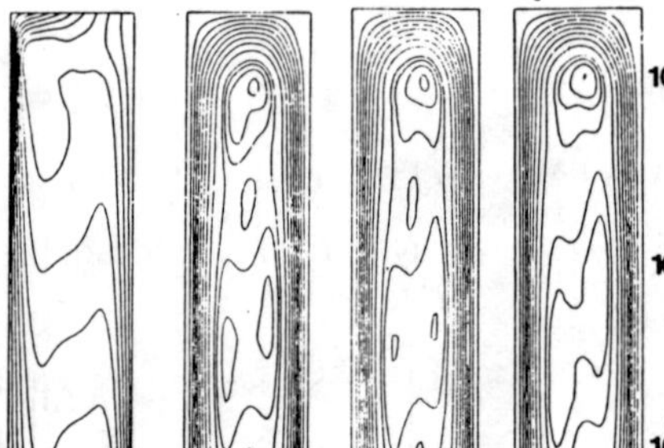

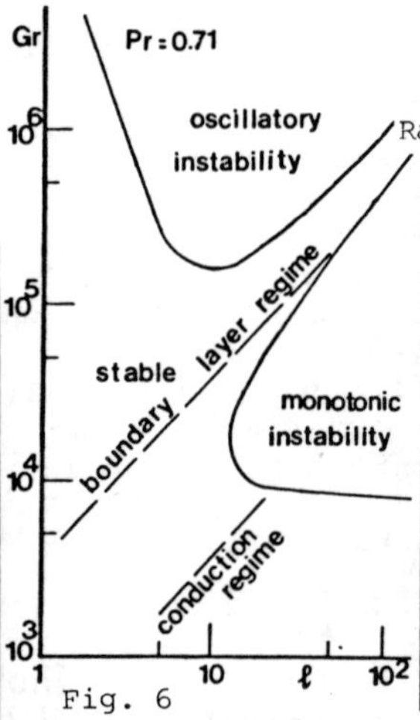

Fig. 6

Fig. 7 : ℓ = 15

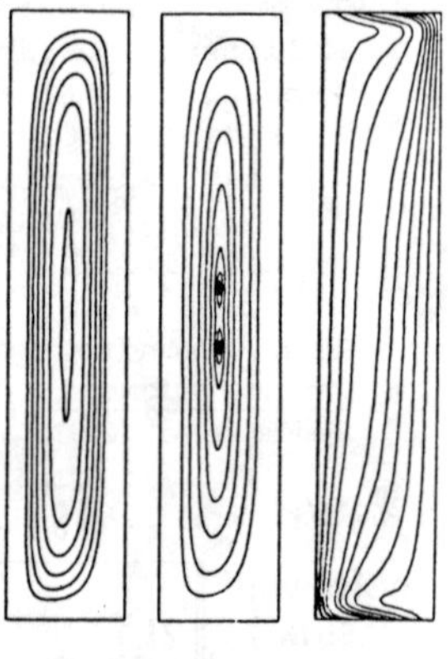

Fig.8: ℓ=5
Ra=$2.5.10^5$

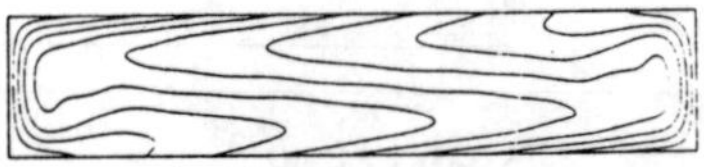

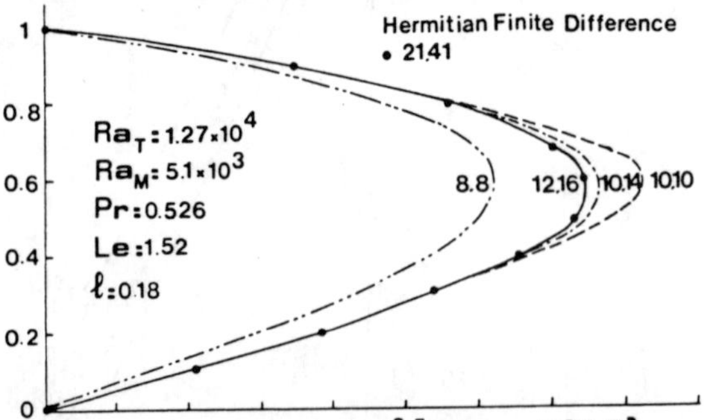

Fig. 9

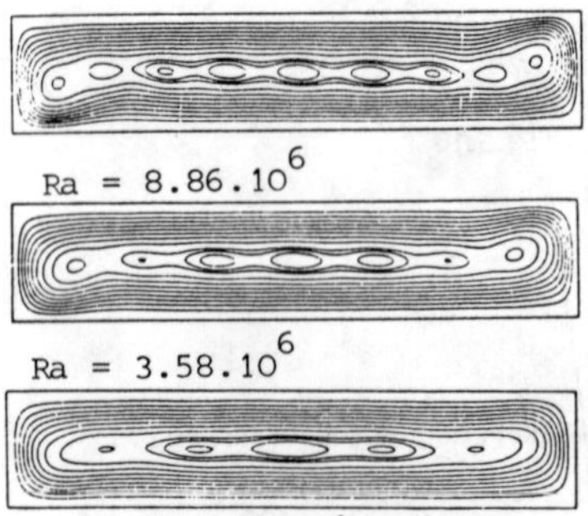

Fig. 10 : ℓ = 0.2

COMPUTATION OF VORTEX FLOW AROUND A CANARD-DELTA COMBINATION

Lars-Erik Eriksson & Arthur Rizzi
FFA, The Aeronautical Research Institute of Sweden
S-161 11 BROMMA, Sweden

SUMMARY

The inviscid, compressible and rotational flow around an isolated 55 degree swept, sharp-edged delta wing and around that same wing closely coupled with a canard is computed at transonic speeds by using meshes of O-O type around the main wing, constructed by transfinite interpolation, and by using a time-marching finite-volume procedure to obtain steady-state solutions to the Euler equations. The canard is represented as a slit in the mesh. Results of these computations, performed on a CYBER 205 vector processor, show that the flow model predicts leading edge vortex separation on the main wing, and in the wing alone case the overall flow agrees qualitatively with that deduced from oil flow pictures for a similar wing. In the canard/delta case, the deflection of the flow due to the canard is verified by velocity vector plots and the influence on the wing pressure distribution is found to be realistic. Crossflow velocity plots show that the canard creates a vortex of its own which interacts with the flow over the main wing.

I. INTRODUCTION

Flows with shed vorticity are important in aerodynamics, especially for wings of large sweep and small aspect ratio. The characteristic feature of such flows is the generation of vorticity at the edges of the wing and the subsequent convection of this vorticity along streamlines. Except for the actual generation of vorticity, which is a very local phenomenon, involving irreversible thermodynamic processes and an entropy production, these flows are effectively inviscid. It has recently been found that the Euler equations admit solutions for these types of flows and this has generated a lot of excitement in computational fluid dynamics because this flow model allows vortex sheets to be captured automatically in the solution. Previous flow models based on potential theory required them to be hand fitted and that process was cumbersome and difficult especially for multiple lifting surfaces.

Although the vortex-generating mechanism in the Euler-equation model is still much debated, a number of solutions to the discretized equations have been presented[1,6] with vorticity being shed from the leading edge of highly swept (around 70 deg.) wings. These speak for a mechanism in at least the discrete model for generating vorticity. Our purpose here is to explore further how the vortex-sheet capturing ability of the Euler-equation model may depend on wing sweep-angle and configuration complexity. We present two calculations for a delta wing at intermediate sweep angle. The first is an isolated delta wing of 55 deg. sweep and the second is that same delta wing closely coupled with a canard ahead of it. As far as we know this latter computation is the first of its kind.

II. OUTLINE OF NUMERICAL METHOD

Our method[4] to generate 3D meshes around wings is based on the concept of transfinite interpolation and gives the desired mesh coordinates in the interior of the flowfield by interpolating among the given boundary meshes. Normal derivatives of the coordinates at the boundaries are used where needed to control the mesh at these surfaces. A novel feature of the method is that it has been generalized to handle slits in the computational domain, which makes it possible to insert additional lifting surfaces in the vicinity of the main wing.

Our method[2] to solve for 3D vortex flowfields is a time-dependent finite-volume approach that uses a multistage explicit time integration scheme together with centered space differences to solve the compressible Euler equations. Significant features of this approach are its integral conservation-law form, important for the correct capturing of shock waves and vortex sheets, its amenability to very general geometry without the need for a global coordinate transformation, and its toleration of mesh singularities because the flow equations are balanced only within the cells of the grid[3], and not at the nodal points.

III. SINGLE DELTA WING

Mesh Generation

The first case presented is a cropped delta wing with a leading edge sweep of 55 deg. and zero trailing edge sweep. A simple 4% thick parabolic-arc profile defines all chordwise sections of the wing and no camber or twist has been added. The mesh used for this wing (Fig. 1) is the standard 0-0 type of mesh previously used for ordinary quadrilateral wings and is thus not the special delta-wing type of mesh presented before[1]. There are two reasons for this choice: 1) The moderate leading edge sweep does not particularly favour any one of the two mesh alternatives when considering mesh density or mesh skewness around the main wing. 2) The standard 0-0 mesh is better suited for modelling the canard in the form of a slit. The overall mesh dimensions used for the flow computations are the following: 56 cells around the chord, 20 in the spanwise direction, 32 in the outward direction.

Results

The results of the flow computations for the single delta wing at a freestream Mach number of 0.70 and 10 degrees angle of attack reveal a number of interesting features in this flow. As the pressure, Mach number and total pressure contours on the upper surface clearly indicate (Figs. 2a-c), leading edge vortex separation is obtained, but seems to start some distance downstream of the wing apex. This behaviour differs from that obtained at higher sweep angles, in which case separation starts immediately at the apex[1]. It is as yet unclear whether this difference is caused by the artificial viscosity in the flow model or by the physics of inviscid flow. Experimental results[5] in the form of oil flow pictures for a similar wing at the same freestream Mach number show that for a certain incidence range, leading edge separation does in fact start downstream of the apex. In view of this, the computed flow pattern appears to be quite realistic. We remark further that the vortex dominates our flowfield much more than it does the corresponding one computed by a singularity method for conical flow.

Another interesting feature of the computed flow is the appearance of a very local trailing edge loading just where the main vortex passes over the trailing edge. Again, there are two possible explanations for this phenomenon, one physical and one numerical. Either it is caused by an interaction between the main vortex and the vortex sheet emitted at the trailing edge or else it is a purely numerical effect due to the sharp trailing edge. However, the fact that such phenomena are not encountered at the trailing edge of ordinary wings with attached flow indicates that it does reflect a real flow mechanism, through admittedly in an imperfect manner due to the coarseness of the mesh.

A third interesting feature of the computed flow is its sensitivity to variations in the applied boundary conditions at the trailing edge of the wing. In formulating the solid wall boundary conditions, local surface curvature is used to improve the accuracy of these boundary conditions. At the trailing edge, this curvature can either be defined as the true local curvature or as an effective curvature based on the assumption that the flow separates at the edge. Using the latter alternative, which can be seen as an implicit Kutta condition, the presented results were obtained, whereas the first alternative resulted in a non-physical type of flow with almost zero lift. Similar numerical experiments for wings with low sweep and attached flow have not shown such dramatic effects.

As a striking demonstration of the ability of the Euler model to simulate vortical flows, the cross-flow pattern on the upper side of the wing is shown in Fig. 2d. The location of the vortex in the computed flow agrees qualitatively with that deduced from the oil flow pictures in the experimental work mentioned above, and even though the planforms are not quite identical, the computed (C_L= .554, C_D= .0914) and measured (C_L= .53) lift coefficients are in good agreement.

IV. CANARD/DELTA COMBINATION

If a computational method is to be useful for aircraft design, it must be able to treat geometries more complicated than a single delta wing. A typical modern fighter is sketched in Fig. 3, and we see that the most simplified configuration that still possesses the fundamental aerodynamic character of the complete plane is a closely coupled nose-wing/delta-wing combination whose shed vortices interact with each other (see schematic Fig. 4). Our solution method is completely general so that once a mesh is constructed around a canard-delta combination like this, no modification other than the added boundary conditions of zero flow through the second wing surface needs to be implemented. The mesh generation for such a geometry, however, is not a trivial matter, as will be evident from the description below.

Mesh Generation

The mesh we choose to use for the canard/delta combination is a standard 0-0 mesh around the main wing supplied with a slit to represent the canard. This mesh topology leads to an H-H type of mesh around the canard, which for reasonable overall mesh dimensions cannot resolve the details of the flow there. However, our primary objective is to obtain the flow-deflection effect of the canard on the main wing and for this purpose the mesh is found to be sufficiently dense.

As before, we use transfinite interpolation to generate the desired mesh. For practical reasons, the mesh generation is accomplished in two steps, first a standard 0-0 mesh is generated around the main wing and then this mesh is deformed in a region around the slit, so that it conforms to the canard surface. Just as for the main wing, normal derivatives of the mesh coordinates are used to control the mesh around the canard. The extension of the transfinite interpolation method to slits is a novel feature of the mesh generation procedure.

As a reasonably realistic example of a canard/delta combination, the 55 deg. swept delta wing presented in the previous section was chosen as the main wing and a smaller similarly swept sharp-edged wing was placed in front of it as a canard. The mesh generated for this geometry (Fig. 5) has the same overall dimensions as the wing-alone mesh ($56\times32\times20$) and the slit is defined so that there are 8×8 cells on both the upper and lower canard surface.

Flow Solution

As mentioned before, the basic solution method is quite general and independent of the mesh. However, the introduction of a slit in the mesh with solid-wall boundary conditions certainly affects the practical coding of the method. Due to the high degree of vectorization in the present code, the changes due to the slit are accomplished as local corrections after each vector statement so that the data streams are not interrupted. This method works well because the number of field points that must be corrected is very small in comparison with the total number of points. For the canard/delta combination presented here the total penalty due to the canard was less than 10%.

Results

The computed flow around the canard/delta combination shown in Fig. 5 at a freestream Mach number of 0.70 and 10 deg. angle of attack is presented in Figs. 6a-e. Comparing these results with the wing-alone results in Figs. 2a-d, it is evident that the canard influences the flow over the main wing to a large degree. The main effect on the pressure distribution is a lift reduction for the inner part of the wing and a small lift increase for the outer part of the wing. As the pressure distribution on the canard shows (Fig. 6a), the mesh is too coarse to resolve the details of the flow there, but the pressure difference between the upper and lower surface indicates that the model does simulate the flow-deflection effect of the canard. Further evidence of this is given by the contours of total pressure (Fig. 6c) which show that the canard generates a "wake" of total pressure loss that is convected downstream over the main wing surface. This indicates that the canard acts as a lifting surface that sheds vorticity into the freestream flow. The final confirmation of this is given by the crossflow velocity vectors plotted in Fig. 6d, which clearly show the vortex generated by the canard.

To demonstrate the effect of the canard's solid-wall boundary conditions on the surrounding flow, velocity vectors around the canard are shown in Fig. 6e.

Acknowledgement

This work has been supported in its entirety by the Air Materiel Department of the Swedish Defence Materiel Administration.

REFERENCES

[1] Eriksson, L.-E., and Rizzi, A.: Computation of Vortex Flow Around Wings Using the Euler Equations, ed. H. Viviand, Proc. 4th GAMM Conf. Num. Meth., Vieweg Verlag, 1982.

[2] Rizzi, A.: Damped Euler-Equation Method to Compute Transonic Flow Around Wing-Body Combinations, AIAA J, Vol. 20, pp. 1321-1328, Oct. 1982.

[3] Eriksson, L.-E.: A Study of Mesh Singularities and Their Effects on Numerical Errors, paper presented at Minisymposium on Analysis of Mesh Effects, SIAM Meeting, July 1982, to be published.

[4] Eriksson, L.-E.: Generation of Boundary-Conforming Grids Around Wing-Body Configurations Using Transfinite Interpolation, AIAA J, Vol. 20, pp. 1313-1320, Oct. 1982.

[5] Sutton, E.P.: Some Observations of the Flow over a Delta-Winged Model with 55 deg. Leading-Edge Sweep, at Mach Numbers between 0.4 and 1.8, Aeronautical Research Council R&M No. 3190, 1960.

[6] Rizzi, A., Eriksson, L.-E., Schmidt, W., and Hitzel, S.: Numerical Solution of the Euler Equations Simulating Vortex Flow Around Wings, in Aerodynamics of Vortical Type Flows in Three Dimensions, AGARD-CPP-342, Paris, 1983.

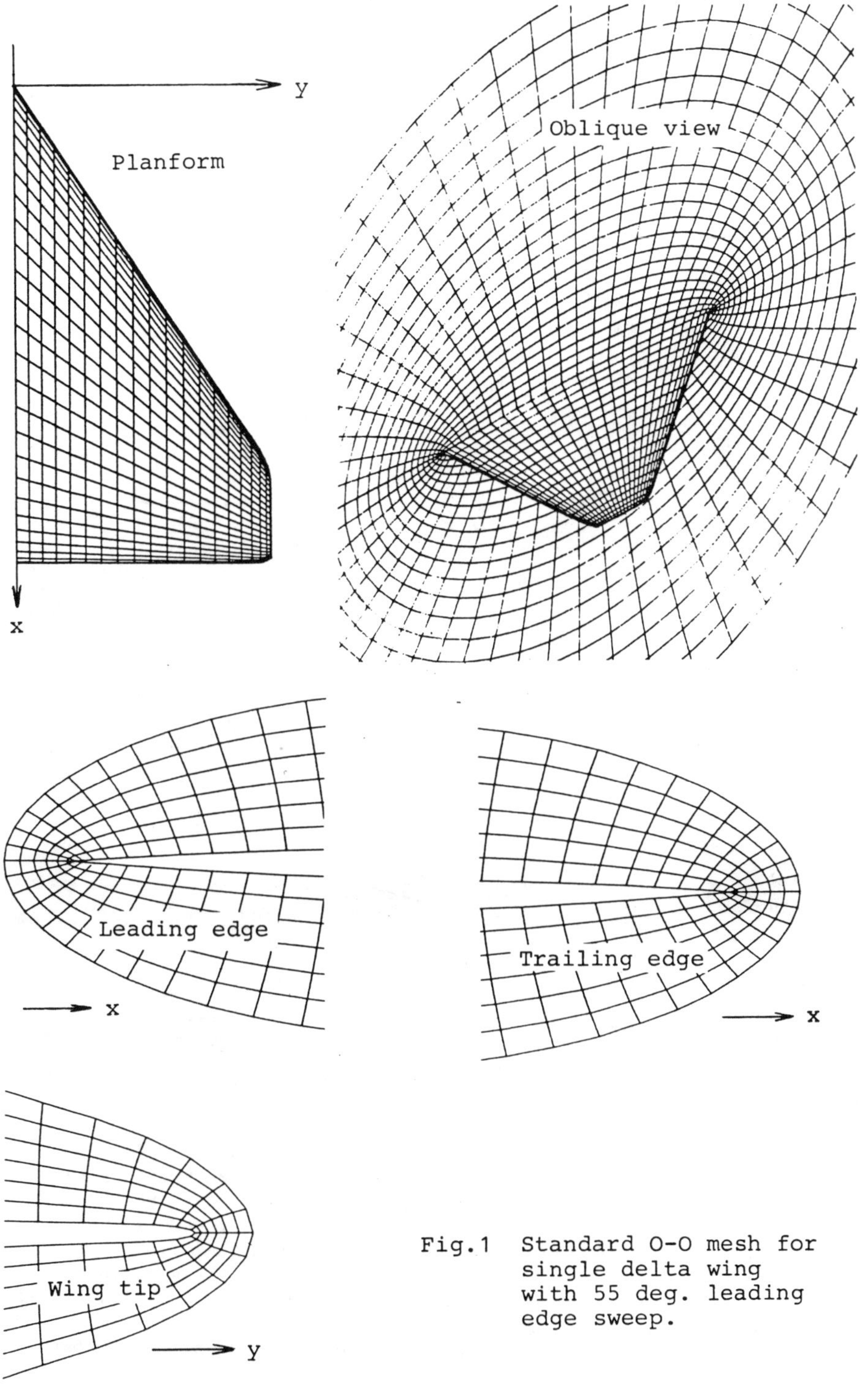

Fig.1 Standard O-O mesh for single delta wing with 55 deg. leading edge sweep.

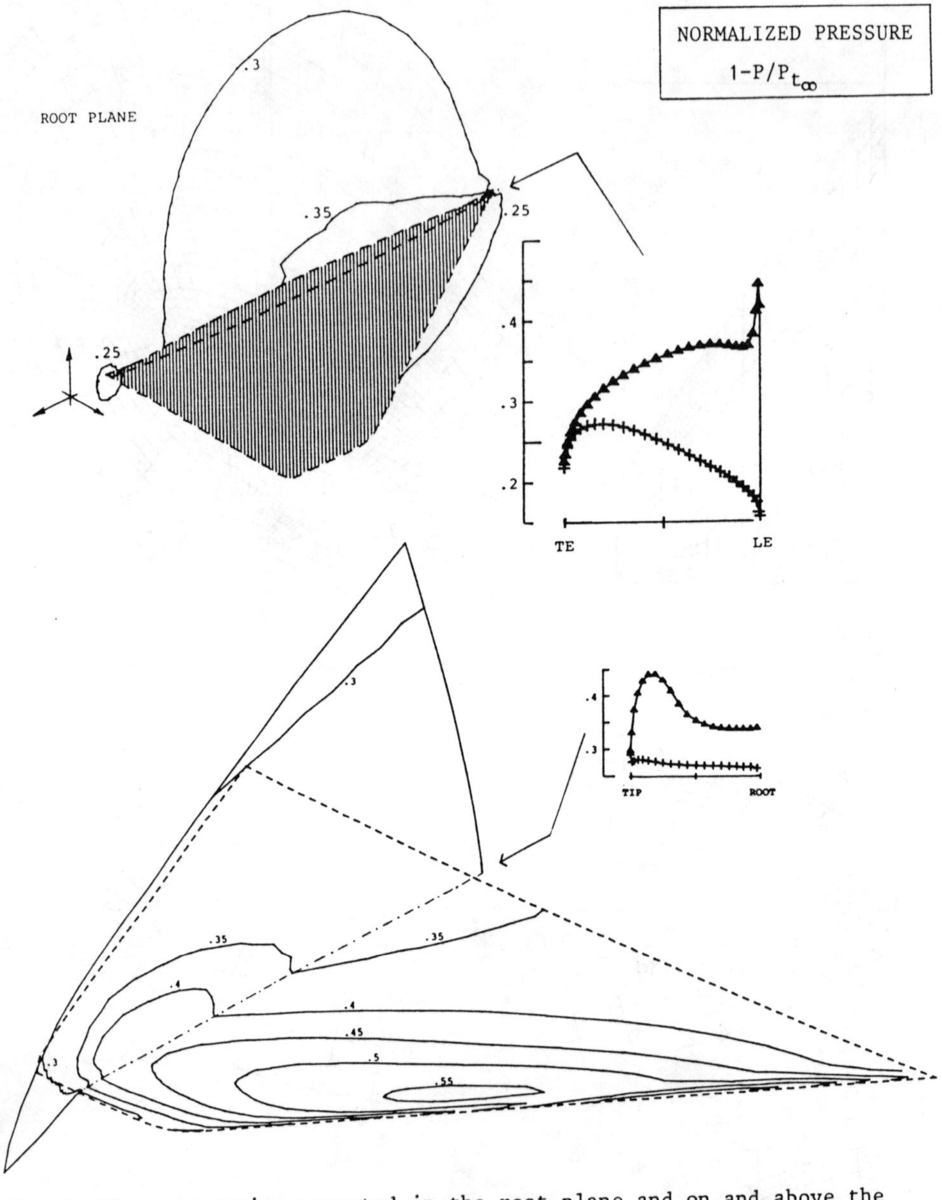

Fig. 2 Flow properties computed in the root plane and on and above the upper surface of the cropped sharp-edged 55 deg swept delta wing for $M_\infty = 0.70$ and $\alpha = 10$ deg. The flowfield is surveyed by isograms 1) in the root plane, 2) on the upper surface, and 3) in an intersecting mesh surface extending outward from the 67.5% chord station of the wing. Along the intersection lines of the wing in the root plane and the wing with the 67.5% chord station the flow properties on the upper and lower surface are plotted versus the local chord and span direction respectively. The displayed properties are:

2(a) Isobars of normalized pressure $1.-p/p_{t_\infty}$

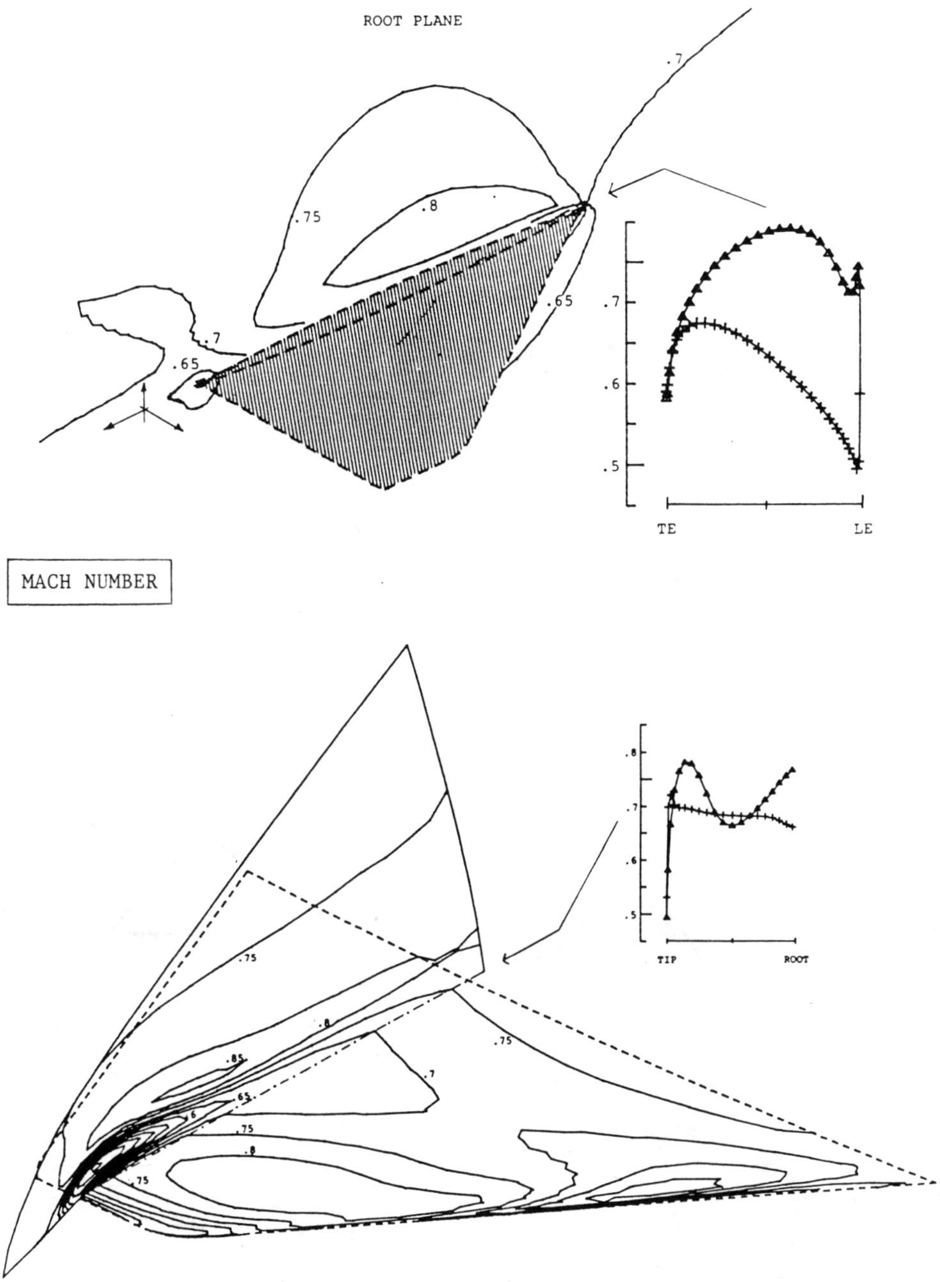

·Fig. 2(b) Lines of constant Mach number.

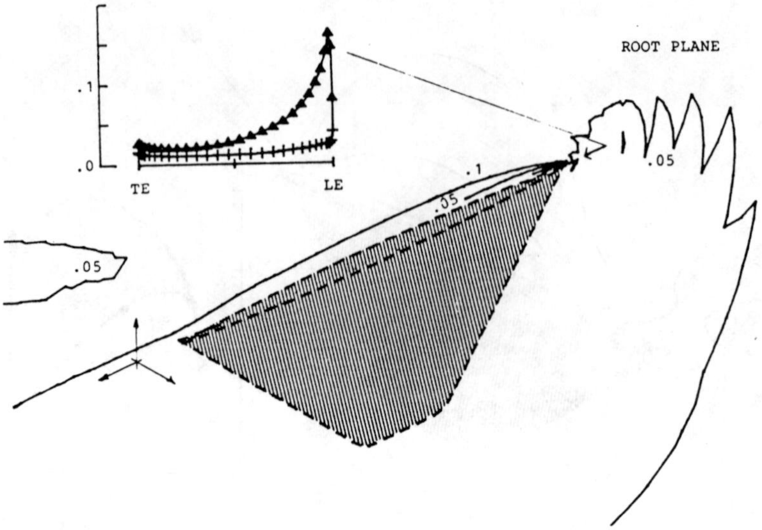

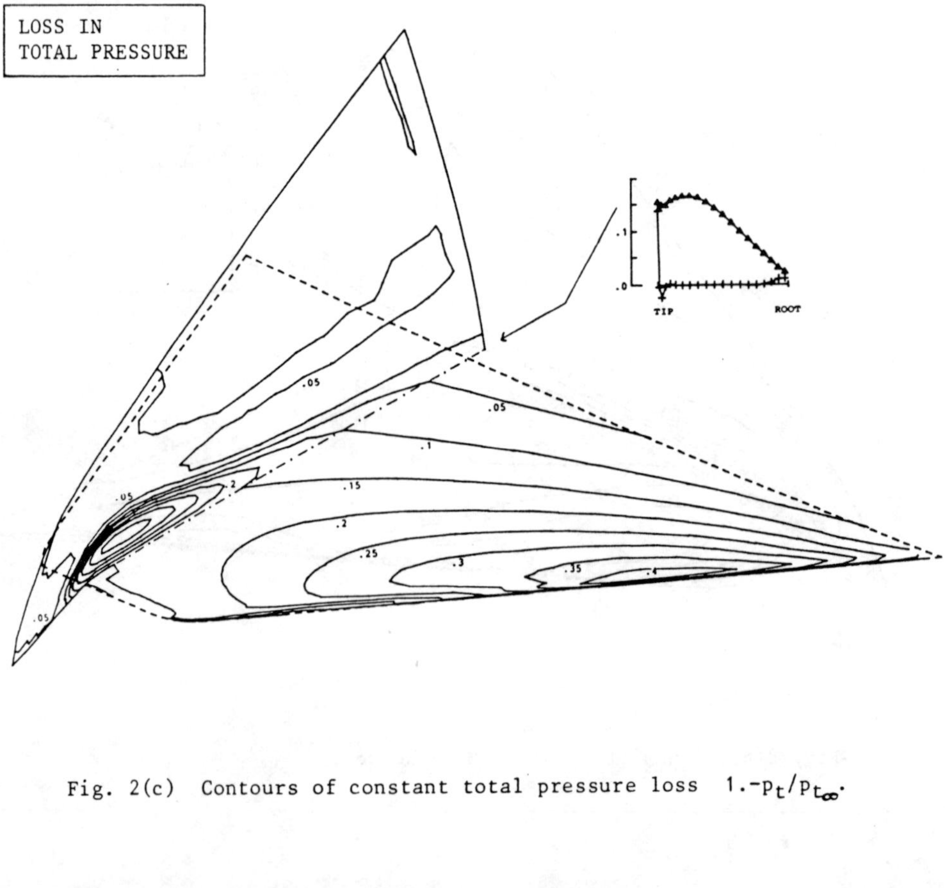

Fig. 2(c) Contours of constant total pressure loss $1.-p_t/p_{t_\infty}$.

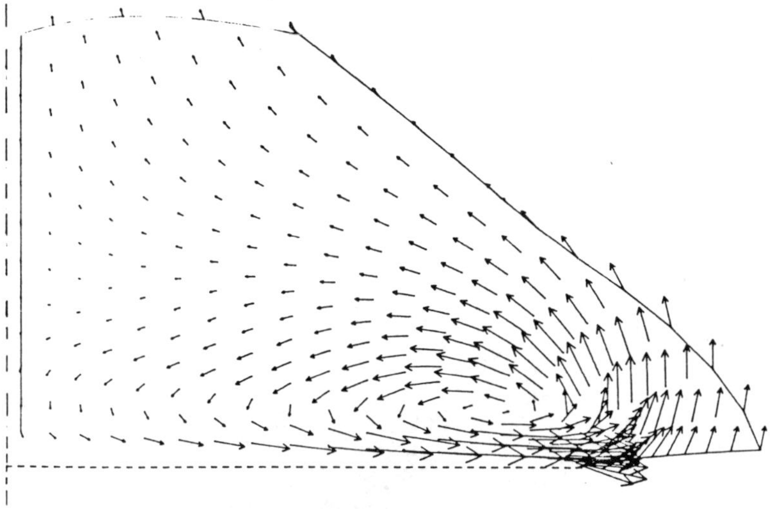

Fig. 2(d) v-w velocity-vector plot in the 67.5% constant chord surface.

Fig. 3 Modern combat aircraft with canard-delta wing combination.

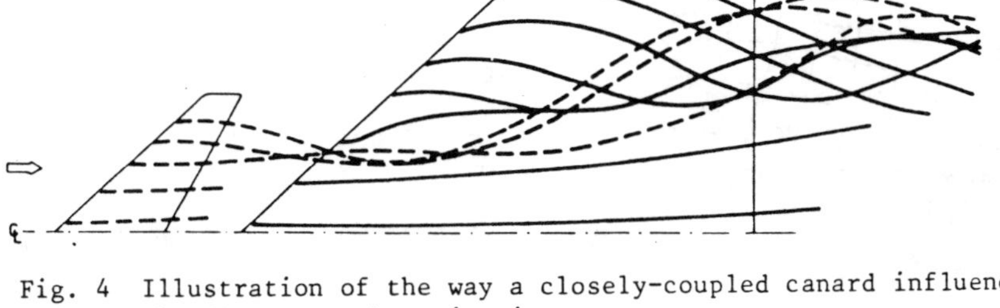

Fig. 4 Illustration of the way a closely-coupled canard influences the flow over the main wing.

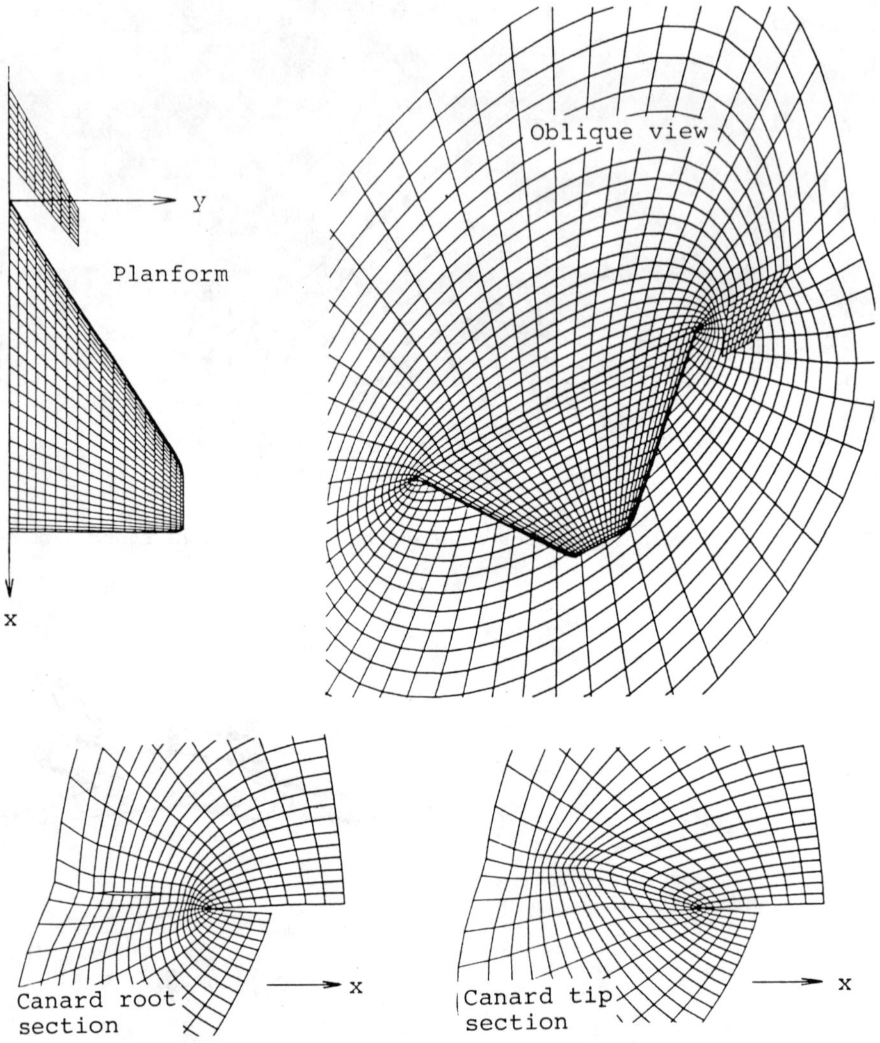

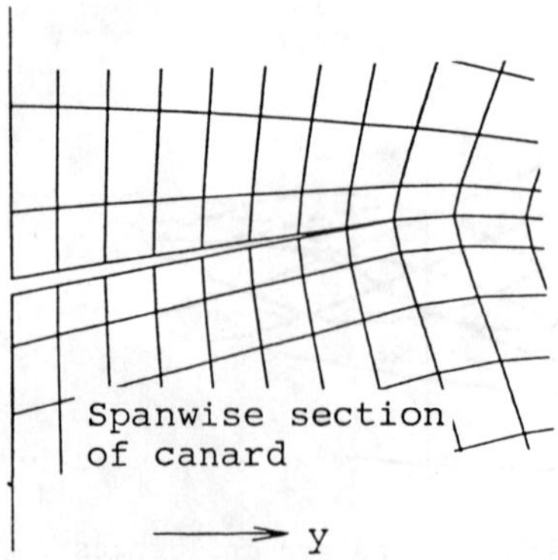

Fig.5

O-O mesh for canard/delta combination. Main wing is the same as that in Fig.1.

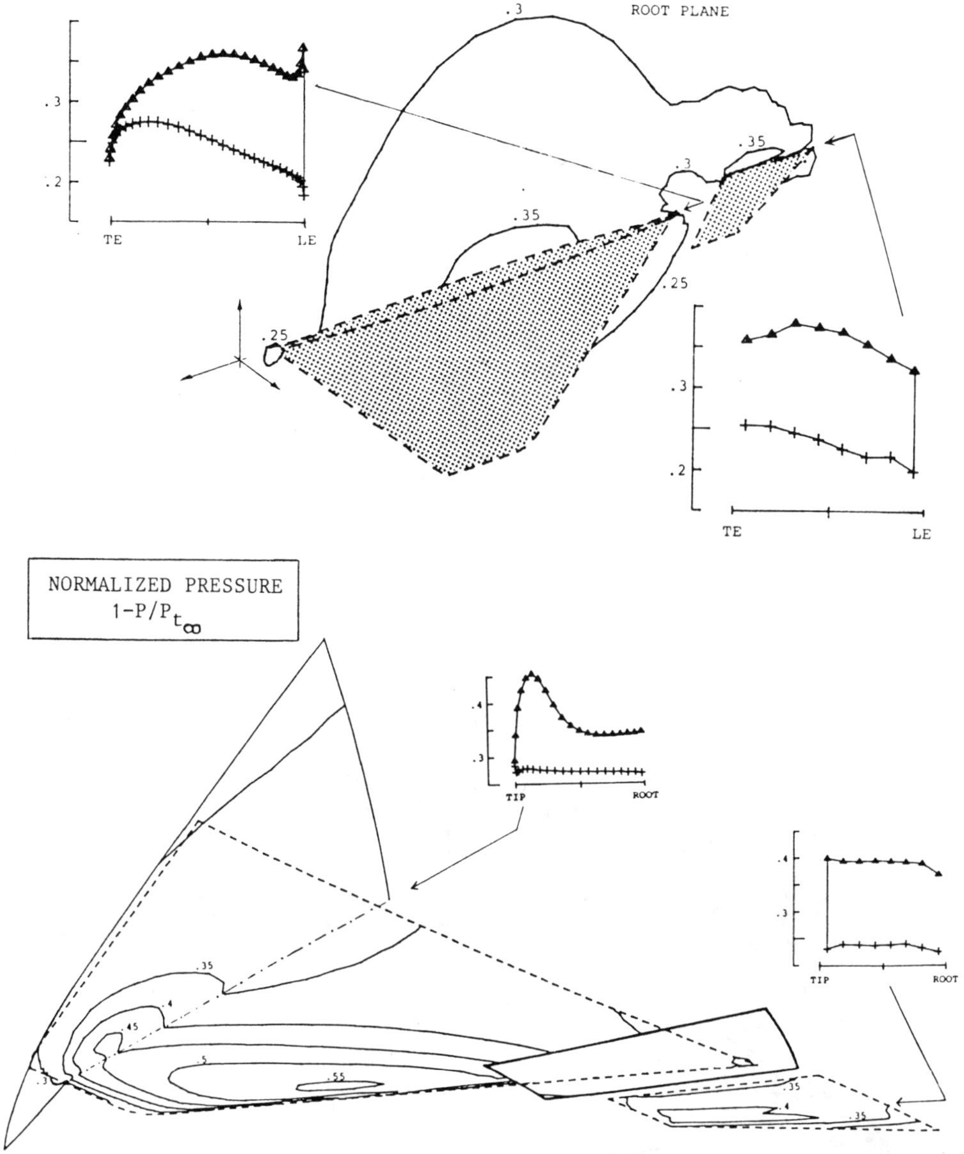

Fig. 6 Flow properties computed in the root plane and on and above the upper surface of the cropped sharp-edged 55 deg swept delta wing with close-coupled canard for $M_\infty = 0.70$ and $\alpha = 10$ deg. The flow-field is surveyed by isograms 1) in the root plane, 2) on the upper surface of the wing and canard, 3) in an intersecting mesh surface extending outward from the 67.5% chord station of the wing, and 4) in a mesh surface just behind and extending above the canard. Along the intersection lines of 1) the wing and canard with the root plane, 2) the wing with the 67.5% chord surface, and 3) the canard with its 44% chord station, the flow properties on the upper and lower surface are plotted versus the local chord and span directions respectively. The displayed properties are: (a) Isobars of $1.-p/p_{t_\infty}$

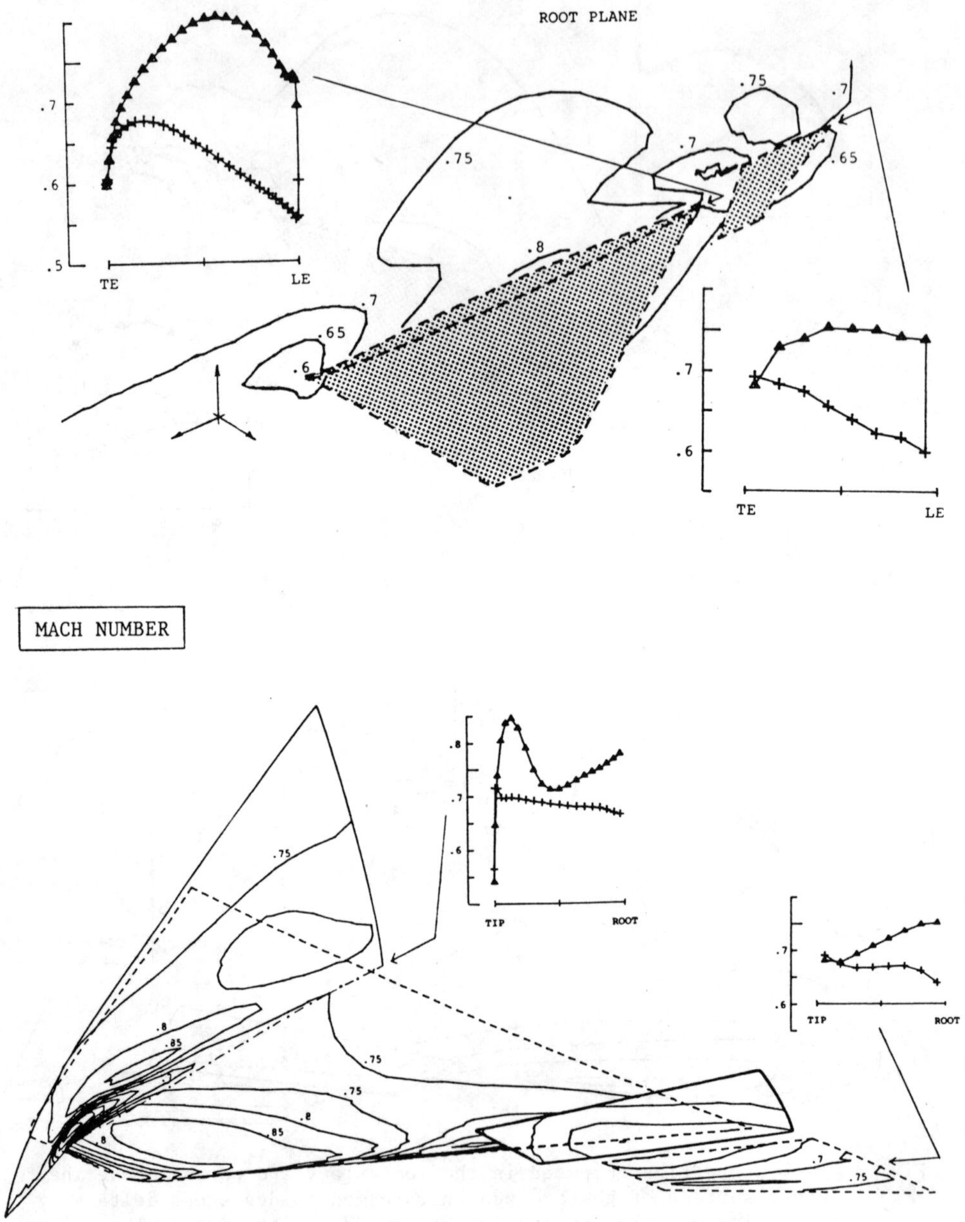

Fig. 6(b) Lines of constant Mach number.

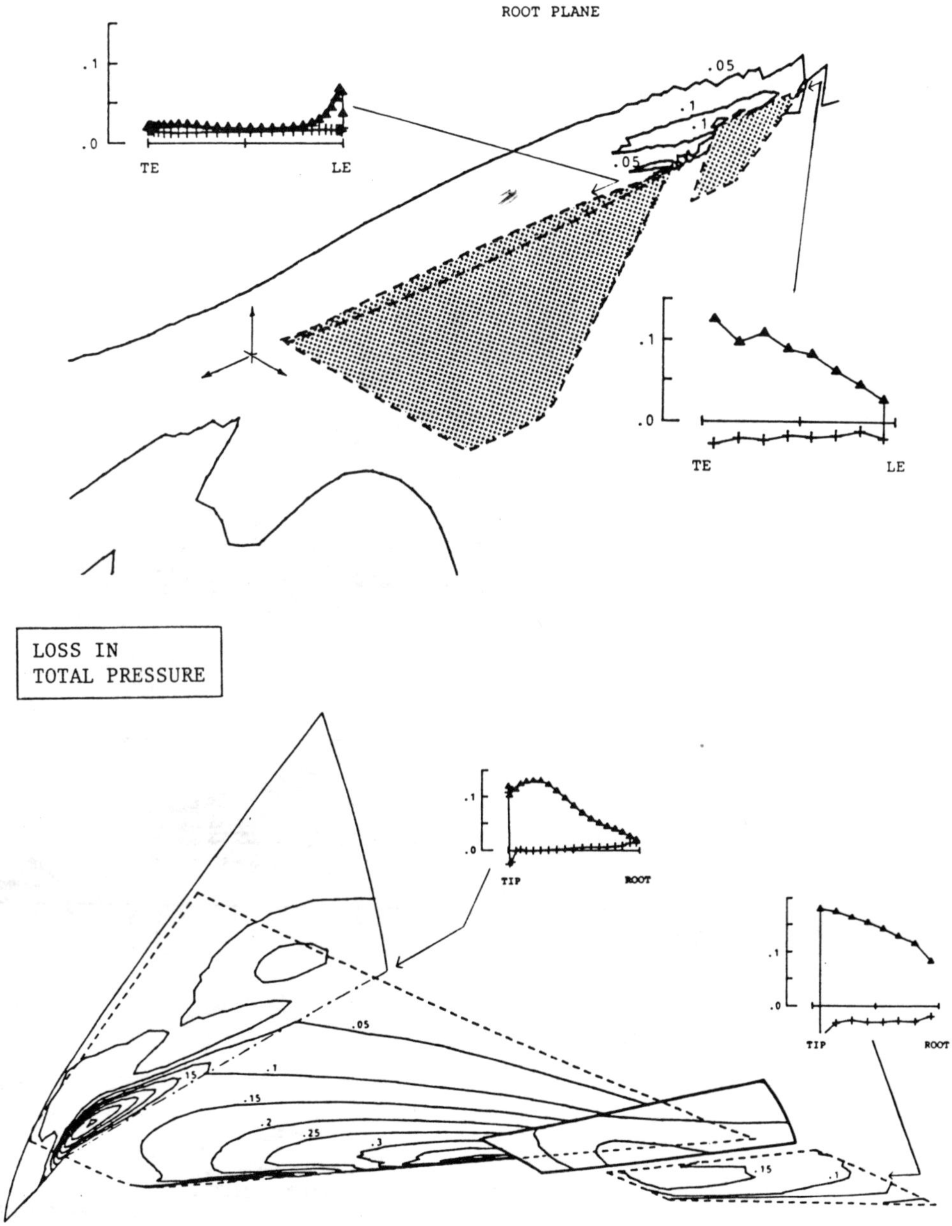

Fig. 6(c) Contours of constant loss in total pressure $1.-p_t/p_{t_\infty}$

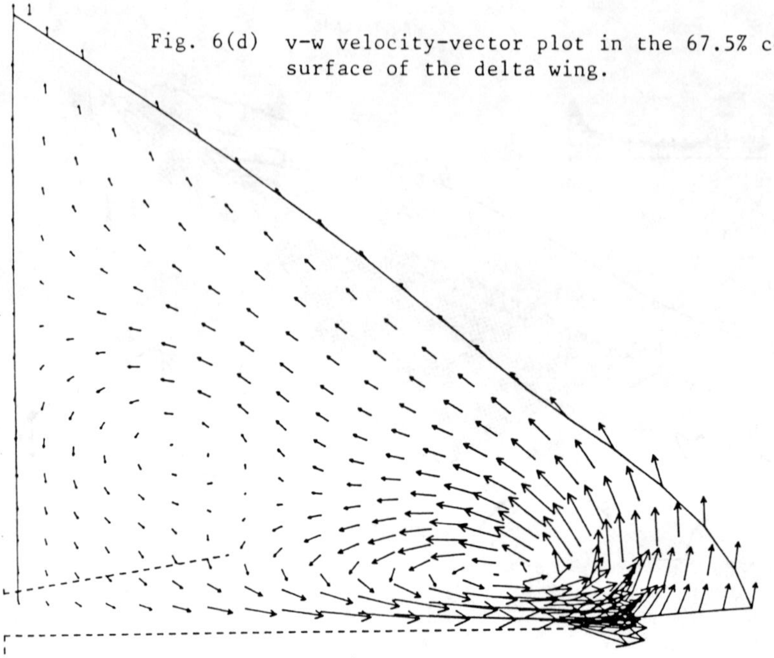

Fig. 6(d) v-w velocity-vector plot in the 67.5% constant chord surface of the delta wing.

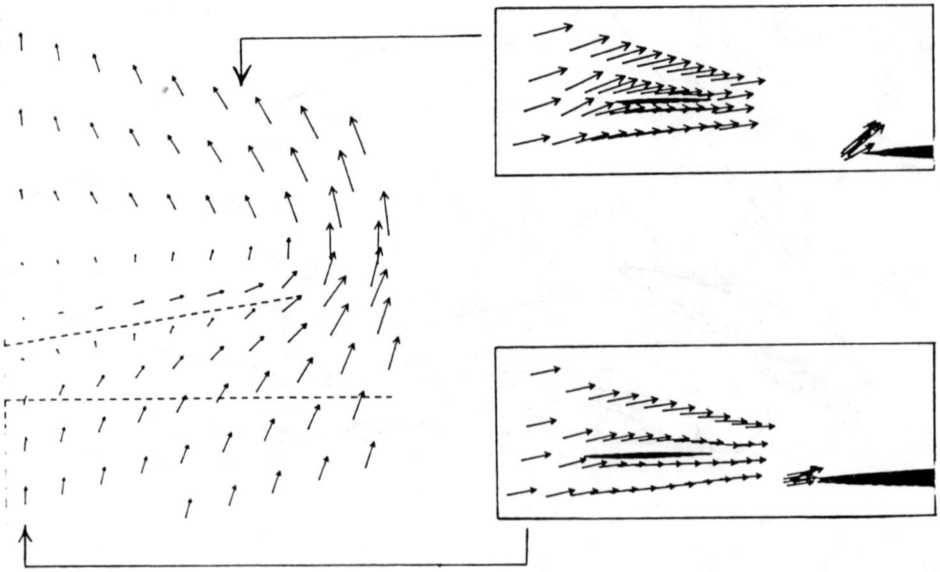

Fig. 6(e) v-w velocity-vector plot (to the left) in the mesh surface that intersects the 80% constant chord station of the canard, together with u-w velocity-vector plots (to the right) in the root plane and in the mesh surface that intersects the 80% span station of the canard.

THE NUMERICAL SIMULATION OF TURBULENT GAS-PARTICLE CHANNEL FLOWS USING A NEW KINETIC MODEL SOLVING COUPLED BOLTZMANN AND NAVIER-STOKES EQUATIONS

J.A. Essers, L. Lourenço & M.L. Riethmuller
von Karman Institute for Fluid Dynamics
Chaussée de Waterloo, 72
B - 1640 Rhode Saint Genèse, Belgium

SUMMARY

A new kinetic model is presented for the numerical simulation of turbulent incompressible gas-particle channel flows. The particulate phase is described by a probability density function governed by an integro-differential Boltzmann equation. The latter is coupled to the Reynolds averaged Navier-Stokes equations governing the gas motion. A space marching approximate factorization technique is used to solve these coupled equations. Results obtained for steady developing flows in a horizontal channel for different loading ratios and particle sizes are found in good agreement with experiments.

INTRODUCTION

Existing mathematical models describing gas-particle flows consider the particulate phase as a continuum, an assumption which is only valid for suspensions with sufficiently large concentrations of very small particles. With the continuum models, it is also difficult to suitably introduce the effect of two phase interactions on the stress tensor, and to simulate the effect of particle interaction with solid walls. Additionally, the particle slip velocity at the wall should be introduced empirically. All these drawbacks can be avoided using the new kinetic model presented below.

In that model, drag and gravity forces, particle-particle collisions as well as wall-particle interactions are accounted for. The effect of particle rotation is neglected. Particles are also supposed to be sufficiently large and heavy. Their random motion (agitation) due to gas turbulence can therefore be neglected. A detailed critical analysis of these approximations is presented in [1]. As opposed to continuum models, the new kinetic model is able to explain why particles do not deposit at the lower wall if the Knudsen number is not small (that number is defined as the inverse of the product $\pi d_p^2 n_p H$, d_p, n_p and H respectively denoting the particle diameter, the particle number density and the channel width). This effect can be explained by the agitation produced during particle-wall interactions [1-3]. That production strongly depends on wall roughness. A correct simulation of these interactions is therefore very important.

THE NEW KINETIC MODEL

For the sake of simplicity, we consider a plane 2D steady, incompressible gas-particle flow between two parallel plates forming a horizontal channel. The x and y axes are respectively chosen parallel and perpendicular to the plates. The particulate phase is represented by a particle distribution function $f(\underline{s},\underline{\xi})$, where $\underline{s}$ is a vector with components x and y defining the position of a particle, and $\underline{\xi}$ is the particle velocity vector with components σ and λ. $f(\underline{s},\underline{\xi})d\underline{s}d\underline{\xi}$ represents the number of particles contained in the phase-space volume $d\underline{s}d\underline{\xi}$. The distribution function is governed by the following transport equation of Boltzmann type :

$$\sigma \frac{\partial}{\partial x} f + \lambda \frac{\partial}{\partial y} f + \frac{1}{m_p} \left[\frac{\partial}{\partial \sigma} (fF_x) + \frac{\partial}{\partial \lambda} (fF_y) \right] - C(f) \tag{1}$$

where m_p is the particle mass, F_x and F_y are the components of the volume forces acting on the particles, and C(f) is a function of f modelling the effect of particle-particle collisions. Equation (1) expresses that for a steady flow, the change in time of the number of particles contained in $d\underline{s}d\underline{\xi}$, which can be due to the fact that particles can leave the space volume $d\underline{s}$ by convection and also to a variation of particle velocities induced by the volume forces and by particle-particle collisions, is equal to zero. Forces F_x and F_y account for gravity effects, particle drag and particle lift due to gas shear. The detailed expressions used for spherical particles are discussed [1-3]. They correspond to complicated non linear functions of σ,λ and of the local gas velocity components.

The collision models we use to express C(f) are fully described in [1]. We consider slightly inelastic collisions between two spherical particles. Our calculations have first been performed with the following sophisticated Boltzmann model rather similar to that usually used in rarefied gas dynamics:

$$C(f) = \int_{\sigma'=-\infty}^{+\infty} \int_{\lambda'=-\infty}^{+\infty} \int_{\theta=0}^{\pi} \int_{\psi=0}^{2\pi} G d_p^2 \sin\theta \cos\theta \; H(\underline{\xi},\underline{\xi}',\underline{\xi}^*,\underline{\xi}'^*) \; d\underline{\xi}' d\theta d\psi \tag{2}$$

with

$$H = \frac{1}{E^3} f(\underline{s},\underline{\xi}^*) f(\underline{s},\underline{\xi}'^*) - f(\underline{s},\underline{\xi}) f(\underline{s},\underline{\xi}')$$

where :

- $(\underline{\xi},\underline{\xi}')$ are the velocities of a couple of particles before collision;
- $(\underline{\xi}^*,\underline{\xi}'^*)$ are the velocities of these particles after collision. They can be expressed as functions of $(\underline{\xi},\underline{\xi}')$ using the collision model;
- d_p is the particle diameter, and θ,ψ are some collision angles;
- G is a function of $(\underline{\xi},\underline{\xi}')$ corresponding to the relative velocity before collision.
- E is a coefficient smaller than 1 accounting for the inelastic character of the collision.

The use of the Boltzmann collision model leads to good results, but is time consuming because it requires the numerical evaluation of a multiple integral at each mesh point. For that reason, another so-called BGK model, using a much simpler expression of C(f), has also been used [1-3]. The latter, which is rather similar to the classical BGK model used in rarefied gas dynamics [4], unfortunately leads to a poor accuracy.

For steady 2D flows, the distribution function f is a function of 4 independent variables x,y,σ and λ. Its numerical calculation performed by solving (1) leads to much more detailed information on the particulate phase than those obtained from a continuum model. In particular, multiple numerical quadratures on σ and λ of functions of f give the local value at each channel point of important quantities such as particle density, velocity components, agitation, kinetic energy of the random motion, etc [1-3].

As F_x and F_y are functions of gas velocity components u(x,y) and v(x,y), equation (1) should be coupled to Navier-Stokes equations describing the gas motion. For incompressible turbulent gas flows, the continuity and Reynolds averaged axial momentum equations can be written as follows using the usual shear flow approximation and assuming that the volume of solid particles is negligible with respect to that occupied by the gas :

$$\frac{\partial u}{\partial x} + \frac{\partial v}{\partial y} = 0 \qquad (3)$$

$$u \frac{\partial u}{\partial x} + v \frac{\partial u}{\partial y} = h + \frac{\partial}{\partial y}\left(\varepsilon \frac{\partial u}{\partial y}\right) + M_{p,x} \qquad (4)$$

where :

- ε is the total effective viscosity field;
- $h = -\frac{1}{\rho}\frac{dp}{dx}$ is an unknown function of x only, which is proportional to the axial pressure gradient;
- $M_{p,x}$ is the momentum sink/source term due to the presence of particles. Its calculation also requires multiple quadratures on σ and λ of functions of f,u and v [1-3] . It can be expressed as :

$$M_{p,x} = \frac{\pi}{8} d_p^2 \iint_{\sigma\lambda} fC_D(\sigma-u)\sqrt{(u-\sigma)^2+(v-\lambda)^2}\; d\sigma d\lambda$$

the particle drag coefficient C_D being expressed as a function of a Reynolds number defined using the local relative gas-particle velocity.

To compute the effective viscosity field ε, we use a generalized version of the Nee-Kovasznay model [5] which is fully described in [1-3] . A convection-diffusion partial differential equation governing ε is solved. That equation exhibits classical Nee-Kovasznay terms modelling the turbulence diffusion, production and dissipation. Some additional sophisticated terms are introduced to model the effect of solid particles on gas turbulence. The latter terms are functions of the gas velocity components, but also of some integrals on σ and λ of the distribution function, thus introducing a coupling with equation (1).

THE NUMERICAL SCHEME

Our steady gas-particle flow calculations have been performed using a space marching technique. The values of u(y), v(y), ε(y) and f(y,σ,λ) are given at the channel inlet x=0. We, e.g. assume that all solid particles are injected in a uniform laminar gas flow parallel to the plates. Turbulent boundary layers then develop progressively in the channel. The flow field is computed successively in cross sections corresponding to increasing values of x until reaching the fully developed regime. The computational domain in the phase plane (σ,λ) is artificially limited by far field boundaries σ = 0, σ = S, λ = -L and λ = +L, S and L being chosen sufficiently larger than any expected particle velocity, i.e., such that f can be assumed to be equal to zero on these boundaries. Equations (1-4) are written in non dimensional form, and a transformation is performed to introduce a stretching of y,σ and λ variables [1-3] . To simplify the presentation, we will however assume here that equations are discretized without stretching, i.e. with a uniform mesh, using their dimensional form (1-4). To ensure numerical stability a fully implicit scheme is used. The following discretized form of equation (1) is used to evaluate $f^{k,\ell}_{m+1,j}$ from the known values of $f^{k\ \ell}_{m,j}$ (for all j,k,ℓ) :

$$\frac{\sigma^k}{\Delta x}\left(f^{k,\ell}_{m+1,j} - f^{k,\ell}_{m,j}\right) + \lambda^\ell \frac{\partial}{\partial y} f^{k,\ell}_{m+1,j} + \frac{1}{m_p}\frac{\partial}{\partial\sigma}\left(f^{k,\ell}_{m+1,j}\tilde{F}^{k,\ell}_{x,j}\right) + \frac{1}{m_p}\frac{\partial}{\partial\lambda}\left(f^{k,\ell}_{m+1,j}\tilde{F}^{k,\ell}_{y,j}\right) = \tilde{C}^{k,\ell}_j \qquad (5)$$

where :

- $f^{k,\ell}_{m,j}$ denotes the value of f at mesh point $x = m\Delta x$; $y = j\Delta y$; $\sigma = k\Delta\sigma$; $\lambda = \ell\Delta\lambda$, m,j,k,ℓ being some integers.

- $\frac{\partial}{\partial y}, \frac{\partial}{\partial \sigma}, \frac{\partial}{\partial \lambda}$ denote some discretized forms of the derivatives. These discretizations are usually fully centered, but hybrid schemes using slightly excentered discretizations should sometimes be used to avoid oscillations.

- The $\sim$ superscript indicates that the corresponding quantities are known. We will first assume that they correspond to mesh points in the plane $x = m\Delta x$. Simple quadrature formulas are used to compute the collision multiple integral from the known values of $f^{k,\ell}_{m,j}$.

Using (5), the values of $f^{k,\ell}_{m+1,j}$ could in principle be obtained from the very time consuming solution of a large linear system. To significantly improve the efficiency, we however performed an approximate factorization of equation (5), and used the following nearly equivalent form successively leading to the solution of several linear systems with tridiagonal matrices:

$$\left(1+\frac{\Delta x}{\sigma^k}\lambda^\ell\frac{\partial}{\partial y}\right)\left(1+\frac{\Delta x}{m_p\sigma^k}\frac{\partial}{\partial\sigma}\tilde{F}^{k,\ell}_{x,j}\right)\left(1+\frac{\Delta x}{m_p\sigma^k}\frac{\partial}{\partial\lambda}\tilde{F}^{k,\ell}_{y,j}\right)f^{k,\ell}_{m+1,j} = f^{k,\ell}_{m,j}+\frac{\Delta x}{\sigma^k}\tilde{C}^{k,\ell}_{m,j} \tag{6}$$

To compute that solution, we need some boundary conditions for the values of $f^{k,\ell}_{m+1}$ on the two plates. Consider e.g. the lower plate. We denote by f^{k,ℓ^-}_{m+1} and f^{k,ℓ^+}_{m+1} the values of the distribution function respectively corresponding to negative and positive values of σ, i.e. to incident and reflected particles. For the incident particles, the boundary condition used just consists in writing equation (6) on the plate using a fully excentered discretization of $\frac{\partial}{\partial y}$. For the reflected particles, the values of f^{k,ℓ^+}_{m+1} should be expressed as functions of the f^{k,ℓ^-}_{m+1} from physical information corresponding to the particle-wall interaction model. Experimental measurements have to be made that are used to construct a transfer matrix W as explained in [1-3]. We can then express the values of f^{k,ℓ^+}_{m+1} from :

$$f^{k,\ell^+}_{m+1} = \sum_{i,n^-} W_{k,\ell^+,i,n^-}\,\tilde{f}^{i,n^-} \tag{7}$$

To compute gas velocity components $u_{m+1,j}$, $v_{m+1,j}$ from the known values of $u_{m,j}$ and $v_{m,j}$ the following discretized forms of equations (3-4) are used:

$$\tilde{u}_j\frac{u_{m+1,j}-u_{m,j}}{\Delta x}+\tilde{v}_j\frac{u_{m+1,j+1}-u_{m+1,j-1}}{2\Delta y} = h_{m+1}+(\tilde{M}_{p,x})_j$$
$$+\frac{1}{\Delta y^2}\left[\tilde{\varepsilon}_{j+1/2}\left(u_{m+1,j+1}-u_{m+1,j}\right)-\tilde{\varepsilon}_{j-1/2}\left(u_{m+1,j}-u_{m+1,j-1}\right)\right] \tag{8}$$

$$v_{m+1,j+1}-v_{m+1,j} = \frac{\Delta y}{2\Delta x}\left(u_{m,j+1}+u_{m,j}-u_{m+1,j+1}-u_{m+1,j}\right) \tag{9}$$

The value of h_{m+1} related to the local pressure gradient is unknown. However, if we assign it any (wrong) value, the values of $u_{m+1,j}$ can be obtained from equation (8) written at each mesh point using no-slip boundary conditions on the plates. That calculation requires the solution of a tridiagonal linear system. Then using (9) and expressing the impermeability of the lower plate (v=0), the values of v can be computed for all mesh points of line (m+1) successively for increasing values of j by marching towards the upper plate. On the latter, v is, however, generally found different from zero. Using another guess for h_{m+1}, the procedure can be repeated to compute another solution for $u_{m+1,j}$ and $v_{m+1,j}$. Then using an original algorithm fully described in [1,3,6], the two solutions can be combined to obtain the correct value of the pressure gradient as well as the correct solution for $u_{m+1,j}$, $v_{m+1,j}$ satisfying the no-slip and impermeability conditions on both plates.

The convection-diffusion Nee-Kovasznay equation can be discretized with a scheme quite similar to that used for the momentum equation. The non linear terms as well as the integrals of the distribution function appearing in the terms corresponding to turbulence production and dissipation are calculated using known values of u,v,f. The value of $\varepsilon_{m+1,j}$ of the effective viscosity field can then be computed by solving a linear tridiagonal system, with the boundary conditions $\varepsilon=0$ on the plates.

Up to now, we have assumed that terms with a $\sim$ superscript appearing in equations (6-9) were computed using known values of u,v,f,ε corresponding to $x = m\Delta x$. That procedure would, however, generally lead to severe non linear instabilities. In practice, the complete procedure described above must be repeated iteratively. For each iteration, terms with a $\sim$ superscript are updated using values of $f^{k,\ell}_{m+1,j}$, $u_{m+1,j}$ $v_{m+1,j}$ and $\varepsilon_{m+1,j}$ obtained at the end of the previous iteration. A satisfactory convergence is usually obtained after a very small number of iterations (typically 2-5).

NUMERICAL RESULTS

The numerical technique has first been tested for 2D turbulent gas flows without particles. Results were found in excellent agreement with experiments [6]. For relatively large Knudsen numbers, i.e. for weak particle loading ratios, the effect of particle-particle collisions is negligible. The new scheme has first been tested by neglecting the collision term. Calculations have then been performed for higher loading ratios using both BGK and Boltzmann collision models. Comparative reliable experimental results have been obtained in a special facility that has been constructed at the von Karman Institute. In that facility, small spherical glass balls are dispersed in air flowing between two parallel plates. Detailed LDV measurements of gas and particle velocities, particle density and distribution function can be performed at different points in the channel [1]. Figure 1 shows typical distribution functions versus the axial particle velocity computed with the Boltzmann model for different values of the y-coordinate in a fully developed flow. Note the bimodal character on the function close to the plates. This character, which is due to the presence of both incident and reflected particles, is also observed in the experiments. Our calculations performed with the Boltzmann model for different Reynolds numbers, loading ratios and particle diameters have been found in good agreement with experiments [1-3]. Figure 2 allows a comparison of computed and measured distribution functions at the center of channel. With the BGK model, a rather poor accuracy is however obtained specially for high loading ratios. A comparison of the distribution functions computed with the two collision models for loading ratios of 0.5 and 3 is presented on figure 3.

All results presented below have been obtained with the Boltzmann model for a fully developed flow. Figures 4 and 5 present a comparison of typical gas velocity and particle density profiles obtained from calculations and experiments for similar values of loading ratios, particle diameter and gas mass flow velocity. Figure 6 shows the variation of particle to gas mean velocity ratio as a function of the loading ratio for two different particle diameters. Experiments indicated that there exists a unique relationship between the relative particle agitation and the Knudsen number for different sizes. A similar conclusion is obtained from the calculations as shown on figure 7. These results are very encouraging.

A typical 2D calculation without any collision term typically requires a CPU time of the order of 3 hours on a VAX 11/780 computer. That time is only increased by about 10% when the BGK model is used. The use of the Boltzmann model is, however, much more time-consuming (typically 5-6 hours).

REFERENCES

[1] LOURENÇO, L.: A kinetic model for a gas particle suspension flow in a horizontal duct and its numerical implementation. Ph.D. Thesis, U. Libre de Bruxelles, 1982.

[2] LOURENÇO,L.; RIETHMULLER, M.L.; ESSERS, J.A.: The kinetic model for gas particle flow and its numerical implementation. Int. Conf. on the Physical Modelling of Multi-Phase Flow, Coventry, England, April 1983, also VKI Preprint 1983-06.

[3] LOURENÇO, L.; ESSERS, J.A.; RIETHMULLER, M.L.: Computation of turbulent gas-particle suspension flow in channels. In: "Computational Fluid Dynamics", VKI LS 1983-04, March 1983.

[4] BATHNAGAR, P.L.; GROSS, E.P.; KROOK, M.: A model for collision processes in gases. I - Small amplitude processes in charged and neutral one component systems. Physical Review, Vol. 96, No. 3, April 1954, pp 511-525.

[5] NEE, V.W. & KOVASZNAY, L.S.G.: The calculation of the incompressible turbulent boundary layer by a simple theory. In: "Computation of Turbulent Boundary Layers", Proceedings of the AFOSR-IFP-Stanford Conference, Vol. 1, 1968, pp 300-319.

[6] LOURENÇO, L. & ESSERS, J.A.: Numerical model for the solution of internal pipe/channel flows in laminar or turbulent motion. VKI TN 141, 1981.

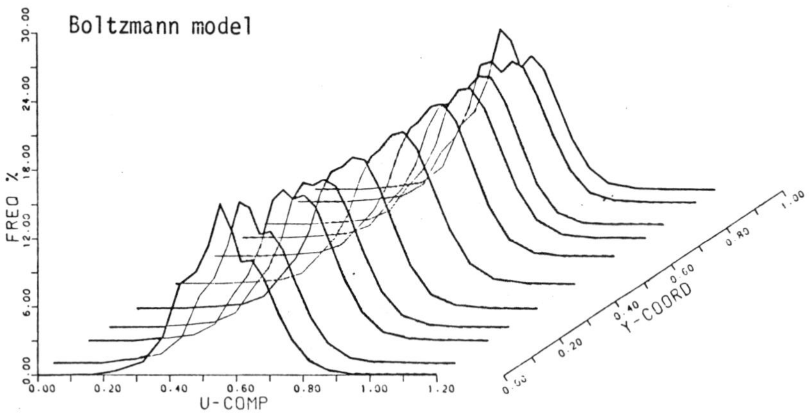

FIG. 1 - TYPICAL DISTRIBUTION FUNCTIONS VERSUS AXIAL PARTICLE VELOCITY COMPUTED ACROSS THE CHANNEL FOR A FULLY DEVELOPED FLOW (LOADING RATIO : 0.5, PARTICLE DIAMETER : 0.5 mm)

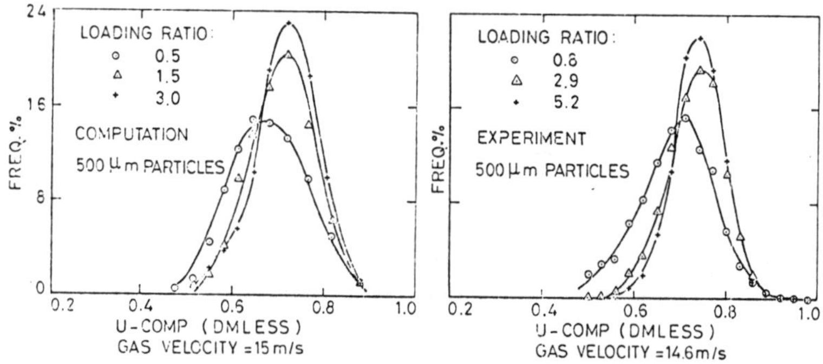

FIG. 2 - DISTRIBUTION FUNCTION OBTAINED AT THE CENTER OF THE CHANNEL FROM EXPERIMENTS AND COMPUTATIONS PERFORMED WITH BOLTZMANN MODEL (FULLY DEVELOPED FLOW)

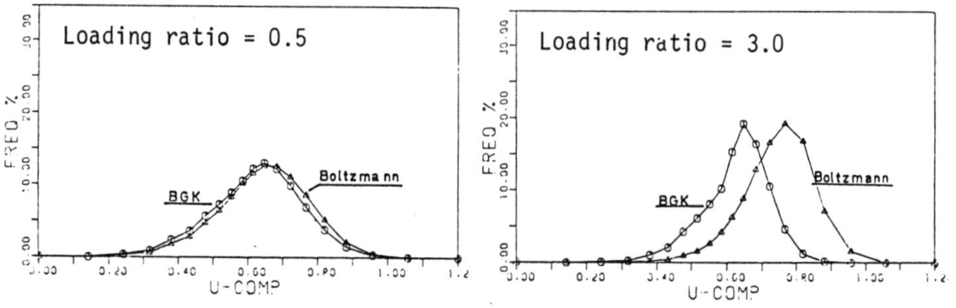

FIG. 3 - COMPARISON OF TYPICAL DISTRIBUTION FUNCTIONS COMPUTED AT THE CENTER OF THE CHANNEL WITH BOLTZMANN AND BGK MODELS (PARTICLE DIAMETER 0.5 mm)

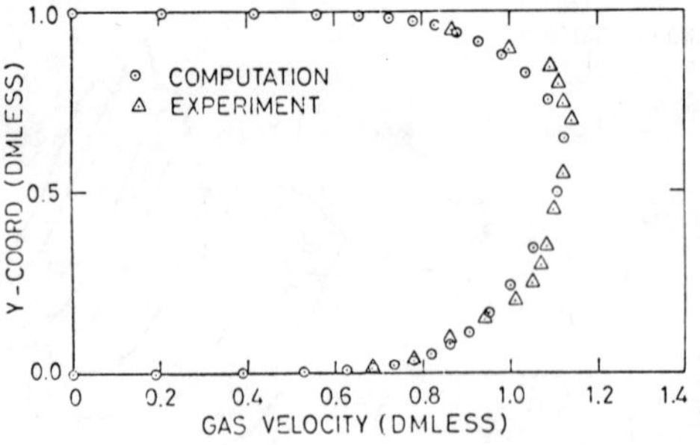

FIG. 4 - COMPARISON OF TYPICAL MEASURED AND COMPUTED FULLY DEVELOPED GAS VELOCITY PROFILES

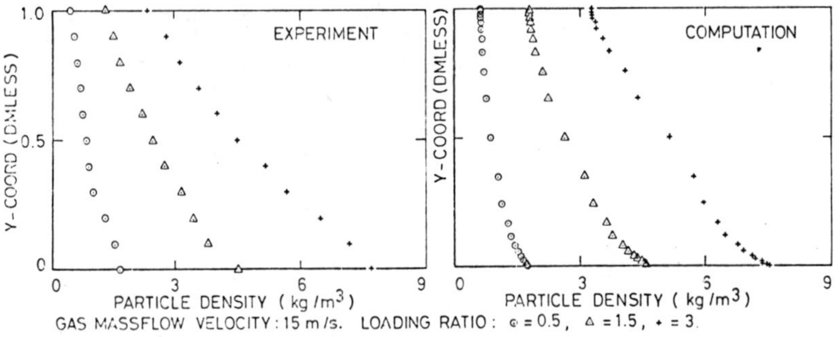

FIG. 5 - COMPARISON OF TYPICAL MEASURED AND COMPUTED PARTICLE DENSITY PROFILES (FULLY DEVELOPED FLOW)

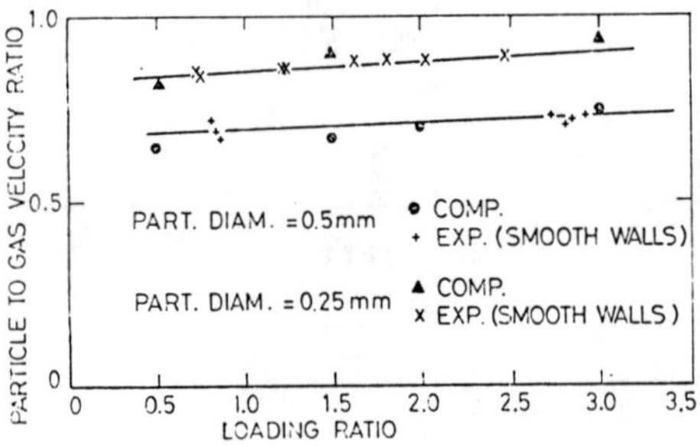

FIG. 6 - PARTICLE TO GAS VELOCITY RATIO VERSUS LOADING RATIO FOR TWO DIFFERENT PARTICLE DIAMETERS (FULLY DEVELOPED FLOW)

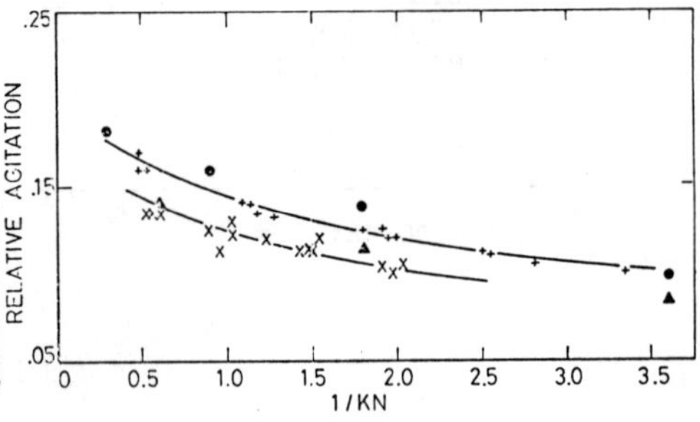

FIG. 7 - RELATIVE AGITATION VERSUS INVERSE KNUDSEN NUMBER FOR TWO DIFFERENT PARTICLE DIAMETERS (FULLY DEVELOPED FLOW)

ACCURATE NUMERICAL SOLUTION OF STIFF EIGENVALUE PROBLEMS

Vijay K. Garg
Department of Mechanical Engineering
Naval Postgraduate School, Monterey, CA 93943

SUMMARY

The principal difficulty in solving stiff eigenvalue problems is the parasitic growth of one or more independent solutions when using a standard shooting technique for integration. Methods to overcome this difficulty either use repeated orthonormalization or increase the total number of first order equations to be solved. Herein we describe a method that avoids orthonormalization and *reduces* the number of first order equations to be integrated.

INTRODUCTION

The general linear boundary value problem may be expressed in the form

$$Dy = A(x)y+f(x), \tag{1a}$$

$$By(0) = c_1, \tag{1b}$$

$$Cy(1) = c_2, \tag{1c}$$

where D denotes differentiation with respect to x, y(x) and f(x) are vector functions with n components, A is an n X n matrix, B is an (n - k) X n matrix of rank n - k, C is a k X n matrix of rank k, c_1 is a vector with n - k components, and c_2 is a vector with k components.

Applying the principle of superposition, any solution of (1a), (1b) can be written as

$$\begin{aligned} y(x) &= y^0(x)+\beta_1 y^1(x)+\beta_2 y^2(x)+\ldots+\beta_k y^k(x) \\ &= y^0(x)+Y(x)\beta. \end{aligned} \tag{2}$$

where β is a vector with k components, $y^0(x)$ is a solution of the nonhomogeneous system

$$\begin{aligned} Dy^0(x) &= A(x)y^0(x)+f(x), \\ By^0(0) &= c_1, \end{aligned} \tag{3}$$

and Y(x) denotes an n X k matrix whose columns $y^1, y^2, \ldots, y^k$ are solutions of the homogeneous system

$$\begin{aligned} Dy(x) &= A(x)Y(x), \\ BY(0) &= o. \end{aligned} \tag{4}$$

The form (2) satisfies the condition (1b) for any choice of the constants $\beta = \{\beta_1, \beta_2, \ldots, \beta_k\}$. If in addition β is chosen so as to satisfy the condition (1c):

$$Cy(1) = Cy^0(1)+CY(1)\beta = c_2, \tag{5}$$

then (2) is the required solution of the linear boundary value problem (1).

It often happens that this mathematically exact procedure for obtaining the solution of (1) leads to very poor or even completely incorrect results when applied as a numerical procedure. This happens when the matrix A in (1) has eigenvalues whose real parts are well separated. In this case the vectors $y^1, y^2, \ldots, y^k$, which form the columns of Y(x), become more and more dependent as integration proceeds from x=0 to 1. When this situation holds, the matrix Y(1) will be poorly conditioned, the vector β determined by (5) will be inaccurate, and the resolution (2) will lead to poor or even absurd numerical results.

Several methods exist that circumvent this difficulty by keeping the columns of Y(x) independent during numerical integration. Four of these methods are critically examined in [1], which recommends the Gram-Schmidt orthonormalization procedure for overall economy and accuracy. Herein we describe a method which is even superior to the Gram-Schmidt orthonormalization method.

THE PROBLEM

The problem in relation to which the present method will be outlined is drawn from hydrodynamic stability theory where a classical example involves the solution of the Orr-Sommerfeld equation [2]

$$L(\phi) \equiv [\{D^2 - \alpha^2 - iRe(\alpha U-\omega)\} (D^2-\alpha^2) + i\alpha ReD^2U]\phi = 0 \qquad (6a)$$

with boundary conditions

$$\phi = D\phi = 0 \text{ when } x = 0 \text{ and } \infty. \qquad (6b)$$

Here $U = U(x)$ is a prescribed function that specifies the laminar velocity profile. This equation governs the spatial stability of a laminar flow (velocity $U(x)$ and Reynolds number Re) subjected to an infinitesimal two-dimensional disturbance of real frequency ω and complex wave number α so that the disturbance is proportional to $\exp[i(\alpha x - \omega t)]$. The range of integration for (6a) may also be $0 \leq x \leq 1$.

In contrast to (1) the system (6) is homogeneous with the complex eigenvalue α and associated eigenfunction $\phi(x)$ that we wish to determine for given $U(x)$, ω and Re. This implies that $y^0(x)$ of (2) will vanish identically, and trouble in determining β from (5) will arise if the characteristic values of the operator L, which are $\pm\alpha$ and $\pm\lambda = \pm[\alpha^2+iRe(\alpha-\omega)]^{1/2}$, differ greatly in their real parts. Thus if α is, say, of order unity, then $Re^{1/2}$ should be at least an order of magnitude larger. This is often the case in hydrodynamic stability problems except when solving for the stability of jet flows that become unstable at Reynolds numbers of order unity.

It is clear that for solution of (6), Y(x) of (2) contains only two columns y^1 and y^2, each containing four elements. The solution $\phi(x)$ of (6) is thus a linear combination of two solutions ϕ_1 and ϕ_2. For integration of (6a) a marching technique such as the Runge-Kutta method can be used with starting values

$$(\phi_1, D\phi_1, D^2\phi_1, D^3\phi_1) = (1, -\lambda, \lambda^2, -\lambda^3)e^{-\lambda x_1} \qquad (7a)$$

$$(\phi_2, D\phi_2, D^2\phi_2, D^3\phi_2) = (1, -\alpha, \alpha^2, -\alpha^3)e^{-\alpha x_1} \qquad (7b)$$

at $x=x_1$, where x_1 is the finite value of x at which boundary conditions, mathematically applicable at $x=\infty$, may be assumed to hold. Since ϕ_1 is strongly increasing in direction of decreasing x, (for large Re), numerical integration of (6a) causes ϕ_2 to be linearly dependent on ϕ_1, thus leading to absurd results.

THE METHOD

Assume we have obtained the rapidly growing solution ϕ_1 (corresponding to the characteristic value λ). We then let

$$\phi_2 = \zeta\phi_1 \qquad (8)$$

Since ϕ_2 is a solution of system (6), substitution of (8) into (6a) leads to a differential equation of *third* order for $D\zeta$:

$$L_1[D\zeta;\phi_1] \equiv \phi_1 D^4\zeta + 4D\phi_1 D^3\zeta + (6D^2\phi_1 - 2\alpha^2\phi_1)\, D^2\zeta$$
$$+ 4(D^3\phi_1 - \alpha^2 D\phi_1)D\zeta - iRe(\alpha U-\omega)$$
$$(\phi_1 D^2\zeta + 2D\phi_1 D\zeta) = 0. \qquad (9)$$

The starting values for (9) at $x=x_1$ can be derived from (7) and (8) as $(\zeta, D\zeta, D^2\zeta, D^3\zeta) = \{1, (\lambda-\alpha), (\lambda-\alpha)^2, (\lambda-\alpha)^3\}e^{(\lambda-\alpha)x_1}$ (10)

Equation (6a) for ϕ_1 and (9) for $D\zeta$ are integrated simultaneously from $x=x_1$ to $x=0$. The function $\zeta(x)$ is then obtained from

$$\zeta(x) - \zeta(0) = \int_0^x D\zeta(\eta)\,d\eta, \qquad (11)$$

where the integration is to be performed in the direction of increasing x. The solution $\phi(x)$ should be such a combination of ϕ_1 and ϕ_2 that both boundary conditions at x=0 are satisfied. The combination

$$\phi(x) = \phi_2(x) - \zeta(0)\phi_1(x)$$
$$= \{\zeta(x) - \zeta(0)\}\phi_1(x) \qquad (12)$$

clearly satisfies $\phi(0) = 0$. Since

$$D\phi(x) = \{\zeta(x) - \zeta(0)\}D\phi_1(x) + D\zeta(x)\phi_1(x),$$

it follows that the second boundary condition $D\phi(0) = 0$ is satisfied only if

$$D\zeta(0) = 0. \qquad (13)$$

This condition (13) corresponds to the fact that we have an eigenvalue problem in (6).

Thus for a given U(x), ω and Re, the guessed value of α is varied using a Newton-Raphson procedure until $D\zeta(0)=0$, yielding the eigenvalue α. A similar procedure holds if the domain for (6) is $0 \le x \le 1$. The above procedure can also be applied to a stiff linear boundary value problem such as (1). In this case, the particular solution $y^0(x)$, (cf.(2)), can be expressed in terms of the rapidly growing solution, say $y^1(x)$, as

$$y^0(x) = \zeta^0(x)\, y^1(x) \qquad (14)$$

in a manner similar to (8). The procedure outlined above is then essentially repeated.

Using this method, we have successfully solved system (6) for very high values of Re much more efficiently and accurately than has been possible heretofore [1]. Some pertinent results will be presented at the conference.

For a three-dimensional disturbance, the operator L in (6a) is of order six, and one more boundary condition at each end of the range of integration is required. In such a case, there are two solutions, say ϕ_1 and ϕ_2, that grow much more rapidly than the third, ϕ_3. Extension of the above method to such cases is underway.

CONCLUSION

An efficient and accurate method for solving stiff eigenvalue and boundary value problems has been developed.

REFERENCES

[1] Garg, V.K., "Improved Shooting Techniques for Linear Boundary Value Problems", Comput. Meth. Appl. Mech. Engrg. 22 (1980) pp. 87-99.

[2] Schlichting, H ., "Boundary Layer Theory", McGraw-Hill, 7th edn. 1979.

STRUCTURAL APPROACH TO TURBULENT MOTION CALCULATION

M.A.Goldshtik, V.N.Shtern
The Institute of Thermophysics
630090 Novosibirsk, USSR

SUMMARY

A new approach to turbulent motion calculation is reported. Existence of coherent structures is essentially used. A solution is constructed by a superposition of these structures. For the structure placing and orientation the probabilistic description is conserved. The method is applied to calculation of stochastic trajectories of the well known Lorenz system. The method advantage is that the integration region is reduced to a correlation radius size. Averaged quantities and pulsation spectrums are found with a saticfactory approximation.

1. INTRODUCTION

Turbulent motion calculation remains to be a great unsolved problem of mechanics. Recently a few hopefull attempts were made of a direct integration of the unsteady Navier-Stokes equations using the Galerkin method [1,2]. But an enormous number of the Galerkin modes and a nescessity of a large time interval for averaging (forces) makes seek an alternative approach.

Our method is concentrated to decrease the time interval to correlation radius order and to use a few "eigen" modes. The main reason is finding the coherent structures in turbulent motion [3,4]. It turns out that turbulence is not a full chaos. It contains typical space-time forms, distributed and orientated by a chance. These forms have repeated and measured scales and patterns. The same things are in dynamical systems having stochastic self-oscillations, when strange attractors appear [5,6]. In this case trajectories consist of a few typical elements, alternating by a chancelike way.

The object is to develop a method which allows on the one hand to calculate the deterministic elements of the motion (structures) and on the other hand to find out a statistical distribution of its placing and alternating. That is to say we want to factorize deterministic and statistic features of the turbulent motion [7].

There were some attempts of a coherent structure calculation for a turbulent motion (see, for example, Lumley [8] and his preceding publications). Unlike these works our approach does not use any empirical data. A disturbation method is applied and a small parameter is the structure overlap. Let us illustrate the disturbation method by a simple example.

2. THE STRUCTURAL REPRESENTATION OF PERIODICAL REGIMES

A periodical function $x(t)$, having a period T, may be expressed by a superposition of structures $\hat{x}(t)$: $x(t)=\Sigma\hat{x}(t-nT)$, moreover by different ways. For instance, one may put $\hat{x}(t)=x(t)$ on a time interval T, and $\hat{x}(t)=0$ outside of it. But it is convenient to choose a structure which is continuous and fast vanishing at $|t| \to \infty$. Any structure is to satisfy conditions

$$\hat{x}(w_k) = 1/\sqrt{2\pi} \int_0^T x(t)\exp(-iw_k t)dt; \; w_k = 2k\pi/T; \; k = 0,1,2,\ldots;$$

where $\hat{x}(w)$ is the Fourier transformation of $\hat{x}(t)$.

Let us consider the equation $\ddot{x} = x - x^2$. It has analytic solutions in terms of the Jacobi elliptic functions. In the range $1 \leqslant x_{max} < 1.5$ the solutions are periodic. On the left end $x_{max} = 1$, $x(t) \equiv 0$, on the right end $x_{max} = 1.5$, $x = 1.5/ch^2(t/2)$. Now we shall find the solutions by the structural approach. We suggest the structure which is placed in the interval $-T/2 \leqslant t \leqslant T/2$, has a maximum at $t = 0$ and vanishes outside the interval exponentially $\sim \exp(\pm t)$, as it follows from the linear part of the equation. Then, taking into account only the neighbouring structures "tails" we have the equation $d^2\hat{x}/dt^2 = \hat{x}(1 - \hat{x} - 4\hat{x}(T/2))$ in the interval $|t| < T/2$ with the boundary conditions: $d\hat{x}/dt \pm \hat{x} = 0$ at $t = \pm T/2$ and the normalizing condition $\hat{x}(0) + 2\hat{x}(T/2) = x_{max}$.

As a result of this nonlinear boundary problem solution we find the structure $\hat{x}(t)$ and the period T and then we construct x(t). The test shows that the structural solutions $\{x(t), T\}$ and exact solutions are equal with an accuracy of one percent or less even in the limit case $x(t) \equiv 1$. It is due to the structure overlap stays to be small, as one can see in the Fig.1.

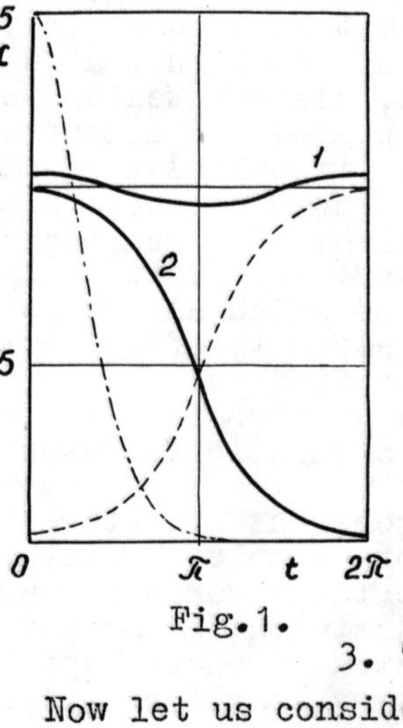

Fig.1.

Fig.1. The structural representation of the periodic regime; curve 1 is a solution at $x_{max} = 1.05$, curve 2 is its structure, the dashed curve is the neighbour structure, the dashed-dotted curve is a solution at $x_{max} = 1.5$.

3. THE STRUCTURALLY-STATISTIC APPROACH

Now let us consider a class of functions, being in a definite sense a generalization of the periodic ones and having the representation: $x(t) = \Sigma\, u_n \hat{x}(t - \Sigma_n)$, where as before $\hat{x}(t)$ is a deterministic function, vanishing rapidly at $|t| \to \infty$ (the structure), but amplitudes u_n and phases Σ_n are chance numbers. This is a model of trajectory form in strange attractors, bifurcating due to a cycle doubling cascade or due to a separatrix loop breaking down [5,6]. In the phase space these trajectories are like infinitly curcuit reelings, placed near the origin cycle or the separatrix loop.

The problem is to find the structure $\hat{x}(t)$ and distributions for u_n and Σ_n. Starting from a base dynamical equation $\dot{x} = F(x)$ (further on only autonomous systems are studied) we shall formulate a bound problem for the structure $\hat{x}(t)$. It is useful for this to divide it on the "core" and "the tails". The core is a central part, where $|\hat{x}(t)|$ is large. It is placed on an interval $-T_n/2 < t < T_n/2$, $T_n = \Sigma_{n+1} - \Sigma_n$, where T_n are distances between the nearest structeres. There are left and right tails. Outside the interval, where $|\hat{x}(t)|$ is supposed to be small enough that $\hat{x}(t)$ to satisfy the linearized dynamical equation $y(t) = dF/dx(0)y$.

Then the left tail may be represented by

$$\hat{x}_L = \Sigma c_j e_j \exp(\lambda_j (t + T_n/2))$$

where λ_j is eigenvalue with Re $\lambda > 0$, and e_j is the eigenvector corresponding to λ_j. It is clear that the left tail vanishes at $t \to -\infty$. Similarly the right tail

$$\hat{x}_R = \Sigma D_k g_k \exp(\mu_k (t - T_n/2)), \text{ where}$$

where μ_k is eigenvalue with Re $\mu < 0$, and g_k is its eigenvector. (Eigenvalues are numerated with decrease of Re λ, μ). The constants $c_j, j = 1,\dots,J$ and D_k, $k = 1,\dots, K$ are to be found. $J + K = N$, where N is an order of the dynamical system. These constants may be excluded and then one has J and K boundary conditions at the interval ends accordingly.

The interval T_n, where the structure is placed may be fixed with some arbitrariness. In the spirit of the present approach the interval T_n may be determined so that $|x^2(t)|$ has minimums at its ends. In this case the linearization is more argumented.

4. SIMPLIFIED STRUCTURALL CALCULATION OF THE LORENZ SYSTEM

In this section the more complex problem to find distributions of u_n and Σ_n is postponed. It is supposed a priori that $\Sigma_n = nT$, where the constant T is unknown and $u_n = (-1)^{\Psi_n}$ where Ψ_n are chance integers, for example 0 and 1 with equal weights, and their moments are $\langle \Psi_n, \Psi_m \rangle = \delta_{nm}$; or for other coordinates $u_n \equiv 1$. This choice (see preceding works [9,10]) is based on a bifurcation analysis of a stochasticlike motion from a separatrix "eight figure".

This simplified approach was applied to nonperiodic trajectories of the well-known Lorenz system [11]:

$$\dot{x} = \sigma (y - x); \; \dot{y} = rx - y - xz; \; \dot{z} = xy - bz \qquad (1)$$

A solution is seeked in the form: $x = \Sigma\, \hat{x}(t - nT)\, (-1)^{\Psi_n}$

$$y = \Sigma \hat{y}(t - nT)(-1)^{\Psi_n} \; ; \; z = \Sigma\, \hat{z}(t - nT)$$

Substituting these expressions in (1) and suitably averaging on Ψ_n, we produce the follouing equations for the structures in the interval

$$\dot{\hat{x}} = \sigma\ (\hat{y} - \hat{x});\quad \dot{\hat{z}} = \hat{x}\hat{y} - b\hat{z} \qquad (2)$$

$$\dot{\hat{y}} = r\hat{x} - \hat{y} - \hat{x}\hat{z} - \hat{x}\hat{z}(T/2)e^{-b(t + T/2)}$$

with boundary conditions: $(\lambda + \sigma)\hat{x} = \sigma\hat{y}$, $\hat{z} = 0$ at $t = -T/2$, and $r\hat{x} + (\lambda + \sigma)\hat{y} = 0$ at $t = T/2$. Here λ is a positive eigenvalue of the linearized system (1). The convention of a solution norm minimum at the interval ends gives the correlation $\hat{x}(-T/2) = \sqrt{b/\lambda}\ \hat{z}(T/2)$. The method of solving is to take tentative values of T and $\hat{z}(T/2) = a$. Then we have all initial data at $t = -T/2$ and integrate the system (2) to $t = T/2$; at this end two conditions are to be fulfiled:

$$F(T,a) \equiv r\hat{x}(T/2) + (\lambda + \sigma)\hat{y}(T/2) = 0$$

$$G(T,a) \equiv \hat{z}(T/2) - a = 0$$

The parameters T,a must be found to vanish F and G.

The problem solution was calculated in [9,10] (with unsignificant modification). The calculation results of the first moment z and spectrums of x(t) are shown in Fig.2,3. We conclude that the structural approach gives a good approximation even in the simplified form.

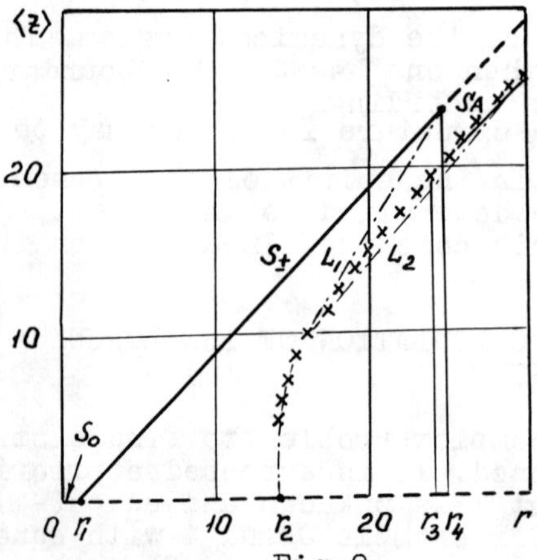

Fig.2.

Fig.2. The dependence of the averaged z on the Rayleigh number r in the Lorenz system. S_0, $S_\pm$ - are stationar solutions: z = 0, z = r-1; L_1, L_2 - are the simpliest cycles, SA is the strange attractor solution, the crosses represent results of the structural approach.

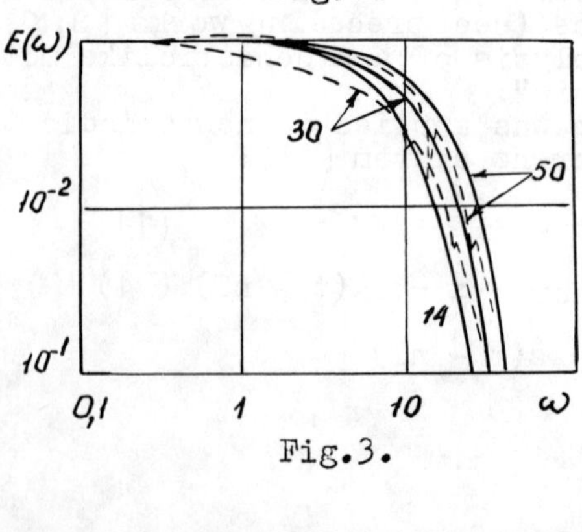

Fig.3.

Fig.3. Pulsation spectrums of x(t) (normalized at $\omega = 0$). Full curves are results of the structural approach, dashed ones are direct calculation. The numbers are values of r.

5. LEADING MAP AND DISTRIBUTION FUNCTIONS

The problem of distribution finding for u_n and Σ_n is difficult enough. But it becomes rather easier near the point of the bifurcation from the separatrix loop ($r = r_2$ in Fig.2).

Here the leading (or switch) one-dimensional map exists. If a supercritic value $\varepsilon = r - r_2 > 0$ is small all curcuits of trajectories pass near the saddle fixed point (the coordinate origin). Moreover they almost tangent the eigenvector e_J (corresponding to the minimum positive value λ).

At $t \to -\infty$ the main term in $\hat{x}_n$ is $c_J\, e_J \exp(\lambda_J t)$ because the other terms decrease more rapidly. In many cases $J = 1$; i.e. there is one eigenvalue with $\mathrm{Re}\,\lambda > 0$. Further on the saddle-node case is considered, when $\mathrm{Im}\,\lambda = 0$.

In this situation a value $c_J(n)$ may be related with the amplitude u_n. Since the loop starting from the origin has a finite amplitude one may define $u_n = c_J(n+1)$. Each circuit transforms u_n into u_{n+1}, so there is a map $u_{n+1} = f(u_n)$.

This map may be called "leading", because it is a basis for finding out the chance parameter distributions. A particular map form depends on a choice of end points of the time interval T_n or (it is the same) the Poincare section in the phase space.

For instance if the section is D_1 = const (at the right end of T_n) the map has a rather simple form (see, for example, [12]): $u_{n+1} = -B + Au_n^{\delta}$, $u_n > 0$;for $u_n < 0$ the map is antisymmetrically continued; $\delta = \mu_1 / \lambda$. Constants A and B may be found by integration the unstable manifold of the saddle point during two circuits. It is convenient to normalize u_n by its maximum value. Then the leading map becomes $u_{n+1} = -1 + au_n^{\delta} \equiv f(u_n)$.

Distribution function $\rho(u)$ is calculated from the Frobenius-Perron equation

$$\rho(u) = \rho(u_1)/f'(u_1) + \rho(u_2)/f'(u_2); \ \{u_1, u_2\} = f^{-1}(u)$$

Some results for variety of the parameters a, are reported in [13] . As a rule $\rho(u)$ is a discontinuous function having infinite breaks or other integrating singularities.

The chance parameter T_n may be correlated with u_n. If the section is D_1 = const this correlation is: $T_n = 1/\lambda \ln 1/u_{n-1} + T_*$ [10]. The constant T_* is seeked as it was made in the section 4.

One produces the boundary problem for $\hat{x}(t)$ by substituting the structural representation in the dynamical equation and averaging on u_n (that means a multiplication by $\rho(u_n)$ and an integration on a positive u_n range). Without losing generality one may put $n = 0$.

Thus in the approach all statistic properties are expressed by a single chance parameter u, and it is enough to know one-dimensional distribution $\rho(u)$, controlled by the leading map.

CONCLUSIONS

A new calculation method for structurally-stochastic motions is suggested. It consists of 1) a formulation of a boundary problem for a structure function in a region of a correlation radius size and 2) finding out a distribution function for the structure spacing and scaling on a basis of a leading map. Application of the method to the Lorenz system shows a satisfactory agreement of moments and pulsation spectrums with direct calculations.

REFERENCES

[1] Orszag S.A., Kells L.G., "Transition to turbulence in plane Poiseuille and plane Couette flow", J. Fluid Mech., 98, pp. 159-205, 1980.

[2] Rozhdestvenskii B.L., Simakin I.N., "Nonstationary secondary flows in a plane channel and the stability of the Poiseuille flow relative to finite disturbances", Soviet Doklady, 266, 6 pp.1337-1340, 1982.

[3] Cantwell B.J., "Organized motion in turbulent flow", Ann. Rev. Fluid Mech. Palo Alto, 13, pp.457-515, 1981.

[4] "The role of coherent structures in modelling turbulence and mixing", Ed. J. Jimenes, Lect. Notes in Phys., 136, 393 p., 1981.

[5] Pikovsky A.C., Rabinovich M.I. "Stochastic oscillations in dissipative systems", Physica 2D, pp.8-24, 1981.

[6] Eckmann J.-P., "Roads to turbulence in dissipative dynamical systems", Rev. Modern Phys., 54, 4, pp.643-654, 1981.

[7] Goldshtik M.A., Shtern V.N., "A method of functional smoothing in the problem of turbulence", Soviet Doklady, 240, 5, pp. 1058-1061, 1978.

[8] Lumley J.L., "Coherent structures in turbulence", Transition and turbulence, N.Y., Academic Press, pp.215-242, 1981.

[9] Goldshtik M.A., Shtern V.N., "On the structural turbulence theory", Soviet Doklady, 257, 6, pp.1319-1322, 1981.

[10] "Structural turbulence", Ed. M.A.Goldshtik, Novosibirsk, Institute of Thermophysics, 166 p., 1982.

[11] Lorenz E.N., "Deterministic nonperiodic flow", J.Atmos. Sci., 20, pp. 130-141, 1963.

[12] Afraimovich V.S., Bykov V.V., Shilnikov L.P., "On attractive noncrude limit sets of the Lorenz attractor type", Trans. Moskow. Math Soc., 44, pp.150-212, 1982.

[13] Shtern V.N., "Onset and properties of a chaos in simple models of a heat convection", Preprint, Institute of Thermophysics, Novosibirsk, 30 p., 1983.

DEVELOPMENT OF A NAVIER-STOKES METHOD BASED ON A FINITE VOLUME TECHNIQUE FOR THE UNSTEADY EULER EQUATIONS

W. HAASE, B. WAGNER

Dornier GmbH, Theoretical Aerodynamics
Postfach 1420, D-7990 Friedrichshafen 1, FRG

A. JAMESON

Princeton University
Princeton, N.J., USA

SUMMARY

Based on the formerly published Runge Kutta Stepping Schemes for solving the unsteady Euler equations in conservation form, a new approach has been developed for solving the unsteady Navier-Stokes equations. The procedure which integrates the governing equations for two-dimensional, compressible, turbulent flow, is based on a finite volume technique and handles arbitrary computational domains associated with body-fitted meshes.

INTRODUCTION

In recent years, finite volume techniques have been developed in order to simulate complex flow phenomena about arbitrary geometries. Associated with Runge-Kutta time stepping schemes, efficient schemes became available integrating the governing equations for inviscid, compressible, unsteady, rotational flows, see e.g. [1,2,3]. Based on the 4-stage scheme a new approach is presented solving the Navier-Stokes equations for two-dimensional turbulent, compressible, time-dependent flow.

This new method is applied to a supersonic, laminar, adiabatic flow about a flat plate with an impinging shock [4], serving as a testcase for numerical simulation of compressible flow [5,6,7]. To evaluate the efficiency of the present method comparisons with the explicit-implicit MacCormack scheme are performed. Furthermore, turbulent, adiabatic flow about the RAE 2822 airfoil is investigated. For transonic flow conditions [8, case 9] the results are compared with those derived from a method [9], based on the hybrid MacCormack scheme [6].

NUMERICAL METHOD

Governing Equations

The physical laws describing conservation of mass momentum and energy are written in integral form for unsteady, compressible, viscous flow in two dimensions

$$\iint_V \frac{\partial \rho}{\partial t}\, dV + \int_S \rho\, \vec{v}\cdot\vec{n}\, dS = 0\ , \tag{1}$$

$$\iint_V \frac{\partial(\rho\vec{v})}{\partial t}\, dV + \int_S \vec{v}(\rho\vec{v}\cdot\vec{n})dS + \int_S p\vec{n}ds - \int_S \overline{\overline{T}}\cdot\vec{n}dS = 0\ ,$$

$$\iint_V \frac{\partial E}{\partial t}\, dV + \int_S (E+p)\vec{v}\cdot\vec{n}dS - \int_S \vec{v}\cdot(\overline{\overline{T}}\cdot\vec{n})dS + \int_S \vec{Q}\cdot\vec{n}dS = 0\ .$$

In two dimensions V denotes the area of an arbitrary flow domain having the boundary S and the local outer normal $\vec{n}$. S is presumed to be fixed in space for the present purposes. ρ denotes the density, p the pressure, $\vec{v}$ the velocity vector, and t the time. The total energy per unit volume, E, reads

$$E = e\,\rho + 0.5\ \ \rho\,\vec{v}\cdot\vec{v} \tag{2}$$

where the specific internal energy per unit mass, e, is related to the pressure through the ideal gas equation of state

$$p = (\gamma - 1)\,\rho\,e \tag{3}$$

with the specific heat ratio γ. The stress tensor $\bar{T}$ consists of the viscous normal stresses

$$\sigma_x = 2\,(\mu + \varepsilon)\,\frac{\partial u}{\partial x} + \lambda \operatorname{div}\vec{v}, \tag{4}$$

$$\sigma_y = 2\,(\mu + \varepsilon)\,\frac{\partial v}{\partial y} + \lambda \operatorname{div}\vec{v},$$

and the shear stresses

$$\tau_{xy} = \tau_{yx} = (\mu + \varepsilon)\,\left(\frac{\partial u}{\partial y} + \frac{\partial v}{\partial x}\right)$$

where u, v are the components of the velocity vector and μ is the molecular viscosity which is derived from the Sutherland formula. The bulk viscosity λ is - related to the Stokes hypothesis - given by $\lambda = -\,2/3\,\mu$. The turbulent viscosity, used for expressing the Reynolds stress terms is derived from an algebraic turbulence model [9] which uses the vorticity distribution instead of boundary layer displacement thicknesses as a scaling parameter. Finally, the heat flux vector $\vec{Q}$ is

$$\vec{Q} = -\,\gamma\,(\mu/Pr + \varepsilon/Pr_t)\ \mathrm{grad}\ e \tag{5}$$

with the Prandtl number $Pr = 0.72$ and the turbulent Prandtl number $Pr_t = 0.9$.

Solution algorithm

Subdividing the computational domain into quadrilateral cells and applying the conservation laws (1) to each cell separately where all physical properties are defined to be constant one obtains a system of ordinary differential equations in time which can be solved by Runge-Kutta time-stepping methods. A class of 4-stage schemes can be written as

$$\begin{aligned} u^{(1)} &= u^{(n)} - \alpha_1\, Pu^{(n)}\, \Delta t, \\ u^{(2)} &= u^{(n)} - \alpha_2\, Pu^{(1)}\, \Delta t, \\ u^{(3)} &= u^{(n)} - \alpha_3\, Pu^{(2)}\, \Delta t, \\ u^{n+1} &= u^{(n)} - \alpha_4\, Pu^{(3)}\, \Delta t, \end{aligned} \tag{5}$$

where n denotes the previous time-level and P represents a central spatial difference operator. This scheme is of fourth order accuracy if $\alpha_1 = 1/4$, $\alpha_2 = 1/3$, $\alpha_3 = 1/2$ and $\alpha_4 = 1$. It is stable up to a CFL-number of $2\sqrt{2}$ and allows latitude in the introduction of dissipation terms. The difference equations (5) are solved by local time-stepping based on the maximum allowable time step for each cell.

To prevent the appearance of oscillations it proves necessary - at least for high Reynolds number flows - to add artificial dissipation to the existing scheme. Following [1] blended second and fourth order artificial dissipation terms have been used.

As already mentioned, an algebraic turbulence model is used in order to model the Reynolds stresses. The BALDWIN/LOMAX model [10] is a two-layer model using the vorticity distribution as a scaling parameter instead of the boundary layer displacement thickness which is extremely difficult to obtain directly from Navier-Stokes calculations.

DISCUSSION OF RESULTS

Interaction of an Oblique Shock Wave with a Laminar Boundary-Layer on a Flat Plate in Supersonic Flow.

From well known measurements [4] a case exhibiting a separation bubble is used for computation which was already a sample case in several publications [5], [6], [7] for demonstrating the computational efficiency of the different MacCormack schemes. The Reynolds number is $Re = 2.96 \cdot 10^5$ based on the distance from the leading edge to the geometric shock impingement point as shown by the insert in Fig. 1. The Mach number of the undisturbed approaching flow is $M_\infty = 1$ and the shock impinges onto the plate with an angle of $\zeta = 32.585^0$ corresponding to a final static pressure rise behind the reflected shock up to $1.4\ p_\infty$.The computational mesh consisted of 32x32 cells and started with 4 cell columns ahead of the leading edge. The mesh is equidistantly spaced in x-direction while the boundary layer is resolved by 16 cell rows of exponentially increasing thickness.

In Fig. 1 the results for the ratio of the wall pressure p_w to the undisturbed pressure p_∞ versus local Reynolds number are compared to the experimental values [4] and to results achieved using the explicit-implicit MacCormack scheme. Small differences can be observed between both numerical results but these differences are generally less than the deviations of both results from the experimental values. The situation is very similar for the skin friction coefficient c_f (related to undisturbed quantities) as shown in Fig. 2. The predicted length of the separation bubble is almost identical.

Convergence for skin friction and velocity profiles was achieved by use of the explicit-implicit MacCormack scheme after 300 time steps. It should be mentioned that the size of the time steps had to be continuously reduced during the last 30 cycles in order to avoid some wiggles which are in fact steady but strongly dependent on the step size. The Runge-Kutta time stepping scheme converged to the same level of accuracy in about 750 time steps without changing the time step size. An implicit residual averaging procedure [2] was used to increase the stability up to a CFL number of 8. The CPU time for the 750 Runge-Kutta steps balances almost exactly the CPU time needed for 300 steps of the explicit-implicit MacCormack scheme. Hence, the computational efficiency is equal for both schemes with respect to this test case.

Transonic Flow about the RAE 2822 airfoil

The flow past the RAE 2822 serves as a testcase with transonic flow conditions. Comprehensive measurements by [8, case 9] are used as a comparison with the calculated data. The calculations are performed on a C-type mesh with 128 x 32 mesh points. The first mesh line near the surface is located at a distance of $\Delta y = 0.083/\sqrt{Re}$. Fig. 3 shows the velocity vector field based on a Mach number of 0.734 (corrected value), a Reynolds number of 6.5 million and an angle of attack of 3.19 degrees. The vectors are scaled according to the magnitude of local velocities. Although boundary layer profiles can be taken from this figure, it is not easy to achieve the boundary layer thickness itself. Therefore, and remembering the features of the eddy viscosity model, i.e. expressing the eddy viscosity in terms of the vorticity, it is more adequate to use the vorticity distribution instead. Fig. 4 gives a contour plot of such lines of vorticity contours. The outermost lines represent a vorticity amount of only 0.01 % of its maximum value and, therefore, can be defined as a boundary between viscid and "inviscid" flow domains. Furthermore, in Fig. 3 and 4 the increase of boundary layer thickness as well as the decrease of vorticity (and skin friction) behind the shock (located at about x/c = 0.6) can be recognized. In Fig. 5 lines of eddy-viscosity contours are plotted as calculated from the turbulence model. The outermost lines represent an eddy viscosity of 10^{-5} which is less than three times smaller than the molecular viscosity based on free stream conditions $\mu_l = 3.65 \cdot 10^{-5}$. The innermost closed line - near the trailing edge - denotes $\mu_t \simeq 8 \cdot 10^{-4}$. The origins of the 10^{-5}-lines located at the airfoil surface and each marked by a small bar are assumed to define the location of transition. Additionally, in Fig. 6 the pressure coefficient distribution is given. The thick solid line encloses the region of supersonic flow conditions. The isobars are plotted based on an increment of 0.05 between two adjacent lines.

A better insight into the accuracy and efficiency of the present scheme is gained from the following four figures. In Fig. 7 the calculated wall pressure coefficient distribution is given as a solid line while the symbols mark the measurements. The agreement is encouraging but could be improved with respect to the suction peak on the upper airfoil surface and the shock resolution, respectively. Without enlarging the total number of grid points and computing time, respectively, an increase in accuracy is expected by means of a solution adaptive grid procedure [11] which redistributes the mesh due to the pressure curvature distribution in order to get smaller grid step sizes in regions of large curvature. Taking this into account, the global accuracy of the skin friction distribution - Fig. 8 - is not too bad although the shock region is poorly resolved. A small separated region very close to the trailing edge is verified by the given skin friction distribution. To investigate the iterative behaviour of the present method, in Fig. 9 the transient behaviour of the lift coefficient is given and compared to that of the formerly used hybrid MacCormack scheme [9]. The advantage of the present method can easily be detected, although one should recognize that a direct comparison - with respect to physical time - is impossible due to the fact that the present scheme uses a variable local time-step while the hybrid scheme bases on constant time-steps in the coarse and the fine mesh, respectively. Additionally, in Fig. 10 the maximal value of the root mean square residuals of $d\rho/dt$ is given. Defining 2000 iteration cycles (ΔC_l=0.0012%) to be a steady state solution for the present scheme and 5000 cycles (ΔC_l=0.18%) for the hybrid scheme the amount of the rms-residuals at this steady state is comparable. The steady state is defined to be achieved for lift- and drag-coefficients tending to constant values. Both methods are applied to the same mesh structure using 128 x 32 mesh points. Therefore, the comparison of normalized CPU-times (with respect to the hybrid MacCormack scheme [9] and based on a IBM 3083 with standard compiler) reads as follows taking into account the time for one iteration as well as the time for reaching the steady state.

IBM 3083	HYBRID CFL=0.75	PRESENT CFL=2.00
CPU-time per iteration	1	0.68
CPU-time for reaching the steady state	1	0.25

Finally, an Euler solution based on the present scheme results in only about a 33 % increase in CPU-time per iteration and mesh point for the corresponding Navier-Stokes run.

REFERENCES

[1] Jameson, A., Schmidt, W., Turkel, E., "Numerical Solutions of the Euler Equations by Finite Volume Methods Using Runge-Kutta Time-Stepping Schemes", AIAA-81-1259 (1981)

[2] Jameson, A., "The Evolution of Computational Methods in Aerodynamics" Princeton University, MAE Report No. 1608 (May, 1983), to appear in: Journal of Applied Mechanics

[3] Schmidt, W., Jameson, A., "Euler Solvers as an Analysis Tool for Aircraft Aerodynamics", in: Recent Advances in Numerical Methods in Fluids, Vol. 4, Habashi, W.G. (Ed.) Pineridge Press, Swansea, U.K. also: Dornier Note BF-P-15/83 (1983)

[4] Hakkinen, R.J., et. al., "The Interaction of an Oblique Shock Wave with a Laminar Boundary Layer", NASA Memo 2-18-59W (1959)

[5] MacCormack, R.W., "Numerical Solution of the Interaction of an Oblique Shock Wave with a Laminar Boundary Layer", Lecture Notes in Physics, Vol.8, Springer Verlag New York, p. 151 (1971)

[6] MacCormack, R.W., "An Efficient Explicit-Implicit-Characteristic Method for Solving the Compressible Navier-Stokes Equations" SIAM-AMS Proceedings, Vol. 11 (1978)

[7] MacCormack, R.W., "A Numerical Method for Solving the Equations of Compressible Viscous Flow", AIAA-81-110 (1981), also: AIAA-Journal Vol. 20, No. 9, pp. 1275-1281 (1982)

[8] Cook, P.H., McDonald, M.A., Firmin, M.C.P., "Airfoil RAE 2822 - Pressure Distributions and Boundary Layer Wake Measurements" AGARD-AR-138 (1979)

[9] Haase, W., "Gitterströmungen", Dornier Report 82 BF 8 B (Dec. 1981)

[10] Baldwin, B.S., Lomax, H., "Thin Layer Approximation and Algebraic Model for Separated Flows", AIAA-78-257 (1978)

[11] Haase, W., Misegades, K., Naar, M., "Adaptive Grids in Viscous Flow" Third Int. Conference on Numerical Methods in Laminar and Turbulent Flow, University of Washington, Seattle (1983)

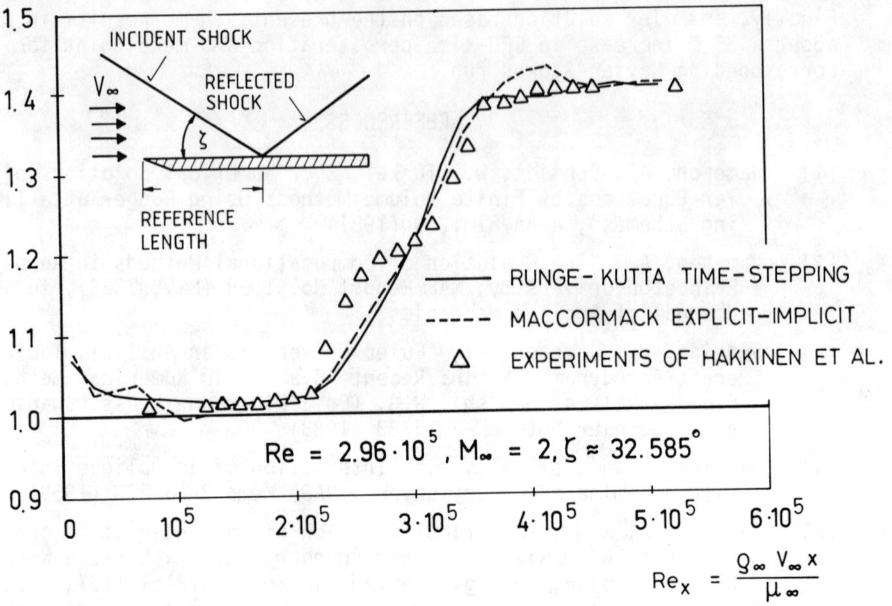

FIG. 1: WALL PRESSURE DISTRIBUTION FOR SHOCK BOUNDARY-LAYER INTERACTION

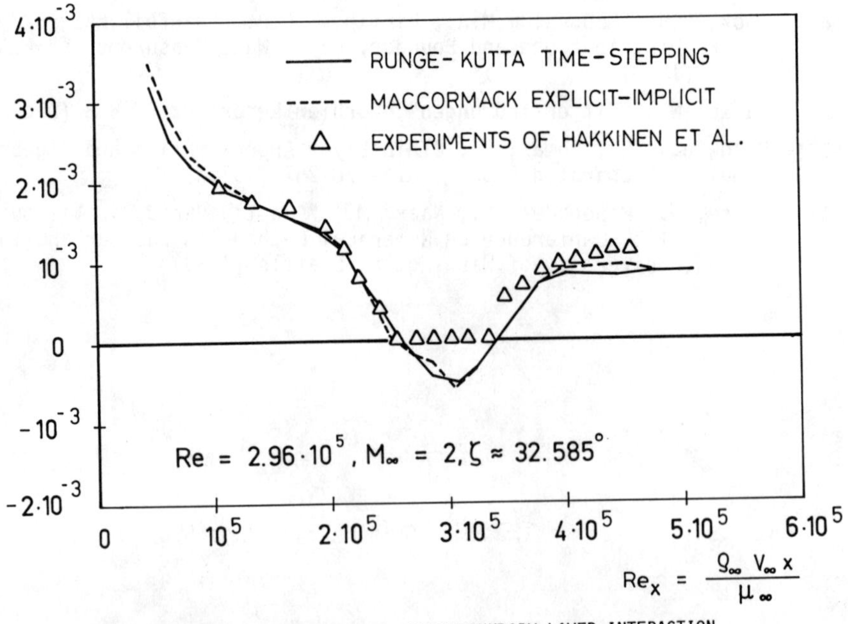

FIG. 2: SKIN FRICTION COEFFICIENT FOR SHOCK BOUNDARY-LAYER INTERACTION

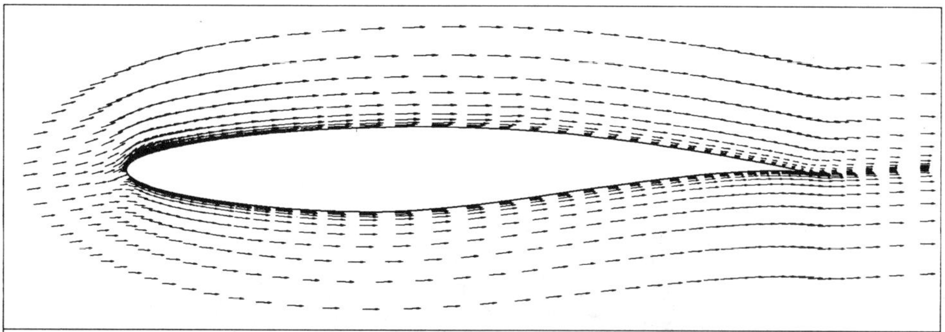

FIG. 3 : RAE 2822 (9) M=0.734, RE=6.5E6, ALPHA=3.19
VELOCITY VECTORS

I=17/112 J=2/18
OFAK=9.000E-01 FLONG=6.000E-01 FMAX= 1.274E+03 ITA=5000 04/10/83 10.08.41

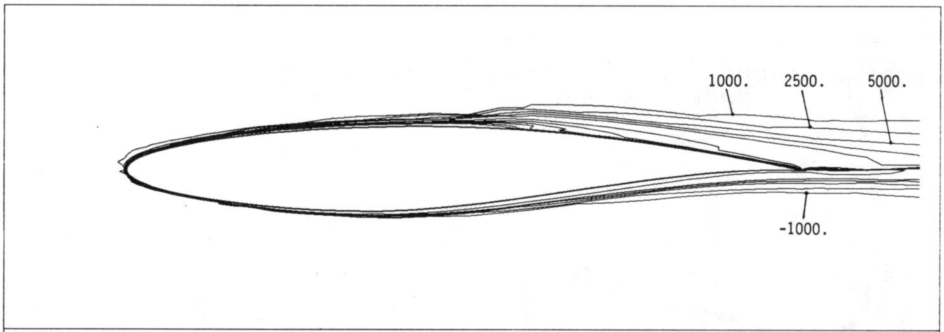

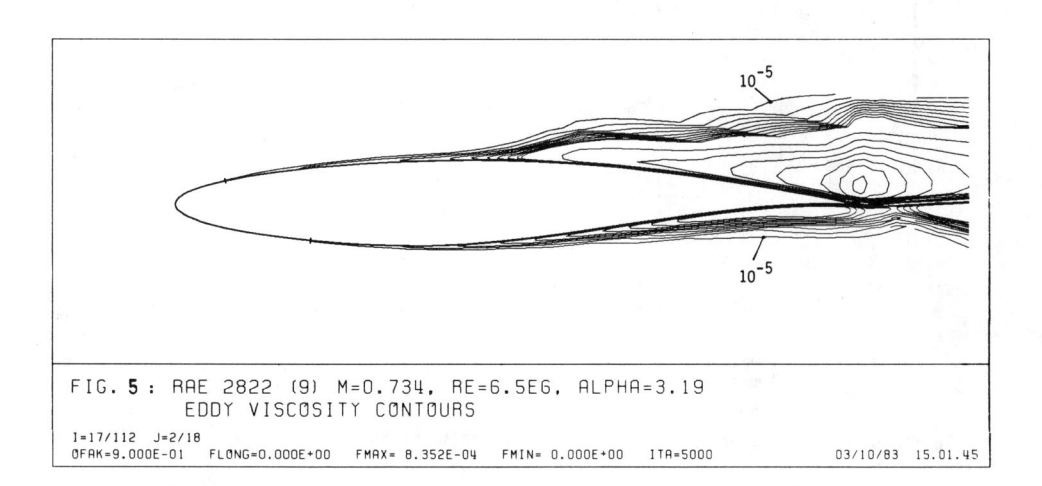

FIG. 4 : RAE 2822 (9) M=0.734, RE=6.5E6, ALPHA=3.19
VORTICITY CONTOURS

I=17/112 J=2/18
OFAK=9.000E-01 FLONG=0.000E+00 FMAX= 9.517E+06 FMIN= -7.142E+06 ITA=5000 03/10/83 15.00.48

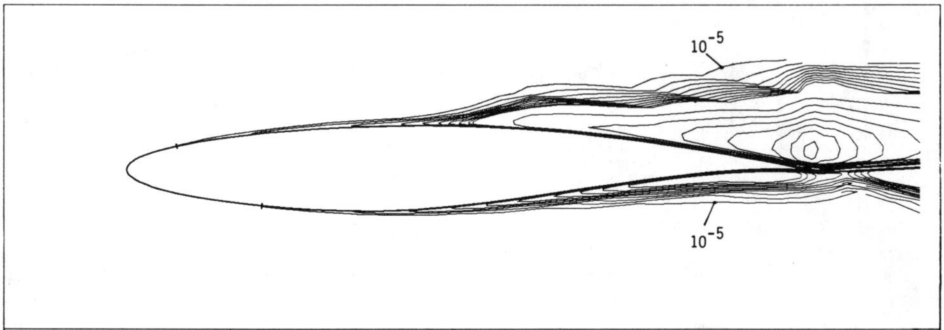

FIG. 5 : RAE 2822 (9) M=0.734, RE=6.5E6, ALPHA=3.19
EDDY VISCOSITY CONTOURS

I=17/112 J=2/18
OFAK=9.000E-01 FLONG=0.000E+00 FMAX= 8.352E-04 FMIN= 0.000E+00 ITA=5000 03/10/83 15.01.45

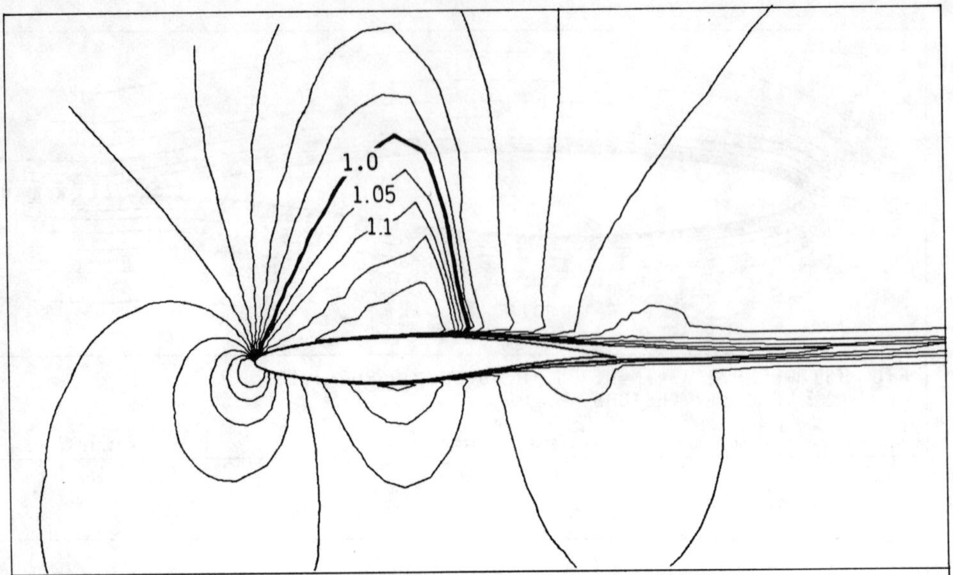

FIG. 6: RAE 2822 (9) M=0.734, RE=6.5E6, ALPHA=3.19
MACH CONTOURS

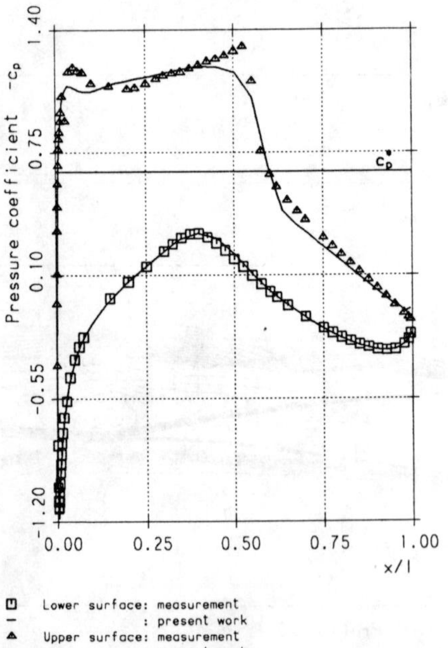

Fig. 7: RAE 2822 (9) M=.734, Re=6.5E6, A=3.19
Wall pressure coefficient distribution

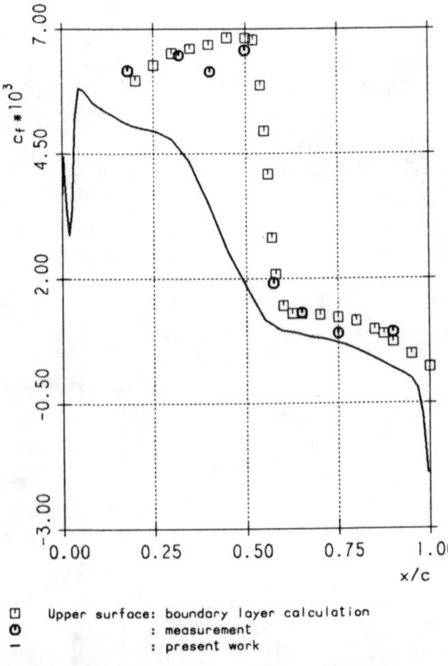

Fig. 8: RAE 2822 (9) M=.734, Re=6.5E6, A=3.19
Skin friction distribution
– normalized with u_∞

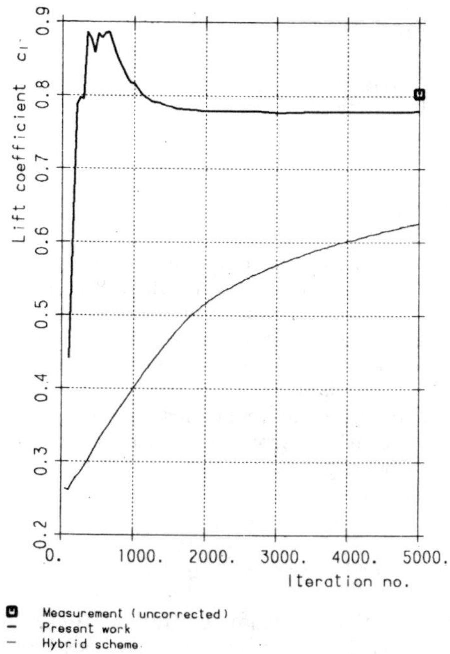

Fig. 9: RAE 2822 (9) M=.734, Re=6.5E6, A=3.19
Transient behaviour of lift coefficient

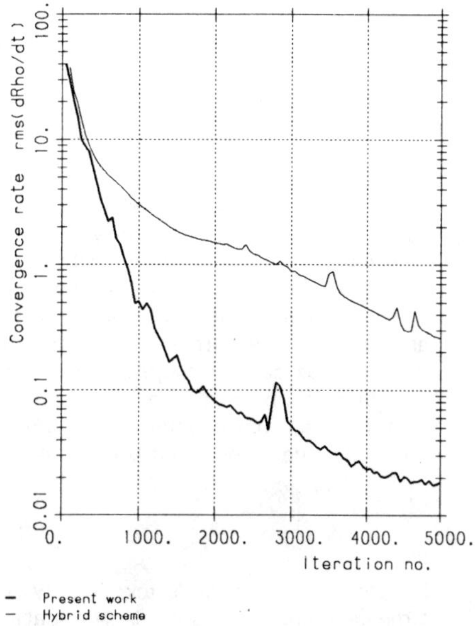

Fig. 10: RAE 2822 (9) M=.734, Re=6.5E6, A=3.19
convergence rate (rms-error(dRho/dt))

THE INFLUENCE OF BOUNDARY CONDITIONS ON THE STABILITY OF APPROXIMATE-FACTORIZATION METHODS

D. Hänel, U. Giese
Aerodynamisches Institut, RWTH Aachen
5100 Aachen, West Germany

SUMMARY

Implicit factorization methods show instabilities for certain finite-difference formulations of the boundary conditions for the intermediate variables. This problem is studied in conjunction with the linear and nonlinear potential equation, solved with different factorization-schemes and several types of boundary conditions for the intermediate variable. Computational instabilities are confirmed by a stability analysis for the difference approximation at the boundary. It was found that the instabilities can be suppressed by satisfying the stability conditions for the boundary or, partially, by use of a sequence of the acceleration parameter α. It is also shown that no stability problems and best rate of convergence are achieved for boundary conditions deduced from the physical boundary conditions.

INTRODUCTION

Factorization methods are widely used in implicit algorithm for the solution of the time-dependent conservation equations, e.g. [1,2] or as relaxation methods like the alternating-direction-implicit method (ADI) [3] or the approximate factorization method (AF), e.g. [4,5,6,7]. All methods have in common that they are fully implicit and unconditionally stable in the sense of the von Neumann analysis. In some cases, however, local instabilities are observed, which lead to a poor rate of convergence or even to divergence. One reason is an unsuitable formulation of the boundary condition of the intermediate variables, introduced through the factorization steps. In the literature little attention was given to this problem, which can completely stabilize the solution. The only other work which is concerned with the analysis of this problem was published recently [8]. In that paper conclusions were reached similar to those obtained in the present one.

THE FINITE-DIFFERENCE PROBLEM

As a model problem the finite-difference formulation of the linear potential equation is considered:

$$L(\Phi) = (r\overrightarrow{\delta}_x\overleftarrow{\delta}_x + s\overrightarrow{\delta}_y\overleftarrow{\delta}_y)\Phi = 0 \tag{1}$$

Herein, $\overrightarrow{\delta}$ and $\overleftarrow{\delta}$ are forward and backward differences and r, s are constant coefficients. For the solution two different approximate-factorization schemes are applied. The first, designated as AF-1 [4], corresponds to the Beam and Warming scheme [1,2] and to the ADI scheme [3]. It is defined as:

$$(\alpha - s\overrightarrow{\delta}_y\overleftarrow{\delta}_y)\,\Delta\Phi^{*} = \alpha\,\omega\,L(\Phi^n) \tag{2a}$$

$$(\alpha - r\overrightarrow{\delta}_x\overleftarrow{\delta}_x)(\Phi^{n+1} - \Phi^n) = \Delta\Phi^{x} \tag{2b}$$

The second scheme is the (AF-2) one, derived for transonic potential flow problems [5,6,7]:

$$(\alpha + s\overleftarrow{\delta}_y)\,\Delta\Phi^* = \alpha\,\omega\,L(\Phi^n) \tag{3a}$$

$$(-\alpha\overrightarrow{\delta}_y - r\overrightarrow{\delta}_x\overleftarrow{\delta}_x)(\Phi^{n+1} - \Phi^n) = \Delta\Phi^* \tag{3b}$$

In eqs. (2) and (3) ω is a relaxation factor and α corresponds to an inverse time step. Generally ω is held fixed at a value between 1 and 2, whereas a sequence of values is used for α [4,5,6] . In both schemes, $\Delta\Phi^*$ is an intermediate variable which is defined only by the factorized operators. It therefore has no real physical meaning.

As a test problem the flow over a thin airfoil is calculated. Dirichlet boundary conditions $\Phi = u_\infty x$ are assumed on three sides of the rectangular integration domain and on the lower boundary (y = 0) a Neumann condition is prescribed:

$$\frac{\partial\Phi}{\partial y} = v_o(x) \quad \begin{cases} v_o = 0 & x < 0 \quad x > 1 \\ v = .2(1-2x) & 0 \le x \le 1 \end{cases} \tag{4}$$

With eq. (4) the stationary difference equation (1) becomes to

$$L_B(\Phi) = (r\,\overrightarrow{\delta}_x\overleftarrow{\delta}_x + \frac{2s}{\Delta y}\overrightarrow{\delta}_y)\,\Phi - \frac{2s\,v_o(x)}{\Delta y} \tag{5a}$$

which uses the relations

$$\overrightarrow{\delta}_y\overleftarrow{\delta}_y\,\Phi = \frac{1}{\Delta y}(\overrightarrow{\delta}_y - \overleftarrow{\delta}_y) \tag{5b}$$

$$\delta_y\,\Phi_{y=0} = v_o(x) = \frac{1}{2}(\overrightarrow{\delta}_y\Phi + \overleftarrow{\delta}_y\Phi)_{y=0} \tag{5c}$$

The first step of the schemes (2) and (3) requires boundary conditions for $\Delta\Phi^*$ at y = 0. Since $\Delta\Phi^*$ depends on the factorization and vanishes in the steady state, the derivation of the boundary conditions is not unique. In this paper, four different formulations are chosen. They are summarized in Table 1.

	B.C. on y=0 (j=1)	α_1	α_2	α_3	α_4
BC 1	$\Delta\Phi^*_{j-1} = 0$	$2(\alpha + \frac{s}{\Delta y^2})$	2α	1	$\alpha/(\alpha + \frac{s}{\Delta y})$
BC 2	$\overleftarrow{\delta}_y\,\Delta\Phi^*_j = 0$	2α	2α	1	1
BC 3	$\overleftarrow{\delta}_y\,\Delta\Phi^*_j = -\overrightarrow{\delta}_y\,\Delta\Phi^*_j$	α	α	1	$\alpha/(\alpha + \frac{2s}{\Delta y})$
BC 4	$\Delta\Phi^*_j = 0,\ \Phi^{n+1}_{j+1} - \Phi^{n+1}_j = \Delta y \cdot v_o$	$2(\alpha + \frac{s}{\Delta y^2})$	2α	$\frac{1}{2}$	$\alpha/(\alpha + \frac{s}{\Delta y})$

Table 1. Boundary conditions for the intermediate variables

A compact formulation of the difference schemes (2) and (3) at the boundary line is for AF-1

$$(\alpha_1 - \frac{2s}{\Delta y}\vec{\delta}_y)\,\Delta\Phi^* = \alpha_2\omega\,(\alpha_3\frac{2s}{\Delta y}\vec{\delta}_y + r\vec{\delta}_x\overleftarrow{\delta}_x)\,\Phi^n \tag{6a}$$

$$(\alpha - r\vec{\delta}_x\overleftarrow{\delta}_x)\,(\Phi^{n+1} - \Phi^n) = \Delta\Phi^* \tag{6b}$$

and for AF-2

$$\Delta\Phi^* = \alpha_4\,\omega\,(\alpha_3\frac{2s}{\Delta y}\vec{\delta}_y + r\vec{\delta}_x\overleftarrow{\delta}_x)\,\Phi^n \tag{7a}$$

$$(-\alpha\vec{\delta}_y - r\vec{\delta}_x\overleftarrow{\delta}_x)\,(\Phi^{n+1} - \Phi^n) = \Delta\Phi^* \quad . \tag{7b}$$

The coefficients α_1 to α_4 are given in Table 1.

The first two conditions in Table 1, where the gradient or the value of $\Delta\Phi^*$ on the dummy points below the boundary $(y = 0)$ vanishes, are justified by the steady-state solution $\Delta\Phi^* = 0$. These conditions were used in [5,6] for transonic flow computation.

The third condition, BC 3, is deduced from the physical boundary condition formulated in eq. (5a). It was implemented in the unfactorized version of eqs. (2) or (3) (with $\Delta\Phi^*$ eliminated) and these equations are factorized afterwards in such a way, that the second step is the same as for inner points.

A further condition, BC 4, set $\Delta\Phi^*$ equal to zero on the boundary and the physical boundary condition for Φ^{n+1} is extrapolated explicitly. This condition is applied very often in algorithm for the conservation equations, e.g. [1,2]. The advantage is that the implicit operator becomes independent of the boundary conditions and therefore it is much easier to implement.

STABILITY ANALYSIS

At the interior points the von Neumann analysis shows unconditional stability for both schemes, eqs. (2) and (3), if $0 \le \omega \le 2$ and $\alpha > 0$. The consideration of boundary conditions requires more rigorous analysis, like [9]. In general, however, their application becomes very tedious. In [10,11] arguments are given which justify the application of the von Neumann analysis also to the finite-difference operators at the boundary.

Therefore, according to the von Neumann analysis, a Fourier solution for Φ was substituted into eqs. (6) and (7). Then the AF-1 scheme, eq. (6), stability is assured, if

$$(\bar{\alpha}_2\omega - 2\alpha)\,[(c_x + c_y)^2 + s_y^2] \le 2[(c_x + c_y)(\bar{\alpha}_1\alpha + (\bar{\alpha}_1 - \alpha)c_x + c_x c_y) + s_y^2 c_x] \quad , \tag{8}$$

where

$$\bar{\alpha}_1 = \alpha_1\alpha_3 \qquad \bar{\alpha}_2 = \alpha_2\alpha_3$$

$$c_x = 2\frac{r}{\Delta x^2}(1 - \cos\xi) \quad , \quad c_y = 2\alpha_3\frac{s}{\Delta y^2}(1 - \cos\eta) \quad , \quad s_y = 2\alpha_3\frac{s}{\Delta y^2}\sin\eta \, . \tag{9}$$

From eq. (9), unconditional stability is indicated if

$$\alpha_2 \omega - 2\alpha \le 0 \qquad \text{and} \qquad \alpha_1 \ge \alpha \quad , \tag{10}$$

which means that

$$\omega \le 2 \qquad \text{for BC 3 and BC 4}$$
$$\omega \le 1 \qquad \text{for BC 1 and BC 2} \tag{11}$$

Thus, in the interesting region $\omega \le 2$ the difference scheme is unconditionally stable, using the conditions BC 3 and BC 4. For BC 1 and BC 2 stability restrictions are found by a numerical solution of eq. (8) for all wave components. The result is shown in Fig. 1, where the dimensionless α-value is plotted against the relaxation factor ω. This stability behaviour, i.e. two stable regions for one ω, is confirmed by test calculation for the linear potential equation, using single α-values.

In a similar way, the stability limits imposed by the boundary conditions can be found for AF-2 scheme,eq. (7):

$$(\alpha_4 \omega - 2)\, c_x (c_x + c_y) + (\alpha_4 \omega - k/\alpha_3)(c_y (c_x + c_y) + s_y^2) \le 0 \tag{12}$$

with eq. (9) and $k = \alpha \dfrac{\Delta y}{s}$.

This inequality is satisfied if

$$\alpha_4 \omega - 2 \le 0$$
and
$$\alpha_4 \omega - k/\alpha_3 \le 0 \quad . \tag{13}$$

For the boundary conditions considered here, the stability regions are shown in Fig. 2 with $\alpha \Delta y/s$ versus ω. Similar to the results of the AF-1 scheme the conditions BC 3 and BC 4 give unconditionally stable results in the interesting region $\omega \le 2$, whereas BC 1 and BC 2 lead to restrictions of the stability. Again, these restrictions could be verified in linear test calculations.

NUMERICAL RESULTS AND DISCUSSION

The influence of the boundary condition of the intermediate variables is studied at first for the linear model problem as described before. The stability restrictions, eq. (8) and (12), could be confirmed in all cases, using single α-value for the calculations. As an example, Fig. 3 shows the convergence history of the AF-2 scheme with the condition BC 2 for single α-values. The behaviour corresponds exactly to that expected from the analysis. The lower curve is calculated with a sequence of two α-values, a stable and a unstable one. Despite of the unstable value, the solution converges, since the frequency range of the unstable modes is small enough. This behaviour also explains, why by using sequences af the α-values the stability restriction do not become so serious.

To avoid the instabilities within a sequence at all, it is possible to satisfy the stability conditions, eq. (11) or (13), by a special choice of α or ω at the boundary line. This was done e.g. in the TAIR code of Holst, as reported in [8] . The present study yielded indeed an improvement of the convergence in that case, but with the condition BC 3 and without manipulation a better rate of convergence was achieved.

In Figs. 4 and 5 the convergence history of the AF-1 and AF-2 schemes is shown for the different boundary conditions, using a geometrical α-sequence. For the AF-1 scheme all conditions show stable results, whereas for AF-2 scheme the condition BC 2 leads to divergence. Best results were achieved with the condition BC 3 in both schemes.

The extrapolation of the boundary values (BC 4) shows a poor rate of convergence, which is in opposite to the experiences made with the Beam and Warming scheme, applied on the full Navier-Stokes equations [2]. In this work, both the conditions BC 3 and BC 4 were tested and have resulted in comparable rates of convergence. A reason may be found in the presence of convective terms, which were not considered here.

Further investigations were made with the transonic nonlinear potential equation. As a test case the supercritical flow around a NACA 0012 airfoil was calculated with the full potential equation in curvilinear coordinates and the AF-2 as method of solution. The mesh arrangement, the pressure distribution and the convergence history for the different boundary conditions were shown in Fig. 6. The rate of convergence behaves very similar to that of the linear problem, shown in Fig. 5. It is noted that the condition BC 3 is superior to others.

The same method of solution was applied to internal potential flow problems, too [7]. The situation was complicated by the fact, that Neumann conditions are present on three boundaries. An effective rate of convergence could be achieved only with the boundary condition BC 3. As an example of the internal flow Fig. 7 shows the isomachs of the flow field and the convergence history for the transonic plane inlet flow.

REFERENCES

[1] T. H. Pulliam, J. L. Steger: Implicit Finite-Difference Simulations of Three-Dimensional Compressible Flow. AIAA Journal, 18, pp. 159-167 (1978).

[2] A. Merten, D. Hänel: Navier-Stokes Solution for Compressible Flow in a Rotating Cylinder. Notes on Num. Fluid Mech., 5, Vieweg Verlag (1982).

[3] D. W. Peaceman, H. F. Rachford: The Numerical Solution of Parabolic and Elliptic Differential Equations. J. Soc. Ind. Appl. Math., 3 (1955).

[4] W. F. Ballhaus, A. Jameson, J. Albert: Implicit Approximate-Factorization Schemes for Steady Transonic Flow Problems. AIAA Journal, 16, pp. 573-579 (1978).

[5] T. L. Holst: Implicit Algorithm for the Conservative Transonic Full-Potential Equation Using an Arbitrary Mesh. AIAA Journal, 17, pp. 1431-1439 (1980).

[6] T. L. Holst: Fast Conservative Algorithm for Solving the Transonic Full Potential Equation. AIAA Journal, 18, pp. 1431-1439 (1980).

[7] U. Giese: Computation of Inviscid Transonic Flow. Proceedings of the 8th Intern. Conf. on Numerical Methods in Fluid Dynamics (Aachen 1982), Lecture Notes in Physics, 170, pp. 217-223 (1982).

[8] J. C. South, M. M. Hafez: Stability Analysis of Intermediate Boundary Conditions in Approximate Factorization Schemes. Pres. on 6th AIAA Comp. Fluid Dyn. Conference, Danvers, July 1983.

[9] B. Gustafsson, H.-O. Kreiss and A. Sundström: Stability Theory of Difference Approximations for Mixed Initial Boundary Value Problems II. Mathematics of Computation, 26, pp. 649-686 (1972).

[10] J. A. Trapp, J. D. Ramshaw: A Simple Heuristic Method for Analyzing the Effect of Boundary Conditions on Numerical Stability. J. of Computational Physics, 20, pp. 238-242 (1976).

[11] M. Goldberg, L.Tadmore: Scheme-Independent Stability Criteria for Difference Approximations of Hyperbolic Initial Boundary Value Problems. Mathematics of Computation, 36, pp. 603-626 (1981).

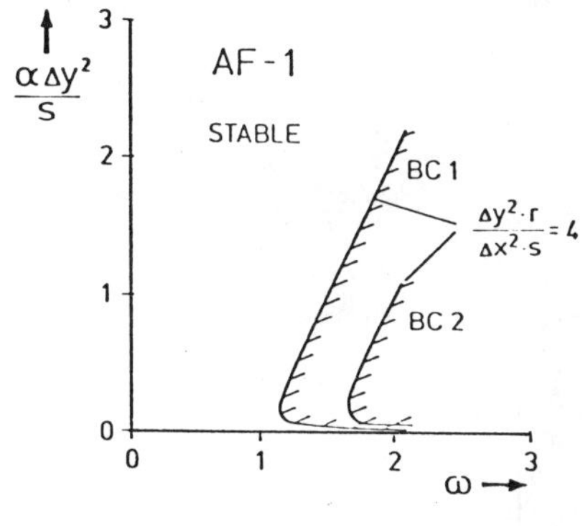

Fig. 1. Stability region of the AF-1 scheme with the boundary conditions BC 1 and BC 2.

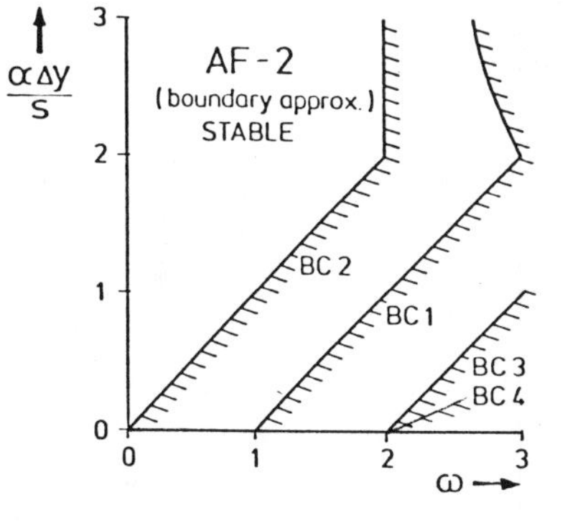

Fig. 2. Stability region of the AF-2 scheme with the boundary conditions BC 1 to BC 4.

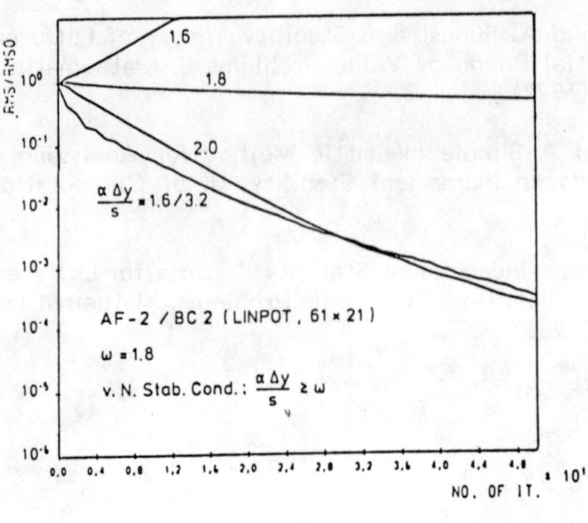

Fig. 3. Convergence of residual (rms) for AF-2 with boundary condition BC 2.

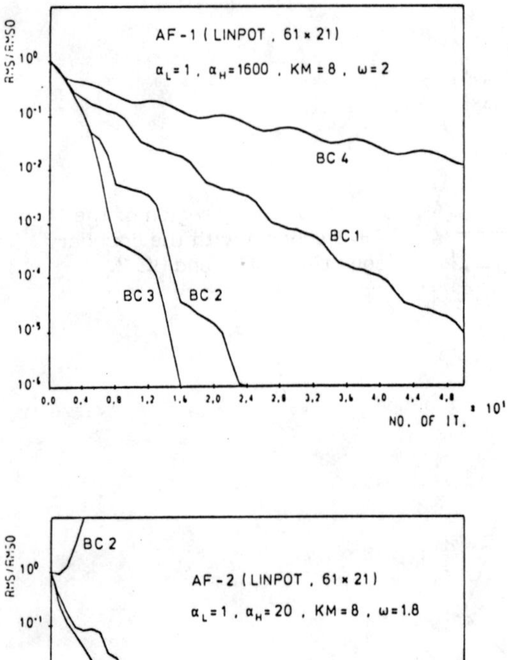

Fig. 4. Convergence of residual (rms) for AF-1 with boundary conditions BC 1 to BC 4 using geometrical sequences of a α-values.

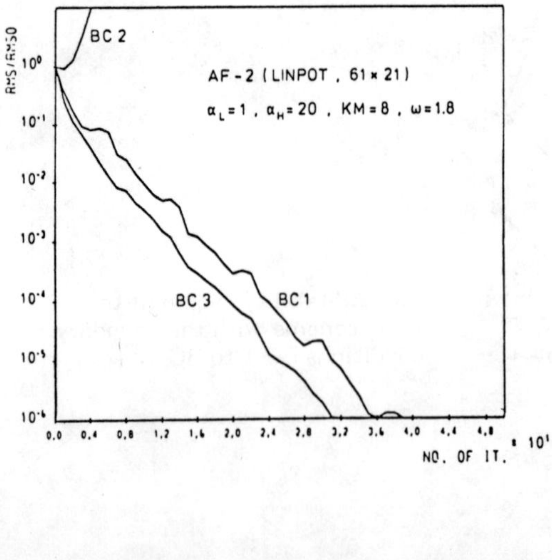

Fig. 5. Convergence of residual (rms) for AF-2 with boundary conditions BC 1 to BC 4 using geometrical sequences of a α-values.

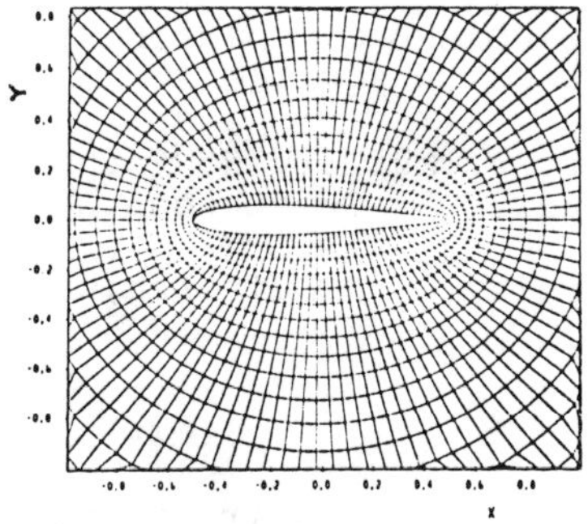

Fig. 6a. Mesh arrangement

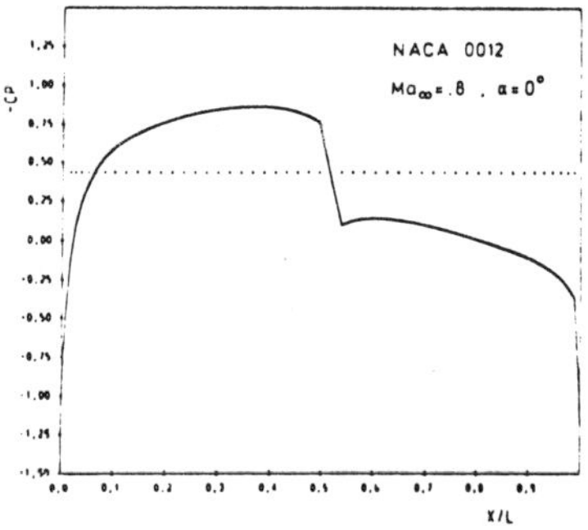

Fig. 6b. Pressure distribution

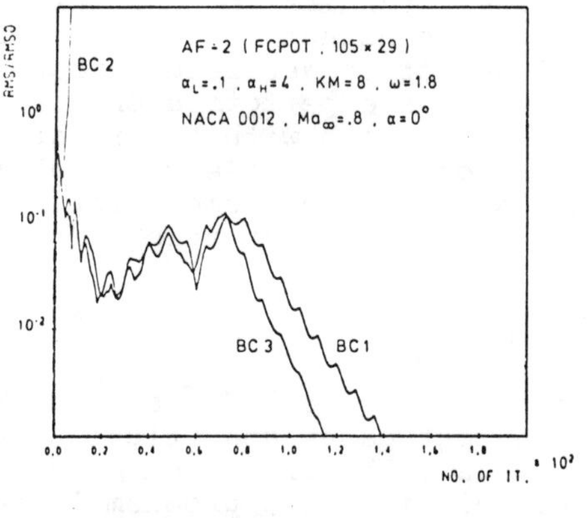

Fig. 6c. Convergence of residual (rms) for AF-2 (full potential equation) with boundary conditions BC 1 to BC 3.
- Transonic flow over a NACA 0012 airfoil -

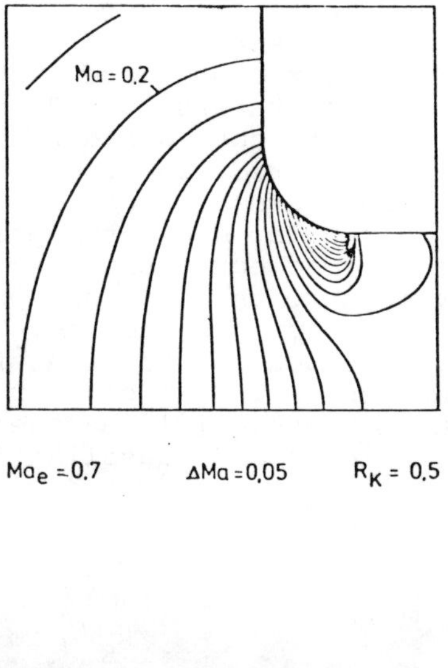

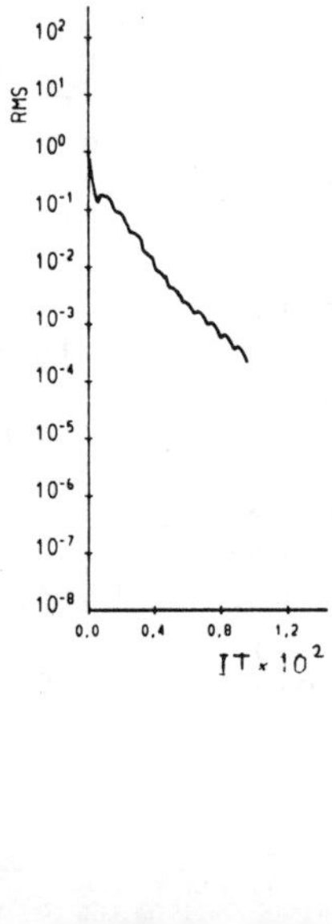

Fig. 7a. Isomachs

Fig. 7b. Convergence of residual (rms) for AF-2 (full potential equation) with boundary condition BC 3
- plane transonic flow through an inlet -

NUMERICAL EXPERIENCES WITH BOUNDARY CONFORMED COORDINATE SYSTEMS FOR SOLUTION OF THE SHALLOW WATER EQUATIONS

J. Häuser, D. Eppel, M. Lobmeyr, A. Müller, H. Paap
Institute of Physics, GKSS-Research Center, 2054 Geesthacht, Germany
F. Tanzer
1. Institute of Physics, Giessen University, 6300 Giessen, Germany

SUMMARY

Many problems of applied oceanography and environmental science demand the solution of the momentum, mass and energy equations on physical domains having curving coastlines. Finite-difference calculations representing the boundary as a stair-step function may give inaccurate results near the coastline where simulation results are of greatest interest for numerous problems of application. This suggests that methods have to be used which are capable to handle the problem of boundary curvature.

This paper presents computational results for the shallow water equations on a circular ring of constant depth employing the concept of boundary fitted grids for an accurate representation of the boundary. All calculations are performed on a rectangle in the transformed plane using a mesh with square grid spacing. Comparisons of the simulations of transient normal mode oscillations and analytic solutions are shown, demonstrating that this technique can yield accurate results in situations involving a curved boundary.

1. INTRODUCTION

Many problems of mathematical physics demand the solution of partial differential equations. In most cases analytical solutions cannot be obtained thus one has to resort to numerical solutions. Very often the numerical solution of a problem is aggravated not only by complex physical processes but also by the irregularity of the solution domain. The present paper is concerned with solutions of the shallow water equations which describe the free surface of a liquid subjected to the gravity field of the earth. The equations serve as an example of the Navier-Stokes equations which are the basic equations for solving the flow field in tidal rivers, bays and estuaries.

Using Thompson's method, the curvilinear coordinate system is constructed by elliptical differential operators (the Poisson equation is solved for each curvilinear coordinate where the right-hand side of the equation is used for grid line control [1]). All calculations are performed in the transformed plane. Normally the solution area in this plane is a rectangle. Hence the problem is reduced to the solution of the transformed equations on a rectangle which is an advantage in comparison to the original problem and substantially facilitates the computer programming. The calculation of the metric coefficients, needed for the transformed equations, can be obtained from Thompson's TOMCAT code or from an improved version by Johnson and Thompson [3]. Another possibility is offered by the code of Coleman and Haussling [2] which uses the segmentation of the solution area.

2. SHALLOW WATER EQUATIONS

For the purpose of this paper we start with the linearized shallow water equations, that is, we assume small oscillations, i.e., $h \ll H$ (see Fig. 1).

$$\frac{\partial u}{\partial t} + g \cdot \frac{\partial h}{\partial x} + f \cdot v = 0 \qquad (2.1)$$

$$\frac{\partial v}{\partial t} + g \cdot \frac{\partial h}{\partial y} - f \cdot u = 0 \qquad (2.2)$$

$$\frac{\partial h}{\partial t} + \frac{\partial (u \cdot H)}{\partial x} + \frac{\partial (v \cdot H)}{\partial y} = 0 \qquad (2.3)$$

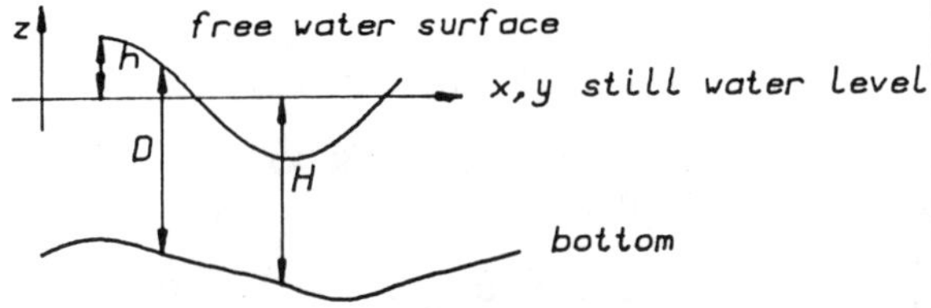

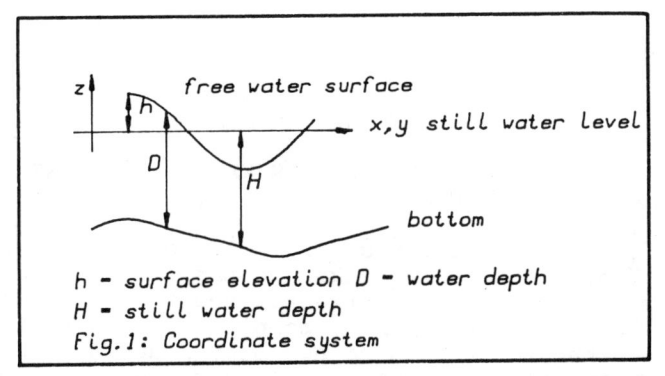

h - surface elevation D - water depth
H - still water depth
Fig. 1: Coordinate system

The above equations describe linear long waves in an inviscid rotating ocean. The Coriolis force is of the form $\vec{f} \times \vec{v}$ where $\vec{f}$ points into the direction of the axis of rotation. Eqs. (2.1) through (2.3) are solved for an annular sheet of water bounded by concentric circles. The still water depth H is assumed to be constant (note that H = D is assumed). This problem can be solved analytically using Lamb's [7] solution for a circle. For the assumption that the concentric circles represent solid walls, it is required that components of the velocity normal to these walls vanish. For the analytical solution harmonic time dependence is assumed. In the following u,v and h depend only on spatial coordinates r, Θ but may be complex quantities. The final solution is obtained by multiplication with $\exp(+ i\omega t)$ and taking the real part. For the harmonic time dependence the spatial part of the analytical solution takes the form:

$$u = \frac{g}{\omega^2 - f^2} \left\{ \frac{\partial h}{\partial r} (i\omega \cos\Theta - f \sin\Theta) - \frac{1}{r} \frac{\partial h}{\partial \Theta} (+ f \cos\Theta + i\omega \sin\Theta) \right\} \qquad (2.4)$$

$$v = \frac{g}{\omega^2 - f^2} \left\{ \frac{\partial h}{\partial r} (i\omega \sin\Theta + f \cos\Theta) + \frac{1}{r} \frac{\partial h}{\partial \Theta} (- f \sin\Theta + i \ \cos\Theta) \right\} \qquad (2.5)$$

$$\frac{\partial^2 h}{\partial r^2} + \frac{1}{r} \frac{\partial h}{\partial r} + \frac{1}{r^2} \frac{\partial^2 h}{\partial \Theta^2} + \frac{\omega^2 - f^2}{gH} h = 0 . \qquad (2.6)$$

Since the normal component of the velocity vanishes at the inner and outer circles, the boundary condition $v_{\hat{n}} = 0$ has to be used for Eq. (2.6). For f equal zero we have $\partial h / \partial r = 0$ and hence contourlines for surface elevation must be perpendicular to the boundaries. Using a separation ansatz in the spatial coordinates,

$h = R(r)\,\phi(\Theta)$, the solution for h is described by $h(r,\Theta,t) = J_n(kr)\cdot \cos(n\Theta - \omega t)$ and the boundary condition takes the form:

$$k_m r_c J'_n (k_m r_c) + \frac{fn}{\omega} J_n (k_m r_c) = 0 \tag{2.7}$$

where $k_m r_c$ (m = 1,2...) denote the set of values satisfying the boundary condition, r_c is the radius of the inner or outer circle and ' denotes differentation with respect to the product kr. Insertion of the Bessel function of order n,$J_n(kr)$, into Eq.(2.6) leads to the dispersion relation:

$$\omega^2 = g H k^2 + f^2. \tag{2.8}$$

The integers n,m identify the normal mode where n is the number of radial nodes (i.e., nodes or zeros along a given value r = constant) while m describes the number of concentric circles where (i.e., in radial direction) the elevation is zero. For f = 0, one simply has the boundary condition $J'_n(k_m r_c) = 0$.

3. TRANSFORMATION OF THE SHALLOW WATER EQUATIONS TO CURVILINEAR COORDINATE SYSTEMS

Since the irregularly shaped solution area in the physical plane is mapped onto a rectangle in the transformed plane, the governing equations must be transformed, too. In general, the transformed equations are more complex, having variable coefficients, but the solution area has been reduced to a rectangle, spanned by an equidistant mesh. However, the solution of the transformed equations on a rectangle is much easier than the original problem. Of course, boundary conditions have to be transformed, too.

Assuming a coordinate system $x = x(\xi,\eta)$, $y = y(\xi,\eta)$ for the two-dimensional case, the transformation of first derivatives in conservative form takes the form (chain rule):

$$h_x = \frac{1}{\sqrt{g}} \left((hy_\eta)_\xi - (hy_\xi)_\eta\right); \quad h_y = \frac{1}{\sqrt{g}} \left(-(hx_\eta)_\xi + (hx_\xi)_\eta\right) \tag{3.1}$$

Using Eqs. (3.1) the transformed shallow water equations read:

$$\frac{\partial h}{\partial t} + \frac{1}{\sqrt{g}} (uHy_\eta - vHx_\eta)_\xi + \frac{1}{\sqrt{g}} (vHx_\xi - uHy_\xi)_\eta = 0 \tag{3.2}$$

$$\frac{\partial (uH)}{\partial t} + \frac{gH}{\sqrt{g}} \left((hy_\eta)_\xi - (hy_\xi)_\eta\right) + fvH = 0 \tag{3.3}$$

$$\frac{\partial (vH)}{\partial t} + \frac{gH}{\sqrt{g}} \left((hx_\xi)_\eta - (hx_\eta)_\xi\right) - fuH = 0 \tag{3.4}$$

where $\sqrt{g}$ is the Jacobi determinant of the transformation (not to be confused with g which is the acceleration of the earth).

In the following the transformed boundary conditions are determined. To this and the shallow water equations are written in the form

$$\frac{\partial v_{\hat{n}}}{\partial t} + g \frac{\partial h}{\partial \hat{n}} + fv_{\hat{t}} = 0 \tag{3.5}$$

$$\frac{\partial v_{\hat{t}}}{\partial t} + g \frac{\partial h}{\partial \hat{t}} - fv_{\hat{n}} = 0\,. \tag{3.6}$$

Since $v_{\hat{n}} = 0$, we obtain from Eqs. (3.5, 3.6)

$$g \frac{\partial h}{\partial \hat{n}} + f\, v_{\hat{t}} = 0 \tag{3.7}$$

$$\frac{\partial v_{\hat{t}}}{\partial t} + g \frac{\partial h}{\partial \hat{t}} = 0\,. \tag{3.8}$$

Eq. (3.8) is used to calculate the tangential velocity while Eq. (3.7) can be used to compute the surface elevation h at the boundary.

If the boundary of the solution area is described by a curve in parameter representation (e.g., spline), $\vec{r} = (x(q), y(q))$, where q is the curve parameter, we find for the unit tangent vector

$$\hat{t} = (\dot{x}^2 + \dot{y}^2)^{-1/2} \cdot (\dot{x},\dot{y}) =: (-\sin\phi, \cos\phi) \tag{3.9}$$

where the dot denotes differentiation with respect to q.
As we are interested only in the cartesian velocity components u and v, the tangential velocity, Eq. (3.8), is separated:

$$\frac{\partial u}{\partial t} - g \frac{\partial h}{\partial \hat{t}} \sin\phi = 0; \qquad \frac{\partial v}{\partial t} + g \frac{\partial h}{\partial \hat{t}} \cos\phi = 0. \tag{3.10}$$

Using the relation

$$\frac{\partial h}{\partial \hat{t}} = -\sin\phi \frac{\partial h}{\partial x} + \cos\phi \frac{\partial h}{\partial y} \tag{3.11}$$

results in the following equations for the boundary points:

$$\frac{\partial u}{\partial t} + g \sin\phi \left(\sin\phi \frac{\partial h}{\partial x} - \cos\phi \frac{\partial h}{\partial y}\right) = 0 \tag{3.12}$$

$$\frac{\partial v}{\partial t} + g \cos\phi \left(-\sin\phi \frac{\partial h}{\partial x} + \cos\phi \frac{\partial h}{\partial y}\right) = 0\,.$$

For the transformation of Eqs. (3.12) onto the computational plane, the conservative form of the first derivatives, Eq. (3.1), is used. Inserting these expressions into Eq. (3.12), gives the final form of the equations valid for the upper and lower sides of the rectangle - which form the solid boundary - in the transformed plane:

$$\begin{aligned} \frac{\partial (uH)}{\partial t} + \frac{gH \sin\phi}{\sqrt{g}} & \{\sin\phi \left((hy_\eta)_\xi - (hy_\xi)_\eta\right) \\ & - \cos\phi \left(-(hx_\eta)_\xi + (hx_\xi)_\eta\right)\} = 0 \\ \frac{\partial (vH)}{\partial t} + \frac{gH \cos\phi}{\sqrt{g}} & \{-\sin\phi \left((hy_\eta)_\xi - (hy_\xi)_\eta\right) \\ & + \cos\phi \left(-(hx_\eta)_\xi + (hx_\xi)_\eta\right)\} = 0\,. \end{aligned} \tag{3.13}$$

Let N and M be the number of points in ξ and η directions, respectively, then the transformed velocity equations (3.3),(3.4) are valid for $\xi = 1(1)N$; $\eta = 2(1)M-1$ whereas the equation for the surface elevation holds for all grid points. The equations derived from the boundary conditions, Eqs. (3.13), hold for $\xi = 1(1)N$; $\eta = 1,M$. Since we have an equidistant mesh in the transformed plane with grid spacings $\Delta\xi = \Delta\eta = 1$, the range of the coordinates ξ and η corresponds to the number of grid points in each direction. No boundary conditions are specified along the reentrant boundaries which form the left and right side of the rectangle in the transformed plane. Rather we have the following correspondence between reentrant grid points: $\xi = N \to 1$ and $\xi = 0 \to N$.

4. NUMERICAL SCHEME FOR THE TRANSFORMED SHALLOW WATER EQUATIONS

For the numerical solution of the system of coupled equations, Eqs. 3.2 - 3.4), two staggered meshed are used; one for the components u,v and the other for h. The meshes are chosen such that velocity components are calculated on the boundary. No calculations have been performed where all three variables are computed at the same grid points. The staggered grids are depicted in Fig. 2. The vertices at the squares are taken as grid points for u and v, marked by crosses, while the dots at the center of the squares represent grid points for h.

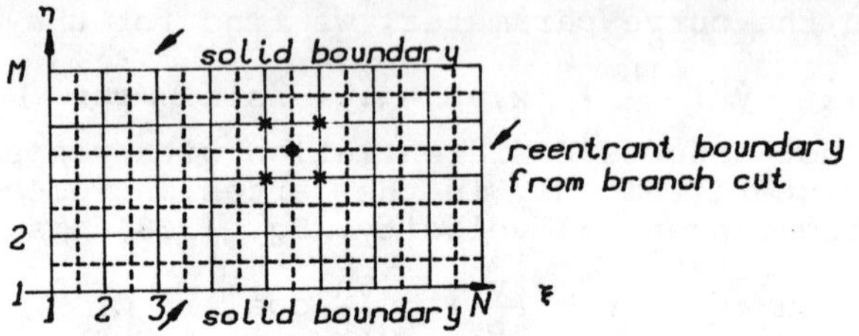

Fig.2: Crosses denote grid points for u,v while dots indicate those for h.

Since the metric coefficients must be known at each vertex of a solid or dashed grid line, it is necessary that the boundary fitted grid, representing this feature, has (2N-1) x (2M-1) grid points. In Fig. 3 the two sets of indices for the staggered grids are shown, where i,j denote positions for u,v and I,J denote those of h.

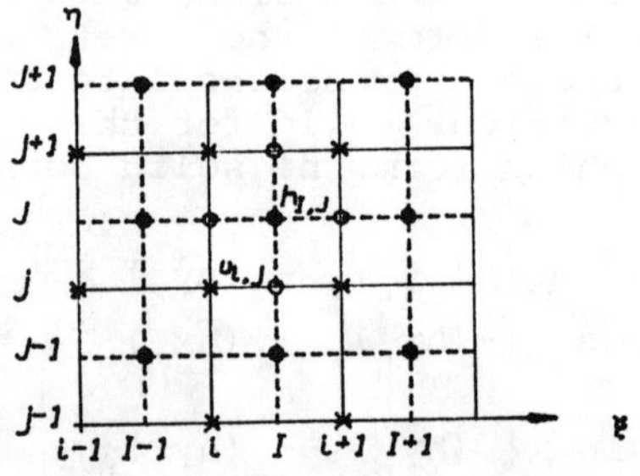

Fig.3: u,v and h have to be interpolated at points denoted by open circles.

The relations between these indices can be seen from Fig. 3 and have the form

$$I = i + \frac{1}{2}; \qquad J = j + \frac{1}{2}. \tag{4.1}$$

In order to satisfy the zero normal flow condition for solid walls, the u and v grid points are situated on the lower and upper boundaries (Fig. 2).

The discretized equations (see below) are then solved for the corresponding set of grid points. For the numerical solution the resulting mathematical system of weakly coupled, ordinary differential equations is integrated in time by the explicit fourth-order Runge-Kutta-Fehlberg method. For the spatial approximation linear interpolation is used. Since the discrete equations are in flux conservative form, flux corrected transport schemes can be used. For this paper no implicit procedure has been used, that is, the time step is limited by the velocity with which the surface elevation is propagated. Using Eq. (4.1) the discretized equations take the form [4 - 6]:

$$\begin{aligned}\frac{\partial h_{I,J}}{\partial t} = -\frac{1}{(\sqrt{g})_{I,J}} \{ & (uHy_\eta)_{i+1,J} - (uHy_\eta)_{i,J} \\ & -(vHx_\eta)_{i+1,J} + (vHx_\eta)_{i,J} \\ & +(vHx_\xi)_{I,j+1} - (vHx_\xi)_{I,j} \\ & -(uHy_\xi)_{I,j+1} + (uHy_\xi)_{I,j} \}\end{aligned} \tag{4.2}$$

$$\begin{aligned}\frac{\partial (uH)_{i,j}}{\partial t} = -\frac{g\,H_{i,j}}{(\sqrt{g})_{i,j}} \{ & (hy_\eta)_{I,j} - (hy_\eta)_{I-1,j} \\ & -(hy_\xi)_{i,J} + (hy_\xi)_{i,J-1} \} - (fHv)_{i,j}\end{aligned} \tag{4.3}$$

$$\begin{aligned}\frac{\partial (vH)_{i,j}}{\partial t} = -\frac{g\,H_{i,j}}{(\sqrt{g})_{i,j}} \{ & (hx_\xi)_{i,J} - (hx_\xi)_{i,J-1} \\ & -(hx_\eta)_{I,j} + (hx_\eta)_{I-1,j} \} + (fHu)_{i,j}\end{aligned} \tag{4.4}$$

The boundary conditions are obtained in the same manner.

From Eqs. (4.2 - 4.4) it can be seen that u,v and h values are needed at locations which are not grid points. For the continuity equation velocity values are needed at points i+1,J; i,J; I,j+1 and I,j. These values are obtained from linear interpolation in the transformed plane:

$$u_{i+1,J} = \frac{1}{2}\left(u_{i+1,j} + u_{i+1,j+1}\right); \quad v_{I,j+1} = \frac{1}{2}\left(v_{i,j+1} + v_{i+1,j+1}\right) \tag{4.5}$$

Since the partial derivatives x_ξ, x_η, y_ξ, y_η and $\sqrt{g}$ are known on the fine grid (formed by both dashed and solid lines, see Fig. 2) no interpolation is necessary. As H is a known function of x and y, all values of H can be calculated on the fine grid prior to the beginning of the numerical calculations. For the momentum equations values for h are needed at points I,j;I-1,j; i,J; and i,J-1. Again linear interpolation is used. For the solid boundaries, that is, j=1 or j=M the following formulas are used

$$\begin{aligned} J=0 &: h_{I,J} = 2h_{I,J=1} - h_{I,J=2} \\ J=M &: h_{I,J} = 2h_{I,J=M-1} - h_{I,J=M-2} \end{aligned} \tag{4.6}$$

5. NUMERICAL RESULTS FOR NORMAL MODE SIMULATIONS

The finite-difference scheme, described in Chap. 4, in conjunction with the boundary fitted grid was used to simulate the normal mode oscillations of the circular ring basin of uniform depth H. For the numerical calculations two modes were simulated (Fig. 4), namely for $n = 0$ with $R = 7.01 - 3.83$, and $n = 4$ (Fig. 5) with $R = 15.96 - 9.28$ where R is the difference of the outer and the inner radius of the ring. Since $k = 1$ was used for all computations, R equals the difference of the corresponding zeros of $J_n'(r_c)$ (see Eq. 2.7; Coriolis force neglected).

When the computations were started it became immediately obvious that the scheme was unstable. The instability was associated with the use of Eq. (3.1), which is a conservative form for the first derivatives. It was found that even for a plane surface (h constant) the time derivative for, e.g. uH, was different from zero, due to the existence of the term $gH/\sqrt{g}\ h((y_\eta)_\xi - (y_\xi)_\eta)$ because second derivatives do not cancel numerically. Hence, this term provides a numerical source of energy. Since it is invariant in time it eventually renders the system unstable. Therefore, a nonconservative form was used.

Second, it seems that the staggered mesh used can cause instabilities, too ($n = 4$), due to the evaluation of the derivatives. This suggests that a mesh where all quantities are calculated at the same grid points is preferable in the case of the shallow water equations.

Third, in most cases it will be advantageous to obtain solutions in components of the curvilinear coordinates (i.e. v_ξ and v_η) in order to reduce numerical diffusion and to facilitate the formulation of the boundary conditions. For the present case this is an advantage for $n = 0$.

Calculations for $n = 4$ poorly simulated the amplitudes and shapes, like frequencies, even after a few cycles. This behavior is still under investigation.

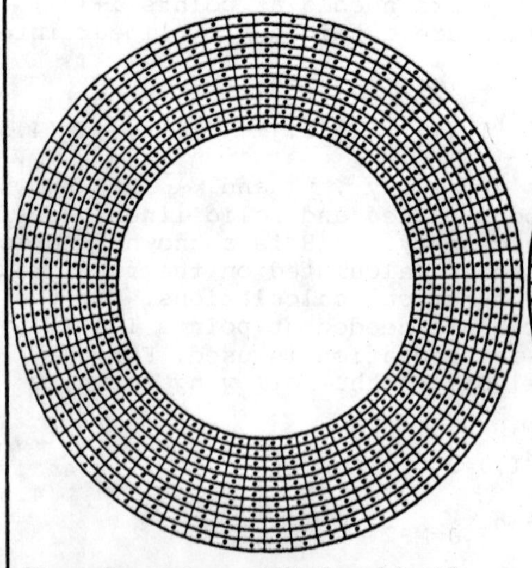

Fig.4: Boundary fitted grid: 119 points in ξ, 23 points in η direction. Vertices of the mesh denote u,v grid points, whereas dots denote those for h.

Fig.5: Surface elevation h obtained from analytical solution, Eq. 2.6, for $n = 4$.

DISCUSSION

The solution of the shallow water equations on a circular ring provides a particularly good basis for the testing of the boundary fitted coordinate system approach, since comparisons with the analytical solution for both velocity components and surface elevation are possible. Moreover, frequencies of the normal mode oscillations can be checked, too. In this paper Coriolis force was not accounted for, i.e. the earth's rotation was neglected. Hence it is not known whether the numerical scheme becomes unstable when this term is included. Although all calculations are performed on a uniform grid, instabilities may be associated to the variable coefficients generated by the transformation. Analytical solutions also exist for a circular ring basin with a parabolic depth profile $H = H_0(1 - r^2/R^2)$. No calculations have been performed for this case. Furthermore,the influence of grid line concentration on the accuracy and stability has not been investigated.

Of great importance is the approximation of the first derivatives since the use of a conservative form may lead to inconsistencies, because second derivatives do not cancel in general, i.e. $\partial^2 h/\partial x\, \partial y \neq \partial^2 h/\partial y\, \partial x$. This is the case for the formulas on the staggered grids. Even for constant h, terms proportional to $h(y_{\xi\eta} - y_{\eta\xi})$ etc. occur which provide a numerical source of energy. Most likely these terms will not cause stability problems in parabolic equations where sufficient damping exists but may cause instabilities in hyperbolic systems. However, the accuracy of the solutions will be affected in both cases. Although the use of boundary fitted grids to problems of curving boundaries is one of the most versatile approaches, further investigations are necessary concerning the stability and accuracy of the solutions; that is, the influence of the transformation on and the use of the numerically calculated metric coefficients for the numerical solution demands additional detailed investigations.

REFERENCES

[1] Thompson,J.F., 1982: Elliptic Grid Generation, in Thompson J.F.: Numerical Grid Generation, pp. 79 - 106.

[2] Coleman, R.M., 1982: Generation of a Boundary Fitted Coordinate System Using Segmented Computational Regions, in Thompson, J.F.: Numerical Grid Generation, pp. 633-652.

[3] Johnson, B.H., 1982: Numerical Modeling of Estuarine Hydrodynamics on a Boundary Fitted Coordinate System, in Thompson, J.F.: Numerical Grid Generation, pp. 409-436.

[4] Zalesak, S.T., 1979: Fully Multidimensional Flux Corrected Transport Algorithm for Fluids, J. of Comp. Physics, 31, pp. 335 - 362.

[5] Book, D.L. (ed), 1981: Finite Difference Techniques for Vectorized Fluid Dynamics Calculations,Springer,Berlin,226 pp.

[6] Häuser, J. et al., 1983: Flux Corrected Transport for One-Dimensional Simulation Models in Rivers and Estuaries Incorporating Three-Dimensional Effects, in Taylor: Laminar and Turbulent Flow, Pineridge Press, Swansea, pp. 433 - 450 (a corrected version can be obtained from the author.)

[7] Lamb, H.,1945: Hydrodynamics,6th Edition,Dover,New York,pp.738.

A BOUNDARY INTEGRAL APPROACH TO COMPUTE THE THREE-DIMENSIONAL OSEEN'S FLOW PAST A MOVING BODY

F.K. Hebeker *)
Fachbereich 17 Mathematik-Informatik
Universität-GHS Paderborn
D-479o Paderborn

SUMMARY

The homogeneous viscous incompressible low Reynolds number flow past a moving body is investigated mathematically and numerically in Oseen's approximation. The classical potential theory for Oseen's equations is developed up to an operational procedure, a boundary integral method, which proves to be well fitted to compute viscous 3D-flows. The numerical approach consists of a collocation-type procedure where the algebraic system is solved effectively by a two-level multigrid method. The results of numerical test computations are presented.

1. INTRODUCTION

Let a body pass through a homogeneous viscous incompressible fluid. Mathematically the fluid flow may be calculated in Oseen's approximation : this is adequate not only at some distance from the body but for low Reynolds number also in its neighborhood.

In the present paper a new numerical approach to compute the steady low Reynolds number flow past a moving body in Oseen's approximation is proposed - to solve by boundary integral methods. They play an increasing role for the numerical solution of mathematical problems in engineering sciences (see Jaswon-Symm [7], Brebbia [1], Hess [5]),e.g. for the computation of potential flows.

Their striking advantages over domain-type methods (FDM,FEM, spectral) in many cases are well-known:

. instead of the threedimensional domain its twodimensional boundary has to be discretized only,
. hence algebraic systems with packed but relatively small matrices arise,
. exterior domains do not cause difficulties.

And in case of homogeneous viscous flow approximate solutions are obtained which

. automatically satisfy the incompressibility constraint.

---

*) This work partly has been supported by the Deutsche Forschungsgemeinschaft in its special program "Finite Approximationen in der Strömungsmechanik".

Oseen [9] constructed a fundamental solution (of the Oseen's equations), a (4.4)-matrix, by means of which the surface potentials (i.e. weakly singular surface integrals) are defined (section 2). For our BVP we introduce a mixed ansatz, containing a free parameter, with potentials of the simple and of the double layer. By means of classical jump relations (Faxén [2]) we are led to a Fredholm integral equations' system of the second kind for the unknown surface source of the potentials.

In section 3 using a collocation-type discretization procedure with piecewise bilinear polynomials as coordinate functions (on the parameter space of the boundary) a linear algebraic system with packed but relatively small matrix is obtained. In order to solve it we propose and compare

- Gauß - Jacobi iteration,
- a two - level multigrid method,
- Gauß' elimination.

Then the potentials (formed with the approximate surface source) easily and speedy may be computed as weakly singular surface integrals, and they form an automatically solenoidal approximation of the fluid flow.

As a numerical test example these procedures have been carried out in case of a ball passing through the fluid - but we point out that the flow around any body of a different shape may be computed in the same manner!

The present author wishes to express his sincere thanks to Dipl.-Math. P. Neugebauer for his valuable help to perform the numerical computations.

2. THE BOUNDARY INTEGRAL EQUATIONS

We consider the steady motion of a homogeneous viscous incompressible fluid flow past a three-dimensional body. In Oseen's approximation this flow may be calculated from the dimensionless Oseen's equations (body-fixed (x_1,x_2,x_3)-frame)

$$\left.\begin{aligned} -\Delta u + R\frac{\partial u}{\partial x_1} + \nabla p &= o \\ \operatorname{div} u &= o \end{aligned}\right\} \quad \text{around the body}, \tag{1}$$

with the boundary conditions $u = -U$ on the boundary B, and $u = o$ at infinity. Here $U = (1,o,o)$ is the velocity of the uniform flow at infinity, R the Reynolds number, $v = U + u$ the velocity field, and p the pressure function of the flow.

Oseen [9] has constructed the fundamental solution of (1). It is given by the tensor Γ,

$$\Gamma_{ij}(x) = \frac{1}{8\pi}\left(\delta_{ij}\,\Delta\phi(x) - \frac{\partial^2\phi}{\partial x_i\,\partial x_j}(x)\right), \tag{2}$$

with the scalar function

$$\phi(x) = \frac{2}{R}\int_o^s \frac{1-e^{-\sigma}}{\sigma}\,d\sigma\ , \quad s = \frac{R}{2}\left(|x| - x_1\right)\ ,$$

and the "Kronecker-delta" $\delta_{ij} = 1$ when $i = j$, $\delta_{ij} = o$ when $i \neq j$.

Any i-th column $\Gamma^{(i)}$ of Γ solves Oseen's equations with the pressure $-\frac{1}{4\pi}\frac{\partial}{\partial x_i}\frac{1}{|x|}$. Let us define the stress tensor T,

$$T_{ij}(u,p,R) = -\delta_{ij}p + \frac{\partial u_i}{\partial x_j} + \frac{\partial u_j}{\partial x_i} - \frac{1}{2}R\,\delta_{1i}\,u_j\ , \quad (3)$$

where $u = (u_1,u_2,u_3)$. If n denotes the normal vector of B pointing to the flow region E, then the stress vector t and its adjoint t' are componentwise given by

$$t_j = \sum_i n_i\,T_{ij}(u,p,R)\ ,\quad t'_j = \sum_i n_i\,T_{ij}(u,-p,-R)\,. \quad (4)$$

Hence arguing as in ordinary potential theory (see [7]) we get the basic representation formula for any solution of (1):

$$u_i(x) = \int_B t'_y\,(\Gamma^{(i)}(x-y))\cdot u(y)\,do_y - \int_B \Gamma^{(i)}(x-y)\cdot t(u(y))\,do_y \quad (5)$$

when $x \in E$. Here t'_y denotes derivation w.r.t. the integration variable. Formulae like (5) in essence are contained in [2], [9]. They suggest to look at the solution of the BVP (1) in terms of <u>Oseen's potentials</u>, the simple-layer potential $V(x,\phi)$

$$V_i(x,\phi) = \int_B \Gamma^{(i)}(x-y)\cdot\phi(y)\,do_y\ , \quad (6)$$

or the double-layer potential $W(x,\phi)$ with components

$$W_i(x,\phi) = \int_B t'_y\,(\Gamma^{(i)}(x-y))\cdot\phi(y)\,do_y\ , \quad (7)$$

with the unknown surface source vector field $\phi(y)$. Both V and W solve Oseen's equations (1) with the corresponding pressure functions

$$P(x,\phi) = \int_B \left(-\frac{1}{4\pi}\nabla_x\frac{1}{|x-y|}\right)\cdot\phi(y)\,do_y\ ,$$

$$\Omega(x,\phi) = \int_B t'_y\left(-\frac{1}{4\pi}\nabla_x\frac{1}{|x-y|}\right)\cdot\phi(y)\,do_y\ ,\ \text{resp.}$$

The potential (6) is continuous in the whole space, but (7) jumps when passing through the boundary B. In fact, we have the following <u>jump relation</u> ([2], analogous to that of ordinary potential theory (see [7])):

$$W^+ - W = W - W^- = \frac{1}{2}\phi \quad \text{on B}\ , \quad (8)$$

where W^+, W^- denote the exterior or interior limit resp. when approaching the boundary, W the direct value on B.

Now we are looking for a solution of BVP(1) in terms of a mixed simple- and double-layer-potential :

$$u(x) = W(x,\phi) + \eta V(x,\phi)\ ,$$

where η is a free parameter. The <u>unknown surface source</u>, the vector field $\phi(y)$, here is determined from the jump relation (8) and V's continuity, which in view of the boundary condition u = - U lead to the following boundary integral equations' system of the second kind:

$$\phi(x) + 2\,W(x,\phi) + 2\,\eta V(x,\phi) = -2U\;, \text{ on } B\;. \tag{9}$$

By potential theoretic reasons system (9) proves to be uniquely solvable in case of $\eta \geq o$. For the classical double-layer potential ansatz (i.e. the limit case $\eta = o$) this has been shown by Faxén [2], but a numerical example (sect. 3) will indicate that in some cases the results can be greatly improved by choosing η positive. Numerically, the singular integral W should be evaluated in the following manner, at least in the neighborhood of the boundary:

$$W_i(x,\phi) = \delta\cdot\phi_i(x) + \int_B t'_y\,(\Gamma^{(i)}(x-y))\cdot(\phi(y)-\phi(x))\,do_y$$

$$-\frac{1}{2}R\,\phi\,(x)\cdot\int_B \Gamma^{(i)}(x-y)\;n_1(y)\;do_y\;, \tag{1o}$$

with $\delta = o$ when $x \in E$ and $\delta = -\frac{1}{2}$ when $x\in B$.

3. THE NUMERICAL PROCEDURE

For simplicity the numerical procedure will be described for the flow around a unit ball - but we point out that the calculations are performed without using any symmetry property of the sphere, hence the method and the results are considered as typical for 3D- flows in more general flow regions.

By means of normed polar coordinates (θ,φ) the sphere is transformed to the unit square $S = [o,1]^2$. S then is divided into N^2 small squares, and as coordinate functions we choose continuous vector fields ϕ_{ij} on S which are bilinear on each of the small squares. Now we are looking for an approximate surface source vector field $\phi_N = (\phi_N^1, \phi_N^2, \phi_N^3)$ on S :

$$\phi_N^k(\theta,\varphi) = \sum_{i,j=1}^{N} \alpha_{ij}^k\;\phi_{ij}^k\;(\theta,\varphi) \quad k = 1,2,3, \tag{11}$$

where the unknowns α_{ij}^k are determined by a collocation-type procedure: the approximate integral equations' system has to be satisfied at N^2 points (θ,φ) of S (here they are chosen as the edges of the small squares):

$$\phi_N^k(\theta_i,\varphi_j) + 2\,W_k(\theta_i,\varphi_j,\phi_N) + 2\,\eta\,V_k(\theta_i,\varphi_j,\phi_N) = -2\,U_k, \tag{12}$$

$i,j = 1,\ldots, N$; $k = 1,2,3$; with $\eta \geq o$ arbitrary. (12) is a linear algebraic system with a nonsparse matrix of $\sim N^4$ elements which has to be calculated numerically using simple formulae of numerical quadrature and cutting -off the singularity.

The algebraic system (12) can be solved iteratively, e.g.

by Gauß-Jacobi iteration. To improve the numerical efficiency we propose to use a two-level multigrid method - for multigrid methods for potential flow see [8],[1o],[12]- which in essence consists of an alternating sequence of a) one Gauß-Jacobi smoothing step, and b) one coarse-grid correcting step. As a coarse-grid we take most simply that of doubled mesh size, if N is odd.

The test series should clarify the following two questions
a) How does the choice of the free parameter $\eta \geq o$ influence the numerical results ?
b) What an accuracy can be attained when solving the algebraic system by means of the iterative procedures ?
Ad a): As a test example we take a simple solution of Oseen's equations (1), namely $u(x) = \Gamma^{(1)}(x)$ outside the unit ball centered at the origin of the frame, and then we compare this flow with the approximate flow which is obtained by solving (12)(with right hand side $2\Gamma_{k1}(x)$ instead of $-2U_k$). Such a test has the advantage of giving the exact error of the procedure. Now at R=2.o the linear system has been solved for N=15(with matrix of size 675x675) by Gauß' elimination with partial pivoting. Figure 1 shows the great influence of η on the numerical results, we can regard η as a "conditioning parameter". (This behavior has been confirmed by another test example for Stokes' equations: see [4].)
Ad b): This question is investigated in case of Stokes' equations (R=o.o). As a test example a simple solution has been constructed, and a comparison was carried out like that of a). The linear system has been solved iteratively for N=17 (with a matrix of size 867x867) at an optimal choice of η, starting at the exact solution of the collocational equations on the coarser grid. Figure 2 shows the multigrid method to be superior to the classical Gauß-Jacobi iteration. With only one multigrid cycle nearly the accuracy of Gauß' elimination was achieved in our test example.

REFERENCES

[1] Brebbia, C.A., ed., "*Boundary element methods in engineering*", Berlin 1982.

[2] Faxén, H., "Fredholmsche Integralgleichungen zu der Hydrodynamik zäher Flüssigkeiten", Ark.Mat.Astr.Fys. 21 A, 14 (1929) pp.1 - 4o.

[3] Fischer, T.M., "An integral equation procedure for the exterior threedimensional slow viscous flow", Integral Equ.Oper.Th.5 (1982) pp. 49o-5o5.

[4] Hebeker, F.K., "A theorem of Faxén and the boundary integral method for threedimensional viscous incompressible fluid flows", preprint, Univ. Paderborn, 1982.

[5] Hess, J.L., "Review of integral equation techniques for solving potential-flow problems with emphasis on the surface-source method", Comp.Meth.Appl.Mech. Engin. 5 (1975) pp. 145 - 196.

[6] Homentcovschi, D., "Three-dimensional Oseen flow past a flat plate", Quart.Appl.Math. 4o (1982) 137-149.

[7] Jaswon, M.A., Symm,G.T., "*Integral Equations in Potential Theory and Elastostatics*", London 1977.

[8] Novak, Z.P., "Use of the multigrid method for Laplacean problems in three dimensions", in Hackbusch,W., Trottenberg, U., eds., "Multigrid methods", Berlin 1982, pp. 576-598.

[9] Oseen, C.W., "*Neuere Methoden und Ergebnisse in der Hydrodynamik*", Leipzig 1927.

[1o] Schippers, H., "Application of multigrid methods for integral equations to two problems from fluid dynamics", J.Comp.Phys. 48 (1982) 441-461.

[11] Wendland, W.,"Die Behandlung von Randwertaufgaben im $\mathbb{R}^3$ mit Hilfe von Einfach- und Doppelschichtpotentialen", Num.Math.11 (1968) 38o - 4o4.

[12] Wolff, H., "Multiple grid method for the calculation of potential flow around 3 D-bodies", preprint, Mathematisch Centrum Amsterdam, 1982.

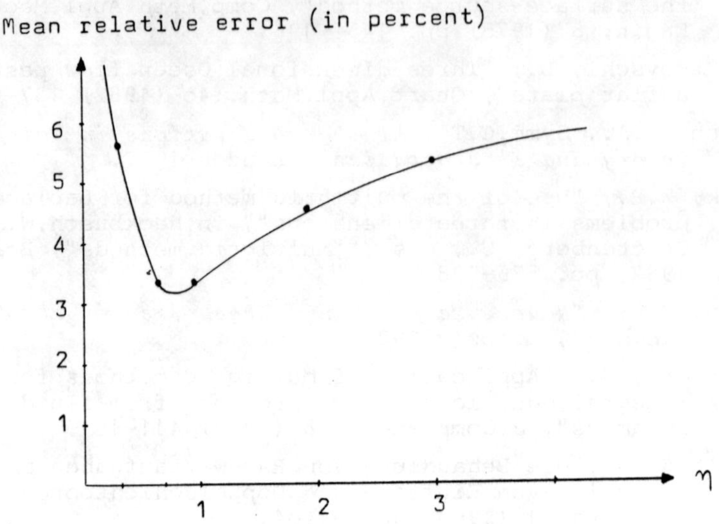

FIG. 1 : Influence of η

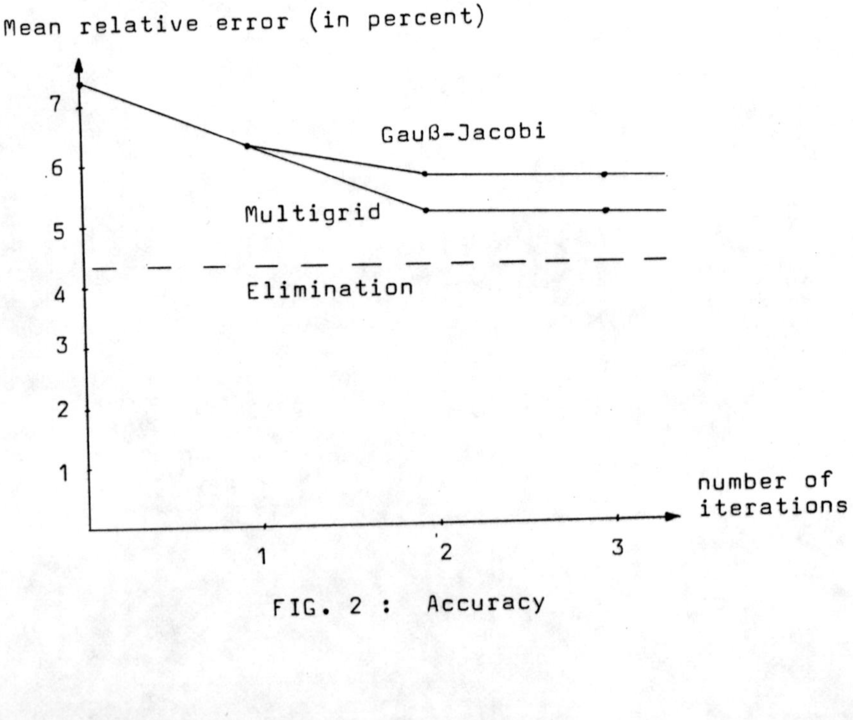

FIG. 2 : Accuracy

NUMERICAL SOLUTION OF MULTI-DIMENSIONAL DIFFUSION-CONVECTION PROBLEMS BY ASYMPTOTIC CORRECTIONS

M. Israeli* and P. Bar-Yoseph**

Technion - Israel Institute of Technology
Haifa, Israel

SUMMARY

The Booster Method for improvement of the numerical solution of partial differential equations by the addition of asymptotic corrections to the right hand side is presented. It is applied here to the diffusion-convection equation for the case of 'small' diffusion. The correction terms were used in finite difference and finite element schemes. The finite element results were used as reference for checking the performance of the finite difference schemes. Excellent results were obtained without the use of upstreaming or artificial diffusion. Theoretical expectations were confirmed.

1. INTRODUCTION

Singularly perturbed initial and boundary value problems for partial differential equations appear in various fields of application such as fluid dynamics, heat transfer, transport of atmospheric pollution, etc. In particular, such equations appear in diffusion-convection processes. Often the (normalized) diffusion coefficient ε becomes small, and thin boundary or interior layers appear within the region of interest. Consequently, these problems become increasingly difficult to solve numerically by discretization methods.

We would like to avoid the use of a prohibitively large number of grid points, as required for resolution by straightforward numerical methods when ε decreases. To this end, several approaches are possible, such as the use of nonuniform meshes, adaptive techniques, positive type schemes, etc. The question of applicability of such schemes to multi-dimensional problems is presently open.

A different approach is motivated by classical singular perturbation methods where 'inner' and 'outer' solutions are combined to give approximate solutions. These solutions become more accurate as the equations become stiffer, however, the error is fixed for a given ε and cannot be improved or estimated reliably in most cases of interest.

The Booster Method attempts to combine the asymptotic approach, with known discretization methods, in order to obtain a numerical method which improves when ε becomes smaller. At the same time, it keeps the property that the error can be made arbitrarily small for any fixed ε by refining the computational mesh (Israeli and Ungarish [1],[2]). For the one-dimensional case, we were able to prove that an improvement by a factor of ε^{n+1} can be obtained where ε is the 'small' parameter and n is the order of the asymptotic approximation used (Israeli and Ungarish [2]). We expect similar behaviour in the multidimensional case [1].

In the present paper, we investigate a multi-dimensional application to diffusion-convection problems.

---

* Department of Computer Science, Technion.
** Department of Mechanical Engineering, Technion.

2. FORMULATION

We consider the transport of a quantity q in a rectangular region. The normalized partial differential equation is

$$L(q) \equiv -\varepsilon\nabla^2 q + \vec{V} \cdot \nabla q = 0 \quad . \tag{1}$$

In the present application the velocity field $\vec{V}$ is assumed to be known and q is specified on the boundaries. This problem was often used as a test case for various finite difference and finite element methods of solution and it is well known that most methods fail as the cell Reynolds number $(|\vec{V}|h)/\varepsilon$ becomes larger than $O(1)$ (here h is a representative mesh size). For example centered schemes develop unphysical oscillations in space, while uncentered schemes have unacceptable artificial diffusion and are of lower order over the same computational stencil.

The Booster Method uses an asymptotic approximation $\tilde{q}(x,y)$ to the solution $q(x,y)$ in order to reduce the truncation error in the numerical scheme.

The 'usual' numerical solution $Q(x,y)$ (defined only at grid points) is obtained from

$$L_N(Q) = f \quad , \tag{2}$$

where L_N is the discrete approximation to the differential operator L. The improved numerical solution $\bar{Q}$ is obtained from

$$L_N(\bar{Q}) = f + L_N(\tilde{q}) - L(\tilde{q}) \quad . \tag{3}$$

Here $L(\tilde{q})$ is the differential operator applied to the approximate solution. Thus the Booster Method applies an asymptotic correction to the right hand side of the equation and therefore requires a negligible amount of extra work. It can be used with any numerical scheme without modification in the method of solution.

The same basic approach of using asymptotic corrections to the right hand side can be used to improve the Standard Finite Element (SFE) method. The resulting Asymptotic Finite Element (AFE) method is described briefly in the following; for details see Bar-Yoseph and Israeli [4],[5].

Suppose that the unit square is divided into elements and that the variation of q within the given region is approximated by

$$Q(\vec{x}) = N_i(\vec{x})Q_i \quad , \tag{4}$$

where Q_i is the value of the approximate solution at the i-th nodal point and N_i is the corresponding global trial function (we use the summation convention, with summation over the nodes within the given region). The Bubnov-Galerkin finite element scheme of eq. (1) is given by

$$\varepsilon(\nabla^T N_j, \nabla N_i Q_i)_h + (N_j, \vec{V}\cdot\nabla N_i Q_i)_h = (N_j, f)_h \quad , \quad j = 1,2,\ldots,m \quad , \tag{5}$$

where m is the number of inner nodal points and $(\cdot,\cdot)$ denotes the usual inner product in $L_2(\Omega)$. The subscript h in $(\cdot,\cdot)_h$ denotes an approximation to $(\cdot,\cdot)$ obtained by a quadrature rule.

Our corresponding asymptotic finite element (AFE) scheme for eq. (1) is the following

$$\varepsilon(\nabla^T N_j, \nabla N_i \bar{Q}_i)_h + (N_j, \vec{V}\cdot\nabla N_i \bar{Q}_i)_h = (N_j, f)_h$$

$$+ \{\varepsilon(\nabla^T N_j, \nabla N_i \tilde{q}_i)_h + (N_j, \vec{V}\cdot\nabla N_i \tilde{q}_i)_h - (N_j, L\tilde{q})\} \ , \quad j = 1,2,\ldots,m \ , \tag{6}$$

where $\tilde{q}_i$ is the value of the asymptotic solution at the i-th nodal point. Here the terms included in the first line coincide with the SFE scheme, eq. (5), while the terms in the second line (in curled brackets) represent the correction term which is the essence of the present AFE scheme. This AFE scheme can also improve the pointwise error estimate of the SFE scheme by a factor of the $O(\varepsilon^{n+1})$ [4].

Usually the finite element solution supplies values everywhere inside the elements via the interpolation (4). Applying the same interpolation using $\bar{Q}_i$ instead of Q_i will not give good results within the elements especially when there are no nodal points inside the boundary layers. One should use instead the interpolation

$$q(\vec{x}) \simeq \tilde{q}(\vec{x}) + N_i(\vec{x})(\bar{Q}_i - \tilde{q}_i)$$

which recovers the proper boundary layer behaviour.

3. ASYMPTOTIC SOLUTION

The approximation $q(x,y)$ can be obtained by the method of matched asymptotic expansions (Cole [3]). Such approximations usually satisfy the boundary conditions and become increasingly accurate as ε decreases. Often the error decreases like some power of ε depending on the number of terms used in the construction of the solution (see Table I).

We first construct the zeroth order asymptotic approximation from the outer solution q_o, where

$$\vec{V}\cdot\nabla q_o = 0 \ , \tag{7}$$

and from the boundary layer q_b satisfying a one-dimensional boundary layer equation in the direction normal to the boundaries (where the flow exits the computational region). A complete zeroth order solution should also include corner regions, tangential regions, and boundary layers developing from discontinuities in boundary conditions. Restricting ourselves for now to continuous boundary conditions and constant velocity field not tangent to any boundary, we find that only the exit corner layer has to be included.

Equation (7) implies that the solution remains constant along streamlines, consequently it carries with it the q-values entering the computational region. The difference between these values and the values encountered at the exit forces the boundary layers.

We consider the flow in the unit square with vertices (0,0), (0,1), (1,0), (1,1). Let the components of $\vec{V}$ be u and v (both positive) and the differences of the exit boundary $x = 1$ and $y = 1$ be $f(y)$ and $g(x)$ respectively. Then the structure of the boundary layers will be

$$q_b = f(y)e^{u(x-1)/\varepsilon} + g(x)e^{v(y-1)/\varepsilon} - pe^{u(x-1)/\varepsilon}e^{v(y-1)/\varepsilon} \ , \tag{8}$$

where $p = f(1) = g(1)$ and $f(0) = g(0) = 0$ by assumption.

The third term in (8) is the corner boundary layer and it satisfies the differential equation exactly (in this particular case). The approximation $\tilde{q}$ is obtained by adding q_o from (7) to q_b from (8). It is a uniformly valid approximation in the square, and is easily adapted to particular cases. In our example $u = v = \sqrt{2}/2$ and the outer solution is

$$q_o = \sin\pi(x-y) \quad , \tag{9}$$

corresponding to the boundary conditions

$$q(x,0) = \sin\pi x; \; q(0,y) = -\sin\pi y \quad . \tag{10}$$

On the other hand we take

$$q(x,1) = \sin\pi(x-1) + x, \quad q(1,y) = \sin\pi(1-y) + y \quad , \tag{11}$$

giving rise to $f(y) = y$ and $g(x) = x$, which satisfy the requirement of continuity on the boundary.

4. RESULTS

We solved the differential equation (1) in the unit square with the boundary conditions (10) and (11) on a net with 6, 12 and 24 equal intervals in the x and y directions. We used centered three-point differences for the first and second derivatives. As we do not have the exact solution for this problem, we used as reference a solution obtained with the AFE scheme employing a uniform mesh of 24×24 biquadratic Lagrangian elements (more details appear in [5]). The range of ε reproduced in the tables was $\varepsilon = 0.01$, $\varepsilon = 0.02$, $\varepsilon = 0.05$.

As a seminorm we used a weighted average of the absolute value of the errors. A proper mix of interior and boundary layer points was obtained by taking only the 25 mesh points contained in a square near the corner (1,1).

We observe that (Table I) the error in the asymptotic approximation $A(\varepsilon,h)$ decreases roughly like ε. The error in the regular (Table III) numerical solution $N(\varepsilon,h)$ decreases very slowly with h and increases as ε decreases.

We note first that the errors in Table II are in all cases smaller than the corresponding errors in Tables I and III. Moreover the analysis of [2] indicates that the error in the improved solution $B(\varepsilon,h)$ should decrease with ε and h (Table II). In fact it should be proportional to the product of the previous errors. Table IV presents the ratio $K(\varepsilon,h) = B(\varepsilon,h)/((A(\varepsilon,h)N(\varepsilon,h)))$. We expect it to approach a constant,the fact that the values in Table IV are not far from unity makes the booster method quite attractive.

We note that the centered method is quite useless by itself. Our experience shows that the Booster Method works equally well with other schemes and other discretization methods.

ACKNOWLEDGEMENT

The first author was partially supported by AFOSR under contract number F49620-83-C-0064.

REFERENCES

[1] Israeli, M. and Ungarish, M.: Proceeding of the ICNMFD7, Stanford, 1980, Lecture Notes in Physics, 141, Springer-Verlag.

[2] Israeli, M. and Ungarish, M.: Numer. Math. 39, 309-324 (1982).

[3] Cole, J.D.: Perturbation Methods in Applied Mathematics, Waltham, Mass., Blaisdell Publ. Co., 1968.

[4] Bar-Yoseph, P. and Israeli, M.: An asymptotic finite element method for improvement of finite element solutions of boundary layer problems (to appear).

[5] Bar-Yoseph, P. and Israeli, M.: An asymptotic finite element method for 2-D convection-diffusion problems (in preparation).

TABLE I

Weighted Average of Absolute Error in Asymptotic Solution

ε/grid	24 × 24	12 × 12	6 × 6
0.01	0.0348	0.0590	0.0466
0.02	0.0504	0.0970	0.0846
0.05	0.0552	0.1440	0.1600

TABLE II

Weighted Average of Absolute Error in Booster Solution

ε/grid	24 × 24	12 × 12	6 × 6
0.01	0.0101	0.0315	0.0295
0.02	0.0057	0.0231	0.0284
0.05	0.0026	0.0106	0.0228

TABLE III

Weighted Average of Absolute Error in Regular Solution

ε/grid	24 × 24	12 × 12	6 × 6
0.01	0.1836	0.2603	0.4585
0.02	0.1369	0.2113	0.3904
0.05	0.1110	0.1618	0.2454

TABLE IV

The Ratio $K(\varepsilon,h)$ (see text)

ε/grid	24 × 24	12 × 12	6 × 6
0.01	1.58	2.05	1.38
0.02	0.83	1.13	0.86
0.05	0.42	0.63	0.58

MARCHING MULTIGRID SOLUTIONS TO THE PARABOLIZED NAVIER-STOKES (AND THIN LAYER) EQUATIONS

Moshe Israeli* and Moshe Rosenfeld*

Technion - Israel Institute of Technology
Haifa, Israel

SUMMARY

An efficient method for the solution of the Parabolized or Thin Layer (PNS or TL) incompressible Navier-Stokes equations is suggested. The efficiency is gained by combining an effective second order marching scheme with a Multi-Grid procedure. The relaxation method takes full account of the reduced order of the PNS equations as it behaves like the SLOR for a single Poisson equation. The proposed algorithm is Reynolds number independent and therefore can be applied to the solution of the subsonic Euler equations (which can be viewed as the large Reynolds number limit of the PNS equations). The convergence rates are similar to those obtained by the Multi-Grid solution of a single elliptic equation; the storage is also comparable as only the pressure has to be stored on all levels. Numerical experiments were performed on a linearized model system of PNS equations.

1. INTRODUCTION

Considerable evidence accumulated recently about the applicability of the Parabolized Navier-Stokes equations for high Reynolds number flows with a principal flow direction, see Rubin [1]. The PNS equations are obtained by neglecting the streamwise viscous terms in the Navier-Stokes (NS) equations. When the viscous terms in the circumferential direction are also neglected, one gets the Thin Layer approximation.

The steady PNS equations still have an elliptic nature and therefore the initial value problem in the marching direction is not well posed [2]. A well posed initial-boundary value problem can be formulated by specifying (for example) upstream and side conditions for the velocities and one downstream condition for the pressure. This coupled system of partial differential equations behaves like a single elliptic equation for the pressure. Therefore the PNS equations must be solved globally and cannot be solved by a single sweep marching. The reduced order of the PNS equation can be exploited by constructing an iterative marching method for updating the pressure field only. Such a multiple sweep iteration method has the advantages that the velocity fields are generated during the marching process and only the pressure field has to be stored from sweep to sweep. A considerable saving in storage results. However, a simple minded marching does not result in good convergence properties and sometimes diverges. Israeli and Lin [3] devised a stable marching scheme that behaves like the Successive Line Over Relaxation (SLOR) method for a single elliptic equation.

Here we have used the good smoothing properties of the above mentioned scheme in a Multi-Grid (MG) framework in order to accelerate the convergence of the solution of the PNS (or TL) equations. The marching scheme is implemented using a new stable algorithm which is second order also in the marching direction. The same method can be used without modification for the subsonic Euler equations as the effect of the Reynolds number on the convergence rate is insignificant. For the sake of simplicity, the detailed presentation is restricted to the two-dimensional case. In two dimensions

* Dept. of Computer Science, Technion.
** Dept. of Aeronautical Engineering, Technion.

the PNS and the TL equations are identical and our analysis apply to both cases.

2. FORMULATION

The steady, incompressible and two-dimensional PNS (or TL) equations in cartesian coordinates [x;y] are:

$$U_x + V_y = 0 \tag{1a}$$

$$(U^2)_x + (UV)_y = -P_x + U_{yy}/Re \tag{1b}$$

$$(UV)_x + (V^2)_y = -P_y + V_{yy}/Re \tag{1c}$$

where x is the mainstream direction, Re is the Reynolds number. U and V are the nondimensional velocity components in the x and y directions, respectively. P is the nondimensional pressure.

The two-dimensional NS equations are elliptic of order four - Brandt and Dinar [5]. The PNS are elliptic only of order two like the Poisson equation (the mathematical nature of several two-dimensional and three-dimensional approximations to the Navier-Stokes equations was analysed in [6]). This ellipticity is due to the pressure gradient terms via the continuity equation. A well posed problem can be formulated by defining the boundary conditions as described in Fig. 1. The following Dirichlet conditions may be specified:

* upstream boundary (AB): $U = U_{in}$; $V = V_{in}$ (2a)
* at a solid wall (AD): $U = U_{wall}$; $V = V_{wall}$ (2b)
* at the outer boundary (BC): $U = U_{out}$; $V = V_{out}$ (2c)
* at the downstream boundary (CD): $P = P_{down}$. (2d)

Other boundary conditions can be used but the same number of conditions on each boundary must be kept.

For the present stage of this research the following linear model of equations (1) was used in order to separate the nonlinear effects from the MG procedure and be able to check the linear convergence rates:

$$U_x + V_y = 0 \tag{3a}$$

$$(aU)_x + (bU)_y = -P_x + U_{yy}/Re \tag{3b}$$

$$(aV)_x + (bV)_y = -P_y + V_{yy}/Re \tag{3c}$$

where a and b are known functions of x and y.

3. DISCRETIZATION

It was previously thought that a stable marching scheme for each global iteration must be of the first order because it requires backward differences for the velocities, but forward differencing of the streamwise pressure gradient (to transmit the upstream influence and preserve the elliptic nature of the system). It turns out that this effect can be achieved by a judicious choice of the placement of the variables to be solved at each station. This choice can be explained most easily by taking $a = 1$, $b = 0$

and $1/Re = 0$ in equation (3b), yielding: $U_x = -P_x$. The first order difference scheme previously used for the m-th column (Fig. 2) was:

$$U_m - U_{m-1} = P_m - P_{m+1} \tag{4a}$$

The unknowns were P_m and U_m. The second order method (suggested by Israeli [7]) is:

$$U_{m+1} - U_m = P_m - P_{m+1} \tag{4b}$$

where the unknowns are now P_m and U_{m+1}. This approach was subsequently used by Rubin and Reddy [4].

In addition, one may stagger the velocity V with respect to the other variables as shown in Fig. 2, where the centering points of the different difference equations are also plotted. The differential equations are approximated by central second-order approximations. Whenever needed averaging was used as is usually done for staggered grids.

4. THE MULTI-GRID ALGORITHM

The Multi-Grid technique is a numerical strategy for substantially improving the convergence rate of an iterative procedure. In order to facilitate comparison with theory the accomodative C-cycle MG algorithm was chosen.*

Each MG process consists of three basic parts: relaxation, restriction and interpolation [5].

The Relaxation Scheme

The overall convergence rate of any MG process is greatly influenced by the smoothing properties of the relaxation scheme. It can be shown analytically and experimentally that the usual multiple sweep marching [1] does not have good convergence and smoothing properties because short wave errors are not efficiently smoothed. Israeli and Lin [3] showed that certain modifications in the streamwise momentum equation, which vanish upon convergence, give rise to an iterative scheme which is equivalent, in the linear case, to the SLOR method for one Poisson equation. In the general nonlinear case the modified iterative process is essentially equivalent to the relaxation of a single nonlinear Poisson-like equation for the pressure. The velocities can be viewed as auxiliary variables needed during the marching since they have no "memory" by themselves.

Furthermore, we have automatically gained the good smoothing properties of the line relaxation scheme of a single Poisson equation. The problems associated with the loss of ellipticity of the difference approximation for the Navier-Stokes equations at high Reynolds number [5] are thus avoided and no upstream-weighting or artificial viscosity are required. There results a considerable saving in storage, as well as a simpler relaxation scheme (compare to the distributive relaxation [5]) where the convergence rate is essentially independent of the Reynolds number. We note that the same marching algorithm can thus be used for the (subsonic) Euler equation with the same favorable convergence rate.

---

* Some of the elements of the present approach were used independently by Rubin and Reddy [4]. Detailed comparisons cannot be made because convergence rates and storage estimates were not presented (but see Section 5).

Restriction and Storage Requirements

Let the finite difference approximation of equations (3) on the finest grid M be represented as in [5]:

$$L_j^M \tilde{W}^M(\tilde{x}) = F_j^M(\tilde{x}) \tag{5}$$

where $\tilde{x} = [x,y]$, $\tilde{W}^M = [U^M, V^M, P^M]^T$ is the exact solution of the difference equations, and j is the number of the differential equation, $j = 1,2,3$.

The problem is transferred from the current level k to a coarser level $k-1$, see Fig. 3, by correcting the right hand side of (5):

$$F_j^{k-1}(\tilde{x}) = L_j^{k-1}(\tilde{I}_{j,k}^{k-1}\tilde{w}^k(\tilde{x})) + I_{j,k}^{k-1}[F_j^k(\tilde{x}) - L_j^k\tilde{w}^k(\tilde{x})] \tag{6}$$

in the Full Approximation Storage (FAS) mode. $\tilde{w}^k(\tilde{x})$ is an approximation to $\tilde{W}^k(\tilde{x})$ in the finer level. $I_{j,k}^{k-1}$ and $\tilde{I}_{j,k}^{k-1}$ are proper restriction operators for equation j.

The term in square bracket in equation (6) is the residual of the j-th equation. For the present marching scheme there is no residual in the continuity and in the y-momentum equations since they are solved exactly in each step. The residual of the x-momentum equation results only from the streamwise pressure gradient term and its computation needs only one substraction. $\tilde{I}_{j,k}^{k-1}$ was chosen to be linear interpolation, which yields for the continuity equation: $L_1^{k-1}(\tilde{I}_{1,k}^{k-1}\tilde{w}^k(\tilde{x})) = 0$. $I_{j,k}^{k-1}$, $j = 1,2$ is computed by averaging in both the x and y directions. $I_{3,k}^{k-1}$ is a simple injection.

In summary, equation (6) takes the following terms:

$$F_1^{k-1}(\tilde{x}) = 0 \tag{7a}$$

$$F_2^{k-1}(\tilde{x}) = L_2^{k-1}(\tilde{I}_{2,k}^{k-1}\tilde{w}^k(\tilde{x})) + I_{2,k}^{k-1}(F_2^k - L_2^k\tilde{w}^k(\tilde{x})) \tag{7b}$$

$$F_3^{k-1}(\tilde{x}) = L_3^{k-1}(\tilde{I}_{3,k}^{k-1}\tilde{w}^k(\tilde{x})) \tag{7c}$$

Two consequences should be emphasized:

(a) Only two corrections $(F_2^{k-1}(\tilde{x}), F_3^{k-1}(\tilde{x}))$ have to be computed and stored.

(b) All the dependent variables must be transferred in order to compute the corrections $(L_j^{k-1}(\tilde{I}_{j,k}^{k-1}\tilde{w}^k(\tilde{x})), j = 2,3)$. Since only the pressure is stored, these corrections must be computed during the marching process.

It follows that in addition to the pressure on all grids, one has to save one correction term for each momentum equation on the coarser grids. Assuming N computational points on the finest grid, a simple-minded estimate gives 32N/7 storage locations for the three-dimensional NS Multi-Grid solution, and 11N/7 for the PNS marching MG solution. For the two-dimensional case the corresponding figures are 13N/3 and 7N/3.

Interpolation

Since the present marching scheme generates the velocity field from the pressure, only the correction to the pressure must be interpolated back to the fine grid.

5. RESULTS

In order to check the MG algorithm we choose the following analytical solution. It satisfies the continuity equation but gives rise to source terms in the momentum equations:

$$U = A + (x+y)^m;\quad V = -(x+y)^m;\quad P = -(E1+E2)(x+y)^m \tag{8a}$$

where a and b from equations (3) are defined by:

$$a = E1 + F(x+y)^n;\quad b = E2 - F(x+y)^n \tag{8b}$$

and $E1 = 1$; $E2 = .2$; $F = .2$; $A = 5$; $Re = 1000$; $m = 4$; $n = 2$. The coarsest grid consists of 4×4 intervals.

Figure 4 compares the MG convergence history of different relaxation schemes. In the MG solutions three levels were involved ($M = 3$). The horizontal coordinate gives the number of Work Units (WU), where each work unit is equivalent to one global iteration on the finest grid. The vertical coordinate gives the logarithm of the dynamic residual ε. The dots show the solution of the equivalent Poisson equation (with the same solution for the pressure but with Dirichlet condition over all the boundaries). The linearized PNS equations were solved with and without the streamwise pressure gradient correction of [3]. The corresponding (17×17 points) single grid convergence history is plotted for comparison. The corrected discrete equations and the Poisson equation exhibit very similar convergence whereas the convergence of the unmodified equations is much worse. Upon increasing the number of grids in the unmodified equations the convergence deteriorates.

In Fig. 5 the convergence history of the corrected linearized PNS equations is shown for different number of levels with the same finest grid (65×65). The Reynolds number independence of the suggested solution method is clearly manifested in Fig. 6.

The MG solution is superior to single grid solutions in all cases. The gain in terms of work units is between one order of magnitudes for a grid of 17×17 points to two orders of magnitude for 65×65 points. Table I summarizes the convergence factor per relaxation work, $\dot{\mu}$ (see [5] for definition), for several solution methods.

From the single figure of reference [4] where MG results were presented, we estimated a MG convergence rate of 0.87 which is worse than our results, however as mentioned before, this comparison may be very inaccurate.

6. CONCLUSIONS

An efficient and storage saving MG procedure for the solution of the PNS, TL and the subsonic Euler equations was presented. The numerical scheme is second order and the convergence rate is Reynolds number independent. Based on current experience [5], it can be assumed that the fully nonlinear version of the present method would have similar convergence properties.

Results of the nonlinear version will be reported elsewhere. We are now working on the extension to three-dimensions and to nonzero Mach number.

ACKNOWLEDGEMENT

The first author was partially supported by AFOSR No. F49620-83-C-0064. The project was supported by Stiftung Volkswagenwerke.

REFERENCES

[1] Rubin, S.G.: Incompressible NS and PNS Solution Procedures and Computational Techniques, Von Karman Institute Lecture Notes, 1982.

[2] Israeli, M., Reitman, V., Salomon, S. and Wolfshtein, M.: On the Marching Solution of the Elliptic Equations in Viscous Mechanics, Proc. of 2nd Int. Conf. on Num. Methods in Laminar and Turbulent Flows, pp. 3-14, 1981.

[3] Israeli, M. and Lin, A.: Numerical Solution and Boundary Conditions for Boundary Layer Like Flows, Proc. of the ICNMFD8, pp. 266-272, Lecture Notes in Physics, 170, Springer-Verlag, 1982.

[4] Rubin, S.G. and Reddy, D.R.: Global PNS Solutions for Laminar and Turbulent Flow, AIAA Paper 83-1911, 1983.

[5] Brandt, A. and Dinar, N.: Multigrid Solution to Elliptic Flow Problems, Numerical Methods for PDE's, pp. 53-147, Academic Press Inc., 1979.

[6] Rosenfeld, M.: An Investigation of the Hierarchy of Approximations to the Multi-Dimensional Incompressible Navier-Stokes equations, TAE Report, Technion, Haifa, 1983.

[7] Israeli, M., NASA Lewis Seminar, July, 1982.

TABLE I

The Convergence Factor ($\dot{\mu}$)

for Several Solution Methods and Number of Grids

M	Finest Grid	Poisson	With Correction	Without Correction	Relaxation Only
2	9 × 9	0.518	0.604	0.656	0.905
3	17 × 17	0.625	0.600	0.784	0.973
4	33 × 33	0.648	0.621	0.894	0.979
5	65 × 65	0.696	0.707	0.906	0.986

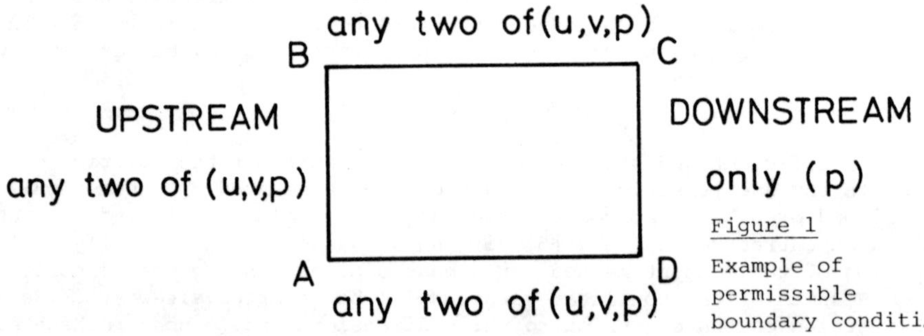

Figure 1

Example of permissible boundary conditions.

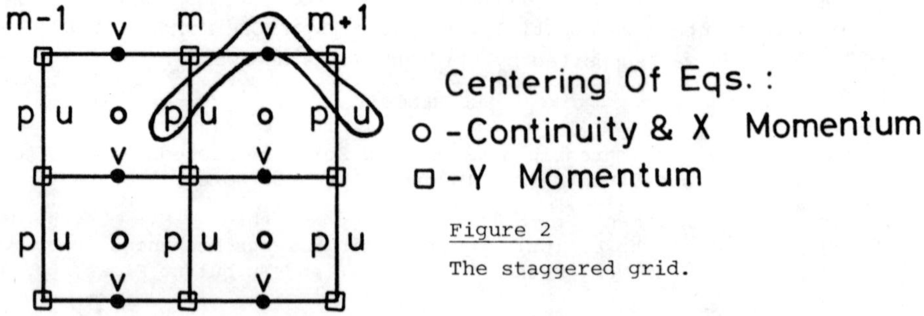

Figure 2

The staggered grid.

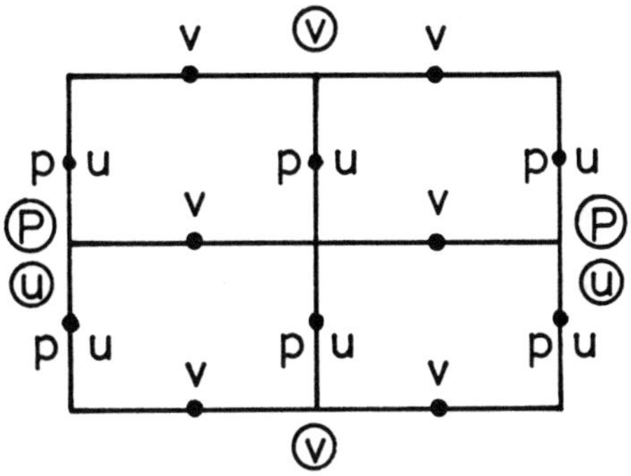

Figure 3

Relative placement of variables on two successive grids.

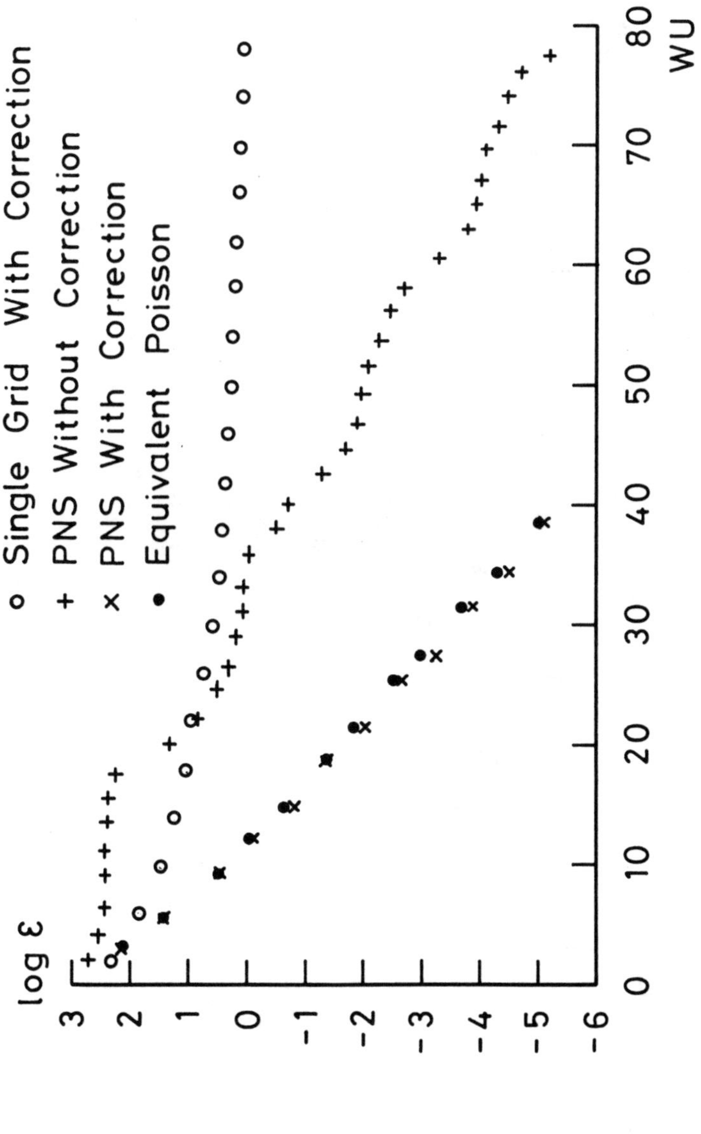

Figure 4 Convergence history for different relaxation schemes (M = 3).

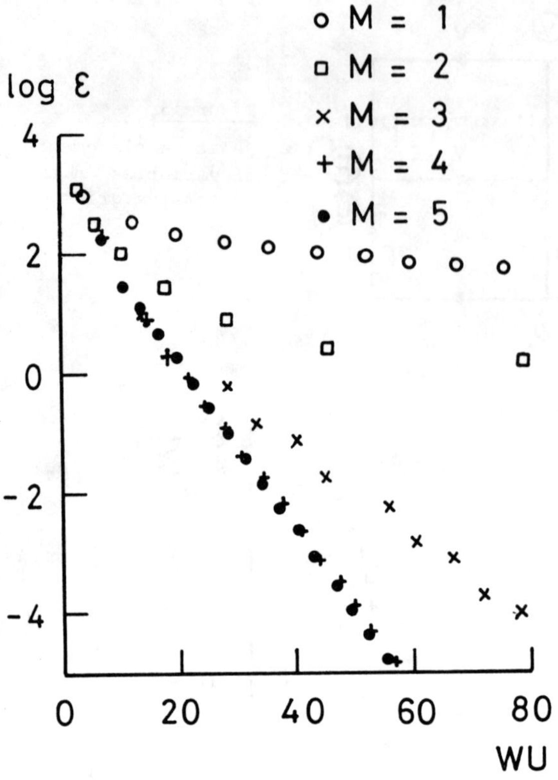

Figure 5

Convergence history for different number of levels (finest grid 65 × 65).

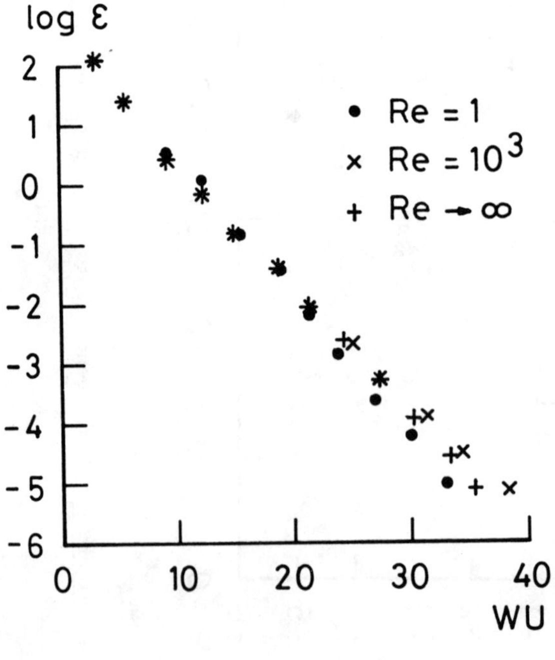

Figure 6

Convergence history for several Reynolds numbers (M = 3).

PSEUDOSPECTRAL CALCULATIONS OF TWO-DIMENSIONAL TRANSONIC FLOW

Wen-Huei Jou
Flow Research Company, Kent, Washington, U.S.A.
Antony Jameson
Princeton University, Princeton, New Jersey, U.S.A.
Ralph Metcalfe
Flow Research Company, Kent, Washington, U.S.A.

SUMMARY

A hybrid pseudospectral-finite difference scheme is used to calculate transonic flow over a two-dimensional object using the Euler equations. The exterior of the object is mapped to the interior of a circle. The flow field variables are discretized using a Fourier series in the circumferential direction, while a central finite difference scheme is used in the radial direction. We used a four-stage Runge-Kutta scheme including a filter and a residue-smoothing process. Transonic flows over a circular cylinder as well as a Karman-Trefftz airfoil were computed. The results are compared to those from finite volume calculations. It is found that the pseudospectral calculations are able to produce shocks with no internal structure, and fewer grid points are needed to obtain the required accuracy.

INTRODUCTION

In recent years, there has been strong interest in computations of transonic flows using the time-dependent Euler equations. This interest stems in part from the possibility of shock-generated vorticity in the flow field and in part from the interest in numerical methods for a nonlinear hyperbolic system. Most of the numerical methods are finite difference in nature and are second order in accuracy. To stabilize the computation and to smooth the dispersive error for unsteady computations, either artificial dissipative terms are added to the equations or a built-in dissipative mechanism is included in the numerical scheme. These dissipative terms usually cause the shock wave to span across three to four grid points. To capture a shock with reasonable accuracy, one is forced to use fairly dense grid distributions over the region where the shock wave is expected to be.

The pseudospectral method is an alternative to the finite difference method. It has been applied successfully to many smoothly varying flows. The numerical analysis of the method has been given in detail by Gottlieb and Orszag in a monograph [1]. Because of its high rate of convergence, the method usually requires relatively few terms of the basis functions for accurate computations. In addition to the spatial accuracy, the dispersive error for unsteady computations is also minimized.

Recently, efforts have been made to apply the pseudospectral method to flows with shock waves. Gottlieb, Lustman, and Orszag [2] have demonstrated the feasibility of the pseudospectral method through the solution of a one-dimensional shock tube problem. By using the shock-capturing technique, they showed that the shock wave can be resolved within one grid point. The Gibbs phenomenon error due to the discontinuity can be filtered to improve the accuracy. Gottlieb, Lustman, and Streett [3] have attempted the two-dimensional problem of the reflection of an oblique shock from a wall. The results from this investigation are encouraging. Using a fairly sparse grid, they showed that the shock wave can be resolved within one grid point. The accuracy of the solution as compared to the exact solution is reasonable considering the sparseness of the grid points.

In the present work, we shall consider steady transonic flows around an airfoil by solving the Euler equations. We shall address problems of applying the pseudospectral method to flows around a complex geometry, including the development of a time-stepping scheme, and enhancement of the stability by residue averaging. Numerical experimentation has been used to confirm convergence with a small number of basis functions, and also the capability to treat shock waves with the aid of filtering.

GOVERNING EQUATIONS AND BASIC APPROACH

The basic approach is to map the exterior of an airfoil to the interior of a circle. Polar coordinates in the mapped plane will be used. Spectral decomposition of the solution can be used in the mapped plane. To serve this purpose, the Euler equations in the mapped plane are written in fully conservative form by using both physical and contravariant velocities. These equations are

$$\frac{\partial}{\partial t}(\vec{q}) + \frac{\partial}{\partial X}(\vec{F}) + \frac{\partial}{\partial Y}(\vec{G}) = 0 \qquad (1)$$

where

$$\vec{q} = J\begin{bmatrix}\rho\\ \rho u\\ \rho v\\ E\end{bmatrix} ; \quad \vec{F} = \begin{bmatrix}\rho U\\ \rho u U + y_Y P/\gamma M_\infty^2\\ \rho v U - x_Y P/\gamma M_\infty^2\\ \rho H U\end{bmatrix} ; \quad \vec{G} = \begin{bmatrix}\rho V\\ \rho u V - y_X P/\gamma M_\infty^2\\ \rho v V + x_X P/\gamma M_\infty^2\\ \rho H V\end{bmatrix} ;$$

$$\begin{bmatrix}U\\ V\end{bmatrix} = \begin{bmatrix}y_Y & -x_Y\\ -y_X & x_X\end{bmatrix}\begin{bmatrix}u\\ v\end{bmatrix} ; \quad H = [(\gamma - 1)P + E]/\rho ;$$

(x,y) are Cartesian coordinates in the physical plane, (X,Y) are polar coordinates in the mapped plane, ρ is the density, (u,v) are velocity components, (U,V) are unscaled contravariant velocity components, E is the energy, H is the specific enthalpy, P is the pressure, J is the Jacobian of the transformation, M_∞ is the free-stream Mach number, and γ is the specific heat ratio. P, ρ and the velocity vector (u,v) are nondimensionalized by their values at the free-stream condition, E and H are nondimensionalized by the free-stream internal energy $\rho_\infty C_V T_\infty$, and other variables at the free-stream condition are computed from these variables.

The physical boundary conditions are defined by the solid-wall condition on the airfoil surface and the fact that the disturbances generated by the airfoil propagate outward to infinity. The numerical implementation of these physical boundary conditions will be discussed later.

This initial effort employed a hybrid spectral-finite difference discretization to gain some experience with the problem. In the circumferential direction, the variables are expanded in a Fourier series because of the periodic nature of the problem. In the radial direction, a central finite difference scheme is used. This hybrid scheme is designed to answer some questions in applying the pseudospectral method to transonic flows. The application of the pseudospectral method to a realistic airfoil shape can be demonstrated by this scheme. Since we expect that the shock wave will be normal to the airfoil surface, the discontinuity is mainly in the X direction. Adequate resolution of the shock wave can be achieved and the question of convergence of the spectral series can be answered using the hybrid scheme. Other properties, such as the time-stepping scheme, convergence acceleration, and the filtering technique can also be studied with the hybrid scheme. We shall use a Karman-Trefftz airfoil for this work because of its simple analytical mapping from the physical plane to the interior of the circle. The method can easily be extended to an airfoil of arbitrary shape by using a truncated complex series to map the profile to a circle.

NUMERICAL SCHEME

A computational mesh is created by equally dividing the (X,Y) coordinates in the mapped plane. In the mapped plane, the spatial derivatives in X at each mesh point are evaluated by application of a fast Fourier transform. The derivatives in Y are evaluated by second-order central finite differences. Evaluation of the elements in the transformation matrices (x_X, x_Y, y_X, y_Y) is performed in the same manner. The singularity of the transformation at the trailing edge is avoided by placing it between two mesh points.

By using this method of evaluating spatial derivatives, the governing equations are converted to a system of ordinary differential equations in time. These equations can be solved numerically by using any of a variety of well-developed techniques for the solution of ordinary differential equations. An approximate fourth-order Runge-Kutta scheme is used in this work. The algorithm is given by the following equations:

$$\vec{q}^{(n)} = \vec{q}^{o} + \frac{1}{(5-n)} \vec{R}^{(n-1)} \quad ; \; n = 1, \ldots, 4 \qquad (2)$$

where $\vec{q}$ represents the flow variables at the mesh points, R represents the terms with spatial derivatives in the equations, and n denotes the Runge-Kutta step. This scheme has been shown to be stable for a CFL number less than 2.8 [4] for a finite difference scheme, and is stable for our hybrid scheme with a CFL number less than or equal to 2. Following Jameson, Schmidt, and Turkel [4], a local time step that is restricted by the CFL number is used. Because of this, no physical interpretation should be given to the transient solutions.

FILTERING

Filtering is required in suppressing the Gibbs error. A Schumann filter used by Gottlieb, Lustman, and Streett [3] and given by the following formula has been applied every 35 time steps at a CFL number of 3.5 (see later section on convergence acceleration).

$$q_K = 0.25(q_{K+1} + 2q_K + q_{K-1}); \; K = i,j \tag{3}$$

where i and j denote indices of the discrete points in the X and Y directions, respectively. At the shock wave, one-sided filtering in the X direction is applied to preserve the sharpness of the shock wave. The Schumann filter is equivalent to a first-order artificial viscosity. However, the filter is applied only every 35 time steps. The order of the error is higher than first order. No high-mode smoothing [3] has been applied.

BOUNDARY CONDITIONS

The numerical implementation of boundary conditions for a hyperbolic system of partial differential equations is an active research subject in itself. Essentially, on a boundary Y_o, there are four characteristics that correspond to the speeds q_n, q_t, $q_n - c$ and $q_n + c$. The respective characteristic variables are $p - c_o^2\rho$, q_t, $p - \rho_o c_o q_n$, and $p + \rho_o c_o q_n$, where the subscript o stands for the quantities at the previous time step. Only the characteristic variables carried on the outgoing characteristics from the interior of the fluid domain can be computed from the governing equations. The characteristic variables carried on the incoming characteristics must be replaced by the appropriate boundary conditions. The flow quantities can then be recovered from the combination of the boundary conditions and the outgoing characteristic variables.

On the solid surface Y = 0, there is only one incoming characteristic. Let (M,N) be the momentum along the surface and normal to the surface, respectively, and ΔQ be the symbol for the temporal change of physical quantities Q as computed by the interior formula, with the subscript c denoting the quantities computed by the interior formula. The following formulae for the physical quantities at the boundary points can then be given.

$$\Delta P = \Delta E + \gamma(\gamma-1)M_\infty^2 \left(-\frac{M}{\rho}\,\Delta M + \frac{1}{2}\,\frac{M^2}{\rho^2}\,\Delta\rho\right) \tag{4}$$

$$P_c = P_o + \Delta P \tag{5}$$

$$\left(\frac{M}{\rho}\right)_c = M_o/\rho_o + (\Delta M - \Delta\rho \cdot M_o/\rho_o)/\rho_o \tag{6}$$

$$\rho_c = \rho_o + \Delta\rho \tag{7}$$

$$P^{(n)} = P_c - \gamma M_\infty^2 \left(\Delta N \cdot c_o\right) \tag{8}$$

$$\rho^{(n)} = \rho_c + \frac{1}{\gamma M_\infty^2} \quad (P^{(n)} - P_c)/c_o^2 \tag{9}$$

$$M^{(n)} = \rho^{(n)} \cdot (M/\rho)_c \tag{10}$$

$$N^{(n)} = 0 \tag{11}$$

$$E^{(n)} = P^{(n)} + \frac{1}{2}\gamma(\gamma-1)\, M_\infty^2 \left[M^{(n)}\right]^2 \Big/ \rho^{(n)} \tag{12}$$

where the superscript n stands for the newly advanced quantities, and all velocities are nondimensionalized by the free-stream velocity.

At the far field boundary, the treatment is essentially the same as that used by Jameson, Schmidt, and Turkel [4] except that the "extrapolated" quantities as defined in that paper are those computed by the interior computations.

CONVERGENCE ACCELERATION

To increase the stability of the time-stepping scheme, an additional "residue-smoothing" process [5] has been implemented. After the residue R has been evaluated at every mesh point, the residues are smoothed by a linear transformation defined as follows:

$$\overline{R} = (1 - \varepsilon\delta_X^2)^{-1}\ (1 - \varepsilon\delta_Y^2)^{-1}\ R \tag{13}$$

where δ_X and δ_Y are conventional finite difference operators in X and Y, and ε is the parameter for the residue-averaging process. The new modified residue field $\overline{R}$ is used to advance the solution in time. This process alters the time-dependent solution without changing its steady state. To bring out the essentials of the residue-averaging process, a simple wave equation is considered:

$$\phi_t + c\phi_x = 0 \quad . \tag{14}$$

The residue-averaging process as described is equivalent, to the lowest order, to adding an additional term to the original simple wave equation and converting it to the following equation:

$$\phi_t + c\phi_x - \varepsilon(\Delta x)^2\ \phi_{txx} = 0 \quad . \tag{15}$$

The dispersion relation for this equation can be given as

$$\frac{\omega}{k} = \frac{c}{1 + \varepsilon k^2 (\Delta x)^2} \tag{16}$$

where ω is the frequency and k is the wave number. By increasing the parameter ε, the wave speed for the high wave number component is substantially increased. This decrease in wave speed for the dangerous short waves contributes to the substantial increase in the time step. In fact, Equation (15) is the linearized form of a model equation for long dispersive waves discussed by Benjamin, Bona, and Mahony [6], who pointed out the numerical advantage of this equation over the Korteweg-de-Vries equation. Other means of manipulating the dispersion relation to gain stability have been suggested (e.g., [7]). However, these methods do not recover the original equation in steady state, although the error is of higher order. The residue-averaging process substantially extends the stability boundary. A CFL number of 3.5 has been used without any difficulty.

COMPUTED RESULTS

For testing the solution algorithm, flows around a circular cylinder are computed. The pressure distribution for a sub-critical flow with $M_\infty = 0.39$ is given in Figure 1. The computation is performed on a 64x24 grid (64 points in the circumferential direction, 24 points radially). It has been computed without filtering. A supercritical case with $M_\infty = 0.45$ is also computed, and the results are shown in Figure 2. Filtering is performed every 35 time steps with a CFL number of 3.5. The results agree with a finite volume calculation by Jameson, Schmidt, and Turkel, [4]. The shock wave has no internal structure and is sharply defined.

A Karman-Trefftz airfoil with the following transformation from the mapped plane ζ to the physical plane z is chosen for calculations.

$$\frac{z}{\kappa L} = \frac{(1+L\zeta)^\kappa + (1-L\zeta)^\kappa}{(1+L\zeta)^\kappa + (1-L\zeta)^\kappa} \quad ; \kappa = 1.9 \tag{17}$$

$$L = (1 - \eta_o^2)^{1/2} - \xi_o \quad ; \zeta_o = (\xi_o, \eta_o) = (-0.1, 0) \tag{18}$$

Supercritical non-lifting flows with $M_\infty = 0.75$ are computed on a 64x24 grid. The results are shown in Figure 3, together with the results from a finite volume calculation. The hybrid calculation shows again a sharply defined shock wave. The agreement between the two calculations is very good. In particular, the positions of the shock as defined by the midpoint of the structure show close agreement. There are discrepancies immediately behind the shock wave, however. The source of these discrepancies is not clear. The pressure ratio across the shock wave using the pseudospectral calculation has been checked against that using the Rankine-Hugoniot relation based on the upstream Mach number. The error is less than 4 percent.

To demonstrate the convergence of the Fourier series, the same case is computed on a 32x24 grid. The results are shown in Figure 4, together with the results of calculations on a denser mesh using a finite volume calculation. The accuracy of the 32x24 calculation is quite good. The shock resolution of the sparse mesh calculation is comparable to that of the 64x24 finite volume calculation. As expected, the finite volume calculation on the sparse grid does not produce acceptable results.

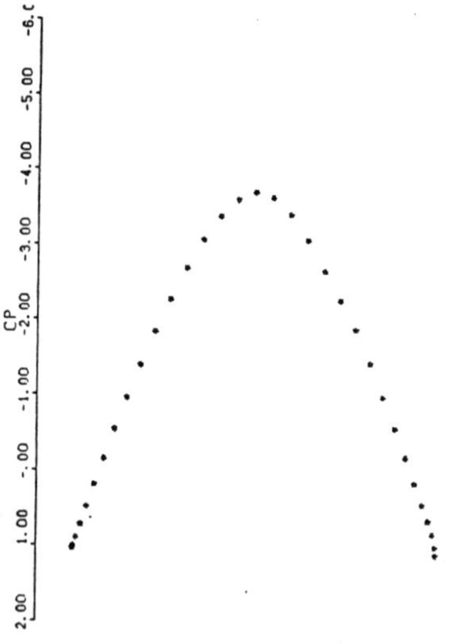

Figure 1. Pressure distribution for a subcritical flow over a cylinder at Mach 0.39

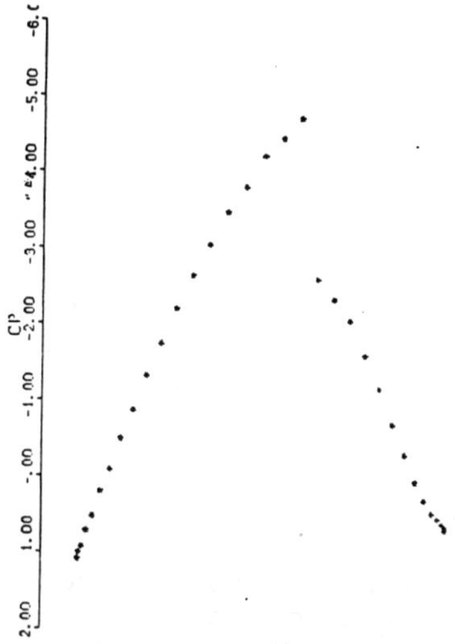

Figure 2. Pressure distribution for a supercritical flow over a cylinder at Mach 0.45

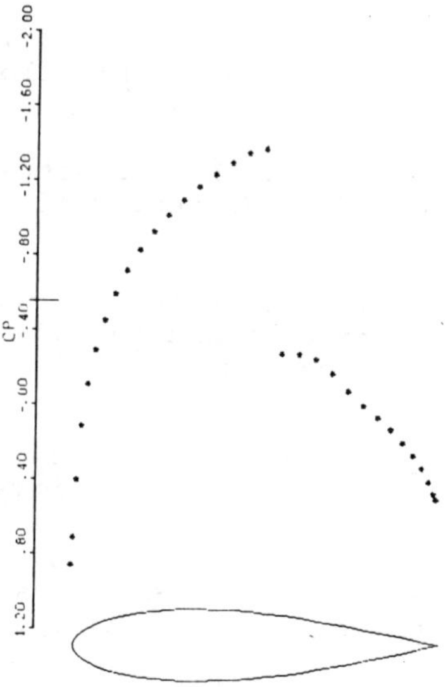

Figure 3. Pressure distribution on a Karman-Trefftz airfoil at Mach 0.75 using a 64x24 grid (350 steps)

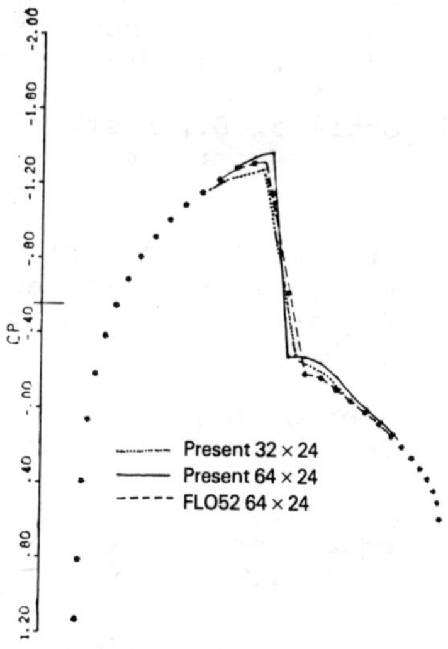

Figure 4. Comparison of results with a finite volume scheme for a Karman-Trefftz airfoil at Mach 0.75

CONCLUSIONS AND ACKNOWLEDGEMENTS

Several conclusions can be drawn from the present investigation.

(1) In computing flows with shock waves, the Gibbs error can be filtered to produce accurate results. Because of the rapid convergence of the Fourier series, fewer grid points are required than with the lower order difference-type scheme.

(2) A shock wave without internal structure can be produced. This capability also contributes to the accuracy of the method in that fewer grid points are required to resolve the shock wave.

(3) Application of the pseudospectral method to flows around a realistic geometry is possible using a mapping technique.

The authors are indebted to Mr. Morton Cooper of Flow Research Company for suggesting the problem. Discussions with Professor Steven Orszag have also been very helpful. This work was supported by the Air Force Office of Scientific Research and the Office of Naval Research under Contract No. F49620-82-C-0022.

REFERENCES

[1] Gottlieb, D., and Orszag, S., Numerical Analysis of Spectral Methods: Theory and Applications, NSF-CBMS Monograph No. 26, Society of Industrial and Applied Mathematics (1977).

[2] Gottlieb, D., Lustman, L., and Orszag, S. A., "Spectral calculations of one-dimensional inviscid compressible flows," SIAM J. Sci. Stat. Comput., 2, 3 (1981) pp. 296-310.

[3] Gottlieb, D., Lustman, L., and Streett, C. L., "Spectral methods for two-dimensional shocks," ICASE rep. 82-38 (1982).

[4] Jameson, A., Schmidt, W., and Turkel, E., "Numerical solutions of the Euler equations by finite volume methods using Runge-Kutta time-stepping schemes," AIAA paper 81-1259 (1981).

[5] Jameson, A., and Baker, T., "Solution of the Euler equations for complex comfigurations," AIAA paper 83-1929 (1983).

[6] Benjamin, T. B., Bona, J. L., and Mahony, J. J., "Model equations for long waves in non-linear dispersive systems," Phil. Trans. Roy. Soc. London Ser. A., 272 (1972) pp. 47-78.

[7] Gottlieb, D., and Turkel, E., "On time discretization for spectral methods," Stud. Appl. Math., 63 (1980) pp. 67-86.

MULTI-GRID SOLUTION OF TIME-DEPENDENT INCOMPRESSIBLE FLOWS

Arne Karlsson and Laszlo Fuchs
Dep. of Gasdynamics
The Royal Institute of Technology
S-100 44 STOCKHOLM, SWEDEN

SUMMARY

We consider the stream function and vorticity form of the governing equations for non-steady incompressible viscous flows. We utilize in our multi-grid scheme the fact that the time scales of the transients are related to the spatial scales of the different components of the transients. In this method both the time step as well as the mesh spacing are varied simultaneously. The method is fast and machine round-off accuracy (reduction of the residuals by 16 orders of magnitude) can be achieved by a computational effort which is equivalent to about 100 succesive point relaxation sweeps on the difference equations. The accuracy of the results with respect to the time-steps and the spatial mesh sizes for the non-steady driven cavity flow is analysed. The flow pattern at some phases of the oscilatory motion are also shown.

INTRODUCTION

The simulation of time-dependent flow problems is more difficult than that of steady-state flows. In this paper we restrict the discussion to flows which either tend, asymptotically, to a steady state or have a periodical solution (in time). For steady-state problems only the length scales must be resolved to get a meaningful solution. In the non-steady state case, the transients may or may not be resolved, but still the steady-state or the periodical solution may be accurate.

The physical transients can be decomposed into Fourier-components in space with time-dependent amplitudes. It has been shown in a previous paper [1], for the heat equation, that the decay of the amplitude of different components of the transient is related to the spatial frequency of that component. Components with short spatial wave-lengths are associated with rapidly decreasing transients, while transients with long spatial wave-lengths decay slowly. Such a behaviour is very suitable for Multi-Grid (MG) methods and is not utilized by standard methods.

Standard methods use, usually, constant or variable time steps but all the computations are done on one, fine spatial grid. From accuracy point of view it is unnecessary to use such fine time-space meshes after the initial elimination of the high-frequency error components. Furthermore, the convergence of standard iterative methods is slower on finer grids than on a coarser one. These facts make accurate solution of time-dependent flow problems with such methods very expensive.

An approach which is suitable for solving slowly varying periodical problems, is to use semi-spectral methods. The time variations are Fourier transformed, and the series is truncated keeping only a limited number of terms. The resulting, coupled-system of partial differential equations has to be solved by a finite difference method. Such a method has been used by Duck [2], where at most 5 terms of the Fourier expansion are used. The method in [2] was applied to the computation of the periodical flow in a driven cavity.

In this work we solve the same problem by using the time-space MG method. This MG method is sketched shortly here. More detailes about the method and its numerical efficiency are given in [1].

Some computed results for the driven cavity flow with a periodically oscilating driving wall are also given. It has been found, that these results differ qualitatively from those which are obtained by Duck [2]. Our results are checked for program errors and the iterative solution has been terminated at machine round-off accuracy.

In the following we analyse the numerical method with regard to its accuracy with respect to the time-step size and the mesh spacing. It has been found that the numerical solution indeed attains the asymptotic accuracy expected from the finite difference approximations. It has been found also that the error is mainly due to the spatial finite difference approximations. Using higher order finite difference approximations would give better accuracy only if fine enough meshes are used. For medium grids, like those we use here, the accuracy level with low order finite differences is about the same as one would obtain using higher order methods (see [3]).

GOVERNING EQUATIONS AND BOUNDARY CONDITIONS

We consider plane incompressible viscous flows. The Navier-Stokes equations are written as a coupled system of a vorticity transport-equation and a Poisson equation. In a cartesian coordinate system (x,y) the governing equations are given by

$$S\omega_t + u\omega_x + v\omega_y = \frac{1}{Re}(\omega_{xx} + \omega_{yy}) \tag{1}$$

$$\psi_{xx} + \psi_{yy} = \omega \tag{2}$$

where Re $(=UL/\nu)$ is the Reynolds number and S $(=L/UT)$ is the Strouhal number, with U, L and T as the characteristic velocity, length and time, respectively, while ν denote the kinematic viscosity of the fluid. The vorticity (ω) and the stream function (ψ) are related by:

$$\omega = v_x - u_y \quad ; \quad u = -\psi_y \quad ; \quad v = \psi_x \tag{3}$$

The following boundary conditions are used:
i) no-slip flow at solid boundaries, and
ii) the velocity components are given on in- and out-flow boundaries.
From these boundary conditions on the velocity vector, the stream function can be computed on all the boundaries. The vorticity on the boundaries satisfies equation (2) and it is computed during the solution process.

In this paper we consider flows which vary in time periodically, with a period T. That is, any dependent variable f, is assumed to satisfy

$$f(x,y,t) = f(x,y,t+T) \tag{4}$$

for all times t. Condition (4) may be used as a convergence test.

FINITE-DIFFERENCE APPROXIMATIONS

Equations (1) and (2) are discretized using second order accurate central differences except in the convective terms of equation (1) where first order up-wind differences are used. The time discretization is done by a Crank-Nicholson scheme, which is also second order accurate.

The boundary values of the vorticity are computed by using equation (2), and the boundary conditions on the velocity. At the boundaries the central five-point formula of the Laplace operator, cannot be used. Instead we use the following relations at the solid walls:
a. For a boundary parallel to the x-axis:

$$\omega_w = \frac{2(\psi_{w+j}-\psi_w)}{(\Delta y)^2} + \frac{2j}{\Delta y}u_w + O(\Delta y) \qquad (5a)$$

where subscript w stands for the wall and subscript $w+j$ stands for the nodal point inside the flow field adjecent to the wall. $j=1$ when the computational domain is above the wall (positive y-direction) and $j=-1$ when the computational domain is below the wall.
b. Similarly, for a wall parallel to the y-axis:

$$\omega_w = \frac{2(\psi_{w+j}-\psi_w)}{(\Delta x)^2} - \frac{2j}{\Delta x}v_w + O(\Delta x) \qquad (5b)$$

where $j=1$ and $j=-1$ when the computational domain is to the right and to the left of the boundary, respectively.

NUMERICAL PROCEDURE

In a previous work [1] we have shown by a Fourier decomposition of the error of evolution in the approximation to the solution that there is a direct relationship between the time constant of the error components and their spatial wave length. High frequency components of the error, which needs a fine spatial mesh to be resolved decay faster than error components with a low frequency. Such a behaviour is very suitable for MG solution procedures where both the time step and the space mesh size are increased, simultaneously.

The time-space MG method has the following basic steps.

i) In each time step the spatial problem is solved by an inner MG solver. (The inner MG solver updates the vorticity transport equation and the Poisson equation for the stream function in each time step. It is similar to the MG solver described in [4].)

ii) When the fast transients are smoothed out one transfers the problem to a coarser grid in space and larger time step.

Steps i) and ii) are repeated until the coarsest grid is reached.

iii) The correction to the approximation is interpolated to the finer grids where step i) is repeated.

Steps iii) and i) are repeated until the finest grid is reached. This MG-cycle is repeated until convergence is obtained. More details are given in reference [1].

COMPUTATIONAL RESULTS

The oscilating flow in a driven cavity (Figure 1) has been used as a model problem. The boundary conditions for this problem are

$$u = v = 0 \qquad \text{on all walls}$$

except

$$u(x,a) = U\cos(2\pi t/T)$$

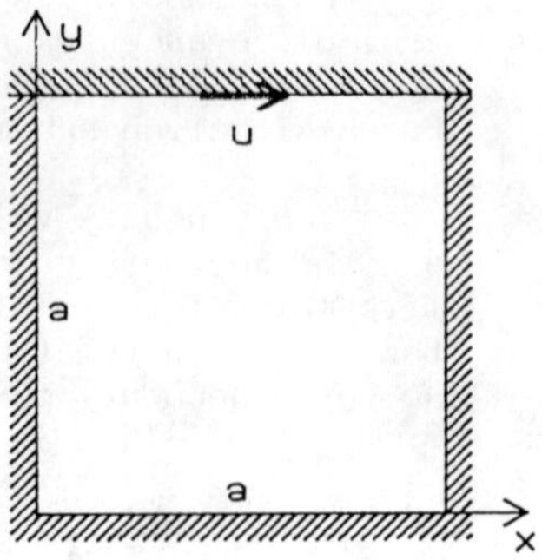

Figure 1

From these conditions on the velocity vector, it is possible to define a (constant) stream function value on all the walls of the cavity. The wall vorticity is computed by equation (5). The Reynolds number is based upon the dimension of the cavity (a) and the maximal speed of the driving wall (U). The Strouhal number ($S=a/UT$) uses the period T. We assume in the following that all the terms in equation (1) has the same importance, and hence we choose $S=1$ for all the cases presented here.

The convergence rate of the MG solver was studied in [1]. It was shown that for Re=200 machine round-off accuracy (reduction of the residuals by 16 orders of magnitude) was obtained after a computational effort equivalent to about 100 succesive point relaxation sweeps on the finest time-space grid. We estimated that standard one-level methods would need several thousands of succesive point relaxation sweeps to obtain a similar accuracy. It was also shown, by comparing the efficiency of the method for Re=20 and 200 that the superiority of the MG method over one-level methods becomes greater for larger Reynolds numbers. For low Reynolds numbers, on the other hand the dissipation is great enough and all the components of the transient are decreasing rapidly.

To check accuracy, we have solved the steady-state driven cavity problem. The solution of the time-dependent program has been compared to a solution computed on a finer mesh (128×128 intervals) using a primitive variable MG-Navier-Stokes solver [5]. The agreement between these solutions was as good as expected due to truncation errors.

The accuracy levels of the solutions obtained by using different time-space grids has also been studied. First a series of computations have been made, using a fixed spatial mesh (with 40×40 mesh intervals) and varying the time step. A mean vorticity has been computed for each of these cases, and an extrapolated value (to $O(\Delta t^5)$ where Δt is the shortest time step in the series of computed solutions) has been computed. The error due to time discretization is defined as the absolute value of the difference between the extrapolated and computed mean vorticities normalized with the extrapolated value. Figure 2 show this error as the function of the number of time steps (N) per period. From this figure it is seen that the asymptotic behaviour of the method, as $\Delta t \downarrow 0$, is as expected from theory, i.e. the method is of second order in time.

A second series of computations have been made, using a fixed time step-size ($\Delta t = 1/8$) and varying the spatial mesh size. Extrapolated values (to $O(h^2)$ where $h = \Delta x = \Delta y$ is the finest mesh size) have been computed at grid points shared by all the grids. A mean relative error due to space discretization is defined in a similar way as above. Figure 3 show this error as the

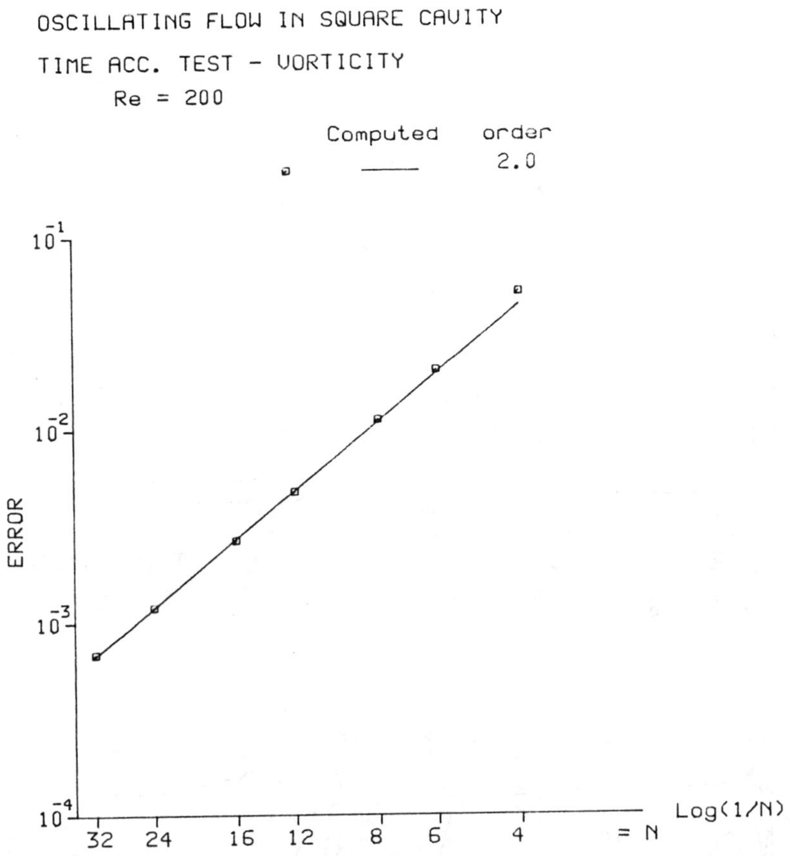

Figure 2: Error vs. number of time steps

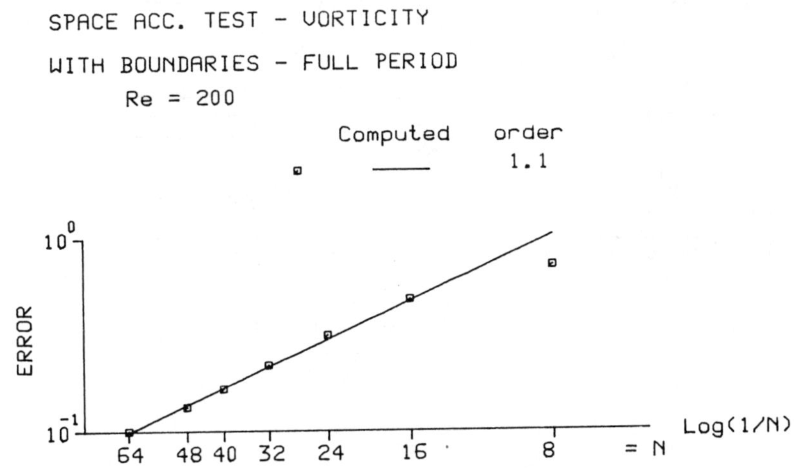

Figure 3: Error vs. number of mesh intervals

VORTICITY STREAMLINES

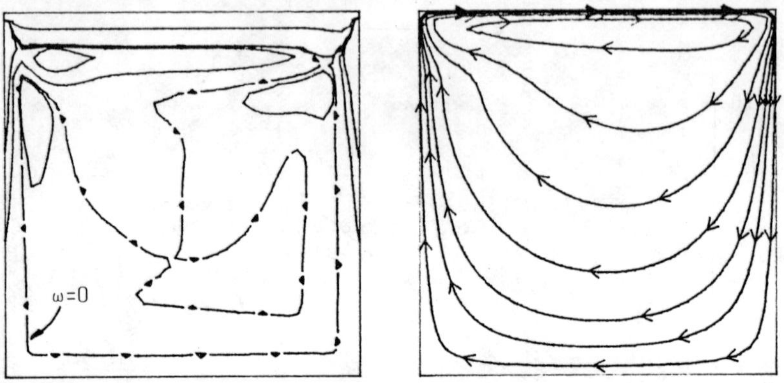

RE = 200 T = 0.00000
DX = 1/40 DY = 1/40 DT = 1/32
RES (V.TR.EQ.) = 6.31E-05

RE = 200 T = 0.00000
DX = 1/40 DY = 1/40 DT = 1/32
RES (V.TR.EQ.) = 6.31E-05

Figure 4.a

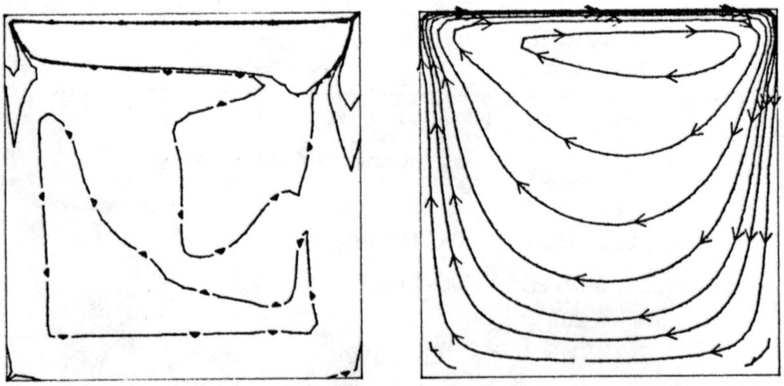

RE = 200 T = 0.25000
DX = 1/40 DY = 1/40 DT = 1/32
RES (V.TR.EQ.) = 6.31E-05

RE = 200 T = 0.25000
DX = 1/40 DY = 1/40 DT = 1/32
RES (V.TR.EQ.) = 6.31E-05

Figure 4.b

Figures 4: Flow patterns at some phases of the periodical motion. Figures 4.a and b corresponds to those cases also given by Duck [2].

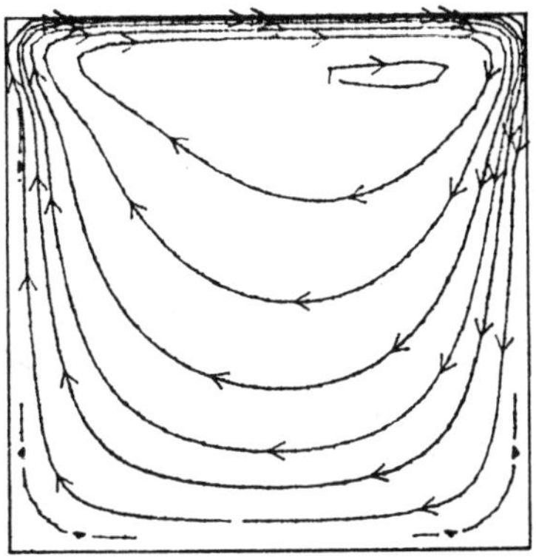

RE = 200 T = 0.28125

Figure 4.c

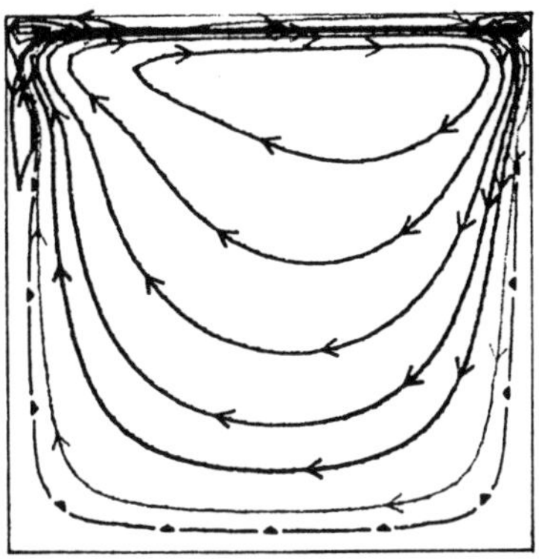

RE = 200 T = 0.31250

Figure 4.d

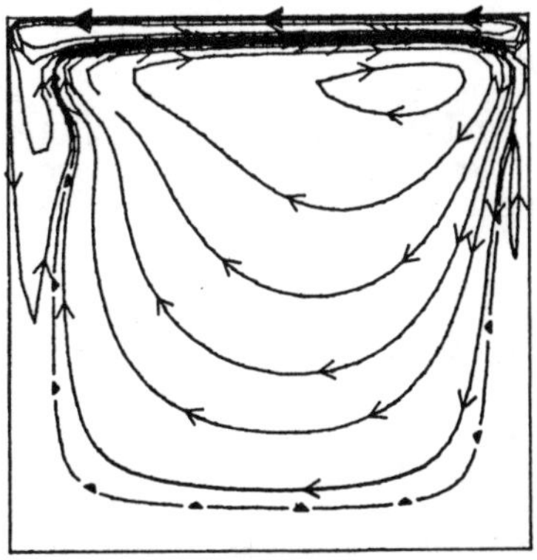

RE = 200 T = 0.34375

Figure 4.e

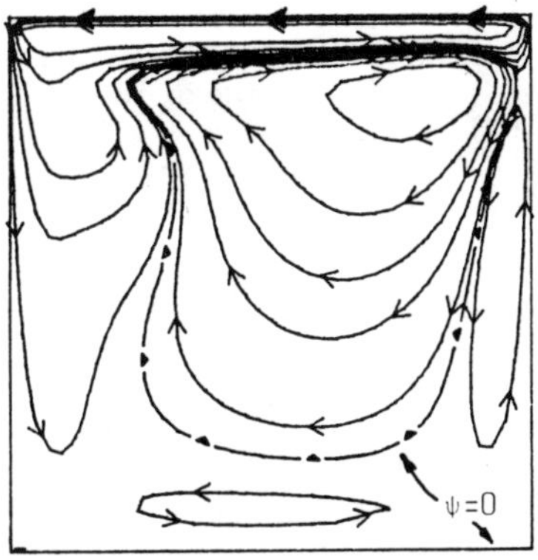

RE = 200 T = 0.37500

Figure 4.f

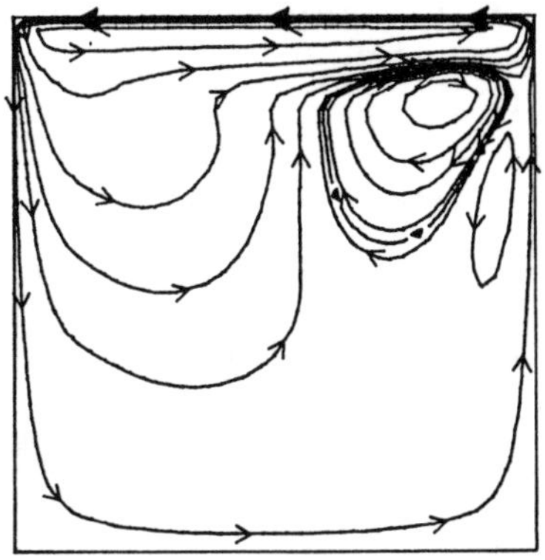

RE = 200 T = 0.40625

Figure 4.g

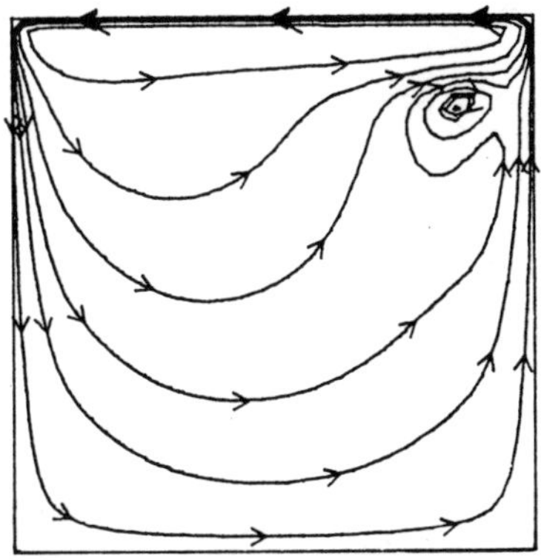

RE = 200 T = 0.43750

Figure 4.h

function of the number of space intervals (N) for the vorticity. We note that the asymptotic behaviour of the method, as $h \downarrow 0$, is of first order, as expected and that for h=1/64 the mean relative error due to space discretization is about 10%. A comparison with figur 2 shows that for medium grids the error due to time discretization is by several orders of magnitude lower than the error due to space discretization.

Figures 4 show the periodical solution of the cavity problem at different phases, for Re=200, h=1/40 and Δt=1/32. In figures (4.a) and (4.b) we display the streamlines and iso-vorticity lines, for t=0 and for t=0.25, respectively.

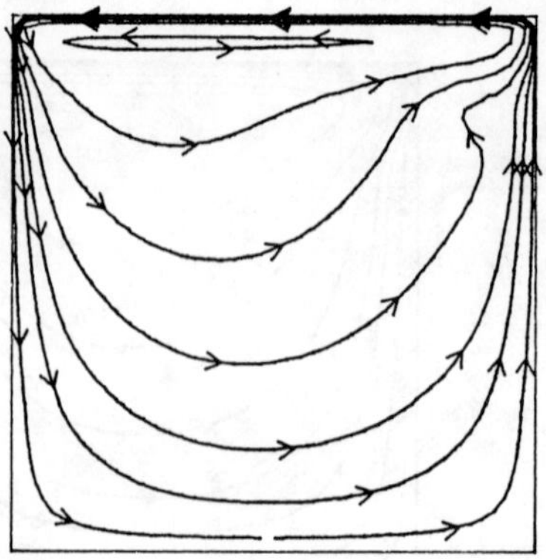

RE = 200 T = 0.46875

DX = 1/40 DY = 1/40 DT = 1/32
RES (U.TR.EQ.) = 6.31E-05

Fgure 4.i

Figures (4.b)-(4.i) show the changes in the flow field when and after that the moving wall changes direction. The large vortex, in the cavity and near the moving wall, at t=0.25 decreases in size and moves toward the right wall as time goes on. At t=0.4375 only a little and well confined vortex is left and at t=0.46875 the remnant of this vortex is seen only as a disturbance in the streamlines. At t=0.5 the picture is the same as the mirror image of the picture at t=0.

CONCLUDING REMARKS

The oscilating motion in the cavity was also computed by Duck [2], using a semi-spectral method. We note a clear qualitative discrepancy between our results and those which are given by Duck. This discrepancy persists in all our computations for different time-space grids. On basis of the accuracy test on our results, we suggest that the results presented in [2] may not be converged and therefore inaccurate. The accuracy of the results with respect to the time-steps and the spatial mesh sizes for the non-steady driven cavity flow have been found to be as good as expected from theory. The MG method itself have been experienced in all the tests to be fast and reliable.

REFERENCES

[1] KARLSSON, A., FUCHS, L. - Fast and Accurate Solution of Time-Dependent Incompressible Flow, Numerical Methods in Laminar and Turbulent Flow, Eds. Taylor, C., Johnson, J.A., Smith, W.R., Pineridge Press, pp. 606-616, 1983.

[2] DUCK, P.W. - Oscillatory Flow Inside a Square Cavity, J. Fl. Mech., vol. 122, pp. 215-234, 1982.

[3] FUCHS, L. - On the Accuracy of Incompressible Flow Simulations, To appear.

[4] THUNELL, T., FUCHS, L. - Numerical Solution of the Navier-Stokes Equations by Multi-Grid Techniques, Num. Meth. in Lam. and Turb. Flow, Eds. Taylor, C., Schrefler, A.B., Pineridge Press, pp. 141-152, 1981.

[5] FUCHS, L. - New Relaxation Methods for Incompressible Flow Problems, Num. Meth. in Lam. and Turb. Flow, Eds. Taylor, C., Johnson, J.A., Smith, W.R., Pineridge Press, pp. 627-640, 1983.

SOLUTION OF THE THREE-DIMENSIONAL, TIME-DEPENDENT NAVIER-STOKES EQUATIONS USING A GALERKIN METHOD

R. Kessler
DFVLR Institute for Theoretical Fluid Mechanics
Bunsenstrasse 10, D-3400 Göttingen

SUMMARY

A Galerkin method is presented for the solution of the three-dimensional, time-dependent Navier-Stokes equations in a rectangular computational region. The particular expansion of the velocity field using nondivergent basis functions allows us to satisfy the equation of continuity exactly. The timestepping is fully implicit and stable, even for large time steps. An efficient selection modus for the basis functions enables us to obtain accurate solutions. Vectorisation of the code leads to high performance on a CRAY-1S computer. We employ the code to investigate nonlinear convective instabilities in a rectangular box heated from below. Aspects of steady three-dimensional flow structures and the onset of oscillatory instabilities are examined. The influence of the Prandtl number and the thermal boundary conditions at the side walls is investigated for steady and time-dependent flows.

INTRODUCTION

The Galerkin method described in this report was developed to simulate the three-dimensional, time-dependent convection flow in a rectangular box heated from below. The convective heat and mass transfer is an important mechanism in technology. There is an increasing number of applications in technical problems, such as energy storage, reactor safety, solar collectors, crystal growth, and microstructures in foundry technology. These thermal convection flows share the common feature of being partly or completely confined.

Besides these applications, there is a great interest in investigating the three-dimensional flow structures and their time dependent behaviour to obtain more information about the transition to turbulent flows. The topological structures appearing in the three-dimensional convection flow are somewhat similar to the structures we can see in seperated flows. Therefore, convection in a box is a rather simple example for studying fundamental phenomena in steady and time-dependent three-dimensional flows.

Figure 1 shows the geometry we have considered and the sketch of a possible steady roll structure. The no-slip conditions at the rigid side

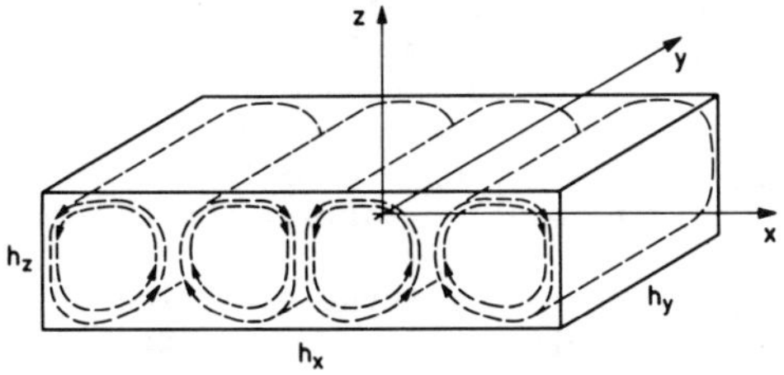

$h_x : h_y : h_z = 4:2:1$

Fig.1: Sketch of the geometry and a possible roll structure

walls cause a complicated three-dimensional flow field, even in the case of steady flow. With the Rayleigh number increasing, the steady convection flow becomes unstable and oscillatory instabilities appear. To determine the influence of the thermal boundary conditions at the side walls, perfectly conducting as well as adiabatic side walls are assumed. Changes in the inertial forces in the flow induced by varying the Prandtl number have a great influence on the steady and time-dependent behaviour of the convection.

For the numerical simulation of the convection, we solve the Boussinesq equation using the Galerkin method, which is a special kind of spectral method. This class of numerical methods has particular advantages in the solving of instability- and transition problems in simple geometries. A survey of the theory of spectral methods and their applications in fluid mechanics is given, for example, in the book by Gottlieb and Orszag [1].

The two-dimensional version of the Galerkin method described here was proven in a comparison problem [2] to provide very accurate solutions. This encouraged us to extend the method to solve three-dimensional, time-dependent problems. The main feature of the two-dimensional version, namely to fulfill exactly the continuity equation and all boundary conditions, can also be realized in the three-dimensional version, using a special expansion for the velocity. The nonperiodic boundary conditions for the velocity vector, caused by the no-slip conditions at the walls, produce additional computational difficulties in the three-dimensional formulation. An effective selection modus for the basis funcions is needed in the three-dimensional code to obtain accurate solutions with a minimum of basis functions.

GALERKIN METHOD

The basic idea of the Galerkin method is to expand the solution g of a differential equation L into complete sets of linearly independent basis functions E_i with the time-dependent coefficients $e_i(t)$.

$$L(x,y,z,t,g,g,....) = 0 \qquad g \cong \sum e_i(t)\; E_i(x,y,z)$$

The basis functions are continuous and defined in the whole computational region V. By inserting the expansion into the differential equation, the error function ε remains on the right hand side.

$$L(x,y,z,t, \sum e_i E_i, ...) = \varepsilon$$

To minimize the residual ε, the inner product of all basis functions with the error function is set equal to zero

$$\int \varepsilon\, E_i\, dV = 0$$

The resulting conditional equations for the coefficients $e_i(t)$ are called Galerkin equations, which are general a system of ordinary differential equations. All differentiations in space can be done analytically with the known basis functions, and therefore numerical dissipation, as it appears in finite difference schemes, is avoided completely. In addition, all basis functions are chosen such as to exactly satisfy the boundary conditions.

By using nondivergent basis functions for the expansion of the velocity field, the pressure term can be eliminated. This procedure is proven to be very efficient in a two-dimensional Galerkin method and yields accurate solutions. To extend this method to three-dimensional problems, we need the formulation of a general three-dimensional nondivergent vector field. According to Joseph [3], such a vector field can be described by the following superposition:

$$\vec{v} = \nabla x \, (\nabla x \, (\Phi \vec{i})) + \nabla x \, (\Psi \vec{i})$$

It is not very efficient to employ this expression in the Galerkin method. As the two terms have a different structure, a great number of integrals have to be calculated. Therefore, we use a formulation that was proven to be equivalent to the above expression [4], and consists of two similar terms

$$\vec{v} = \nabla x \, (\Psi^{(1)} \vec{j}) + \nabla x \, (\Psi^{(2)} \vec{i})$$

By using this formulation to describe the velocity field, the no-slip conditions cause a coupling of the two scalar functions $\Psi^{(1)}$ and $\Psi^{(2)}$.

$$\vec{v} = (-\Psi^{(1)}_z, \Psi^{(2)}_z, \Psi^{(1)}_x - \Psi^{(2)}_y) = 0 \qquad \text{at the walls}$$

Therefore, an additional term is necessary in the expansion of the velocity, in order to satisfy the boundary condition exactly. The expansion of the general three-dimensional velocity field can be written as follows:

$$\vec{v} = \nabla x \left\{ \sum_{ijk} a^{(1)}_{ijk}(t) \cdot \Psi^{(1)}_{ijk}(x,y,z) \cdot \vec{i} \right\}$$
$$+ \nabla x \left\{ \sum_{ijk} a^{(2)}_{ijk}(t) \cdot \Psi^{(2)}_{ijk}(x,y,z) \cdot \vec{j} \right\}$$
$$+ \sum_{ij} a^{(3)}_{ij}(t) \left\{ \nabla x \, \Psi^{(3\,1)}_{ij}(x,y,z)\vec{i} + \nabla x \, \Psi^{(3\,2)}_{ij}(x,y,z)\vec{j} \right\}$$

With the boundary conditions:

$$\Psi^{(1)}_z = \Psi^{(1)}_x = \Psi^{(2)}_z = \Psi^{(2)}_y = \Psi^{(3\,1)}_z = \Psi^{(3\,2)}_z = 0$$
$$\Psi^{(3\,2)}_x = \Psi^{(3\,1)}_y \qquad \text{at the walls}$$

The expansion of the temperature is rather simple, even in three dimensions

$$T = \sum_{lmn} b_{lmn}(t) \cdot T_{lmn}(x,y,z)$$

All the scalar functions appearing in the expansions consist of products of three functions, one function for each coordinate in space. According to the quasi-periodic behaviour of the convection flow in x-direction, in this direction we choose trigonometric functions and the so-called "Beam-functions" which consist of trigonometric and hyperbolic functions [5]. In the x- and y-directions we employ special sets of orthogonal polynomials. These polynomials yield a high resolution of the flow field near the walls and fulfill all boundary conditions at the walls.

The modus of selection is carried out for each sum of the expansion separately. This leads to a complicated but efficient system of basis function selection. Therefore, it is possible to get accurate three-dimensional solutions, even for high Rayleigh numbers, with only 400 basis functions for the velocity and the temperature field.

Applying the expansions in the Boussinesq equations leads -together with the orthogonality condition- to the Galerkin equations. As mentioned above, this is a system of ordinary differential equations in time for the unknown coefficients of the expansions. Discretisation of the time-derivatives reduces the equations to a system of nonlinear algebraic equations. Using explicit methods, numerical oscillations can be avoided only by applying short time-steps. We therefore employ an implicit method, called the "one-leg method" and described by Dahlquist [6].

$$\frac{\partial a}{\partial t} = F(a) \qquad \rightarrow \qquad \frac{a^{n+1} - a^{n}}{\Delta t} = F\left(\frac{a^{n+1} + a^{n}}{2}\right)$$

This timestepping method is of second order accuracy and yields stable solutions, even for large time steps. The resulting system of nonlinear algebraic equations is solved by using the Newton method. Test calculations with various time steps show that the truncation error of the time discretisation is always greater than the truncation error resulting from the Newton method after the first iteration. The good initial value for the Newton iteration obtained by extrapolation of the last time steps is responsible for this behaviour. Therefore, we can restrict ourselves to the use of only one iteration per time step. The matrix of the resulting linear system is full and the application of iterative linear solvers is not effective. We therefore use the Gaussian elimination to solve the linear system.

As more than 98% of the algorithm of the Galerkin code can be vectorised very effectively, we gain a high performance on a vector computer such as the CRAY-1S. Considering the IBM 3081K for comparison, a speed-up factor of 25 was obtained with the vectorised code on the CRAY-1S. Approximately 1.7 seconds per time step are required when using the best approximation with 400 basis functions. Details of the vectorisation of the Galerkin method are described in [7].

DISCUSSION OF RESULTS

At first some results of the calculated steady flow structures are discussed. All calculations presented in this paper were carried out for the geometry shown earlier in figure 1. The influence of the thermal boundary conditions at the side walls can be seen in figure 2. Calculated streamlines in the symmetry plane of the box are plotted for perfect conducting and adiabatic side walls. In three-dimensional convection flows, the streamlines are no longer closed in the symmetry plane. Starting in the centers of the three convection rolls, the streamlines spiral out and all converge in the two small vortices in the corners of the box, in the case of conducting side walls. For adiabatic side walls, the vortices in the corners disappear and the streamlines converge in the outer region of the middle convection roll.

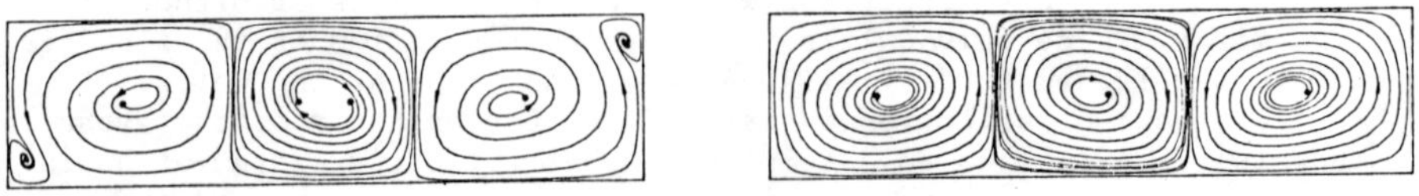

Fig. 2: Streamlines of steady convection in the symmetry plane, Pr = 0.71
left: conducting side walls, right: adiabatic side walls

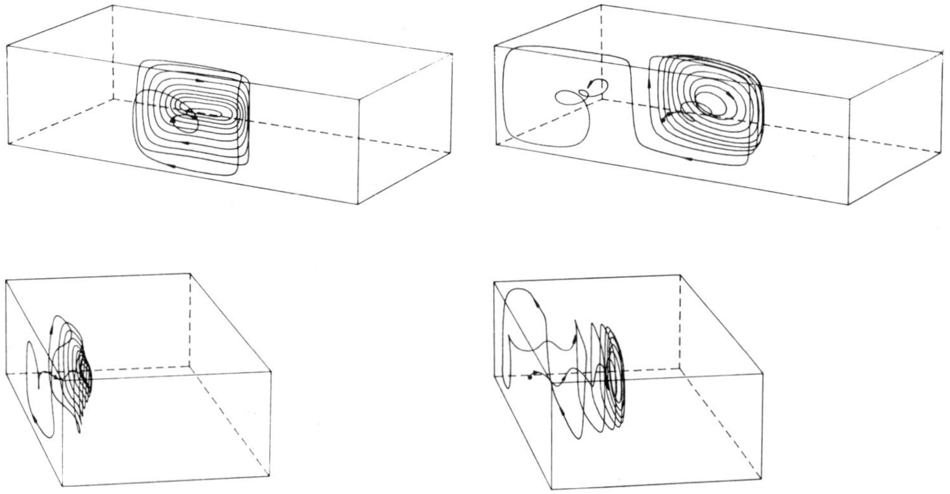

Fig. 3: Streamlines of steady convection, adiabatic side walls, left: Pr = 7, right : Pr = 0.71

Changing the inertial forces in the flow by varying the Prandtl number has a considerable influence on the topological structure of the steady convection. For adiabatic side walls, calculated streamlines are shown in figure 3; on the left side for a Prandtl number of 7, and on the right side for the Prandtl number of 0.71. For water the flow is nearly two-dimensional in the middle region of the box. The three-dimensional region near the wall is caused by the thermal end effect. In air, the inertial forces cause a three-dimensional flow throughout the whole box. We can observe an exchange of mass between the convection rolls, which causes a complicated topological structure of the flow.

By increasing the Rayleigh number, the steady flow becomes unstable and we obtain oscillations of the velocity and the temperature field. Figure 4 shows the time dependent behaviour of the temperature in air at a fixed point in the box with the corresponding power spectra for Rayleigh numbers up to 70,000. In addition to the mean frequency, we obtain higher harmonics of the oscillation. Nonharmonic frequencies appear with an increasing Rayleigh number and seem to indicate the onset of an aperiodic regime.

In figure 5 the calculated mean frequency is compared with experimental data obtained for the same geometry [8]. We recognize a reasonable agreement in the dependence of the frequency on the Rayleigh number. So far, the subharmonics that appear in the experiment have not been found in numerical simulations. We attribute this to the symmetry conditions employed in the Galerkin method. As the occurence of subharmonics and also aperiodic flow is related to a loss of symmetry in the flow field, our simulation cannot describe these phanomena accurately.

The structure of the oscillations in the box can be seen in figure 6. Isothermal lines in two midplanes of the box show the upward and downward motions in the flow. Half a period of the oscillations in water and air can be seen. The structure of the oscillations depends heavily on the

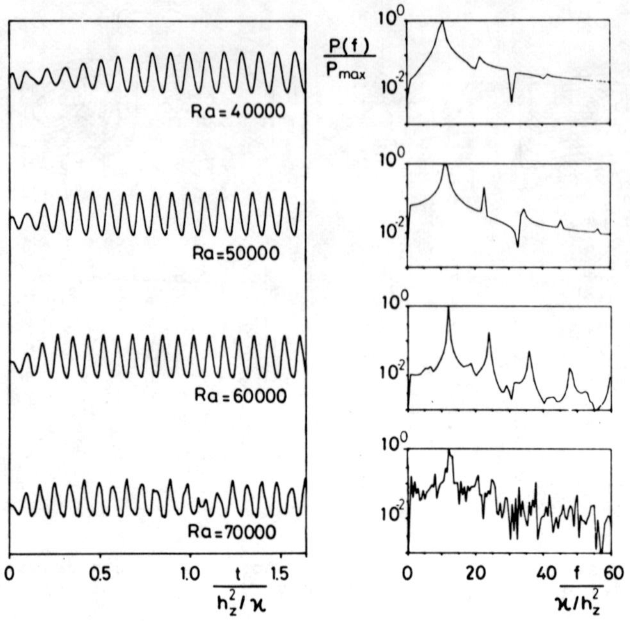

Fig. 4: Calculated temperature oscillations in air. Spectral evolution

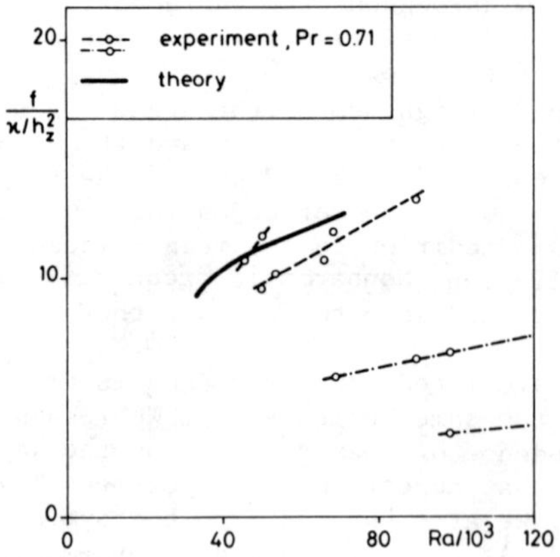

Fig. 5: Frequency of oscillations in air (— • — subharmonics)

Fig. 6: Isothermal lines of oscillatory convection, left: air, right: water

Prandtl number. In air, with a Prandtl number of 0.71, the calculations yield large periodic contractions of the rolls with the maximum deformation in the middle of the box. For a Prandtl number of 7 the maximum amplitude of the upward and downward motions moves along the otherwise almost permanent rolls.

CONCLUDING REMARKS

With the numerical simulation of the nonlinear convective instabilities in a closed box, it has been shown that we can obtain accurate solutions with the Galerkin method even for three-dimensional, time-dependent flows. Using nondivergent basis functions for the velocity, we avoid difficulties related to the calculation of the pressure. This leads to an efficient method of calculating incompressible flows in simple geometries. A special selection modus is necessary to obtain accurate solutions with the rather low number of 400 basis functions. Effective vectorisation of the code yields a high performance on the CRAY-1S computer.

The physical results show that steady and time-dependent convection flow in a box is quite different from the behaviour in infinitely extended fluid layers. The vertical walls have a considerable influence on the flow structures and the transition to time-dependent convection. Three-dimensionality is the key to understanding the physics of the convective instabilities in a box. Further investigations of convection flows can give us important information about the transition to turbulence and the topological structures in three-dimensional, time-dependent flows.

REFERENCES

[1] D. Gottlieb and S.A. Orszag, Numerical analysis of spectral methods: theory and applications. NSF-CBMS Monograph No. 26, SIAM, Philadelphia 1977

[2] G. De Vahl Davis and I.P. Jones, Natural convection in a square cavity: a comparison exercise. Report 1982/FMT/3 University of N.S.W., Kensington, 1982

[3] D.D. Joseph, Stability of fluid motions I, II. Springer Verlag, Heidelberg, 1976

[4] R. Kessler, Oszillatorische Konvektion, Dissertation, Universität Karlsruhe, 1983

[5] D.F. Harris and W.H. Reid, On orthogonal functions which satisfy four boundary conditions. Astrophys. J.S. Series 3, 429-453, 1958

[6] G. Dahlquist, Error analysis for a class of methods for stiff nonlinear initial value problems. Numerical analysis Dundee 1975, Lecture Notes in Mathematics 506, A. Dold and B. Eichmann eds., Springer Verlag Berlin, 60-72, 1976

[7] R. Kessler, Vectorisation of the Galerkin method on the CRAY-1S, in W. Gentzsch, Vectorisation of computer programs with the application to computational fluid dynamics, to appear in "Notes on Numerical Fluid Mechanics", Vieweg Verlag, Wiesbaden, 1984

[8] W. Jäger, Oszillatorische und turbulente Konvektion, Dissertation Universität Karlsruhe, 1982

COMPUTATION OF LAMINAR FLOW IN A PIPE OF MULTIPLY-CONNECTED CROSS-SECTION

M.E.Klonowska, W.J.Prosnak
Institute of Fluid-Flow Machines
of the Polish Academy of Sciences
ul. J.Fiszera 14, 80-952 Gdańsk, Poland

SUMMARY

The paper deals with steady laminar flow in a cylindrical pipe of arbitrary, multiply-connected cross-section, and infinite length. The flow may be regarded, therefore, as generalisation of the well-known Poiseuille flow in a circular cylindrical pipe. The problem of determination of the flow considered reduces to the Dirichlet problem. It is solved in the paper by means of representing the unknown function, i.e. the only remaining velocity component, as the real part of a series containing rational functions. The constant coefficients of the series follow from a set of linear algebraic equations, equivalent to the boundary conditions. This set of equations is presented in the paper as well as some solutions. The analogy with the problem of elastic torsion of multihole circular cylinders is reminded.

INTRODUCTION

Let us regard steady, laminar flow of viscous, incompressible liquid - in a cylindrical pipe, the pipe containing (K-1) cylindrical rods. The generatrices of the pipe and the rods are parallel, and infinite. The cross-section of such a pipe is shown schematically in the Fig.1: the liquid flows perpendicularily to the plane of the Figure through the unshadowed region, bounded by the wall of the pipe and those of the rods. Introducing a rectangular coordinate system x_1, x_2, x_3, the x_3 - axis being parallel to the generatrices of the pipe and the rods, one can write the continuity as well as the Navier-Stokes equations for steady, incompressible flows, as follows:

$$\frac{\partial v_1}{\partial x_1} + \frac{\partial v_2}{\partial x_2} + \frac{\partial v_3}{\partial x_3} = 0 \; ;$$

$$\rho \left(v_1 \frac{\partial v_i}{\partial x_1} + v_2 \frac{\partial v_i}{\partial x_2} + \frac{\partial v_i}{\partial x_3} \right) = - \frac{\partial p}{\partial x_i} + \mu \Delta v_i ; \qquad i=1,2,3$$

the field of mass forces being neglected. The symbols v_1, v_2, v_3 denote the velocity components in direction of the corresponding coordinate axes, and the symbols p, ρ, μ stand for the pressure, the density, and the dynamic viscosity, resp. The two last mentioned functions are assumed to be constant. The symbol Δ denotes the Laplace operator.

The present investigation is confined to laminar flows. Consequently, it is postulated:

$$v_1 = v_2 = 0$$

which leads to the following conclusions:

$$v_3 = v_3 \, (x_1, x_2); \qquad p = p(x_3);$$

$$\frac{1}{\mu} \frac{\partial p}{\partial x_3} = \text{const} = P; \qquad p = p_o + \mu P \cdot x_3;$$

$$\Delta \, v_3 = \text{const} = P.$$

The velocity distribution is governed, therefore, by the Poisson equation:

$$\Delta \, v_3 = P \, . \tag{1}$$

As far as boundary conditions for the only unknown function v_3 are concerned, the slip will be allowed for the sake of generality. Hence, the boundary condition for the velocity component v_3 assumes the following form:

$$v_3 \, (x_{D_k} \, , \, y_{D_k} \,) = v_3^{[k]} \, ; \tag{2}$$

where x_{D_k}, y_{D_k} denote coordinates of a point at the contour D_k, the contour representing either the inner boundary of the cross-section of the pipe (k=1) or the outer boundary of the cross-section of any rod (k=2,3,... K; see Fig.1). The symbol $v_3^{[k]}$ denotes the given slip-velocity of the liquid at the k-th contour.

The constant P is not essential. One can easily get rid

of it by dividing (1) and (2) by $(\frac{1}{4}P)$, and introducing a new function:

$$V = \frac{4\ v_3}{P}.$$

The problem of determination of the flow velocity reduces therefore to the following form:

$$\left.\begin{aligned} &\Delta V = 4\ ; \\ &V(x_{D_k},\ y_{D_k}) = V_k\ ; \qquad k=1,2,\dots,K, \end{aligned}\right\} \quad (3)$$

which is - mathematically - identical with that concerning elastic torsion of cylindrical bars [1,2].

THE METHOD OF SOLUTION

The problem (3) for the Poisson equation can be reduced to another one, for the Laplace equation. In order to do so, the unknown function must be expressed as the sum:

$$V(x,y) = \varphi(x,y) + (x^2 + y^2). \qquad (4)$$

Indeed, insertion of (4) into (3) yields:

$$\left.\begin{aligned} &\Delta\varphi(x,y) = 0\ ; \\ &\varphi(x_{D_k},\ y_{D_k}) = -\ (x^2_{D_k} + y^2_{D_k}) + V_k\ ; \qquad k=1,2,\dots,K, \end{aligned}\right\} \quad (5)$$

i.e. simply the Dirichlet problem for the function $\varphi(x,y)$.

We propose to solve this problem in two steps, which are:

1. application of conformal mapping of the multiply-connected region of flow onto an auxiliary region bounded by circles;
2. construction of the solution in the auxiliary region.

The cross-section shown in Fig.1 may be thought of as laying in the complex z-plane. Consequently, it can be assumed, that a known holomorphic function

$$z = z\ (\zeta)\ , \qquad (6)$$

where

$$z = x + iy\ ; \qquad \zeta = \xi + i\eta\ ;$$

and

$$x = x\ (\xi, \eta)\ ; \quad y = y\ (\xi, \eta)\ , \qquad (7)$$

transforms the given multiply-connected domain of flow in the z-plane onto the multiply-connected auxiliary domain in the ζ-plane (Fig.2). The contours $D_1, D_2, \dots D_K$ transform onto circles of known centres determined by complex numbers ζ_k, and known radii denoted by real numbers a_k ; $k=1,2,\dots K$.

The problem of construction of the mapping function (6) has been dealt with elsewhere [3,4], and will not be considered in the present paper. Hence, we shall confine ourselves to discussion of the second of the abovementioned two "steps".

As consequence of performing the change of independent variables, indicated by (7), a new unknown function will be introduced:

$$\phi(\xi,\eta) = \varphi(x(\xi,\eta),\ y(\xi,\eta))$$

and also a new notation:

$$x_{D_k} = x(\xi_{D_k}, \eta_{D_k}) ;$$

$$y_{D_k} = y(\xi_{D_k}, \eta_{D_k}) , \qquad k=1,2,\dots, K ,$$

where

$$\xi_{D_k} = \xi_k + a_k \cos\theta_k ; \qquad \eta_{D_k} = \eta_k + a_k \sin\theta_k$$

denote coordinates of a point at the k-th circle.

Because the Laplace equation is invariable with respect to the change of independent variables corresponding to conformal mapping, the Dirichlet problem (5) may be rewritten as:

$$\Delta\phi(\xi,\eta) = 0 ; \tag{8}$$

$$\phi(\xi_{D_l}, \eta_{D_l}) = -\left[x(\xi_{D_l}, \eta_{D_l})\right]^2 - \left[y(\xi_{D_l}, \eta_{D_l})\right]^2 + V_l =$$

$$= \sum_{n=o}^{\infty} A_{ln} \cos n\theta_l + B_{ln} \sin\theta_l ; \tag{9}$$

where $l=1,2,\dots,K$ and $x(\xi,\eta)$, $y(\xi,\eta)$ denote known functions, representing the real and the imaginary part, resp., of the mapping function (6). The coefficients of the Fourier series can be always computed numerically, however, for some special classes of this function they can be represented by analytic expressions, containing parameters of this function.

The mapping function dealt with in [3,4] is an example of such a class.

The function $\phi(\xi, \eta)$ being harmonic represents the real or the imaginary part of a suitable holomorphic function $W(\zeta)$. Chosing the first possibility one can write:

$$\phi(\xi, \eta) = \mathrm{Re}\left[W(\zeta)\right] .$$

On the other hand, the holomorphic function $W(\zeta)$ can be represented by means of suitable series. Basing on our former considerations [3] we assume the following most general form:

$$W(\zeta) = \sum_{n=0}^{\infty} C_{1n}\left(\frac{\zeta}{a_1}\right)^n + \sum_{k=2}^{K} C_{ko} \ln \frac{\zeta - \zeta_k}{a_1} + \sum_{k=2}^{K} \sum_{n=1}^{\infty} C_{kn} \left(\frac{a_k}{\zeta - \zeta_k}\right)^n , \tag{10}$$

where

$$C_{kn} = \mu_{kn} + i\,\nu_{kn} ; \qquad k=1,2,\ldots,K; \quad n=0,1,2\ldots \tag{11}$$

denote constant complex coefficients.

With respect to its physical meaning the function $\phi(\xi,\eta)$ must be one-valued. Consequently, the coefficients C_{ko} must be real:

$$\nu_{ko} = 0 ; \qquad k=2,3,\ldots,K. \tag{12}$$

Summarising the considerations of the present section one can say, that the original problem has been reduced to determination of the coefficients (11) of the function $W(\quad)$.

SYSTEM OF EQUATIONS FOR THE COEFFICIENTS

In order to arrive at the system of equations for the said coefficients, our formerly developed method [5,6] will be applied. It consists - in the present case - in expanding the functions appearing at the left-hand side of (9) into Fourier series with respect to Θ_k, and equalising their coefficients with those of the series appearing at the right-hand side.

The desired expansion of $\phi(\xi_{D_1}, \eta_{D_1})$ into Fourier series can be done simply by introducing

$$\zeta_{D_1} = \xi_{D_1} + i\eta_{D_1} = \zeta_1 + a\,e^{i\Theta_1} ; \qquad \zeta_1 = 0$$

into (10), and extracting the real part from the expression obtained. The calculations differ slightly for l=1 and for l≠1, therefore the two cases have to be treated separately. The system of equations so obtained has the following form in the case l=1:

$$\mu_{10} = A_{10} \; ;$$

$$\mu_{1t} + \sum_{k=2}^{K}{}' \Big[-\frac{1}{t}\mu_{ko}R_{kt} + \sum_{p=0}^{t-1} \left(\frac{a_k}{a_1}\right)^{t-p} u_{t-p,p} \left(\mu_{k,t-p}R_{kp} - \nu_{k,t-p}S_{kp}\right) = A_{1t} \; ;$$

$$-\nu_{1t} + \sum_{k=2}^{K}{}' \Big[-\frac{1}{t}\mu_{ko}S_{kt} + \sum_{p=0}^{t-1} \left(\frac{a_k}{a_1}\right)^{t-p} u_{t-p,p} \left(\nu_{k,t-p}R_{kp} + \mu_{k,t-p}S_{kp}\right) = B_{1t} \; ,$$

and - in the case l=2,3,...,K :

$$\mu_{l0} + \mu_{l0} \ln \frac{a_l}{a_1} + \sum_{n=1}^{N} \left(\mu_{1n}R_{1n} - \nu_{1n}S_{1n}\right) + \sum_{k=2}^{K}{}' \left[\mu_{k0}X_{kl} + \sum_{n=1}^{N} \left(\mu_{kn}r_{lkn}^{[0]} - \nu_{kn}s_{lkn}^{[0]}\right)\right] = A_{l0} \; ;$$

$$\mu_{lt} + \sum_{n=t}^{N}{}' v_{nt}\left(\frac{a_l}{a_1}\right)^{t} \left(\mu_{1n}R_{1,n-t} - \nu_{1n}S_{1,n-t}\right) + \sum_{k=2}^{K}{}' \left[\mu_{ko}w_{lkt}r_{lkt}^{[0]} + \sum_{n=1}^{N}\left(\mu_{kn}r_{lkn}^{[t]} - \nu_{kn}s_{lkn}^{[t]}\right)\right] = A_{lt} \; ;$$

$$\nu_{lt} - \sum_{n=t}^{N}{}' v_{nt}\left(\frac{a_l}{a_1}\right)^{t} \left(\nu_{1n}R_{1,n-t} + \mu_{1n}S_{1,n-t}\right) - \sum_{k=2}^{K}{}' \Big[\mu_{ko}w_{lkt}s_{lkt}^{0} + \sum_{n=1}^{N}\left(\nu_{kn}r_{lkn}^{[t]} + \mu_{kn}s_{lkn}^{[t]}\right) = B_{lt} \; .$$

In the both cases t=1,2,...,N, so that altogether the system consists of (2N+1)K equations with the unknowns:

$$\mu_{ko},\ \mu_{kn},\ \nu_{kn};\qquad k=1,2,\dots,K;\quad n=1,2,\dots,N; \tag{13}$$

The symbols appearing in the system denote:

$$u_{nq} = \frac{n(n+1)(n+2)\dots(n+q-1)}{q!};\quad u_{no} = 1;\quad n,q=1,2,\dots$$

$$v_{nq} = \binom{n}{q};\qquad v_{no} = 1;$$

$$w_{lkp} = \frac{(-1)^{p+1}}{p}\left(\frac{a_l}{a_k}\right)^p;\qquad k,l=1,2,\dots,K;\ p=1,2,\dots$$

$$R_{kq} + iS_{kq} = \left(\frac{\zeta_k}{a_1}\right)^q;$$

$$r_{lkn}^{[q]} + is_{lkn}^{[q]} = \left(r_{lkn}^{[q-1]} + is_{lkn}^{[q-1]}\right)\left[-\frac{n+q-1}{q}\,\frac{a_l}{\zeta_l-\zeta_k}\right];$$

with:

$$r_{lkn}^{[0]} + is_{lkn}^{[0]} = \left(\frac{a_k}{\zeta_l-\zeta_k}\right)^n;$$

$$X_{kl} = \ln\frac{\sqrt{(\xi_l-\xi_k)^2+(\eta_l-\eta_k)^2}}{a_1};\qquad k\neq l.$$

The system can be conveniently solved by means of the Gauss-Seidel method.

NUMERICAL EXAMPLES

Two examples have been selected in order to illustrate the method presented. The first one concerns a region bounded by circles (Fig.3), and the second one - a region bounded by a certain outer curve [4] and an inner ellipse (Fig.4). The respective mapping functions (6) have the following form:

$$z=\zeta;\quad z = a_- + a_+\zeta + 0.5a_-(\zeta-0.5)\ \text{with}\ a_{\mp} = \frac{5\mp\sqrt{17}}{40}.$$

The velocity distributions in some sections of the regions of flow are shown in the same Figures.

REFERENCES

[1] L.Landau, E.Lifszic, Mechanics of Continuous Media [in Polish]. PWN, 1958, 726-732.

[2] A.K.Nahdi, Torsion of multihole Circular Cylinders.Transactions of the ASME, Vol.49, June 1982, 432-435.

[3] W.J.Prosnak, A Theory of Radial-flow Hydraulic Machines, Bull.Acad.Polon.Sci., Ser.Sci.Techn., 27 (1979),25 - 273 .

[4] W.Prosnak, Z.Kosma, J.Szymański, Algorithms for Conformal Mapping of Multiply-Connected Domains [in Polish]. Zeszyty Naukowe IMP-PAN w Gdańsku, 116/1044/81.

[5] W.J.Prosnak, On the Flow around an Arbitrary Multiplane. Bull.Acad.Polon.Sci., Ser.Sci.Techn., 24, (1976),31 - 611.

[6] W.J.Prosnak, M.P.Strojnowski, Computation of Flow about an Arbitrary System of Discs. IFTR-Reports, 73/1977.

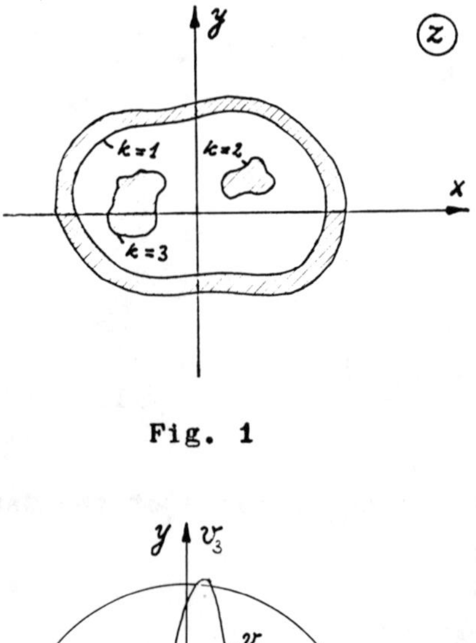

Fig. 1

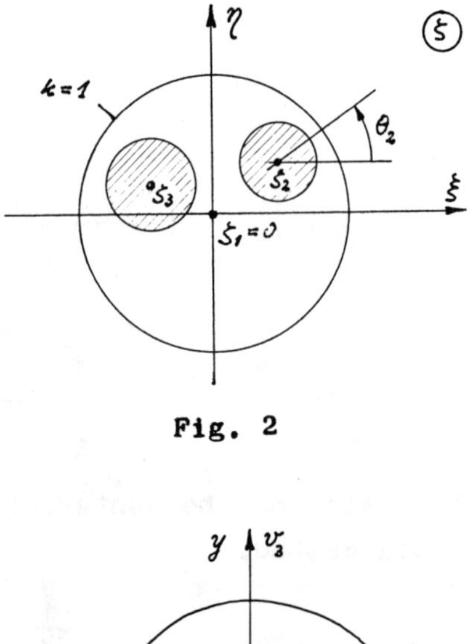

Fig. 2

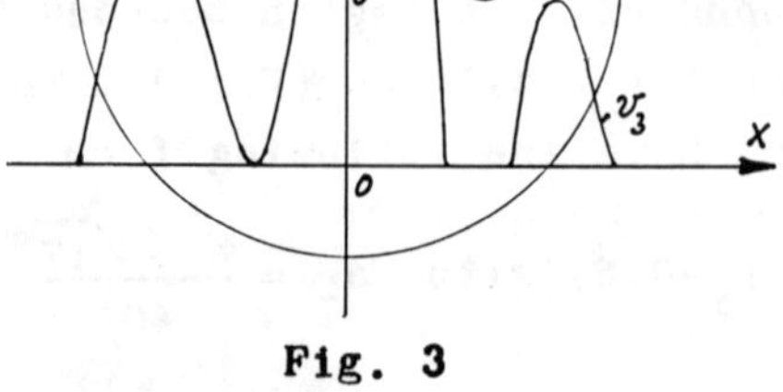

Fig. 3

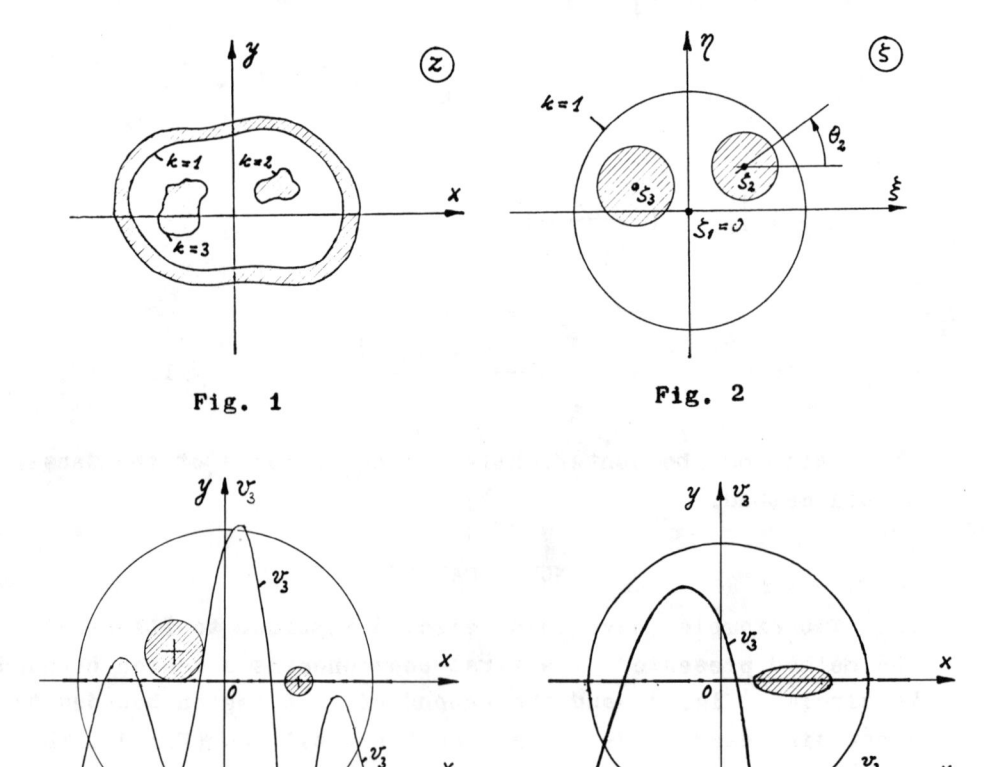

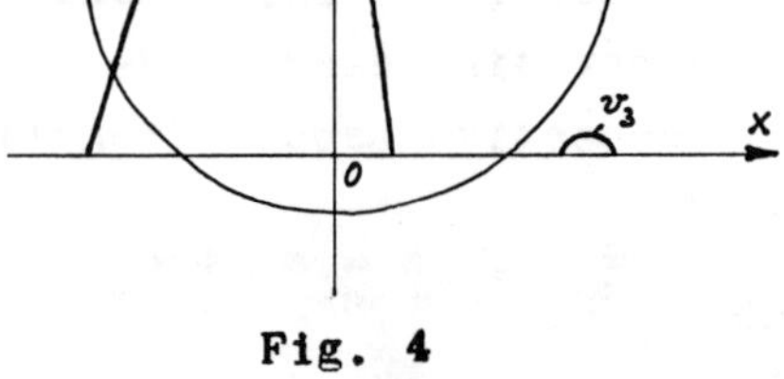

Fig. 4

COMPUTATIONS OF THREE-DIMENSIONAL TRANSONIC INVISCID FLOWS ON A WING BY PSEUDO-UNSTEADY RESOLUTION OF THE EULER EQUATIONS.

Ch. Koeck and M. Néron
MATRA 78140 VELIZY-VILLACOUBLAY (FRANCE)
Office National d'Etudes et de Recherches Aérospatiales
92320 CHATILLON (FRANCE)

SUMMARY

A pseudo-unsteady method developed at ONERA for solving the Euler equations with an isoenergetic assumption is applied to the numerical computation of three-dimensional transonic inviscid flows past a wing. Unlike the methods based on the potential equation, this method yields exact solution of the steady Euler equations, in the case of isoenergetic flows, including weak solutions with shocks and vortex sheets. The continuity and momentum equations are integrated step by step in time using an explicit predictor-corrector scheme with a local time-step technique. This scheme is discretized directly in the physical space and stabilized by second and fourth order dissipative terms. The boundary conditions are treated using the compatibility relations theory. A mesh system of C-O type has been developed, which takes into account the existence of a rounded wing tip. Computations of flows on the ONERA M6 wing have been made in both lifting and non lifting cases. They are compared to full potential results, and also to experiment. Computations made with two different artificial viscosity models are compared.

INTRODUCTION

Computations of inviscid transonic three-dimensional flows on wings have been made at ONERA by methods using the potential flow theory. However, this type of method presents two types of limitations : the first one, which concerns strong shock waves is not too limitative for flows on wings of transport aircraft, in which the shock strength is usually reasonable. The second one, which is its inability to capture contact discontinuities such as vortex sheets, demands to fix arbitrarily their positions. The evolution of the vortex sheets, and especially their roll-up cannot be represented. Nevertheless, it can be noted that, with the assumptions of potential flow theory, i.e. an isoenergetic and isentropic flow, it is possible to use a system which captures correctly the contact discontinuities, based on the momentum equation instead of the potential equation [1].

On the contrary, the pseudo-unsteady method using the Euler equations with an isoenergetic assumption, presented in this paper, is not subject to these two types of limitations. It yields exact solutions of the steady Euler equations in the case of isoenergetic flows, including weak solutions with shocks and vortex sheets. This method has been used at ONERA for various applications in transonic aerodynamics [2][3][4] including three-dimensional flows [5,6](nozzles, rotor cascades, blunt body flows). More recently, it has been extended for the computation of flows past wings, as a result of a cooperation between MATRA and ONERA [7]. This last development is used here for computations on the ONERA-M6 wing. Attention is focused on the influence of the artificial viscosity model and of the mesh on the solution.

Computations of three-dimensional transonic flows using the Euler equations have also been made by Rizzi and Eriksson [8], Godunov et al [9], Denton and Singh [10], Jameson, Schmidt and Turkel [11], Chen, Yu, Rubbert and Jameson [12].

One of the main features of the Euler equations consists in the ability to capture vortex sheets. Therefore, it becomes possible to obtain complex vortex structures such as a double vortex core issued from the leading edge of a delta wing (Hitzel and Schmidt [13]). These methods are able to give a satisfactory overall description of such vortex structures. Yet, they cannot describe precisely phenomena such as the roll up of wakes, as it can be approached by free vortex methods [14], or furthermore, the structure itself of their core, as it can be obtained by multiscale methods[15]. Moreover, if the trailing edges of wings appear to be naturally a possible place for emission of wakes, without any special treatment, the problem of their emission on a smooth surface is still to be solved.

EQUATIONS, NUMERICAL SCHEME, AND ARTIFICIAL VISCOSITY TERMS

The system of equations is simply composed of the unsteady continuity and momentum equations (1), in which the pressure p is a function of the density ρ and the velocity $\bar{V}$, through Bernoulli's relation (1c).

$$\left|\begin{array}{ll} \frac{\partial \rho}{\partial t} + \operatorname{div} \rho \bar{V} = 0 & \text{(1a)} \\ \frac{\partial \rho \bar{V}}{\partial t} + \overline{\operatorname{div}}(\rho \bar{V} \otimes \bar{V} + p \bar{\bar{I}}) = \bar{0} & \text{(1b)} \\ p(\rho, \rho\bar{V}) = \frac{\gamma - 1}{\gamma}\left[\rho H - \frac{1}{2}\frac{(\rho\bar{V})^2}{\rho}\right] & \text{(1c)} \end{array}\right.$$

The replacement of the unsteady energy equation by (1c) allows a reduction of the number of the scalar equations from 5 to 4, and therefore, a reduction of the computational cost. The pseudo-unsteady system (1), which can be written :

$$\frac{\partial f}{\partial t} + \frac{\partial F_\ell}{\partial x_\ell} = 0 \qquad (2)$$

is integrated step by step with respect to time, by means of an explicit predictor-corrector finite-difference scheme, derived from the MacCormack scheme, the discretization of the equations being performed directly in the physical space, using an arbitrary curvilinear mesh [2]. The predictor $\tilde{f}^{n+1}$ and the corrector $f^{(n+1)_0}$ are computed by :

$$\left|\begin{array}{l} \tilde{f}^{n+1} = f^n - \frac{\Delta t}{2}[\delta_\ell^+ F_\ell(f^n)] \\ f^{(n+1)_0} = f^n - \frac{\Delta t}{2}[\delta_\ell^+ F_\ell(f^n)] - \frac{\Delta t}{2}[\delta_\ell^- F_\ell(\tilde{f}^{n+1})] \end{array}\right. \qquad (3)$$

where δ_ℓ^+ and δ_ℓ^- are the upward and backward finite difference operators relative to each index ℓ (cf [5]).

Artificial viscosity terms are added after the computation of the corrector, in order to insure the stability of the scheme. Two different models have been used, which use both a non linear second order term, and the same fourth order term. They can be written : $f^{n+1} = f^{(n+1)_0} + D^2(f^n) + D^4(f^n)$ D^4 being given, in the one dimensional case, by :

$$D^4(f) = \mu_0 \left[f_{i+2} - 4 f_{i+1} + 6 f_i - 4 f_{i-1} + f_{i-2} \right] \qquad (4)$$

In the first model, the second order term which is, for one dimension :

$$D^2(f) = \mu_1 \left| \frac{\rho_{i+1} - 2\rho_i + \rho_{i-1}}{\rho_i} \right| (f_{i+1} - 2f_i + f_{i-1}) \qquad (5)$$

is applied to $\rho, \rho u, \rho v, \rho w$.

In the second model, D^2 is expressed by :

$$D^2(f) = \mu_1 \left[+ \frac{\rho_i + \rho_{i+1}}{2} \left| g_{i+1} - g_i \right| (g_{i+1} - g_i) - \frac{\rho_i + \rho_{i-1}}{2} \left| g_i - g_{i-1} \right| (g_i - g_{i-1}) \right] \qquad (6)$$

for : $g = (0, u, v, w)$

which is analogous to : $D^2(f) = \mu_1 \frac{\partial}{\partial x}\left(\rho \left|\frac{\partial g}{\partial x}\right| \frac{\partial g}{\partial x} \right)$ (7)

PROPERTIES OF THE PSEUDO-UNSTEADY SYSTEM, BOUNDARY CONDITIONS AND NUMERICAL STABILITY

The properties (hyperbolicity, compatibility relations and characteristic cone) which have been presented in 2-D in [2], and in 3-D in [5] will be only briefly recalled. Let consider the system (1) written at the current point P as

$$\frac{\partial f}{\partial t} + \bar{\bar{A}} \frac{\partial f}{\partial \xi} = - \bar{\bar{A}}_2 \frac{\partial f}{\partial \xi_2} - \bar{\bar{A}}_3 \frac{\partial f}{\partial \xi_3} \qquad (8)$$

where $\frac{\partial f}{\partial \xi}$ represents the partial derivative of the unknown vector $f = (\rho, \rho u, \rho v, \rho w)$ according to the particular direction $\bar{\xi} = \bar{\xi}_1$. Directions $\bar{\xi}_2$ and $\bar{\xi}_3$ belong to the plane orthogonal to $\bar{\xi}_1$. Let us consider the eigenvalues λ of the matrix $\bar{\bar{A}}$, solutions of the characteristic equation $\det(\bar{\bar{A}} - \lambda \bar{\bar{I}}) = 0$ and their respective left eigenvectors $\bar{\alpha}$, such as : $\bar{\alpha}.\bar{\bar{A}} = \lambda\bar{\alpha}$. The expressions of the four eigenvalues are, λ_0 being a double eigenvalue :

$$\lambda_0 = \bar{V}.\bar{\xi} \quad ; \quad \lambda_+ = \frac{\gamma+1}{2\gamma}\,\bar{V}.\bar{\xi} + \sqrt{\left(\frac{\gamma-1}{2\gamma}\,\bar{V}.\bar{\xi}\right)^2 + \frac{a^2}{\gamma}\,\bar{\xi}^2} \quad ; \quad \lambda_- = \frac{\gamma+1}{2\gamma}\,\bar{V}.\bar{\xi} - \sqrt{\left(\frac{\gamma-1}{2\gamma}\,\bar{V}.\bar{\xi}\right)^2 + \frac{a^2}{\gamma}\,\bar{\xi}^2} \qquad (9)$$

The main property of the system(8) is to be hyperbolic, i.e the four eigenvalues λ are real and the four left eigenvectors are linearly independent, and so, constitute a basis of the space where the solution is searched. The hyperbolicity with respect to time allows the time integration from any given initial conditions of the system (8). The equations (8), expressed in the basis of the left eigenvectors, are called compatibility relations. They can be written :

$$\bar{\alpha}\left(\frac{\partial \bar{f}}{\partial t} + \lambda\frac{\partial \bar{f}}{\partial \xi}\right) = \bar{\alpha}\left(-\bar{\bar{A}}_2\,\partial\bar{f}/\partial\xi_2 - \bar{\bar{A}}_3\,\partial\bar{f}/\partial\xi_3\right) \qquad (10)$$

or locally as :

$$\left(\frac{\partial}{\partial t} + \lambda\frac{\partial}{\partial \xi}\right)\bar{\alpha}_0.\bar{f} = \bar{\alpha}\left(-\bar{\bar{A}}_2\,\partial\bar{f}/\partial\xi_2 - \bar{\bar{A}}_3\,\partial\bar{f}/\partial\xi_3\right) \qquad (11)$$

where λ and $\bar{\alpha}_0$, evaluated in P at the instant t_0, are fixed quantities. The systems (8) and (10) are equivalent. If the partial derivatives of $\bar{f}$ relative to $\bar{\xi}_2$ and $\bar{\xi}_3$ are known at the time t_0, each equation (10) appears locally as a transport equation (11) of the quantity $\bar{\alpha}_0.\bar{f}$ along the straight line of slope $\frac{1}{\lambda}$ in the plane ($\bar{\xi}$, t).

The treatment of the boundary conditions uses these compatibility relations. Let P be the current point of a boundary element, $\bar{\xi}$ being a unit vector, normal to the boundary element, and directed from the inside to the outside. Each equation (11) represents, for increasing times, a transport of information from the inside to the outside if λ is positive, and inversely if λ is negative. Therefore, the number of necessary boundary conditions is equal to the number of negative eigenvalues. So, at each point of the boundary, the system (8) can be replaced by a system composed of m compatibility relations, associated to the m positive eigenvalues, the ($4-m$) others being replaced by as many boundary conditions. A boundary being called supersonic when the normal velocity is supersonic (id. for a subsonic boundary), it can be seen from the expressions (9) that the number of negative eigenvalues may be 4(for an upstream supersonic boundary), 3 (upstream subsonic), 1 (downstream subsonic) or 0 (downstream supersonic). In the present problem of a transonic flow past a wing, the outer boundary will be chosen far enough of the wing in order to be subsonic everywhere. The three specified conditions at the upstream boundary are the velocity direction (two conditions) and the total pressure. At the downstream boundary is used the equation : $\frac{\partial p}{\partial t} + \rho_0 c_0^- \frac{\partial \bar{V}.\bar{\xi}}{\partial t} = 0$, with : $c_0^- = \lambda^- - \frac{\bar{V}.\bar{\xi}}{\gamma}$, which is an extension of the "non-reflecting" condition proposed in 1D by Hedstrom [16]. On the wall, as well as on the symmetry plane is used the simple slip condition : $\bar{V}.\bar{\xi} = 0$. Let us note that there is no special treatment at the trailing edge. A local time step technique is used, the time step at the current point being computed from a maximal value, Δt_{max} , obtained by the CFL criterion, which specifies, as well known, that the physical dependence domain at P at the time t1 must be within the numerical dependence domain. The physical dependence domain, exactly known, is limited, for a time step Δt , by the trace of the characteristic cone in a hyperplane $t = t_0 = t_1 - \Delta t$, which is here (see fig. 1) an axysymmetric ellipsoïd determined by : $\overline{PI} = -\frac{\gamma+1}{2\gamma}\,\bar{V}\,\Delta t$; $IA = \sqrt{\left(\frac{\gamma-1}{2\gamma}\bar{V}\right)^2 + \frac{a^2}{\gamma}}\;\Delta t$; $IB = a/\sqrt{\gamma}\;\Delta t$. The numerical dependence domain is approached by a sphere centered in P and whose radius R is equal to the smallest distance between this point and the neighbouring points which are used in the discretized equations. The maximum time step at each point P being : $\Delta t_{max} = \frac{R}{|\lambda max|} = R/\left(\frac{\gamma+1}{2\gamma}V + \sqrt{\left(\frac{\gamma-1}{2\gamma}V\right)^2 + \frac{a^2}{\gamma}}\right)$, the CFL criterion gives :

$\Delta t \leq \min_P \left[\Delta t_{max}(P)\right]$. The method used here goes far beyond this criterion by using at each point P a local value of Δt , $\Delta t = c_t\,\Delta t_{max}$, where c_t is an empirical coefficient ($c_t \simeq 0.8$). This technique allows important computing time savings, the time steps not being limited by the size of the smallest mesh cell. This limitation would be especially severe here, because of the small mesh cells used near the wing tip.

MESH

The computations of 3-D flows past a wing present a difficulty, related to the modelization of the flow around the wing tip. The passing round of the wing tip by the flow cannot be taken into account without a sufficiently fine mesh. To this end, a mesh system of C-O type has been developed (Fig. 2). The mesh, as the wing, is symmetrical with respect to the ($z=0$) plane. It is built in three steps : first, two 2-D C type meshes are defined in the vertical symmetry plane and in the horizontal cut half-plane. Then the 3-D mesh is built, by joining them by arcs of ellipses (f.e. around the wing tip) and straight line segments. Finally, the mesh is exactly fitted to the shape of the wing (Fig. 3). The constant-I and constant-J surfaces turn respectively around the leading edge and the tip (Fig. 4, 5). As the basic numerical scheme (3) is non symmetric, symmetrical discretisations of this scheme are used in the upper and lower half-meshes, in lifting cases, on each side of the horizontal cut half-plane. This would permit to obtain a symmetrical solution for a symmetrical wing at a zero angle of attack. Two meshes have been used, a (51x21x17) half-mesh, with (37x21) points on the wing in a non-lifting case, and a (57x50x19) mesh, with (42x50) points on the wing, in a lifting case (Fig. 3 to 7).

NUMERICAL RESULTS

A non-lifting case ($M_\infty=0.92$, $\alpha=0°$) has been computed on the symmetrical ONERA-M6 wing in a (51x21x17) half-mesh. 3000 iterations were performed with viscosity 1, the flow being initially uniform, and 6000 with viscosity 2, the mean residuals $\frac{1}{N}\sum_1^N|\rho^{n+1}-\rho^n|$ having the same value after 4200 iterations ($\simeq 1.6\ 10^{-3}$) than with viscosity 1 after 3000 iterations. With viscosity 2, the mean residuals value is about 0.5 10^{-3} after 6000 iterations. Slight oscillations appear near the shock (Fig. 9), which is located quite downstream, after a large supersonic region. The computations made with the two viscosity models are compared to computations obtained by a method based on the full potential equation in non conservative form [17] and also to experiment (Fig. 8). These results are very close on the most part of the wing. Compared to the potential flow solution, the solutions of the Euler equations present a higher peak in the leading edge region, and a shock stiffer and located further downstream, especially for the second viscosity model. Yet, the solutions obtained with this second model need more computing time. The calculated total pressure Pi, which should be constant, except through the shock, varies very slightly in most of the field (Fig. 10), except in the leading edge region, because of the coarse mesh used , and in the shock region, as expected. A lifting case ($M_\infty = 0.84$, $\alpha=3.06°$) has been computed with the first viscosity model in the (57x50x19) points mesh. The mean residuals after 3000 iterations are less than 10^{-3}. A strong shock can be observed on the upper surface of the wing, near the tip, more upstream than in the non-lifting case, at about 25% of the local chord (Fig. 11). The good coupling of the solutions computed in the upper and the lower half space appear on fig. 12 and 13. Both a strong Mach jump and a good pressure continuity can be observed in the wake downstream of the profile. The evolution of the Pi/Pi_∞ ratio (Fig. 14, 15) shows that the total pressure Pi stays rather constant except in the shock region, on the upper surface, and in the leading edge region, the oscillations of the dark lines on fig 14 ($Pi = Pi_\infty$) being non significant. The gap between the Euler solution, the potential flow solution, and the experiment is less important than in the non-lifting case (Fig. 16). Compared to the potential flow solution, the solution of the Euler equations presents lower C_p values on the upper surface, upstream of the shock, the shock itself being stiffer and located slightly downstream. On fig. 17 are represented the streamlines issued from the leading edge and the tip, and on fig 18, the velocity fields on two mesh surfaces : _ a, at the trailing edge (mesh surface I = 42) ; _ b, further downstream (mesh surface I = 46, for $x/c = 1.18$). We can see the streamlines turning around the wing tip, and their roll up, downstream of the trailing edge tip.

REFERENCES

[1] Viviand, H., "Pseudo-Unsteady Methods for Transonic Flow Computation". Lecture Notes in Physics, Vol. 141, Springer-Verlag, Berlin, (1980) pp. 44-54.

[2] Viviand, H., and Veuillot, J.P., "Méthodes pseudo-instationnaires pour le calcul d'écoulements transsoniques". ONERA, Publication n° 1978-4, (1978).

[3] Laval, P., "Schémas explicites de désintégration du second ordre pour la résolution des problèmes hyperboliques non linéaires : Théorie et Application aux écoulements transsoniques", Thèse de Doctorat d'Etat Paris VI et N.T. ONERA 1981-10.

[4] Lerat, A., Sides, J. and Daru, V., "An Implicit Finite-Volume Method for solving the Euler Equations", Lecture Notes in Physics, vol.170, (1982), pp 286-295.

[5] Brochet, J., "Calcul numérique d'écoulements internes tridimensionnels transsoniques". La Recherche Aérospatiale, pp. 301-315, n° 1980-5. English translation ESA-TT 673.

[6] Enselme, M., Guiraud, D. and Boisseau, J.P. "Contribution à l'évaluation de l'intérêt de calculateurs à structure parallèle pour la résolution de systèmes d'équations aux dérivées partielles". ONERA-TP 1980-160, (1980).

[7] Chattot, J.J., Boschiero, M. and Koeck, C. "Méthodes numériques de prédiction de l'aérodynamique des missiles", AGARD-CP 336, (1982).

[8] Rizzi, A., and Eriksson, L.E. "Computation of Vortex Flow around Wings using the Euler Equations", proceedings of the IVth GAMM Conference, Vieweg-Verlag, (1981).

[9] Godounov, S., Zabrodine, A., Ivanov, M., Kraiko, A. and Prokopov, G., "Résolution numérique des problèmes multi-dimensionnels de la dynamique des gaz", Editions de Moscou, (1979).

[10] Denton, J.D. and Singh, V.K., "Time Marching Methods for Turbomachinery Flow Calculation", VKI Lecture Series 1979-7.

[11] Jameson, A., Schmidt, V., and Turkel, E., "Numerical Solutions of the Euler Equations by Finite Volume Methods using Runge-Kutta Time Stepping", AIAA paper 81-1259, Fluids and Plasma Conf. (1981).

[12] Chen, H.C., Yu, N.J., Rubbert, P.E. and Jameson, A. "Flow Simulations for General Nacelle Configurations using Euler Equations", AIAA paper 83-0539, (1983).

[13] Hitzel, S.M., and Schmidt V., "Slender Wings with Leading-Edge Vortex Separation. A challenge for Panel-Methods and Euler Codes", AIAA paper 83-0562, (1983).

[14] Huberson, S. "Etude numérique et asymptotique de nappes tourbillonnaires enroulées", thèse de 3ème cycle,Paris VI, (1979).

[15] Guiraud, J.P., and Zeytounian, R. "A Double Scale Investigation of the Asymptotic Structure of rolled up Vortex Sheets", Journal of Fluid Mechanics, 79, (1977) , pp 93.

[16] Hedstrom, G.W., "Non-reflecting Boundary Conditions for Non linear Hyperbolic Systems", J. of Comp. Physics n° 30 (1979).

[17] Chattot, J.J., Coulombeix C. and Tome C. "Calcul d'écoulements transsoniques autour d'ailes", Recherche Aérospatiale, n° 1978-4 English translation ESA TT 61.

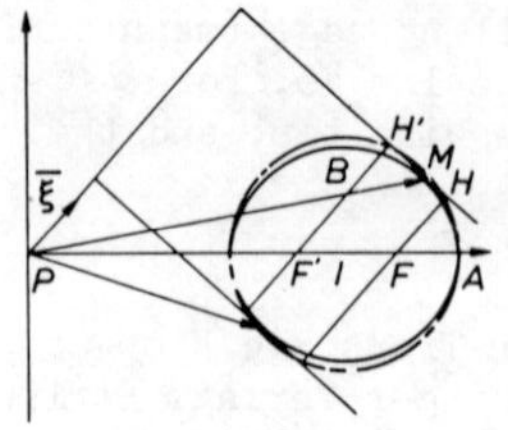

Fig. 1 — Trace of the characteristic cone in a hyperplane $t = t_0$.

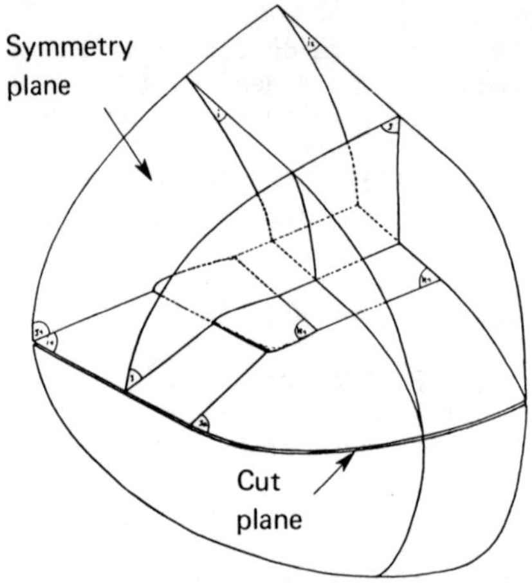

Fig. 2 — C-O type mesh.

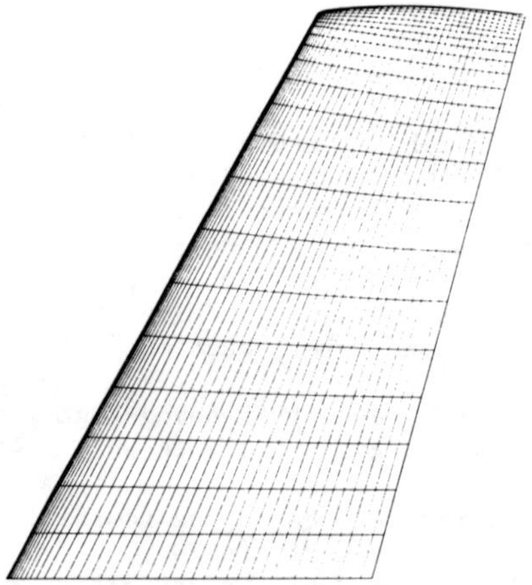

Fig. 3 — Mesh on the M6 wing surface.

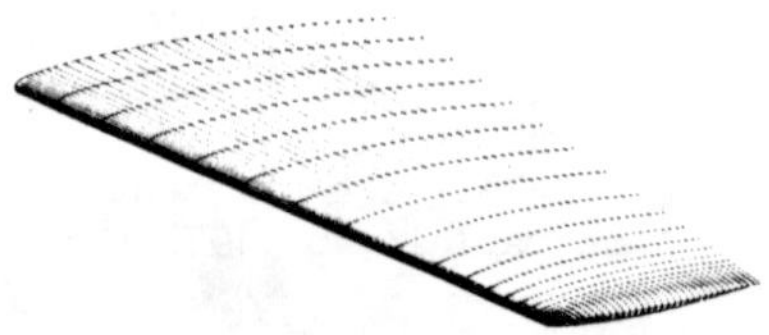

Fig. 6 — M6 wing.

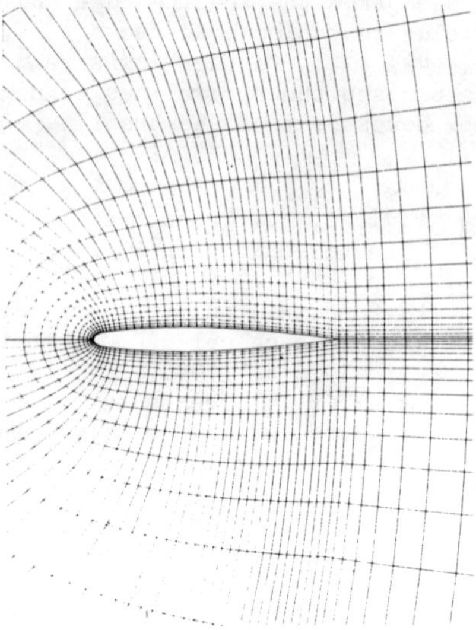

Fig. 4 — M6 wing. Mesh in the plane of symmetry (Y = 0).

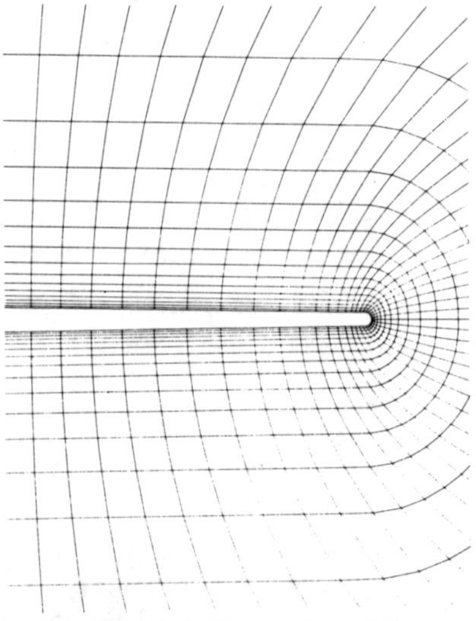

Fig. 5 — M6 wing. Planar view of the mesh surface I = 22.

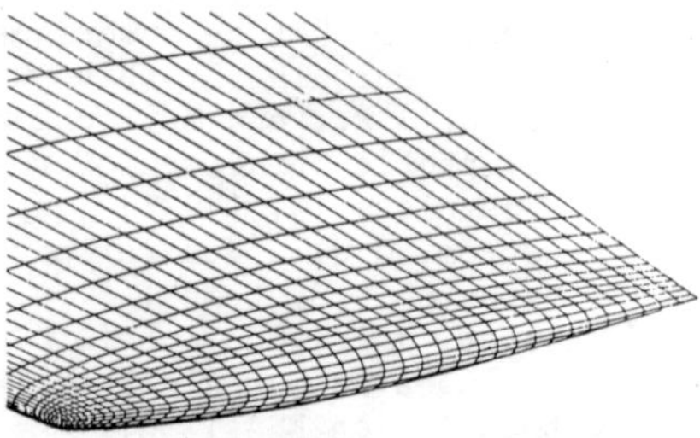

Fig. 7 — M6 wing. Wing tip mesh.

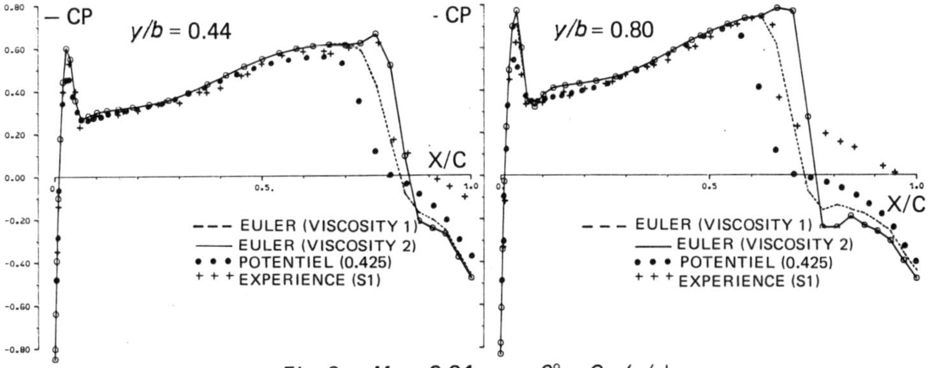

Fig. 8 — $M_\infty = 0.84$; $\alpha = 0°$; Cp (x/c).

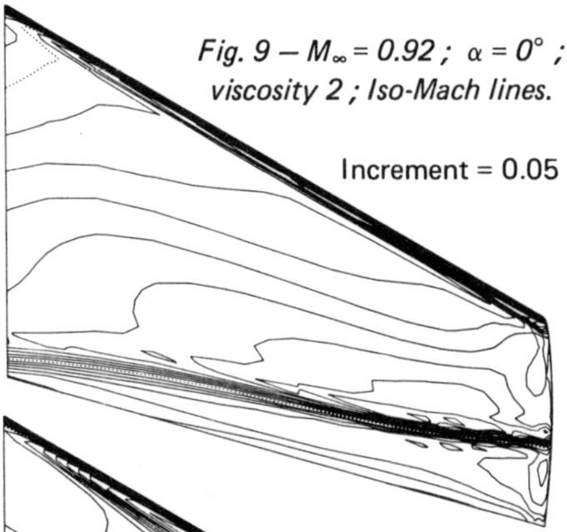

Fig. 9 — $M_\infty = 0.92$; $\alpha = 0°$; viscosity 2; Iso-Mach lines.

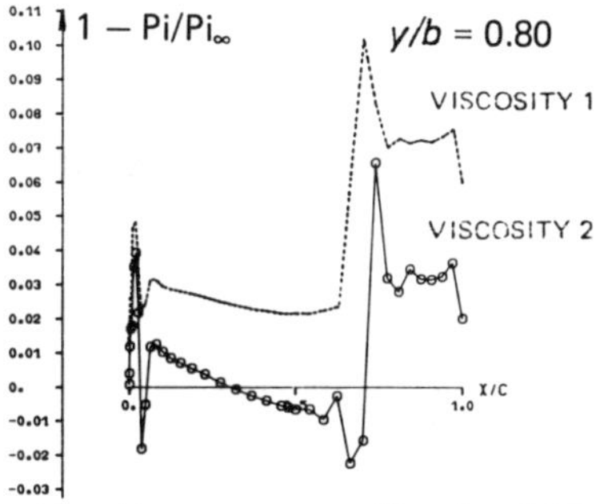

Fig. 10 — $M_\infty = 0.92$; $\alpha = 0°$; $y/b = 0.80$; $1 - Pi/Pi_\infty$ (x/c).

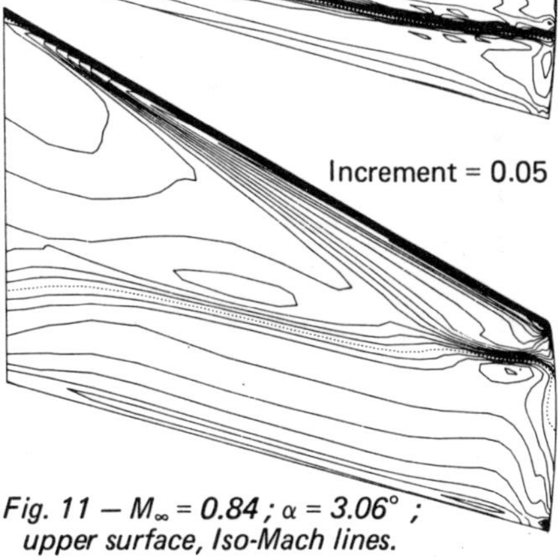

Fig. 11 — $M_\infty = 0.84$; $\alpha = 3.06°$; upper surface, Iso-Mach lines.

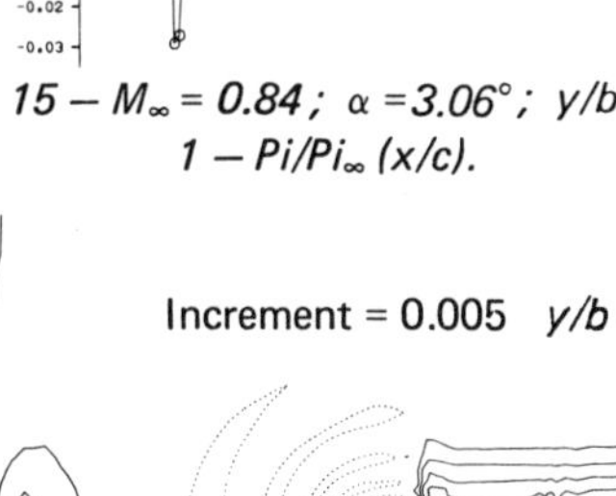

Fig. 15 — $M_\infty = 0.84$; $\alpha = 3.06°$; $y/b = 0.80$; $1 - Pi/Pi_\infty$ (x/c).

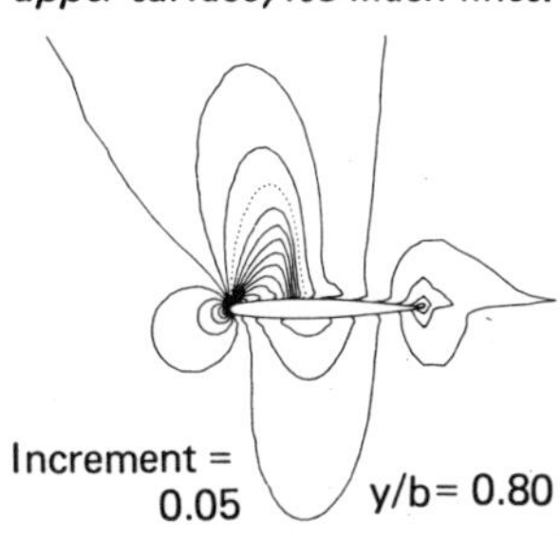

Fig. 12 — $M_\infty = 0.84$; $\alpha = 3.06°$; Iso-Mach lines.

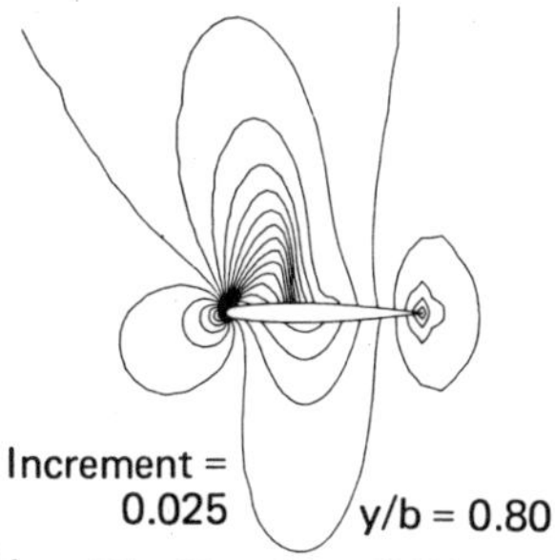

Fig. 13 — $M_\infty = 0.84$; $\alpha = 3.06°$; Iso-p lines.

Increment = 0.005 y/b = 0.80

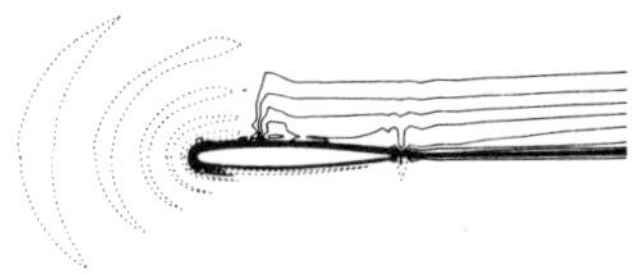

Fig. 14 — $M_\infty = 0.84$; $\alpha = 3.06°$; Iso Pi/Pi_∞ lines.

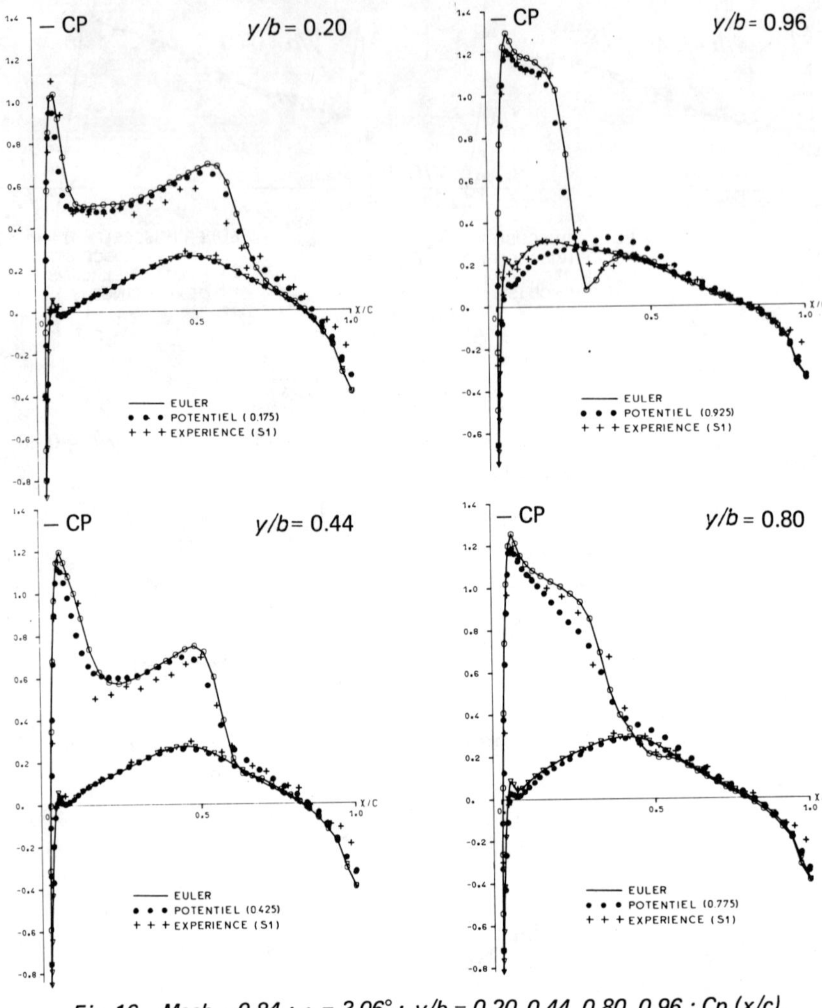

Fig. 16 – Mach = 0.84 ; α = 3.06° ; y/b = 0.20, 0.44, 0.80, 0.96 ; Cp (x/c).

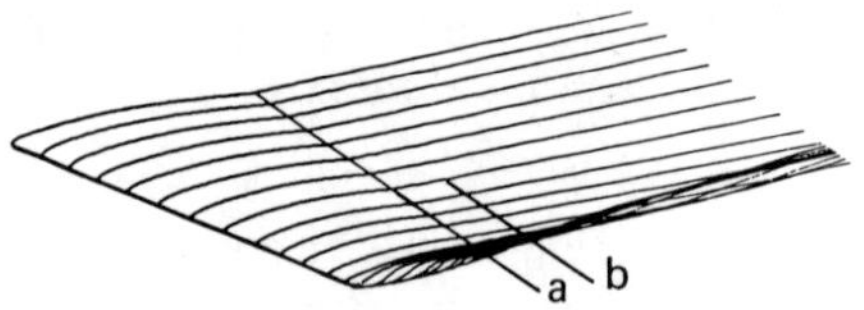

Fig. 17 – Mach = 0.84 ; α = 3.06° ; Computed streamlines.

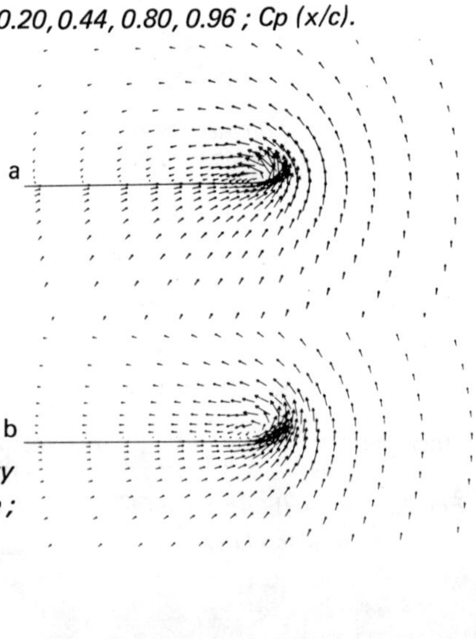

Fig. 18 – Mach = 0.84 ; α = 3.06° ; velocity field on mesh surfaces (a : I = 42. Trailing edge ; b : I = 46).

PSEUDOSPECTRAL SOLUTION OF TWO-DIMENSIONAL GAS-DYNAMIC PROBLEMS

David A. Kopriva*
Institute for Computer Applications in Science and Engineering

Thomas A. Zang
NASA, Langley Research Center

M. D. Salas
NASA, Langley Research Center

M. Y. Hussaini*
Institute for Computer Applications in Science and Engineering

SUMMARY

Chebyshev pseudospectral methods are used to compute two-dimensional smooth compressible flows. Grid refinement tests show that spectral accuracy can be obtained. Filtering is not needed if resolution is sufficiently high and if boundary conditions are carefully prescribed.

1. INTRODUCTION

Pseudospectral approximations to the compressible flow equations have recently been studied as an alternative to second or fourth order finite differences [1], [2]. The motivation is to obtain the superior accuracy characteristic of spectral solutions to smooth incompressible flows. For simple linear hyperbolic models it is easy to demonstrate that spectral approximations are indeed far superior to finite difference approximations. (See, for example, Hussaini, Salas and Zang [3].)

Hussaini, et. al. [4] have shown that spectral approximations do work in a variety of compressible flows. However, it has not been demonstrated that spectral accuracy is obtained in actual problems. Typically, more complicated flows have shocks and shock capturing spectral approximations introduce global oscillations which must be smoothed. In such cases a definite advantage of spectral over finite difference approximations has not yet been established [4]. For this reason we examine the use of spectral methods in conjunction with shock fitting algorithms. In the region where the solution is smooth there is hope of obtaining spectral accuracy.

In this paper we examine the accuracy of spectral methods when applied to smooth compressible flows. Even in such

*) Research supported by the National Aeronautics and Space Administration under NASA Contract No. NAS1-17130.

cases, spectral solutions sometimes exhibit "wiggles". We look at the need for smoothing the solutions to such problems. Two benchmark problems which are non-trivial and two-dimensional are used. The first is the interaction of a plane wave with a shock. The second is the classical Ringleb flow.

2. PSEUDOSPECTRAL METHOD

The problems in the next two sections use the Euler equations in non-conservation form

$$Q_t + BQ_X + CQ_Y = 0 \qquad X,Y\epsilon[0,1],\ t \geq 0 \tag{1}$$

where Q is the column vector of the unknowns and $B(Q)$, $C(Q)$ are square matrices. For the shock/plane wave interaction problem of Section 3, the Y direction is periodic. Q is approximated by a Chebyshev-Fourier expansion

$$Q(X,Y,t) = \sum_{p=0}^{M} \sum_{q=-N/2}^{N/2-1} Q_{pq}(t)T_p(\xi)e^{2\pi iqY}, \tag{2}$$

where $\xi = 2X-1$. The derivatives of the interpolant are

$$Q_X = 2\sum_{p=0}^{M} \sum_{q=-N/2}^{N/2-1} Q_{pq}^{(1,0)}(t)T_p(\xi)e^{2\pi iqY} \tag{3}$$

$$Q_Y = 2\pi\sum_{p=0}^{M} \sum_{q=-N/2}^{N/2-1} Q_{pq}^{(0,1)}(t)T_p(\xi)e^{2\pi iqY} \tag{4}$$

The coefficients $Q_{pq}^{(1,0)}$ are computed with the standard recursion formula [3] and

$$Q_{pq}^{(0,1)} = iqQ_{pq} \tag{5}$$

For the Ringleb problem a double Chebyshev approximation is used and the solution is approximated by

$$Q(X,Y,t) = \sum_{p=0}^{M} \sum_{q=0}^{N} Q_{pq}(t)T_p(\xi)T_q(\eta), \tag{6}$$

where $\eta = 2Y-1$. The derivatives in both directions are evaluated in a manner analogous to equation (3).

While the approximation of derivatives at boundaries often requires points outside the mesh, this is not the case for the spectral approximations. The derivatives use only points within the mesh and hence do not require special treatment.

The time discretization used is the second order modified Euler. Let $L(Q)$ denote the spatial discretization of $B\partial_X + C\partial_Y$ and let $t = n\Delta t$. Then

$$\tilde{Q} = [1-\Delta t L^n]Q^n$$
$$Q^{n+1} = \tfrac{1}{2}\,[Q^n+(1-\Delta t\tilde{L})\tilde{Q}] \tag{7}$$

where $\tilde{L} = L(\tilde{Q})$.

3. SHOCK/PLANE WAVE INTERACTION

The first benchmark problem is the time-dependent interaction of a plane wave with an infinite normal shock. We use this to demonstrate the appearance of wiggles in a case where the relevant features are not resolved. A detailed discussion of the problem and a comparison of spectral and finite difference computations with linear theory predictions can be found in Zang, et. al. [5]. For low amplitude waves whose wave fronts are nearly parallel to the moving shock the linear theoretical solutions are quite accurate.

The computational domain lies between some arbitrarily chosen left boundary x_L and the shock, x_S, on the right. The y direction is periodic, $-\infty < y < \infty$. This domain is mapped to the unit square by

$$X = (x-x_L)/(x_S-x_L), \quad Y = y/y_\ell \tag{8}$$

where y_ℓ is the period in y. The dependent variables are $Q = (P\ u\ v\ S)^T$ where P is the logarithm of the pressure, u and v are the velocities in x and y, and S is the entropy divided by the specific heat at constant volume.

The boundary conditions at $Y = 0$ and 1 are periodic. The right side is bounded by the moving shock and a shock fitting algorithm is used to determine the flow variables and move the shock. The left boundary is supersonic inflow so all variables are specified.

Table I shows the RMS error for the acoustic transmission coefficient of an incident 10° pressure wave for $A = 0.001$ with an $M_S = 3$ shock on three different Chebyshev grids. The error is defined by $e^2 = \sum_{p=0}^{M} (A'_p-A'_e)^2/M$. The transmission coefficient A'_p is taken as the fundamental Fourier amplitude in the Y direction at each grid point in X. The linearized solution is A'_e.

Figure 1 shows A' as a function of x behind the shock for the $N = 16, 32$ Chebyshev grids. The solid line shows the linear theory results for a constant amplitude wave. The numerical wave is started up smoothly. Because the 16 point mesh cannot resolve the startup rise, large oscillations are present. (Note that at the time chosen the beginning of the wave occurs near the coarsest grid spacing.) Without smoothing, if the solution is adequately resolved as in the 32 point calculation, the oscillations are almost eliminated. This conclusion is also reached by Gottlieb, et. al. [6] for both Fourier and Chebyshev approximations.

TABLE I

RMS Error in Acoustic Transmission Coefficient

Number of Chebyshev Modes	RMS Error
8	13.0
16	2.4
32	0.062

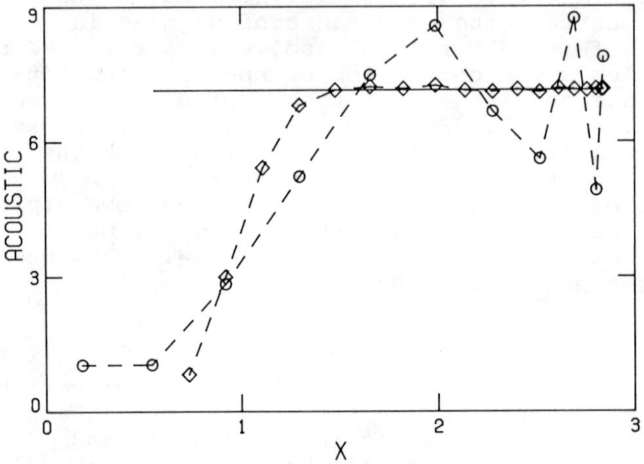

Figure 1: Acoustic transmission coefficient computed with a 16 (circles) and 32 (diamonds) point Chebyshev grid. The solid line is the linear theory prediction.

4. RINGLEB FLOW

The classical Ringleb flow is used for the second benchmark problem. The dependent variables $Q = (P\ u\ v)^T$ were chosen where u,v are the Cartesian velocity components. Since the flow is homentropic we simply set the entropy constant. The flow geometry is computed from the exact solution by specifying the Mach numbers at the inflow and outflow along the lower streamline and the outflow Mach number along the upper streamline. Figure 2 shows the Mach contours of the transonic problem used. The geometry is mapped onto a square by a transformation to (ϕ,ψ), the potential-streamfunction coordinates, which are computed from the exact solution. A double Chebyshev grid is then used in these coordinates. Figure 3 shows a 17×9 point grid for the flow of fig. 2. The supersonic flow uses only the exit portion of the channel.

In the mapped coordinate system (ϕ,ψ) correspond to (X,Y) in eq. (1). The coefficient matrices are

$$B = \begin{bmatrix} U & \gamma\phi_x & \gamma\phi_y \\ \frac{a^2}{\gamma}\phi_x & U & 0 \\ \frac{a^2}{\gamma}\phi_y & 0 & U \end{bmatrix} \qquad C = \begin{bmatrix} V & \gamma\psi_x & \gamma\psi_y \\ \frac{a^2}{\gamma}\psi_x & V & 0 \\ \frac{a^2}{\gamma}\psi_y & 0 & V \end{bmatrix} \tag{11}$$

where a is the sound speed and γ is the ratio of specific heats. The contravariant velocity components are

$$\begin{aligned} U &= u\phi_x + v\phi_y \\ V &= u\psi_x + v\psi_y \end{aligned} \tag{12}$$

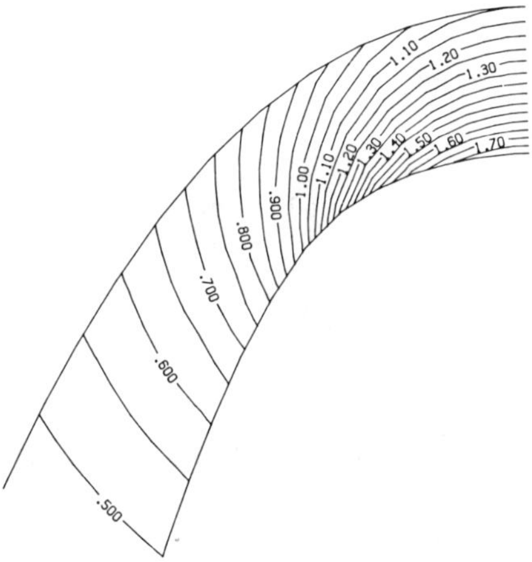

Figure 2: Mach number contours for the exact Ringleb solution.

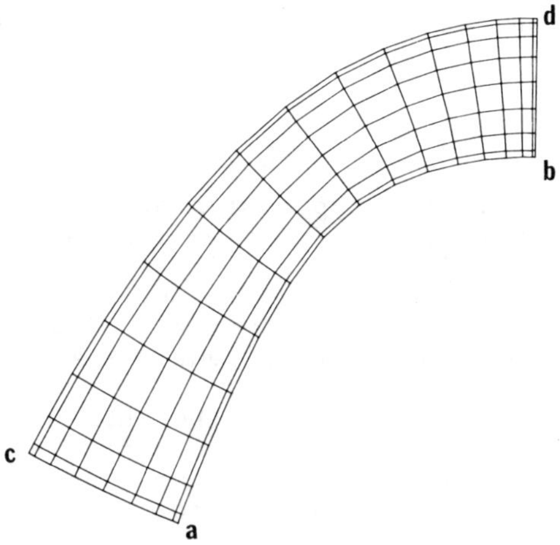

Figure 3: The 17 x 9 Chebyshev-Chebyshev grid for the flow in fig. 2.

The specification of the boundary conditions has turned out to be a most important aspect of computing the Ringleb problem. Reference [7] details several studies applied to finite difference methods.

The Ringleb problem requires several types of boundary conditions. First, the upper and lower streamlines (**ab** and **cd** in fig. 3) are treated as impermeable boundaries, hereafter referred to as walls. The outflow boundary **bd** is chosen to be supersonic. Finally, the inflow boundary **ac** can be either a subsonic or supersonic boundary depending on where it is placed along the channel.

For the wall boundaries the tangential momentum equation can be written as

$$U_t + U\left(u_\phi \phi_x + v_\phi \phi_y\right) + \frac{a^2}{\gamma}|\nabla\phi|^2 P_\phi = 0 \tag{13}$$

The equation is left in this form without explicitly writing the ϕ derivative of the contravariant velocity, U, because the derivatives of u and v are available from the Chebyshev interpolant. The time discretization is performed as in eq. (7). From the fact that the contravariant velocity, V = 0 along the wall, u and v can be determined.

Particular care must be used in specifying the wall pressure when using spectral methods. An example of the disastrous results which can occur when boundary conditions are overspecified can be seen in reference [4]. Computing the pressure from the enthalpy or directly from the pressure equation are also unsatisfactory. Such boundary conditions produce wiggles even for finite difference computations. The only approach which works effectively is to use the compatibility relation for the characteristics intersecting the wall from the interior of the flow.

By combining the pressure and normal momentum equations, an equation for the pressure is

$$P_t = \mp a|\nabla\psi| P_\psi - \Big[UP_\phi + \gamma(u_\psi\psi_x + u_\phi\phi_x + v_\psi\psi_y + v_\phi\phi_y) \pm \frac{\gamma U}{a|\nabla\psi|}(u_\phi\psi_x + v_\phi\psi_y)\Big] \tag{14}$$

where the upper sign applies to the lower wall and the lower sign to the upper wall boundary. The equation is updated according to eq. (7).

The supersonic outflow and inflow boundaries pose no difficulties. At the inflow all the quantities are specified. The outflow requires no boundary condition, either physical or numerical.

Finally, for the subsonic inflow we specify the total enthalpy and the angle of the flow. Typically this leads to a faster approach to the steady state. A compatibility condition combining the normal momentum equation and the pressure equation is

$$P_t + (U - a|\nabla\phi|)P_\phi - \frac{\gamma}{|\nabla\phi| a}\big[U_t + U(u_\phi\phi_x + v_\phi\phi_y)\big] = -\gamma(u_\psi\psi_x + u_\phi\phi_x + v_\psi\psi_y + v_\phi\phi_y) \tag{15}$$

Since the total enthalpy is taken to be a constant along the inflow boundary, another relation between P and U can be obtained by differentiating the total enthalpy equation in time

$$P_t = -\frac{U}{|\nabla\phi|^2}\exp\Big(-\frac{\gamma-1}{\gamma}P\Big)U_t \tag{16}$$

Solving eq. (15) and (16) allows both P_t and U_t to be computed. From the computed U and the fact that V = 0, the Cartesian velocities are calculated.

The fully supersonic flow is a relatively easy problem to compute. The exact solution was chosen as the initial condition and the computations were run long enough for errors to propagate out of the mesh. The time steps were kept small so that the errors would be dominated by the spatial discretization. A grid refinement study is presented in Table II where the maximum error in the pressure from the spectral calculations are compared to second order MacCormack finite

TABLE II
Maximum Error in p for MacCormack and Spectral Computation of Supersonic Ringleb Flow

Grid	MacCormack	Spectral
5 x 5	2.2×10^{-2}	7.5×10^{-4}
9 x 9	4.1×10^{-3}	1.1×10^{-6}
17 x 17	1.0×10^{-3}	6.6×10^{-11}

difference results. The superior error convergence for the spectral computations is clear.

The computation of the transonic flow depicted in fig. 2 is more difficult than the supersonic flow. The reason is the presence of the sonic line and the rapid expansion to sonic conditions along the inner wall. The computations were started with the exact solution and run for approximately the same length of physical time. The slow, explicit time integration method used does not allow relaxation to convergence. The Mach contours of a 17 x 9 point calculation are shown in fig. 4 and can be compared directly with fig. 2. The largest errors occur near the high curvature section of the lower wall, near the sonic line, and at the lower inflow corner. A grid refinement study is shown in Table III. The results are not as spectacular as the supersonic case, but the spectral still outperforms the finite difference computations.

Finally, no filtering was needed for the Ringleb problem either for the supersonic or the transonic cases. Solutions with wiggles result from boundary conditions other than the ones which we described. Application of the compatibility relations at the boundaries appears to be the best approach.

TABLE III
Maximum Error in p for MacCormack and Spectral Computation of Transonic Ringleb Flow

Grid	MacCormack	Spectral
9 x 5	2.6×10^{-2}	2.2×10^{-2}
17 x 9	1.1×10^{-2}	1.9×10^{-3}
33 x 17	3.2×10^{-3}	5.0×10^{-5}

CONCLUSIONS

We have shown that for two-dimensional smooth flows it is possible to obtain spectral accuracy characteristic of more simple problems. The shock/plane wave interaction needed no smoothing for stability. Oscillations were significant only if the flow was not well resolved. The Ringleb problem provided a more general boundary value test. Careful specification of the boundary conditions allowed the computation to be performed without smoothing.

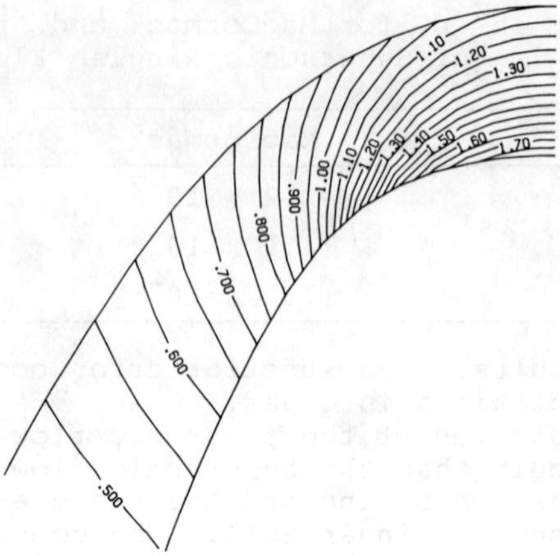

Figure 4: Computed Mach number contours of the transonic flow for the 17 x 9 grid.

REFERENCES

[1] Turkel, E., "On the practical use of high order methods for hyperbolic systems", J. Comp. Phys., Vol. 35, (1980) pp. 319-340.

[2] Zang, T. A., Hussaini, M. Y., "Mixed spectral/finite difference approximations for slightly viscous flows," Lecture Notes in Phys., 141 Springer-Verlag (1980) pp. 461-466.

[3] Hussaini, M. Y., Salas, M. D., Zang, T. A., "Spectral methods for inviscid compressible flows", in Advances in Computational Transonics, G. Habashi, ed., Pineridge Press, Swansea, United Kingdom (1983).

[4] Hussaini, M. Y., Kopriva, D. A., Salas, M. D., Zang, T. A., "Spectral methods for the Euler equations," AIAA paper 83-1942.

[5] Zang, T. A., Kopriva, D. A., Hussaini, M. Y., "Pseudospectral calculation of shock turbulence interactions," Third Intl. Conf. on Numerical Methods in Laminar and Turbulent Flow., C. Taylor, ed., Pineridge Press, (1983).

[6] Gottlieb, D., Orszag, S. A., Turkel, E., "Stability of pseudospectral and finite difference methods for variable coefficient problems", Math. Comp., 37 (1981) pp. 293-305.

[7] K. Förster, ed., Boundary Algorithms for Multidimensional Inviscid Hyperbolic Flows, Vieweg, Braunshweig, (1978).

THE COMPUTATION OF THREE-DIMENSIONAL TRANSONIC FLOWS WITH AN EXPLICIT-IMPLICIT METHOD

W. Kordulla
DFVLR Institute for Theoretical Fluid Mechanics
Bunsenstrasse 10, D-3400 Göttingen

SUMMARY

Recently MacCormack presented an explicit-implicit method to integrate the time-dependent Reynolds-averaged Navier-Stokes equations. The present paper reports on the extension of the method to three dimensions based on an integral formulation for general, surface-oriented coordinates. Computations of the laminar transonic flow past a hemisphere-cylinder configuration are used to validate the code, and comparisons are made with both, experimental pressure data and other theoretical results as obtained with a fully implicit finite-difference method. The code has been vectorised for use on a CRAY-1S computer.

INTRODUCTION

In 1981, MacCormack presented an explicit-implicit method to integrate the time-dependent Navier-Stokes equations, and applied the scheme to a two-dimensional shock wave boundary-layer interaction problem in the frame work of Cartesian coordinates [1]. Since then, several authors report on the application of that method. First, in order to calculate two-dimensional supersonic internal flows, based on a finite-difference formulation, the method was extended to general coordinates although the actual applications used near-Cartesian coordinates [2,3]. A finite-volume approach for general coordinates for transonic flows past airfoils was carried out by the present author in collaboration with MacCormack [4]. For such flows modifications to the orginally proposed method were necessary to enhance stability for large time steps. These modifications included the introduction of additional numerical diffusion terms, of intermediate steps to enable a smooth blending between explicitly and implicitly treated computational domains, and of the immediate cancellation of the wall normal fluxes in the implicit predictor sweeps at the wall. Horstman [5] has predicted separated turbulent trailing-edge flows at transonic Mach numbers, investigating the effect of various turbulence models, including higher order models. More recently Lawrence et al. [6] have used the method to solve the steady parabolized Navier-Stokes equations for supersonic boundary layers and two-dimensional hypersonic corner flow.

The present method extends MacCormack's explicit-implicit scheme in an integral (finite-volume) formulation to simulate three-dimensional transonic flow problems using general coordinates. For validation purposes a geometrically simple configuration, a hemisphere-cylinder combination, has been chosen. Both, experimental as well as theoretical data are available to compare with. As the use of the Baldwin-Lomax turbulence model for the Pulliam-Steger solution resulted in solutions close to those obtained with inviscid-flow assumptions (Pulliam, private communication), mainly laminar flows are considered. Furthermore this is justified as the flow is indeed laminar on the hemisphere portion and becomes turbulent somewhere downstream[7]. Within this paper the main interest is anyway in the correct simulation of the flow on the nose portion of the configuration, because the complete flow field cannot be resolved properly. The test case is for $M_\infty = 0.9$, $\alpha = 5^\circ$, and Re = 212500, formed with the sphere's radius [7]. Pulliam and Steger were the first to simulate Hsieh's experiments by means of the numerical integration of the three-dimensional Navier-Stokes equations [8]. Similar calculations were performed later on the Illiac IV computer exploiting the parallel architecture [9]. All these calculations are based on the fully implicit finite-difference scheme of Beam and Warming. Using the code for the serial computer Hsieh carried out more applications to a variety of flow cases [10,11]. More recently Reddy modified the code by introducing pseudospectral approximations based on Fourier series to compute appropriate derivatives in the circumferential direction [12]. Thereby a 40 % increase in efficiency with respect to a vectorised version of the Pulliam-Steger code is estimated for use on a CRAY computer.

MACCORMACK'S EXPLICIT-IMPLICIT ALGORITHM

MacCormack's explicit-implicit scheme is based on his well-proven second-order accurate explicit predictor-corrector method [13,14]. An implicit predictor-corrector sequence, following the corresponding explicit sweeps, enables to take large time steps, comparable to those used with fully implicit schemes [1]. As with the latter schemes additional diffusive terms are advantageous to enhance stability for large time steps. Note that due to the predictor-corrector sequence of the implicit sweeps the scheme requires only the inversion of bidiagonal systems instead of the costly inversion of the usually tridiagonal ones of fully implicit methods. Furthermore, the scheme becomes implicit only where it is really necessary. If the scheme becomes implicit, however, only the Eulerian Jacobians need to be considered [1], and the viscous terms are approximated by representative viscous coefficients, these being added to the Eulerian eigenvalues. The modifications, needed for transonic airfoil flow predictions [4], are not necessary for the type of flow considered here. A thorough discussion of the three-dimensional scheme will be given in a forthcoming paper (see also [15] for a time-split version). Here the algorithm is briefly sketched due to lack of space.

The explicit portion of the scheme is based on the integral form of the governing equations for Newtonian fluids, while the implicit portion can be viewed as an approximation to the corresponding partial differential equation, after having been differentiated with respect to time [1,4,15, 16]. The latter equation can be interpreted to allow the locally obtained changes of the explicit solution to convect in a physically appropriate manner. Due to its implicit formulation large time steps can be employed.

MacCormack's predictor-corrector scheme then takes the following form

PREDICTOR $$\Delta \underline{U}^n = -\Delta\tau \, (\Delta_+ \, {}^{\ell}\hat{\underline{F}}/\Delta x^{\ell}), \qquad \ell = 1(1)3,$$

$$(L_1 L_2 L_3)_+ CFL_2 \delta\overline{U}\_ = CFL_2 \Delta \underline{U}^n,$$

$$CFL_1 \equiv \min\,[1.0,\ 0.5CFL/CFL_{i,j,k}],\quad CFL_2 \equiv 1-CFL_1,$$

$$\underline{\overline{U}} = \underline{U}^n + CFL_2\,\delta\underline{\overline{U}} + CFL_1\,\Delta\underline{U}^n.$$

CORRECTOR

$$\Delta\underline{\bar{U}} = -\Delta\tau\ (\Delta_{-}\ {}^{\ell}\underline{\hat{\bar{F}}}/\Delta x^{\ell}), \qquad \ell = 1(1)3,$$

$$(\bar{L}_1\bar{L}_2\bar{L}_3)_{-}\ \overline{CFL}_2\ \delta\underline{U}^{n+1} = \overline{CFL}_2\ \Delta\underline{\bar{U}},$$

$$\underline{U}^{n+1} = \frac{1}{2}\ [\underline{U}^{n} + \underline{\bar{U}} + \overline{CFL}_2\ \delta\underline{U}^{n+1} + \overline{CFL}_1\ \Delta\underline{\bar{U}}].$$

Here $\Delta\tau$ is the time step, $\underline{U}$ the numerical solution vector, ${}^{\ell}\underline{\hat{F}}$ the flux across the cell surface $x^{\bar{\ell}}$ = const. and L_{ℓ} the bidiagonal one-dimensional factor of the implicit operator with respect to the direction x^{ℓ}, see Figures 1 and 2. The plus and minus subscripts indicate the forward and backward two-point differences in positive or negative direction x^{ℓ}. The overbars in the corrector sweeps indicate that the quantities are determined with predictor values. Note that the sweeping directions for the surface-tangential x^1- and x^2-sweeps are reversed every other time step such that four time steps constitute one cycle. Thus symmetry is enforced, this being reflected in the fact that the number of implicitly treated cells, for each direction considered, remains the same for predictor and corrector sweeps.

As the present method employs an integral or finite-volume formulation, the determination of the volume of the mesh cell is required. A novel formula is used here which is of particular importance in the case of moving meshes. Instead of the usual non-symmetric decomposition of a cell into five tetrahedrons, the cell is cut into six tetrahedrons, such that gaps between neighbouring cells cannot occur. The determination of the volume requires at most the computation of three volumes of tetrahedrons, and of just the scalar product of a vector, representing the diagonal of a cell, with the sum of three surface normals, if the latter are stored [17]. For the calculation of the Eulerian fluxes across the surfaces of the control volumes the knowledge of surface normals is sufficient in finite-volume formulations. For the determination of the viscous terms in the Navier-Stokes equations the distance between representative cell centers is required to approximate derivatives. Here the representative distances are obtained by dividing volumes by surfaces (see also [15]).

DISCUSSION OF RESULTS

The test case is the transonic flow past a hemisphere-cylinder combination [7] as was mentioned earlier. The configuration is sketched in Figure 2, which also indicates the axisymmetric coordinate system used in the calculations. A typical mesh is depicted in Figure 3. Note that the mesh shown has been constructed by linking the cell centers of the finite-volume grid, in order to show the locations where the solution is obtained. While the mesh points are sufficiently clustered in the hemisphere region, this is obviously not the case downstream of the hemisphere-cylinder juncture. Also the downstream boundary is only 10 radii away from the juncture. Hence the main interest is in simulating the front portion of the flow field. The first step size away from the wall is rather small, i.e. 0.00005 times the sphere's radius and 0.005 for the inviscid-flow simulation. In finite-difference computations a special procedure becomes necessary because of the mesh singularity along the nose ray, this is not the case for the finite-volume formulation where a surface simply degenerates to a line without any flux across it. The mesh size employed consists of 31 x 20 x 31 cells including boundary cells, in x^1 (i), x^2 (j) and x^3 (k) directions, see Figure 2. Using in-core storage the largest mesh possible was 42 x 20 x 31, for which no results will be presented as those do not show dramatic improvements of the solution. The flow chosen has the free stream Mach number of 0.9, an angle of attack of 5 degrees and a free stream Reynolds number based on the sphere's radius of 212500.

The code used has been vectorised nearly completely for a CRAY-1S computer as is described in [16]. Note that the vectorisation does not require sophisticated programming, instead it is essentially based on the autovectorisation features of the CRAY compiler. The comparison with the vectorised fully implicit Pulliam-Steger code shows for laminar flows that the present code needs 0.000060 seconds per cell and time step versus 0.000086 for the finite-difference solution, both using the thin layer approximation and identical time steps. Note that the finite-difference solution employs the corresponding full viscous terms on the left-hand, implicit side of the difference equations, and that the solution is implicit throughout the entire computational domain. If a scalar turbulence model is included the times increase by roughly 0.00001 second for both schemes. Considering the computation times required on an IBM 3081K computer, which were found to be roughly 30 times larger than those on a CRAY-1S, it is quite natural to ask for supercomputers of the CRAY class, if one has to tackle three-dimensional problems in reasonable times. The vectorised MacCormack code for the above problem in the laminar thin-layer version requires roughly 55 minutes of computation time for 3000 time steps. All the calculations have been carried out using the fast in-core storages. It is, however, necessary for practical global applications or detailed basic investigations of the flow structure [18], with the correspondingly larger number of mesh cells, to use external storage devices, and hence, one needs fast input/output procedures. Also the code must be re-coded because one will have in core only three to four data planes simultaneously.

Figure 4 shows the shadowgraph picture (courtesy of T. Hsieh) of the flow field considered. Pointed out are the shear layer separating from the body on the leeward plane of symmetry and the shock on the windward side. Figure 4 shows also computed curves of constant Mach number, plotted in the plane of symmetry, the sonic line being denoted by crosses. The agreement in shock location on the windward plane is good, the same holds for the separation location in the leeward plane. The lower picture has been obtained by combining two graphic plots, hence the discrepancies along the nose ray. Figure 5 compares calculated with measured pressures in terms of Cp for the leeward, horizontal and windward planes. Note that for the front portion the agreement is very good as well as for the flow far downstream, this in spite of the coarse mesh. However, the flow region in the leeward plane, where separation occurs (as can be seen from velocity profiles, not shown here) and where transition from laminar to turbulent flow is expected to take place, that region is not well simulated, because of the laminar-flow assumption. In the windward plane where the shock is located, the grid resolution is clearly not sufficient (compare Figure 3). A comparison with the Pulliam-Steger solution on the basis of the data in [8] show a better agreement of predicted with measured pressure data for the Pulliam-Steger solution in the windward plane. In the horizontal and leeward planes the present solution gives somewhat better results. In the leeward symmetry plane Pulliam and Steger report a smaller supersonic bubble than is presented in Figure 4. A final discussion of the merits of both methods ramains to be made for results based on identical meshes and time steps. The reason for the discrepancy must not be sought in the fact that the present solution uses the complete viscous terms, because the thin-layer assumption did give identical results.

In order to investigate the flow structure, which is known not to be uniquely determined by surface properties [18], the complete flow field has to be explored. This is a non-trivial task in three dimensions. The result of the starting effort is presented in Figure 6. Shown are lines of constant Mach number for several computations surfaces i = const. and k = const. (see Figure 2). The plots of constant Mach number furnish an idea of how the supersonic bubble develops in the downstream and wall normal direction, and that it encompasses the body (see also Figure 4). A more detailed investigation will be pursued in order to understand the structure of reversed flow, disclosed by various velocity plots.

Before the concluding remarks it is mentioned that an inviscid and a turbulent flow solution have also been obtained, using the same grid as shown above.

CONCLUDING REMARKS

With the transonic flow past a hemisphere-cylinder configuration as test case it has been shown that the explicit-implicit method, presented recently by MacCormack, can be an accurate and efficient tool to predict three-dimensional transonic flow fields. For supersonic flows past blunt fins this has been shown earlier [15]. The judgement is based on a comparison with experimental data as well as with other theoretical results as obtained with a fully implicit finite-difference solution. Thereby the main emphasis was put on the prediction of the laminar flow on the spherical nose of the configuration. The scheme is easily vectorisable for a CRAY-1S computer, indeed, the present code has been nearly completely vectorised by using just the compiler's autovectorisation features. The physical results presented are encouraging, indicate, however, that basic investigations of separated flows in the sense of [18,19] would require to pursue probably one of the following approaches: either to use the concept of zonal methods by overlaying finer meshes where needed, if the arrays are to be stored in-core, or to use external data storages to enable finer global meshes which, however, will require advanced high-speed input/output devices. As usual, of course, the modeling of turbulence and, in particular, transition would then be the major pacing item. For the present test case, for example, one knows that the flow is laminar on the hemisphere, but it is not known where it turns turbulent.

ACKNOWLEDGEMENTS

The author thanks T. Hsieh for the provision of his data in tabular form and for the copy of a shadowgraph picture. He also thanks M. Vinokur for helpful discussions concerning finite-volume formulations and T.H. Pulliam for using his code while the author spent some time as NRC Senior Research Associate at the NASA Ames Research Center. The help of M. Neher and H. Graf, DFVLR Göttingen, in preparing the graphic package used is also greatly appreciated.

REFERENCES

[1] R.W. MacCormack, A numerical method for solving the equations of compressible viscous flow, AIAA paper 81-110, 1981 (see also AIAA Journal, Vol. 20, No. 9, Sept. 1982, pp 1275-1281).

[2] E. von Lavante and W.T. Thompkins, An implicit bi-diagonal numerical method for solving the Navier-Stokes equations, AIAA Paper 82-63,1982.

[3] A. Kumar, Some observations on a new numerical method for solving the Navier-Stokes equations, NASA TP-1934, 1981.

[4] W. Kordulla and R.W. MacCormack, Transonic-flow computations using an explicit-implicit method, Proceedings, 8th Int. Conf. Numerical Methods in Fluid Dynamics, Lecture Notes in Physics, Vol. 170, Springer-Verlag, 1982, pp 420-426.

[5] C.C. Horstman, Prediction of separated asymmetric trailing-edge flows at transonic Mach numbers, AIAA Paper 82-1021, 1982.

[6] S.L. Lawrence, J.C. Tannehill and D.S. Chaussee, Application of the implicit MacCormack scheme to the PNS equations, AIAA Paper 83-1956, 1983.

[7] T. Hsieh, An investigation of separated flow about a hemisphere-cylinder at 0- to 90-deg incidence in the Mach number range from 0.6 to 1.5, AEDC-TR-76-112, July 1976.

[8] T.H. Pulliam and J.L. Steger, On implicit finite-difference simulations of three dimensional flow, AIAA Paper 78-10, Jan. 1978.

[9] T.H. Pulliam and H. Lomax, Simulation of three-dimensional compressible viscous flow on the Illiac IV computer, AIAA Paper 79-0206, Jan. 1979.

[10] T. Hsieh, Calculation of Viscous, sonic flow over hemisphere-cylinder at 19 degree incidence: The capturing of nose vortices, AIAA-81-0189, Jan. 1981.

[11] T. Hsieh, An investigation of surface flow pattern and pressure distribution for viscous, sonic flow over hemisphere-cylinder at incidence, in Proceedings, ed. H.H. Fernholz and E. Krause, "Three-Dimensional Turbulent Boundary Layers", IUTAM Symposium Berlin, 1982, Springer-Verlag, 1982.

[12] K.C. Reddy, Pseudospectral Approximation in a Three-Dimensional Navier-Stokes Code, AIAA Journal, Vol. 21, 1983, pp 1208-1210.

[13] R.W. MacCormack, The effect of viscosity in hypervelocity impact cratering, AIAA paper 69-354, 1969.

[14] R.W. MacCormack, The numerical solution of the interaction of a shock wave with a laminar boundary layer, Proceedings, 2nd Int. Conf. Numerical Methods in Fluid Dynamics, Lecture Notes in Physics, Vol. 8, Springer Verlag, Berlin, pp 151-163, 1971.

[15] C.M. Hung and W. Kordulla, A time-split finite-volume algorithm for three-dimensional flow-field simulation, AIAA paper 83-1957, 1983.

[16] W. Kordulla, MacCormack's methods and vectorisation, in: W. Gentzsch, Vectorisation of Computer Programs with the Application to Computational Fluid Dynamics, to appear in "Notes on Numerical Fluid Mechanics", Vieweg Verlag, Wiesbaden, 1984.

[17] W. Kordulla and M. Vinokur, Efficient computation of volume in flow predictions, AIAA Journal, Vol. 21, pp 917-918, 1983.

[18] U. Dallmann, Topological structures of three-dimensional vortex flow separation, AIAA Paper 83-1735, July 1983.

[19] U. Dallmann, Three-dimensional vortex separation phenomena - a challenge to numerical methods -, 5th CAMM-Conference on Numerical Methods in Fluid Mechanics, Oct. 1983, Rome.

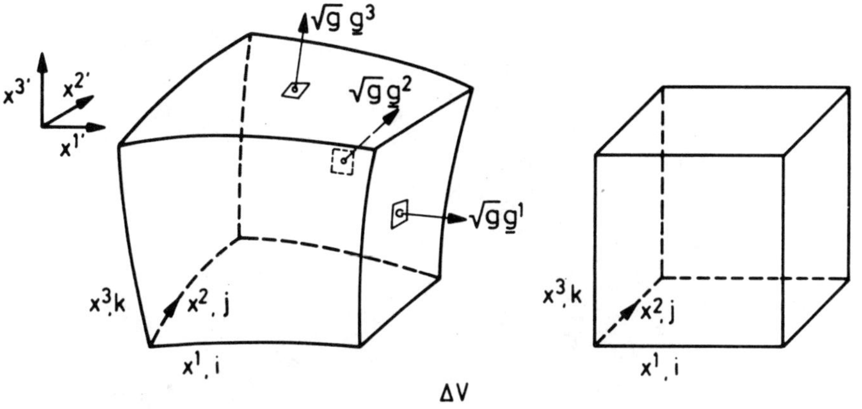

Fig. 1: Sketch of the coordinate system.

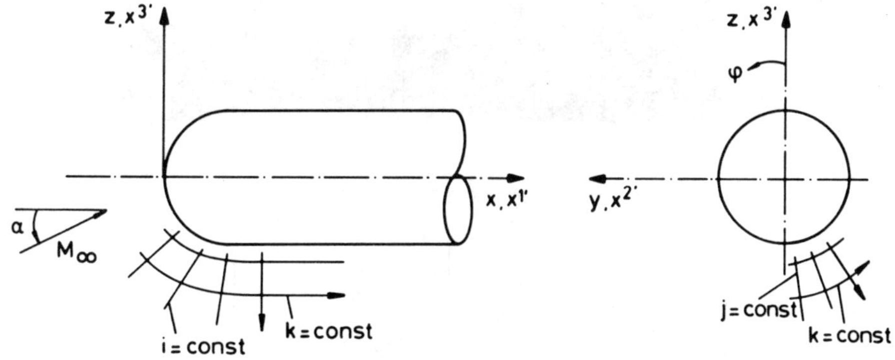

Fig. 2: Sketch of the hemisphere-cylinder configuration.

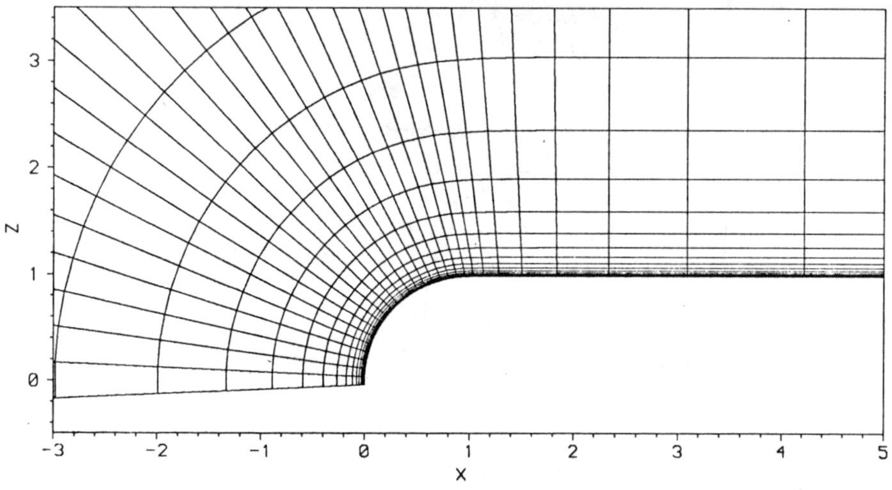

Fig. 3: A typical mesh for j = const. (volume centers).

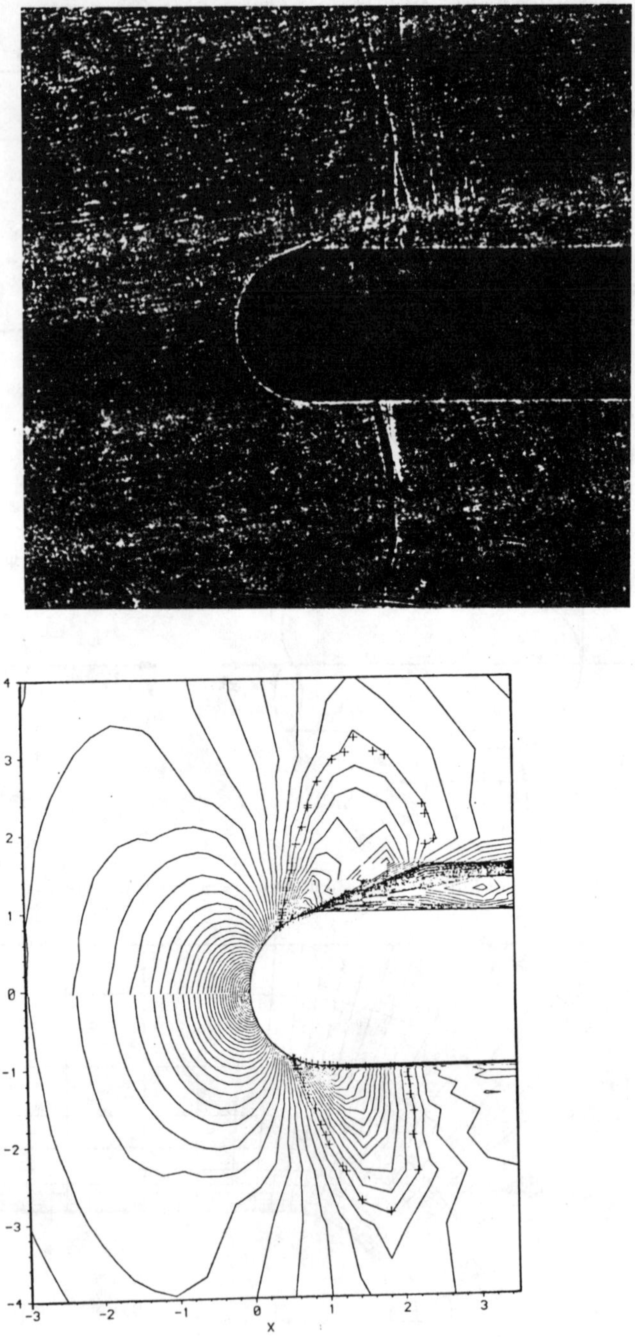

Fig. 4: Shadowgraph picture (top, courtesy of T. Hsieh) and plot of Mach number contours (+ sonic) in the symmetry plane (M_∞ = 0.9, α = 5°, Re = 212500).

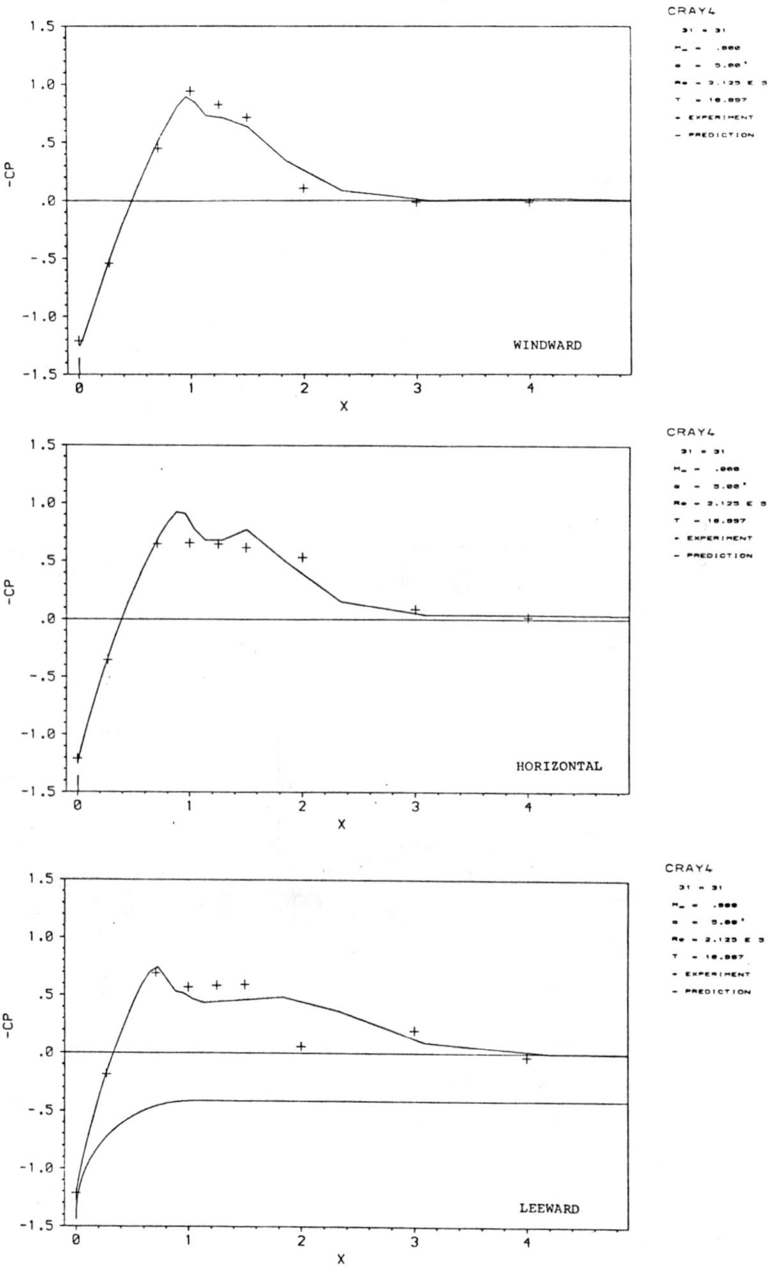

Fig. 5: Comparison of predicted with experimentally obtained C_p value.

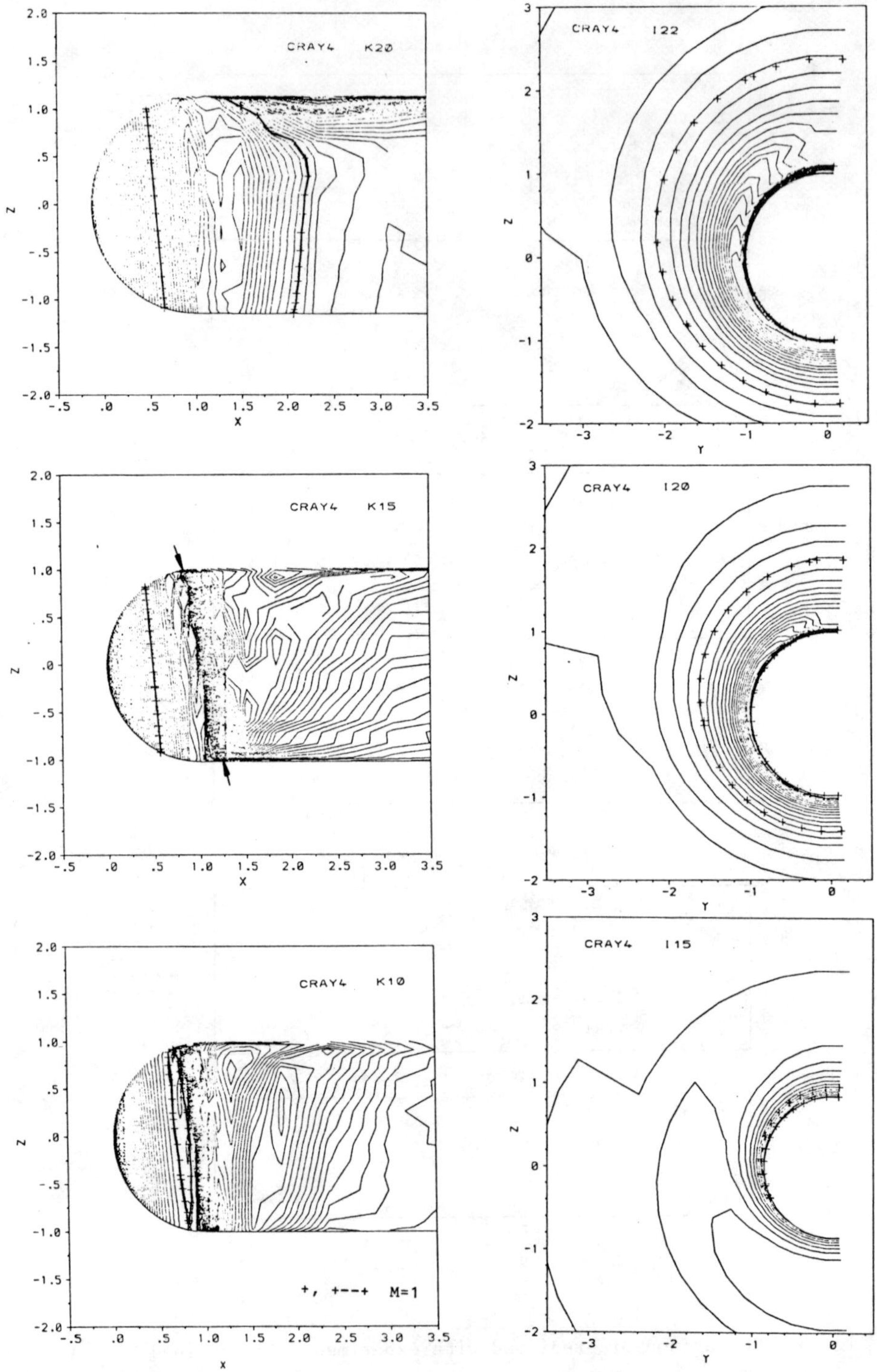

Fig. 6: Mach number contours for several surfaces
i = const. and k = const. (M_∞ = 0.9, α = 5°, Re = 212500

A SUBDOMAIN DECOMPOSITION TECHNIQUE AS AN ALTERNATIVE FOR TRANSONIC POTENTIAL FLOW CALCULATIONS AROUND WING-FUSELAGE CONFIGURATIONS

by T.H. Lê
Office National d'Etudes et de Recherches Aérospatiales (ONERA)
92320 Châtillon (France)

ABSTRACT

As an extension of the author's previous work on potential transonic flow calculations about wing-body configurations using a subdomain decomposition technique, this paper deals with the reliability of this approach to perform computations with more severe flow conditions and more realistic geometries. Two test cases are selected : the W_AB_2 configuration of the RAE at a free-stream Mach number of 0.9 and at an angle of attack of 1° and the DFVLR/Garteur F.4 wing-body at a free-stream Mach number of 0.75 and at an angle of attack of 0.1°.

1. INTRODUCTION

Successful computational algorithms for the transonic potential equation have evolved to solve three-dimensional problem [1-5] with increasing complex geometries up to an aircraft [6].

Commonly, computations around wing-fuselage configurations are performed in a single domain using contour conforming meshes for the discretization. With such a single grid system, control of grid point distribution, skewness, and clustering are difficult to achieve for a complex configuration [7].

Indeed, simultaneous adaptation of the grid to each aircraft component and high gradient zone becomes an almost impossible task. This is a strong motivation to study a solution method with subdomain decomposition techniques.

Multi-domain or zonal approaches have been also investigated in recent years by several authors [8-13] as practical techniques for analysis of complex flow field problems.

As an extension of the author's previous work [10] on transonic potential flow calculations about wing-body configurations using a subdomain decomposition technique, this paper deals with the reliability of this approach to perform computations with more severe flow conditions and more realistic geometries.

1.1 Subdomain decomposition technique

The basic idea underlying this method is to construct grid systems which are fitted for each component of the configuration. In this approach, the computational domain is divided into several overlapping subdomains, which are defined according to the different components of the configuration. Figure 1 shows a sketch of the decomposition of the flow field into two overlapping subdomains, a fuselage subdomain S_f and a wing subdomain S_W. The overlapping region is limited by two plane surfaces R_W

and R_f which belong to the boundaries of the wing subdomain and of the fuselage subdomain respectively.

In each subdomain surface-fitted grids are generated separately with the minor restriction that R_W and R_f are mesh surfaces for both subdomains but with different distributions of mesh points for each subdomain.

The full potential equation is solved in non conservative form by a finite difference SLOR algorithm, which is implemented independently in each subdomain [14].

In the present method using the Schwarz's alternating algorithm [15] the solution process is performed in cycles : it starts by calculating a solution for the flow field in wing subdomain. After a number of SLOR iterations, the distribution of velocity potential in R_f boundary is used as Dirichlet condition for calculation in the fuselage subdomain. Calculation of a solution in the fuselage subdomain is then started and a number of SLOR iterations are performed. This distribution of the velocity potential is used as a Dirichlet condition for the wing subdomain. This constitutes one cycle and the whole process is then repeated. Convergence is achieved when the error norm is below some given tolerance.

1.2 Grid Generation

Grid generation for the two subdomains is based on two completely separate coordinate transformations. Second order accurate finite differences are used to evaluate the partial derivatives of the metrics so that the solution procedure is disconnected from the grid generation step.

- Wing Subdomain Grid

Grid generation in wing subdomain is based on the so-called parabolic transformation which gives a two-dimensional grid of "C" type. The three-dimensional mesh is constructed by the usual way of stacking the two dimensional grids generated around selected spanwise sections of the wing. More details on the method can be found in ref. [1].

- Fuselage Subdomain Grid

In the present calculations, the fuselage is extended to infinity in the upstream and downstream directions by two circular cylinders which can be chosen with a small radius. (fig. 3)

A grid of "H" type is constructed in each cross-section (constant-x plane) by a sequence of transformations which take into account the change in the geometric shapes. Several cases are shown on fig. (2, a, b, c), corresponding to sections a, b, c of the F4 configuration (fig. 5) :

- a cylindrical section (circular or non circular) with a cut or a sheet (fig. 2a)

- a section containing the fuselage and the leading edge of the wing, with a cut (fig. 2b).

- a section with a portion of the wing reaching the lateral boundary R_F (fig. 2c).

In each cross section, the grid is constructed with an upper and lower part and a connection is made in order to achieve the continuity of the mesh.

The figure 2d is a partial perspective view of the mesh system around the F4-Airbus wing/fuselage configuration. The figure shows the meshes on the wing and fuselage, a part of the parabolic grid in a constant-y plane in the wing subdomain and the mesh in the plane of symmetry of the fuselage subdomain.

1.3 Numerical Results

Preliminary numerical results around a wing/fuselage combination using this approach were presented in ref. [10] . In order to illustrate the reliability of this approach computations are performed with more severe flow conditions and more realistic geometries.

The first test case consists of W_AB_2 configuration (fig. 3) of the RAE [16] at a free-stream Mach number of 0.9 and at an angle of attack of 1°.

The computational grid for the fuselage subdomain comprises 71 x 20 x 36 = 51, 120 points with 59 x 14 = 826 points on the fuselage surface.

The computational grid for the wing subdomain has 170 x 20 x 32 = 108, 800 points with 102 x 13 = 1326 points on the wing surface.

For a residual error less than 10^{-6}, about 80 cycles of Schwarz's method are necessary. For each cycle we perform, in each subdomain, about 10 iterations of SLOR algorithm at the beginning, and about 40 at the end of the computation.

On fig. (4a, b) are shown comparisons between present numerical results, numerical results of ARA [4] , and experiments from RAE [16] at two stations on the wing, including one located near the overlapping region (fig. 4a).

In general, the agreement is rather good, particularly for the upper surface where the general change in the pressure distributions in the spanwise direction is fairly well predicted. The position and strength of the shock appear also to agree reasonably well although there is a minor deviation between the two numerical results.

Fig. 4c shows comparisons between the present calculations and the experiments in the symmetry plane on the top and bottom lines of the fuselage. Again there is a fairly good agreement except in a zone on the bottom line around $x'/l = 0.5$.

The second case presented here (fig. 5) is the DFVLR/Garteur F.4 wing-body configuration [17]. Although the wing is low-mounted, the size and topology of the grid used for this case are similar to those of previous one. Iteration strategy is also similar.

The numerical results for a free-stream Mach number of 0.75 and an angle of attack of 0.1° are presented in figure 6 and compared with the experimental results obtained at ONERA [18].

On the wing, fig. 6a, b, the numerical results reflect the overall evolution of experimental data, particularly on the lower surface.

Viscous effects are known to be important on the rear part of the wing [18-19], where they tend to diminish the load, thus accounting for the present discrepancies in this area. The lack of resolution for the suction peak in the inner section may be due to coarseness of the grid near the leading edge.

On the fuselage (fig. 6c) the agreement is quite good. The presence of the cockpit can be noticed in the calculated pressures.

CONCLUSION

Using a subdomain decomposition technique to perform transonic potential either with severe flow conditions or around a realistic configuration, it has been demonstrated that this multi-domain approach is robust.

In future developments, additional subdomains could be included, to describe the nose region and the tail region, to take advantage of cartesian meshes in the far field and to take into account local viscous effect.

Acknowledgement

We gratefully acknowledge the very efficient help of D. Destarac in developing the F4-Airbus geometry pre-procesor and in generating the corresponding three-dimensional mesh.

REFERENCES

[1] Chattot, J.J., Coulombeix, C., Manie, F., Schmitt, V. "Calcul d'écoulements transsoniques autour d'ailes". D.G.L.R. Symposium Transonic Configuration Bad Harzburg, June 1978 ONERA T.P. n° 1978-67.

[2] Jameson, A., Caughey, D.A., "Progress in Finite Volume Calculations for Wing-Fuselage combinations", AIAA Journal, Vol. 18, Nov. 1980, pp. 1281-1288.

[3] Holst, T., Thomas, S. "Numerical Solution of Transonic Wing Flow Fields", AIAA Paper 82-010S, 1982.

[4] Baker, T.J., Forsey, C.R., "A Fast algorithm for the calculation of transonic flow over wing/body combinations. AIAA paper 81-101S, 1981.

[5] Chen, L.T. "A More accurate Transonic Computational Method for Wing body Configurations", ICAS Paper n° 82-0162, 1982.

[6] Perrier, P., "Computational Fluid Dynamics Around Complete Aircraft Configurations", ICAS Paper n° 82 6.1.1, 1982.

[7] Yu, N.J. "Transonic Flow Simulation for Complex Configurations with surface fitted grids", AIAA paper 81-1258, 1981.

[8] Lee, K.D. "3-D Transonic Flow computations using Grid Systems with Block Structure" AIAA Paper n° 81-0998, June 1981.

[9] Atta, E.H., Vadyak, T., "A Grid Interfacing Zonal Algorithm for Three Dimensional Transonic Flow About Aircraft Configurations", AIAA Paper n° 82-1017, June 1982.

[10] Lê, T.H., "Transonic Potential Flow Calculation about complex bodies by a technique of overlapping subdomains 8ICNMFD Aachen 1982.

[11] Cambier, L., Ghazzi, W., Veuillot, J.P., Viviand, H., "A Multi-Domain approach for the computation of viscous transonic flows by unsteady type methods". Recent Advances in Numerical Methods in Fluids, Vol. 3.

[12] Dinh, Q.V., Glowinski, R., Mantel, B., Periaux, J., Perrier, P. "Sub-domain Solutions of Non-Linear Problems in Fluid Dynamics on Parallel Processors", Computing Methods in APplied Sciences and Engineering, 1982.

[13] Morchoisne, Y., "Résolution des équations de Navier-Stokes par une méthode spectrale de sous-domaines", Numerical Methods in Engineering, 1983.

[14] Lê, T.H., "Calcul d'écoulements transoniques autour d'une configuration aile-fuselage dans une approche par domaines" La Recherche Aérospatiale, 1983, n° 3 (Mai-Juin).

[15] Schwarz, H.A., Gesammelte Mathematische Abhandleingen, Vol. 2 (1960).

[16] Treadgold, D.A., Jones, A.F., Wilson, K.H., "Pressure distribution measured in the RAE 8 ft x 6ft transonic wind tunnel on RAE wing "A" in combination with an axi-symmetric body at Mach numbers of 0.4, 0.8 and 0.9."Experimental Data Base for Computer Program Assessment', AGARD-AR-138.

[17] Redeker, G., Schmidt, N., Müller, R., "Design and experimental verification of a transonic wing for a transport aircraft" AGARD CP-285 Paper n° 13, 1980.

[18] Schmitt, V., "Aérodynamique d'un ensemble voilure fuselage du type "Avion de Transport" 18ème Colloque d'Aérodynamique Appliquée de l'AAAF, Poitiers, Novembre 1981. T.P. n° 1981-122.

[19] Lazareff, M., Le Balleur, J.C., "Calculs des écoulements visqueux tridimensionnels sur ailes transsoniques par interaction fluide parfait couche limite". La Recherche Aérospatiale n° 3 (1983).

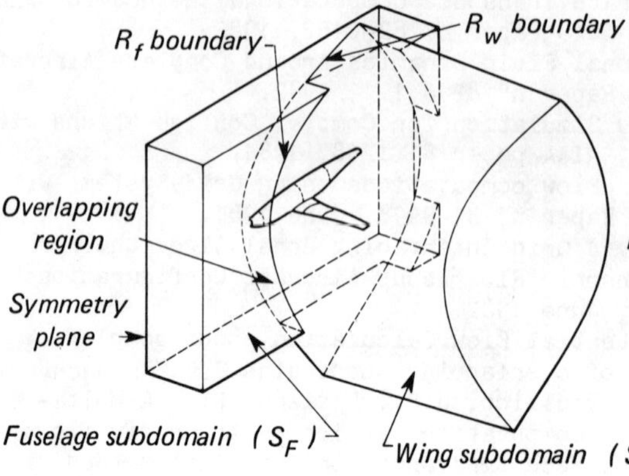

Fig .1 –Sketch of the decomposition into wing and fuselage subdomains

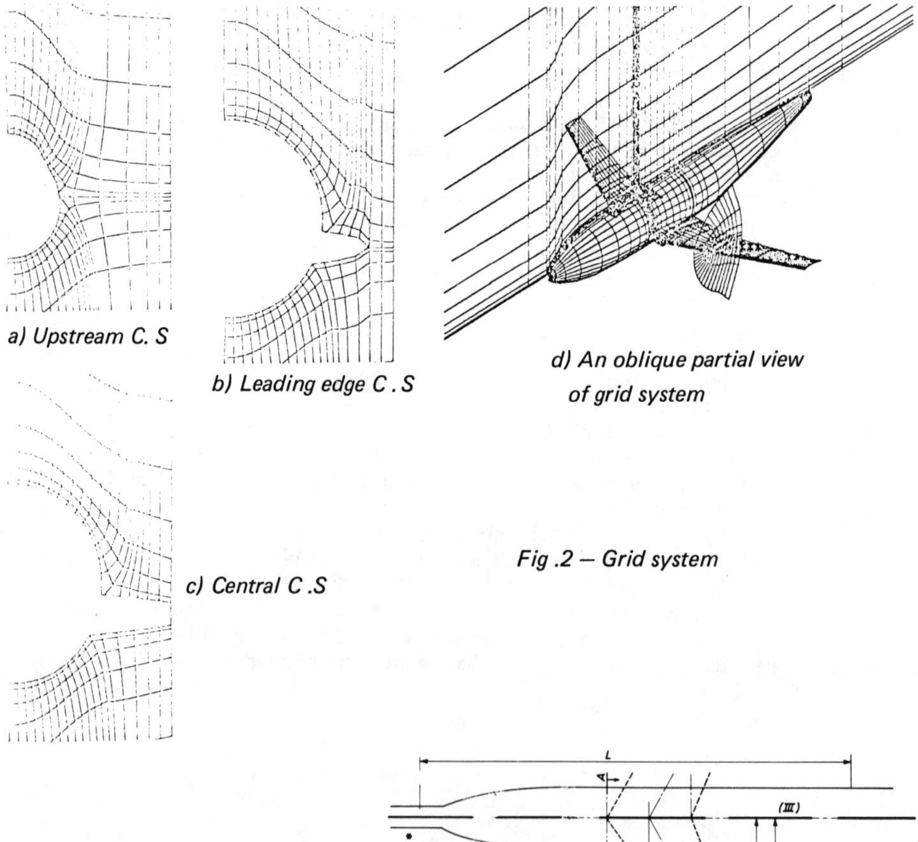

a) Upstream C. S

b) Leading edge C . S

c) Central C .S

d) An oblique partial view of grid system

Fig .2 – Grid system

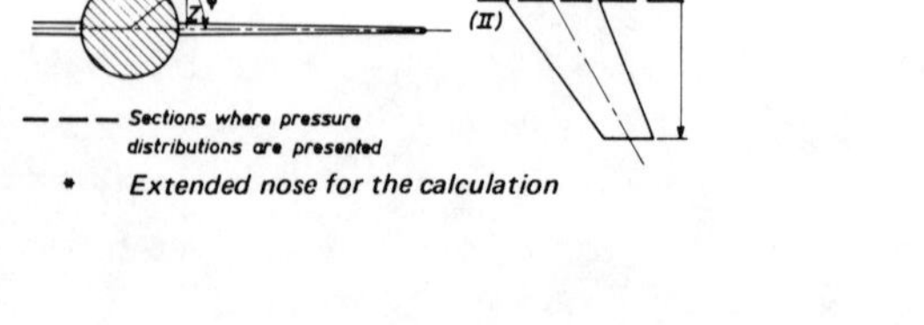

Fig .3 – Sketch of wing A mounted on body B_2; configuration W_A B_2

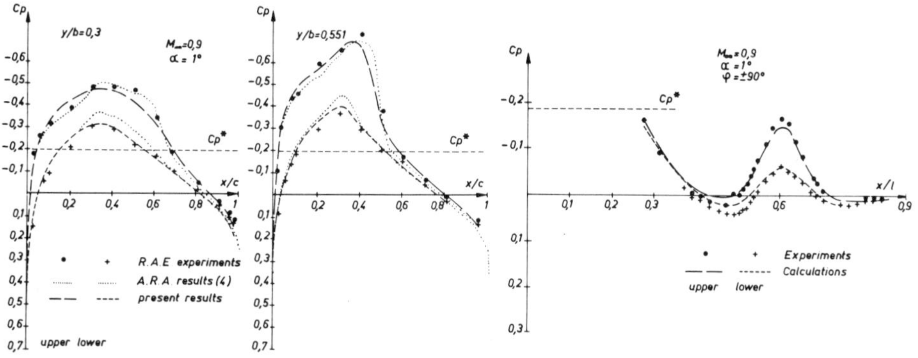

a) Wing section (I) b) Wing section (II) c) Fuselage section (III) in symmetry plane

Fig.4 – Pressure distributions

Fig.5 – Transport Aircraft configuration: DFVLR/Garteur F.4 wing Body

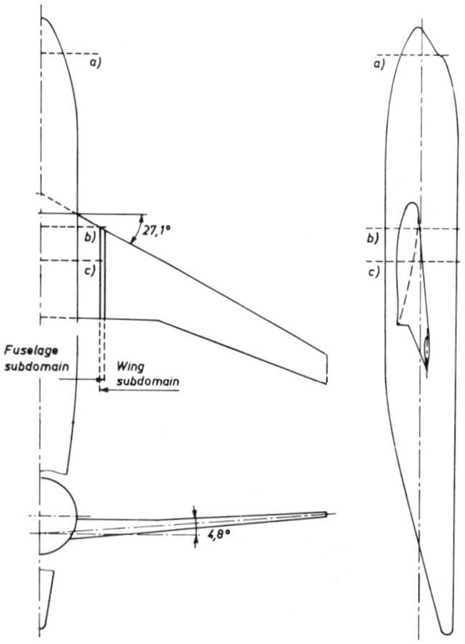

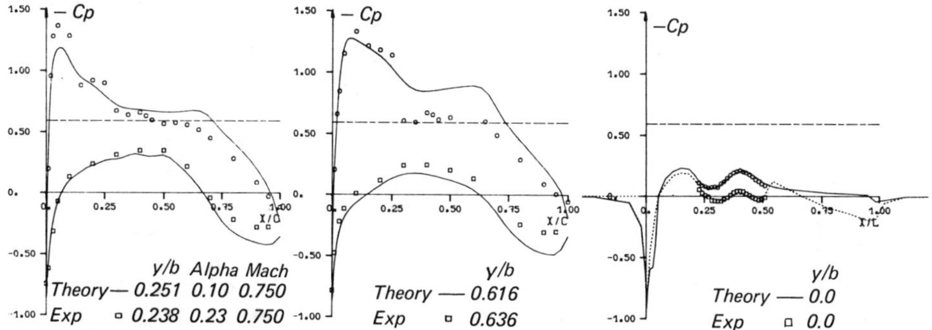

a) Wing section (I) b) Wing section (II) c) Fuselage section in symmetry plane

Fig.6 – Pressure distributions

NUMERICAL ANALYSIS OF FINITE-AMPLITUDE PERISTALTIC FLOW AT SMALL WAVELENGTH

G.V.Levina
Institute of Continuum Mechanics
of the Ural Research Center USSR Academy of Science
Acad.Korolev Street 1
614013 PERM', The USSR

SUMMARY

Consideration is given to the plane flow due to a sinusoidal peristaltic wave propagating along the walls of uniform channel. The Navier-Stokes equations in the curvilinear moving coordinate system translating longitudinally at a constant velocity equal to the wave speed are solved by the finite-difference method based on the stream function/vorticity formulation. The pressure distribution is of particular attention due to significance of the pressure drop over the wavelength as an essential characteristic of peristaltic pumping. In this paper, the compatible pressure marching method, based upon compatibility between vorticity and pressure finite-difference schemes, is applied to studies of peristaltic flows at considerable wall curvature (i.e. large wave amplitude and small wavelength) and finite Reynolds numbers. It is found, that at small wavelengths a new type of trapping can be observed, not associated with splitting of the axial streamline (non-axial trapping).

INTRODUCTION

Peristaltic pumping, being the form of fluid transport with the aid of a progressive contraction/expansion wave along the walls of the channel, is widely employed in biological systems as it is in the human body. The studies of peristaltic flows are of particular importance for better understanding of many physiological transport processes, as well as for construction of medical and engineering pumping devices (peristaltic transport used for the pumping of as blood, slurries, corrosive fluids when it is desirable to prevent transported material from coming into contact with the mechanical parts of the pump).

In the last two decades the transport phenomena created by peristalsis has become the subject of numerous investigations. Some of them are summarized in the review by Jaffrin & Shapiro [1]. In the majority of early works the problem of peristaltic flows was being solved by application of perturbation methods, using the assumption of small magnitude of one or simultaneously two constitutive parameters (Reynolds number, the wavenumber or the wave amplitude).

The more recent numerical investigations (see, for example, [2-4]), accounted for the inertia and finite deformation effects of the channel walls, introduced only some quantitative corrections for the boundaries description of the previously investigated flow regimes, and revealed no new effects.

The present paper reports the results of numerical investigations of the flow structure at large deformations of channel boundaries and finite Reynolds numbers. A new type of trapping is found to occur at small wavelengths, followed by the formation of vortex flow region without splitting of the central streamline.

GOVERNING EQUATIONS AND BOUNDARY CONDITIONS

Let us consider the plane flow of a viscous incompressible fluid in an infinite channel, the walls of which are deformed by the wave, travelling along the X-axis with the velocity C (see figure 1):

$$Y_B = \pm \left[d + b\sin\frac{2\pi}{\lambda}(X - Ct) \right] \quad , \tag{1}$$

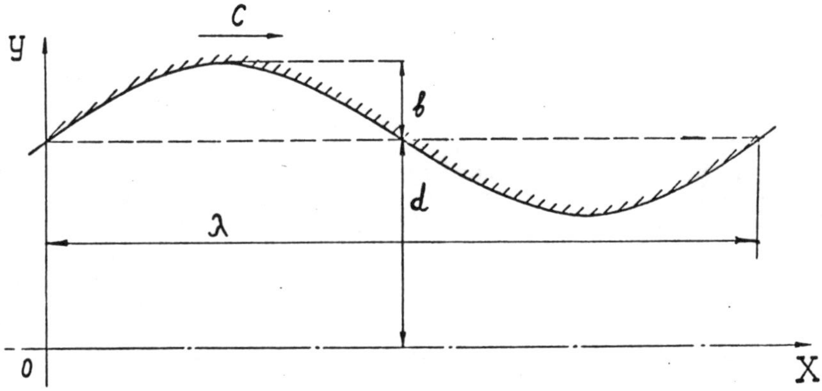

Figure 1. Geometry of peristaltic channel.

The analysis is carried out in the moving co-ordinates (x,y) (the wave frame), in which the developed flow will be treated as steady since the configuration of the boundaries appears to be stationary. Co-ordinate and velocity transformation between the laboratory frame (X,Y) and wave frame (x,y) can be described in terms of the following relations:

$$x = X - Ct \ , \quad y = Y \ ; \qquad v_x = V_X - C \ , \quad v_y = V_Y \ ; \tag{2}$$

where V_X, V_Y and v_x, v_y denote the velocity components in the laboratory and the wave frame respectively.

Then we introduce the curvilinear co-ordinates(ξ, η), enabling to flatten the walls of the channel and employ the uniform rectangular mesh in finite-difference calculations

$$\xi = x \quad , \qquad \eta = \frac{y}{1 + \zeta\sin\alpha x} \quad . \tag{3}$$

After rearrangements of (2) and (3), the Navier-Stokes equations and boundary conditions in dimensionless variables can be specified in terms of stream function Ψ and vorticity φ

$$\frac{\partial \varphi}{\partial t} + \mathrm{Re}\,a_2\left(\frac{\partial \psi}{\partial \eta}\frac{\partial \varphi}{\partial \xi} - \frac{\partial \psi}{\partial \xi}\frac{\partial \varphi}{\partial \eta}\right) = \frac{\partial^2 \varphi}{\partial \xi^2} + 2a_1\frac{\partial^2 \varphi}{\partial \xi \partial \eta} + a_3\frac{\partial^2 \varphi}{\partial \eta^2} + a_4\frac{\partial \varphi}{\partial \eta}, \quad (4)$$

$$\frac{\partial^2 \psi}{\partial \xi^2} + 2a_1\frac{\partial^2 \psi}{\partial \xi \partial \eta} + a_3\frac{\partial^2 \psi}{\partial \eta^2} + a_4\frac{\partial \psi}{\partial \eta} = -\varphi\ ; \quad (5)$$

$$\psi = \varphi = 0\ (\eta = 0);\quad \psi = q-1,\quad \frac{\partial \psi}{\partial \eta} = -(1 + \zeta \sin \alpha \xi)\ (\eta = 1);(6)$$

$$\mathrm{Re} = \frac{Cd}{\nu}\ ,\quad \zeta = \frac{b}{d}\ ,\quad \alpha = \frac{2\pi d}{\lambda}\ .$$

Here

$$a_1 = \frac{\partial \eta}{\partial x}\ ,\quad a_2 = \frac{\partial \eta}{\partial y}\ ,\quad a_3 = \left(\frac{\partial \eta}{\partial x}\right)^2 + \left(\frac{\partial \eta}{\partial y}\right)^2\ ,\ a_4 = \frac{\partial^2 \eta}{\partial x^2} \quad (7)$$

are coefficients, arising from curvilinear transformation of co-ordinates (ξ,η); ν is the kinematic viscosity of fluid, Re - the Reynolds number, ζ - the relative wave amplitude, α - the wavenumber, q - the dimensionless time-mean flow defined in fixed co-ordinates; $d, d^2/\nu$, C and Cd are taken to be the units of length, time, velocity and stream function respectively.

We shall consider the space-periodic flows with the period of $2\pi/\alpha$ equal to the peristaltic wave period:

$$\psi(\xi,\eta,t) = \psi(\xi + \tfrac{2\pi}{\alpha},\eta,t)\ ,\quad \varphi(\xi,\eta,t) = \varphi(\xi + \tfrac{2\pi}{\alpha},\eta,t). \quad (8)$$

Due to conditions (8) we may choose the size of the computation region lying on the ξ-axis to be equal to one wavelength.

COMPUTATIONAL METHOD

While searching for appropriate numerical algorithm a particular attention is concentrated on pressure field definition, since the relation between the time-mean flow and pressure drop over the wavelength is an important characteristic of peristaltic pumping. Works in computational fluid dynamics [5-6] describe the difficulties, encountered in numerical pressure calculations at finite Reynolds numbers. In case of the iterative method, which involves solving a Poisson equation with Neumann boundary conditions, convergence is slow that leads to a considerable cost of computer time. The approach, based on step integration of Navier-Stokes equations (the marching schemes) for obtaining pressure results, appears to be more efficient and easier in application. However, results, produced by these methods in case of increasing Reynolds numbers, are most unsatisfactory, since variations in marching route lead to considerable deviations in pressure values. Richards & Crane [6] introduced the concept of compatible pressure and vorticity schemes and showed that the lack of compatibility would be the principal reason for the poor results obtained by using the marching schemes.

In the present paper in accordance with [6] we are first to devise a suitable pressure scheme and then to work out the vorticity finite-difference scheme by imposing the compatibility requirement.

To construct the pressure marching scheme it is often more convenient to work in terms of total head or stagnation pressure H [6], defined as

$$H = p + \frac{1}{2} V^2 , \tag{9}$$

where p is the pressure measured in units of $\rho C\nu/d$ and ρ is the fluid density. Using relation (9) the Navier-Stokes equation for stationary flow in cartesian rectangular co-ordinates (x,y) takes the form:

$$\nabla H = \mathrm{Re}(\vec{V} \times \vec{\varphi}) - \mathrm{rot}\vec{\varphi} , \tag{10}$$

where $\vec{\varphi} = \{0,0,\varphi\}$. In orthogonal curvilinear co-ordinates the ξ and η components of (10) give:

$$\frac{\partial H}{\partial \xi} = - \mathrm{Re}\,\varphi \frac{\partial \psi}{\partial \xi} - \frac{a_1}{a_2} \frac{\partial \varphi}{\partial \xi} - \frac{a_3}{a_2} \frac{\partial \varphi}{\partial \eta} , \tag{11}$$

$$\frac{\partial H}{\partial \eta} = - \mathrm{Re}\,\varphi \frac{\partial \psi}{\partial \eta} + \frac{1}{a_2} \frac{\partial \varphi}{\partial \xi} + \frac{a_1}{a_2} \frac{\partial \varphi}{\partial \eta} . \tag{12}$$

Further treatment of the problem requires the adoption of pressure scheme that is compatible with the vorticity finite-difference scheme employed. This is most easily achieved if pressure values are sought at node points of an interlaced mesh system, as shown in figure 2. In curvilinear coordinate

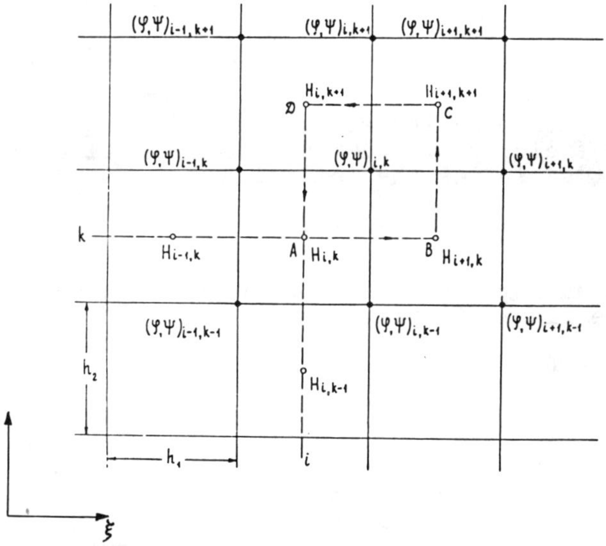

Figure 2. Fragment of mesh system.

system (ξ,η) the calculations are performed in uniform rectangular mesh. Finite-difference approximations of the resulting equations (11),(12) are chosen to be:

$$\frac{H_{i+1,k} - H_{i,k}}{h_1} = - \mathrm{Re}\left(\varphi \frac{\partial \psi}{\partial \xi}\right)_{i,k-1/2} - \frac{1}{(a_2)_i} \left[\left(a_1 \frac{\partial \varphi}{\partial \xi}\right)_{i,k-1/2} + \right.$$

$$\left. + (a_3)_{i,k-1/2} \frac{\varphi_{i,k} - \varphi_{i,k-1}}{h_2}\right] , \qquad (13)$$

$$\frac{H_{i,k+1} - H_{i,k}}{h_2} = - \mathrm{Re}\left(\varphi \frac{\partial \psi}{\partial \eta}\right)_{i-1/2,k} + \frac{1}{(a_2)_{i-1/2}} \left[\frac{\varphi_{i,k} - \varphi_{i-1,k}}{h_1} + \right.$$

$$\left. + \left(a_1 \frac{\partial \varphi}{\partial \eta}\right)_{i-1/2,k}\right] . \qquad (14)$$

Function products at some chosen interior node points (i,k--1/2), (i-1/2,k) of the interlaced grid can be readily obtained from the following formulae:

$$\left(f \frac{\partial g}{\partial \xi}\right)_{i,k-1/2} = \frac{1}{2}\left(f_{i,k-1} \frac{g_{i+1,k} - g_{i-1,k}}{2h_1} + f_{i,k} \times \right.$$

$$\left. \times \frac{g_{i+1,k-1} - g_{i-1,k-1}}{2h_1}\right) , \qquad (15)$$

$$\left(f \frac{\partial g}{\partial \eta}\right)_{i-1/2,k} = \frac{1}{2}\left(f_{i-1,k} \frac{g_{i,k+1} - g_{i,k-1}}{2h_2} + f_{i,k} \times \right.$$

$$\left. \times \frac{g_{i-1,k+1} - g_{i-1,k-1}}{2h_2}\right) . \qquad (16)$$

With this marching scheme one can uniquely determine the pressure field, provided that the integral over the closed circuit L satisfies the condition:

$$\oint_L \nabla H d\vec{l} = 0 . \qquad (17)$$

The grid analogue of this condition for the closed circuit ABCD (see figure 2) takes the form

$$F_{i,k} = (H_{i+1,k} - H_{i,k}) + (H_{i+1,k+1} - H_{i+1,k}) +$$

$$+ (H_{i,k+1} - H_{i+1,k+1}) + (H_{i,k} - H_{i,k+1}) = 0 . \qquad (18)$$

Substituting the expressions (13) and (14) into (18) and performing some simple algebraic manipulations one gets the approximation of steady-state vorticity equation that is compatible with the pressure marching scheme (13)-(14). Making use of stabilization method we find the stationary distributions of the vorticity and stream function

$$\frac{\varphi^{n+1}_{i,k} - \varphi^{n}_{i,k}}{\tau} = F^{n}_{i,k} , \qquad (19)$$

where the superscript n indicates the time step number and τ is the time step. The stream function field is determined at each time step by solving the finite-difference analogue of the equation (5) using the upwind succesive-over-relaxation (SOR) method.

Computations are carried out according to the following procedure:

1. A numerical solution $\varphi^{n+1}_{i,k}$ for the vorticity equation (19) is found.
2. The stream function field $\psi^{n+1}_{i,k}$ from the finite-difference analogue of the eq.(5) is calculated.
3. The boundary values for vorticity on the wall $\varphi|^{n+1}_{B}$ and values of functions ψ and φ on the side boundaries (see, for example, [5]) are computed.
   The stages 1-3 are repeated at each time step.
4. On getting the steady distributions of ψ and φ , by the stabilization method, one computes the function H distribution using the marching scheme (13)-(14): a computational cycle begins with computing the H values on the initial marching line of η = const. It is evident that the employment of the marching scheme (13)-(14) is valid only for the steady-state solution ($\partial\varphi/\partial t = 0$).
5. The pressure is computed from the known values of ψ and H

$$p_{i,k} = H_{i,k} - \frac{1}{2}\left[\left(-\frac{\partial\psi}{\partial x}\right)^2_{i-1/2,k-1/2} + \left(\frac{\partial\psi}{\partial y}\right)^2_{i-1/2,k-1/2}\right]. \quad (20)$$

It should be noted, that the applied computational method provides a perfect agreement between the pressure drop results over the whole cross-section of the channel; i.e. at various y-values we get pressure drop results, that generally agree to 6-7 decimal figures, even in the case of finite Reynolds numbers.

RESULTS AND DISCUSSION

The regime of peristaltic flow is controlled by four dimensionless parameters: the relative amplitude ζ, the wavenumber α , the Reynolds number Re and the time-mean flow q.

In order to test the reliability of the considered finite-difference method a comparison was made between the obtained numerical results and available analitical solutions [1]. For Stokes approximation (Re = 0) and large wavelength ($\alpha \leq 0.5$) the numerical data are consistent with predictions of linear analysis (Re = 0, α = 0) reported elsewhere [1]. Thus our computations justify with reasonable accuracy the validity of the analitically derived condition for initiation of wave trapping of fluid particles, followed by the splitting of the central streamline (axial trapping):

$$q > \frac{2 - \zeta}{3} \quad . \quad (21)$$

Present findings were compared with the theoretical results [7,8] , based upon perturbation method, for the finite

Reynolds numbers and finite wavenumbers. It has been found out (see, also [4]) that the applicability of these asymptotical solutions is restricted to a much narrower range compared with the predicted ones [7,8].

The employed numerical method has no restrictions, in principle, on the magnitudes of wave amplitude, wavelength and Reynolds number. This fact enables us to observe a new flow regime, existing only in case of sufficently large wavenumbers (i.e. small wavelengths) and considerable wave amplitudes.

To discuss the main results of our numerical investigation for large wavenumbers and finite amplitudes of wall deformation we consider the case of creeping flow (Re = 0). We shall describe the flow regimes, existing at various values of the time-mean flow q and the wavenumber α, for the fixed wave amplitude $\zeta = 0.4$. Figure 3 shows the diagram of flow regimes on the plane (q,α).

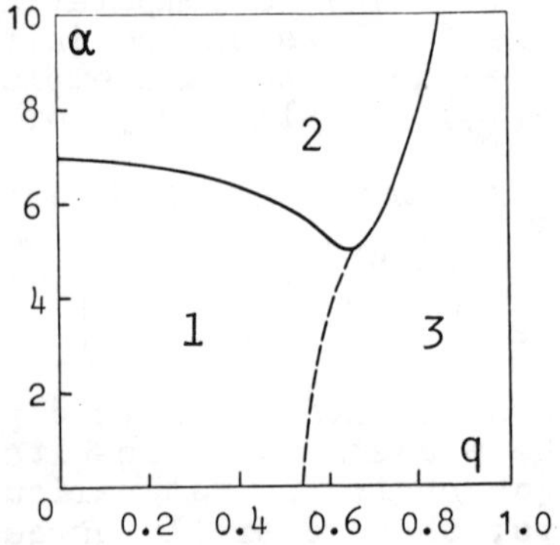

Figure 3. Regime diagram Re = 0, $\zeta = 0.4$.

In the region of $\alpha < 5$, adjacent to the axis $\alpha = 0$, one can observe the flow patterns, which are in qualitative agreement with those, observed at infinitely large wavelengths: for small values of time-mean flow q, the streamlines repeat the shape of the wall and their curvature decreases as they approach the axis of the channel (region 1); the increase in the time-mean flow shows the splitting of the streamline from the axis and the sequent flow trapping into vorticity zone in the core of the channel (axial trapping), travelling with the top of the wave (region 3). The boundary of the axial trapping region is depicted by the dotted line. The analysis of longitudinal velocity profiles shows that with increase of the wavenumber they diverge from the parabolic profile, considered by linear theory for Re = 0, $\alpha = 0$ [1].

At wavenumber $\alpha > 5$ the discrepancy from the local Poiseuille flow structure has become so substantial, that it results in the formation of a new flow regime. This flow pattern is characterised by a formation of the closed contours of the streamlines far from the axis of the channel (non-axial trapping) and the existence of contraflows: the flow velocities in the neighbourhood of the axis and in the adjacent section of the trapping region have the opposite directions. The flow contour corresponding to a non-axial trapping is shown in figure 4; on the flow regime diagram the boundary of the new flow pattern (region 2) is represented by the solid line. As indicated in figure 3, the axial trapping occurs only on exceeding some critical value of the time-mean flow, whereas the new non-axial trapping appears at arbitrary small time-mean flow values, provided the wavenumber is higher than the threshold.

The transition of one flow structure to another is readily

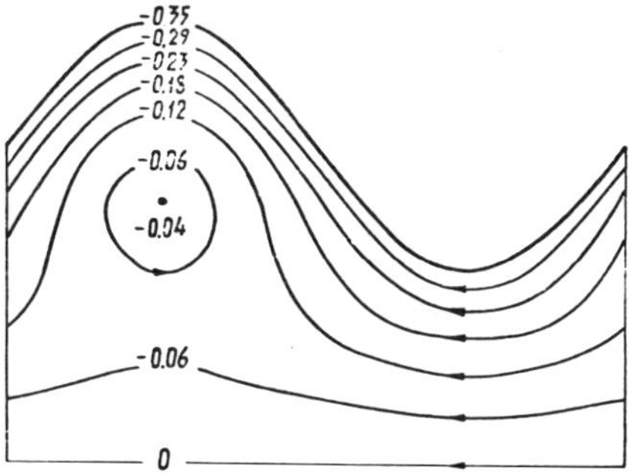

Figure 4. Flow structure: $\alpha = 6.5$, q = 0.65.

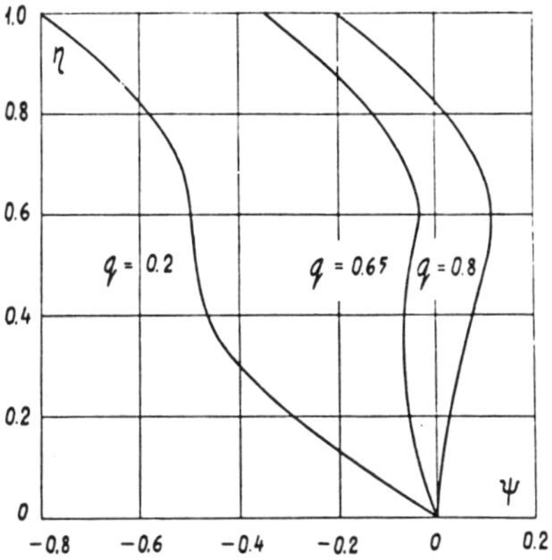

Figure 5. Dependences $\Psi(\eta)$ $\alpha = 6.5$.

seen from the variation of the shape of the stream function dependence on the transverse co-ordinate in the wide cross-section of the channel. In figure 5 the flow with unclosed streamlines is shown by the monotonous curve $\Psi(\eta)$ (q = 0.2), the non-axial trapping - by the curve with minimum and maximum (q = 0.65), and the axial trapping - by the curve with a maximum only, the latter being placed in the region of positive values of Ψ (q = 0.8).

The numerical data, obtained for the wave amplitudes $\zeta \leq 0.8$ and Reynolds numbers Re $\leq$ 100 are in the qualitative accordance with the above conclusions concerning the possible regimes of the flow and the character of transitions between them, but the location of boundary lines on the diagram of regimes is naturally varied.

The author wishes to thank Dr.G.I.Burde and Prof.E.M.Zhukhovitski for his helpful comments and discussions.

REFERENCES

[1] Jaffrin M.Y. & Shapiro A.H. "Peristaltic pumping", Ann.Rev. Fluid Mech., 3, pp. 13-36 (1971).

[2] Brown T.D. & Hung T.-K. "Computational and experimental investigations of two-dimensional nonlinear peristaltic flows", J.Fluid Mech., 83 (2), pp.249-272 (1977).

[3] Zaiko V.M.,Starobin I.M.,Utkin A.V. "Numerical simulation of motion of viscous fluid (blood) in a tube with actively deforming walls", Mekh.Komp.Materialov, pp.515-523 (1979).

[4] Takabatake S. & Ayukawa K. "Numerical study of two-dimensional peristaltic flows", J.Fluid Mech., 122 (1982).

[5] Roache P.J. Computational Fluid Dynamics. 1976.

[6] Richards C.W. & Crane C.M. "Pressure marching schemes that work", Int.J.For Num.Meth.In Eng., pp. 599-610 (1980).

[7] Zien T.F. & Ostrach S. "A long wave approximation to peristaltic motion", J.Biomech., 3 (1), pp. 63-75 (1970).

[8] Jaffrin M.Y. "Inertia and streamline curvature effects on peristaltic pumping", Int.J.Eng.Sci.,11,pp.681-699 (1973).

APPLICATION OF A NEARLY ORTHOGONAL COORDINATE TRANSFORMATION FOR PREDICTING VISCOUS FLOWS WITH SEPARATION

A. Lippke, D. Wacker
Fachbereich Mathematik

F. Thiele
Hermann-Föttinger-Institut

Technische Universität Berlin
Straße des 17. Juni 135
D - 1000 Berlin 12, West-Germany

SUMMARY

The prediction of viscous two-dimensional flows with regions of separation is investigated by using a nearly-orthogonal coordinate transformation. The numerical method for generating the nearly-orthogonal grid applies a two-step algorithm based on a general non-orthogonal transformation of the physical plane to the rectangular domain of computation. To solve the Navier-Stokes equations written in terms of vorticity and stream function a coupled strongly implicit procedure is used. The calculations performed for various flow problems indicate that the application of nearly-orthogonal coordinates yields to a more stable and efficient solution procedure.

INTRODUCTION

In recent years curvilinear coordinate systems have been extensively applied to study solutions of flow problems in fluid mechanics [1]. For two-dimensional geometries the boundary-fitted coordinates are, in general, numerically generated by the solution of two elliptic differential equations [2]. Although such coordinate systems overcome the limitations of conformal mappings these transformations are not, in most cases, orthogonal and can have a profound influence on the stability of Navier-Stokes solvers and the accuracy of the solutions. However, nearly-orthogonal curvilinear coordinates are less restrictive and retain most of the flexibility of general transformations.

Haussling and Coleman [3] proposed a method to generate a grid with arbitrary distribution of the coordinates on all boundaries. However, the application of the procedure to highly curved geometries can lead to numerical difficulties. Such a problem does not arise by using the ideas of Mobley and Stewart [4] or Visbal and Knight [5]. This may be due to the fact that the coordinates are only prescribed on some of the boundaries.

In this paper the method of Visbal and Knight [5] is applied to the calculation of viscous flows in two-dimensional geometries where regions of separation can occur. The emphasis of the present investigation is on evaluating to what extent the use of a nearly-orthogonal transformation has an influence on the behaviour of the Navier-Stokes solver.

COORDINATE TRANSFORMATION

Boundary-fitted non-orthogonal coordinate transformations are able to transform general geometries into a rectangular computational domain. The transformed grid is a cartesian one. Therefore the approximation of the difference molecules on or in the vicinity of a body is no longer necessary.

This is of advantage for the solution procedure of differential equations. On the other hand first and mixed second derivatives will arise due to the transformation. The behaviour of these additional terms in regard to the Navier-Stokes solvers depends on the non-orthogonality of the grid. With the definitions $B := (x,y)$ and $B^* := (s,t)$ of the physical and transformed plane, respectively, the general transformation can be given by

$$\begin{pmatrix} s \\ t \end{pmatrix} = \begin{pmatrix} f(x,y) \\ g(x,y) \end{pmatrix} . \tag{1}$$

In the transformed plane the Navier-Stokes equations are written in terms of vorticity and stream function as

$$\alpha \frac{\partial^2 \omega}{\partial s^2} - 2\beta \frac{\partial^2 \omega}{\partial s \partial t} + \gamma \frac{\partial^2 \omega}{\partial t^2} + \sigma \frac{\partial \omega}{\partial t} + \tau \frac{\partial \omega}{\partial s} = \mathrm{Re} \cdot J \left(\frac{\partial \psi}{\partial t} \frac{\partial \omega}{\partial s} - \frac{\partial \psi}{\partial s} \frac{\partial \omega}{\partial t} \right) , \tag{2}$$

$$\alpha \frac{\partial^2 \psi}{\partial s^2} - 2\beta \frac{\partial^2 \psi}{\partial s \partial t} + \gamma \frac{\partial^2 \psi}{\partial t^2} + \sigma \frac{\partial \psi}{\partial t} + \tau \frac{\partial \psi}{\partial s} = - J^2 \omega . \tag{3}$$

Here, the coefficients α, β, γ, σ, τ and J depend on s and t, and Re is the Reynolds number. Let equation (1) be a bijective transformation then the coefficients read

$$\begin{cases} \alpha = \left(\frac{\partial x}{\partial s}\right)^2 + \left(\frac{\partial y}{\partial s}\right)^2 , & \gamma = \left(\frac{\partial x}{\partial t}\right)^2 + \left(\frac{\partial y}{\partial t}\right)^2 , \\ \beta = \frac{\partial x}{\partial s}\frac{\partial x}{\partial t} + \frac{\partial y}{\partial s}\frac{\partial y}{\partial t} , & J = \frac{\partial x}{\partial s}\frac{\partial y}{\partial t} - \frac{\partial x}{\partial t}\frac{\partial y}{\partial s} , \\ \sigma = J^{-1}\left(\frac{\partial y}{\partial s} Dx - \frac{\partial x}{\partial t} Dy\right) , & \tau = J^{-1}\left(\frac{\partial x}{\partial s} Dy - \frac{\partial y}{\partial t} Dx\right) , \end{cases} \tag{4}$$

where the operator is defined by

$$\begin{pmatrix} Dx \\ Dy \end{pmatrix} = \left(\alpha \frac{\partial^2}{\partial s^2} - 2\beta \frac{\partial^2}{\partial s \partial t} + \gamma \frac{\partial^2}{\partial t^2} \right) \begin{pmatrix} x \\ y \end{pmatrix} . \tag{5}$$

The scaling factors α, σ and τ will interfere with the stability of the solver. This is not surprising as most solution procedures for large linear systems, e.g. for 30×30 to 70×70 mesh points, are unstable in absence of diagonal dominance. This is closely related to the condition of the coefficient matrix. Another unfavourable fact is that the truncation error of second order approximation will increase by $\sin^{-2}\varphi$ where φ denotes the angle between two intersecting coordinate lines (see [2]).

The nearly-orthogonal grids are able to overcome these problems without loosing most of the advantages of numerically generated boundary-fitted curvilinear coordinate systems. If we generate an orthogonal coordinate system the molecules of the finite difference representation will reduce from 9 to 5 points. This can be easily deduced from the Cauchy-Riemann equations; with $\partial x/\partial s = \partial y/\partial t$ and $\partial x/\partial t = - \partial y/\partial s$ follows from equation (4) that $\beta = \sigma = \tau \equiv 0$. With respect to the solution of the equations (2) and (3) the storage requirement is reduced.

GENERATION OF NEARLY-ORTHOGONAL GRIDS

In recent years a set of two coupled Poisson equations [1]

$$\left(\frac{\partial^2}{\partial x^2} + \frac{\partial^2}{\partial y^2}\right)\begin{pmatrix} s \\ t \end{pmatrix} = \begin{pmatrix} P \\ \Omega \end{pmatrix} \tag{6}$$

has been applied to generate general non-orthogonal transformations. To differentiate between analytically and numerically generated orthogonal grids the following definition is suggested. A grid is called orthogonal if β is identically zero whereas all nearly-orthogonal grids satisfy the criterion

$$|\beta| \leq \varepsilon(\alpha + \gamma) \;, \qquad \varepsilon \ll 1 \;. \tag{7}$$

Based on this definition a procedure for the generation of nearly-orthogonal grids can be proposed. We consider the inverse transformation of equation (1)

$$\begin{pmatrix} x \\ y \end{pmatrix} = \begin{pmatrix} \tilde{f}(s,t) \\ \tilde{g}(s,t) \end{pmatrix} \tag{8}$$

where $g^{(ij)}$ is the first fundamental form

$$g^{(11)} = \left(\frac{\partial \tilde{f}}{\partial s}\right)^2 + \left(\frac{\partial \tilde{g}}{\partial s}\right)^2 \;, \qquad g^{(22)} = \left(\frac{\partial \tilde{f}}{\partial t}\right)^2 + \left(\frac{\partial \tilde{g}}{\partial t}\right)^2 \;, \tag{9}$$

$$g^{(12)} = g^{(21)} = \frac{\partial \tilde{f}}{\partial s}\frac{\partial \tilde{f}}{\partial t} + \frac{\partial \tilde{g}}{\partial s}\frac{\partial \tilde{g}}{\partial t} \;. \tag{10}$$

To ensure an orthogonal transformation it is required that $g^{(12)} = g^{(21)} = 0$. Using equation (6) in the transformed plane we obtain

$$\Delta\begin{pmatrix} s \\ t \end{pmatrix} = \left\{ J^{-1}\left(\frac{\partial}{\partial s}\left(\frac{J}{g^{(11)}}\frac{\partial}{\partial s}\right) + \frac{\partial}{\partial t}\left(\frac{J}{g^{(22)}}\frac{\partial}{\partial t}\right)\right)\right\}\begin{pmatrix} s \\ t \end{pmatrix} \tag{11}$$

with $J^2 := g^{(11)} \cdot g^{(22)}$. Equation (11) yields the coupled system of differential equations

$$\Delta s = J^{-1}\frac{\partial}{\partial s}\left(\frac{J}{g^{(11)}}\right) \;, \qquad \Delta t = J^{-1}\frac{\partial}{\partial t}\left(\frac{J}{g^{(22)}}\right) \;. \tag{12}$$

Instabilities arise by solving these equations simultaneously. For this reason we choose a two step procedure according to Visbal and Knight [5]:

<u>Step 1:</u> Solve the inverted equations to

$$\Delta s = \frac{1}{J}\frac{\partial}{\partial s}\left(\frac{J}{g_{11}}\right) \;, \qquad \Delta t = 0 \tag{13}$$

with the boundary conditions

$$\Gamma_1: \quad s = 0 \;, \qquad \partial t/\partial n = 0 \;; \qquad \Gamma_2: \quad s = 1 \;, \qquad \partial t/\partial n = 0$$

$$\Gamma_3: \quad s = f_3(\cdot) \;, \qquad t = 0 \;; \qquad \Gamma_4: \quad \partial s/\partial n = 0 \;, \qquad t = 1$$

where $(\cdot)$ denotes the arclength of Γ_3.

Step 2: Solve the inverted equations to

$$\Delta s = \frac{1}{J}\frac{\partial}{\partial s}\left(\frac{J}{g_{11}}\right) , \qquad \Delta t = Q \qquad (14), (14b)$$

with the boundary conditions

$$\Gamma_1: \quad s = 0 , \qquad t = f_1(\cdot) ; \qquad \Gamma_2: \quad s = 1 , \qquad \partial t/\partial n = 0$$

$$\Gamma_3: \quad s = f_3(\cdot) , \qquad t = 0 ; \qquad \Gamma_4: \quad s = f_4(\cdot) , \qquad t = 1$$

where $(\cdot)$ denotes the arclength of Γ_1.

The function Q has to be determined from local orthogonality. f_4 is the distribution of boundary grid points obtained in step 1.

The boundary conditions indicate that the distribution of the boundary grid points on Γ_1 and Γ_3 can be arbitrarily specified whereas the distribution on the other boundaries Γ_2 and Γ_4 are automatically choosen according to orthogonality. A good choice for the forcing function Q in equation (14) seems to be $Q = S/J^2$ with

$$S = \gamma T - \left(\frac{\partial y}{\partial t}\frac{\partial s}{\partial x}\right)R \qquad \text{if} \quad \partial x/\partial s \neq 0 \qquad (15)$$

$$S = \gamma T + \left(\frac{\partial x}{\partial t}\frac{\partial s}{\partial y}\right)R \qquad \text{if} \quad \partial y/\partial s \neq 0 \qquad (16)$$

$$T = -\left(\frac{\partial x}{\partial t}\frac{\partial^2 x}{\partial t^2} + \frac{\partial y}{\partial t}\frac{\partial^2 y}{\partial t^2}\right)/\alpha \qquad R = \frac{\partial x}{\partial s}\frac{\partial^2 y}{\partial s^2} - \frac{\partial y}{\partial s}\frac{\partial^2 x}{\partial s^2} . \qquad (17)$$

These equations are solved using successive overrelaxation (SOR) with uniform or pointwise acceleration. The investigations indicate that a uniform parameter is sufficient in most cases. The r.h.s. of equation (14a) may be updated during the solution procedure, although linear interpolation from the first step is sufficient in "well behaved" domains. The function Q is a function of t alone and should be evaluated at boundary points by a second order non-symmetric approximation to avoid instabilities.

SOLUTION OF THE NAVIER-STOKES EQUATIONS

The Navier-Stokes equations are solved using the strongly implicit procedure of Stone [6]. This method has been extendet to the non-orthogonal transformations for multiple connected regions by Wacker [7]. The coupled arbitrary equations of second order with the nine point molecule can be solved. The procedure solves the equation

$$M * W = G . \qquad (18)$$

Here, M is the coefficient matrix of the discrete form of the Navier-Stokes equations and W is a vector with the components (ω_i, ψ_i).

The main idea of Stone is to add a matrix N to equation (18) which will ensure a factorisation in an upper and lower triangular matrix with band-structure. This is obtained by the following modifications:

$$(M + N) * W = (M + N) * W - (M * W - G) . \qquad (19)$$

Following Stone we set

$$(M+N) * W^{n+1} = (M+N) * W^{n} - (M * W^{n} - G) \ . \tag{20}$$

Then defining

$$\delta^{n+1} := W^{n+1} - W^{n} \ , \qquad R^{n} := G - M * W^{n} \tag{21}$$

equation (20) becomes

$$(M+N) * \delta^{n+1} = R^{n} \ . \tag{22}$$

A factorisation of the modified coefficient matrix in equation (22)

$$M + N = L * U \tag{23}$$

yields a two sweep algorithm for the solution of equation (22):

$$L * V = R^{n} \text{ (forward sweep)} \ , \quad U * \delta^{n+1} = V \text{ (backward sweep)}. \tag{24}$$

This algorithm allows a full implementation of the vorticity equation on all boundaries. As the coupled strongly implicit procedure (CSIP) can handle both 9 and 5 point molecules we can apply the solution procedure to either non-orthogonal or nearly-orthogonal grids with the same parameters.

RESULTS AND CONCLUSIONS

The following flow problems have been chosen to investigate the behaviour of the solution procedure derived above:

1. Driven Cavity,
2. Backward facing step,
3. Channel with symmetric contraction (diffuser).

The numerical calculations have been performed for various Reynolds numbers. There the Reynolds number is defined in 2 and 3 by the characteristic values of the channel height and the mean velocity, both evaluated at the channel entrance.

The driven cavity problem has been selected to show that the procedure results in a trivial cartesian coordinate system. Figure 1 shows the number of iterations required to reduce a maximum error to 10^{-4}. The plot indicates that the number of iterations can be decreased by about 30 % using the nearly-orthogonal grid. The calculation procedure converged for Re up to 400 and 800 using the non-orthogonal and nearly-orthogonal grid, respectively. The value of the stream function at the vortex center in Table 1 agree well with the results of Burggraf cited by Tuann and Olson [8].

TABLE 1
Streamfunction at vortex center for driven cavity

Re	CSIP	Burggraf
0	-0.09985	-0.0998
10	-0.09982	-0.1000
50	-0.10051	-
100	-0.10218	-0.1015
200	-0.10541	-0.1032
300	-0.10682	-
400	-0.10695	-0.1017

For a geometry with sharp corners a backward facing step problem has been selected. The nearly-orthogonal grid and a streamline plot for Re = 300 are given in Figure 2. The number of iterations required (Table 2) indicate that the nearly-orthogonal grid is of benefit for higher Reynolds number cases (Re > 50).

The diffuser has a similar geometry but smooth boundaries than the step flow problem. Figure 3 presents the nearly-orthogonal grid and a streamline plot for Re = 500.

TABLE 2
Number of iterations
for backward facing step

Re	Non.-orth.-grid	Nearly-orth.-grid
0	184	253
10	73	96
50	153	147
100	449	174
150	1025	177
200	682	248

TABLE 3
Number of iterations
for diffuser

Re	Non.-orth.-grid	Nearly-orth.-grid
0	100	112
10	52	52
50	247	156
100	205	147
200	386	190
300	782	360

TABLE 4
Results for the diffuser
at Re = 350

case	Stream function at vortex center	reattachment length
1	- 8.249E-03	4.82
2	- 8.442E-03	4.83
3	- 9.066E-03	4.80

Although the diffuser problem does not have sharp corners the results in Table 3 show a similar behaviour in the convergence acceleration as obtained for the backward facing step.

The nearly-orthogonal transformation can only ensure that the coefficients β, σ and τ in the equations (2) and (3) are small compared with α and γ . For this reason it is of interest to what extend the accuracy of the solution is influenced by omitting the coefficients β, σ and τ. Preliminary investigations have been carried out for the diffuser problem where the Navier-Stokes equations were selected for the following three cases:

1. Coefficients β, σ and τ are taken into account.
2. Coefficients β, σ and τ are set to zero if $|\beta| < 5\cdot 10^{-4}\ (\alpha+\gamma)$.
3. Coefficients β, σ and τ are set to zero everywhere.

Table 4 contains some of the characteristic flow properties. Even though the values for the stream function differ by about 10 % the length of the reverse flow region is predicted quite accurately.

From the present investigations we can conclude that the use of nearly-orthogonal grids is beneficial in the numerical solution of the Navier-Stokes equations for flow problems with regions of separation. The main feature of the procedure proposed is that the application of nearly-orthogonal grids results in a better stability and convergence behaviour. In addition the procedure could be extended to calculate flows with higher Reynolds numbers. Furthermore it seems worthwhile to extend the investigation of the influence of certain transformation coefficients on the accuracy of the solution procedure.

REFERENCES

[1] Thompson, J.F., "Proceedings of the Symposium on the Numerical Generation of Curvilinear Coordinate Systems and use in the Numerical Solution of Partial Differential Equations", Nashville, Tennessee, April 1982.

[2] Warsi, Z.U.A., "Basic Differential Models for Coordinate Generation", Symposium on the Numerical Generation of Curvilinear Coordinate Systems, Nashville, Tennessee, April 1982.

[3] Haussling, H.J. and Coleman, R.M., "A Method for Generation of Orthogonal and Nearly Orthogonal Boundary-Fitted Coordinate Systems", Journal of Computational Physics, 43, pp. 373-381, 1981.

[4] Mobley, C.D. and Stewart, R.J., "On the Numerical Generation of Boundary-Fitted Orthogonal Curvilinear Coordinate Systems", Journal of Computational Physics, 34, pp. 124-135, 1980.

[5] Visbal, M. and Knight, D., "Generation of Orthogonal and Nearly Orthogonal Coordinates with Grid Control near Boundaries", AIAA Journal, 20, pp. 305-306, 1982.

[6] Stone, H.L., "Iterative Solution of Implicit Approximation of Multidimensional Partial Differential Equations", SIAM Journal of Numerical Analysis, 5, pp. 530-558, 1968.

[7] Wacker, D., "Ein streng implizites Verfahren zur Lösung der Navier-Stokes-Gleichungen mit Koordinatentransformation", Hermann-Föttinger-Institut, IB-01/83, TU Berlin, 1983.

[8] Tuann, S.Y. and Olson, M.D., "Review of Computing Methods for Recirculating Flows", Journal of Computational Physics, 29, pp. 1-19, 1978.

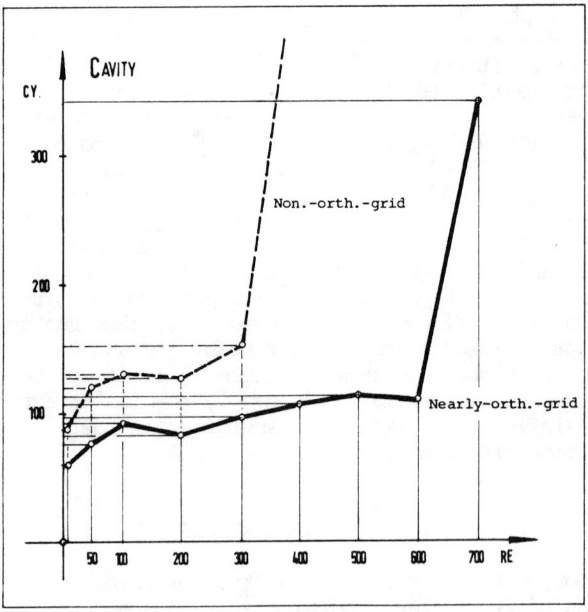

Fig. 1 - Number of iterations for the driven cavity problem

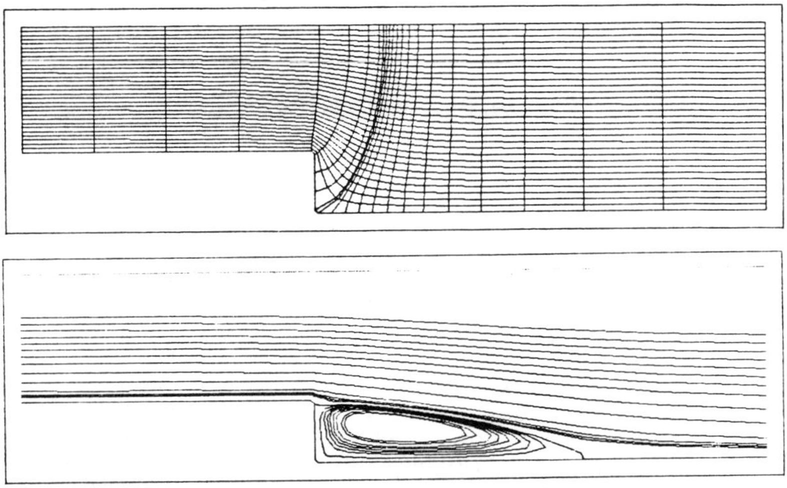

Fig. 2 - Grid configuration and streamline plot
for the backward facing step at
Re = 300

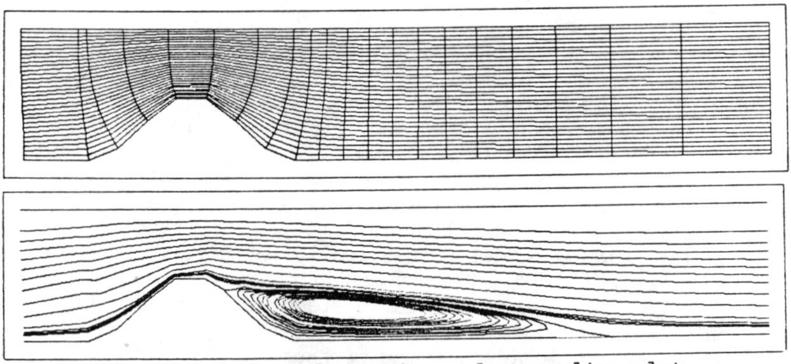

Fig. 3 - Grid configuration and streamline plot
for the diffuser flow at Re = 500

HYDRODYNAMIC INSTABILITY MECHANISMS IN MIXING LAYERS*

P. Mele, M. Morganti, A. Di Carlo

University of Rome "La Sapienza"
Via Eudossiana, 18 - 00184 - Rome, Italy

SUMMARY

A numerical approach, based on a finite element scheme, was developed in order to study the fluid dynamic instability of mixing layers according to the linear theory. For flows definited on unbounded domain as the mixing layer the asymptotic behaviour allows to derive the shape of the trial/test functions over the infinite elements from the inviscid solution of the reduced equation. In the present case a continuous spectrum of the eigenvalues formed by two half straight-lines is shown, while the neutral stability curve generated by the discrete spectrum was calculated. The disturbed flow field obtained by a computer simulation is compared with the experimental results.

INTRODUCTION

Interest in the instability phenomena has been renewed by recent observations in the remarkable coherence of large scale eddies in turbulent shear flows (Laufer, 1975 [11], Roshko, 1976 [15]). In particular, the mixing layer between two uniform streams is endowed with a well documented large scale coherent structure (Konrad, 1976 [10], Breidenthal, 1978 [3], Jimenez, Martinez-Val and Rebollo, 1979 [9]). The hypothesis that these structures may somehow be deterministically forecast from an analysis of the unstable basic flow, has been recently risen (Corcos, 1980 [4]). According to this hypothesis the early stages of formation of these structures should be traced by the linear stability theory.

By superimposing a velocity and pressure perturbation of the kind:

$$\vec{u}'(\vec{x},t) = \varepsilon\,\vec{u}(y)\exp\left[i(\alpha x + \beta z - \omega t)\right] \tag{1}$$

$$p'(\vec{x},t) = \varepsilon\,p(y)\exp\left[i(\alpha x + \beta z - \omega t)\right] \tag{2}$$

upon a plane basic flow of a viscous incompressible fluid, defined on the domain I , and by linearizing the Navier-Stokes equation with respect to ε , the Orr-Sommerfeld equation is obtained:

$$L\phi = 0 \tag{3}$$

the linear fourth-order differential operator L is given by:

$$L: = (D^2+\alpha^2) + i\,\alpha\,Re\left[\left(U(y)-c\right)\left(D^2+\alpha^2\right) + d^2U/dy^2\right]$$

where:

$$D: = -i\,\frac{d}{dy} \tag{4}$$

Re is the Reynolds number, $U(y)$ is the basic velocity field, and $c = \frac{\omega}{\alpha} = c_R + i\,c_I$ is the eigenvalue with c_R representing the phase velocity and c_I a measure of damping ($c_I < 0$) or amplification ($c_I > 0$). The complex eigen-

---

*This study was supported by C.N.R. under contract No. 82.02652.07 .

function ϕ defines the two components of the two-dimensional perturbation velocity*:

$$u' \equiv D\phi\, i \exp[i\alpha(x-ct)] \quad ; \quad v' \equiv -\phi\, i\alpha \exp[i\alpha(x-ct)] \qquad (5)$$

ASYMPTOTIC BEHAVIOUR FOR THE MIXING LAYER

In the case of mixing layers, the definition domain is unbounded $I :=]-\infty,+\infty[$ and the basic velocity profile can be considered constant for $|y|$ great enough, that is:

$$U(y) = u^*_\infty \Rightarrow \frac{d^2U}{dy^2} = 0 \qquad \text{for} \qquad |y| > y_\infty \qquad (6)$$

with:

$$u_\infty = u^1_\infty \;,\; y > y_\infty \quad ; \quad u_\infty = u^2_\infty \;,\; y < -y_\infty \quad ; \quad u^1_\infty \,,\, u^2_\infty \text{ and } y_\infty \in \mathbb{R} \qquad (7)$$

Where the basic velocity profile is assumed to be constant, the behaviour of the eigenfunction ϕ is described by the reduced equation:

$$(D^2+\alpha^2)^2\phi = -i\alpha\, Re\left[(u^*_\infty - c)(D^2+\alpha^2)\right]\phi \qquad (8)$$

This equation with four constant coefficient has four linearly independent solutions; hence the general integral can be expressed by:

$$\phi = \sum_{i=1}^{4} A_i \exp(\lambda_i y) \qquad (9)$$

with:

$$\lambda_1 = -Q^{\frac{1}{2}} \;,\; \lambda_2 = Q^{\frac{1}{2}} \;,\; \lambda_3 = -\alpha \;,\; \lambda_4 = \alpha \quad \text{and} \quad Q = i\alpha\, Re(u^*_\infty - c) + \alpha^2 \qquad (10)$$

The first two members of (9) corresponding to λ_1 and λ_2 represent "viscous solutions", and other two corresponding to λ_3 and λ_4 represent "inviscid solutions". The complex constants A_i of (9) must satisfy the boundary conditions. For $y = y_\infty$ or $y = -y_\infty$, the (9) must fit and match the behaviour of ϕ for respectively $y < y_\infty$ or $y > -y_\infty$. The conditions at the infinity must be considered with some care.

For bounded interval I the boundary conditions impose the vanishing of ϕ together with its first derivative,

$$\phi = 0 \;,\; D\phi = 0 \quad \text{on} \quad \partial I \qquad (11)$$

Boundary conditions impose in fact that both disturbance velocity component u' and v' vanish at the walls.

The weaker condition:

$$\phi \;,\; D\phi \quad \text{bounded for} \quad |y| \to \infty \qquad (12)$$

contains (11) and allows us to consider disturbances originating from the external flow. Grosh and Salwen [8] considering the weaker boundary condition (12) found out the continuous spectrum of eigenvalues for the boundary layer flow and for the free jet: it forms a half straight-line on the complex plane $(c_R, i c_I)$. In the mixing layer case the continuous spectrum is formed by two half straight-lines on the complex plane:

$$c = u^1_\infty - i(1+k^2)\,\alpha/Re \qquad (13_1)$$

and:

$$c = u^2_\infty - i(1+k^2)\,\alpha/Re \qquad (13_2)$$

with k real.

The reduced equation (8) holds for $y \in [y_\infty, +\infty[$ and $y \in]-\infty, -y_\infty]$.

---

*Squire transformations assure that the limitation to two-dimensional perturbation is not restrictive.

Over one of the two intervals, the eigenfunctions corresponding to the modes of the continuous spectrum possess a periodic component (viscous solution) in addition to the exponential inviscid solution; over the other interval the eigenfunction tends to zero (with modulus exponentially damped). In fact for the modes (13_1), the quantity Q assumes the values:

$$Q = -k^2\,\alpha^2 \qquad \text{for} \quad y > y_\infty$$

and:
$$Q = i\,\alpha\,\mathrm{Re}(u_\infty^2 - u_\infty^1) - k^2\,\alpha^2 \qquad \text{for} \quad y < -y_\infty \tag{14}$$

The eigenfunction ϕ becomes for $y > y_\infty$:

$$\phi = A_1\,e^{-i\alpha k y} + A_2\,e^{i\alpha k y} + A_3\,e^{-\alpha y} + A_4\,e^{\alpha y} \qquad \text{for} \quad y > y_\infty \tag{15}$$

The boundary conditions (12) impose that only A_4 must vanish, while A_1, A_2 and A_3 can be different from zero showing a periodic component. The modes (13_2) present an analogous behaviour.

The modes of the continuous spectrum representing disturbances somehow generated in the external flow and coming from it, are stable. The eigenvalues have negative imaginary part:

$$c_I = -(1 + k^2)\,\alpha/\mathrm{Re} \tag{16}$$

They tend to zero when $\alpha/\mathrm{Re} \to 0$, i.e. when the ratio between the disturbance wave number and Reynolds number tends to zero. As a consequence, the half straight-line $\alpha = 0$ for $\mathrm{Re} > 0$ can be considered a limit curve of stability in (α, Re) plane. But this curve cannot be overcrossed. The limit of stability is therefore determined by an eigenvalue of the discrete spectrum.

APPROXIMATED SOLUTIONS OF THE O.S. EQUATION FOR THE MIXING LAYER

A numerical approach has been considered in order to obtain the eigenvalues and the corresponding eigenfunctions. The velocity profile is expressed by:

$$U = \tanh y \tag{17}$$

The approximated numerical solution of the Orr-Sommerfeld equation is obtained by devising a standard Galerkin procedure based on Hermite cubics as trial/test functions in the interval $[-y_\infty, y_\infty]$, plus two special "bubble" functions over the infinite elements $[y_\infty, \infty[$ and $]-\infty, -y_\infty]$ derived from the reduced Orr-Sommerfeld equation.

The discretization procedure is obtained by defining the set of mesh points $\{-y_\infty, y_1, y_2 \,\ldots\ldots\, y_n, +y_\infty\}$ such that:

$$-\infty < -y_\infty < y_1 < y_2 \ldots < y_n < y_\infty < +\infty \tag{18}$$

Two d.o.f. are attached to each internal mesh point $\{y_1, y_2, \ldots\ldots y_n\}$: the value of the function and its derivative. One further d.o.f. is attached to each of the mesch points y_∞ and $-y_\infty$. For y_∞ the shape function is formed by Hermite cubics for $y_n < y < y_\infty$ and the exponential inviscid solution of the reduced Orr-Sommerfeld equation for $y > y_\infty$ such that the function and derivative are continuous. The details of the numerical model can be found in [12]. The adoption of the "bubble" functions over the infinite elements allows us to obtain the eigenvalues of the continuous spectrum in addition to those of the discrete spectrum, if the value of y_∞ is great enough ([5],[13]).

For the sake of illustration, the eigenvalues obtained for $\mathrm{Re} = 5.02$ and $\alpha = 0.494$ with $y_\infty = 40$ are shown in the fig. 1. The eigenvalues corresponding to the continuous spectrum lie along the half straight-lines:

$$\omega \equiv \alpha c = 0.494 - i(1+k^2)0.0486$$
$$\omega \equiv \alpha c = -0.494 - i(1+k^2)0.0486 \qquad (19)$$

The eigenfunction corresponding to one of the obtained eigenvalue, i.e. $c = 0.9944 - i\,0.2641$, belonging to the continuous spectrum is shown in fig. 2. The periodic behaviour is well approximated (the wave lenght should be 9.802). Fig. 3 shows the corresponding streamlines, representing disturbances coming from the external region upon the basic flow.

THE NEUTRAL CURVE OF STABILITY FOR THE MIXING LAYER

In addition to the eigenvalues of the continuous spectrum, isolated eigenvalues (discrete spectrum) are obtained. They present real part equal to zero, and imaginary part higher than the imaginary part of the eigenvalues forming the continuous spectrum. The neutral curve of stability is obtained starting with particular values of α and varying Re till imaginary part of the less stable eigenvalue vanishes. The results of the calculation are plotted in fig. 4.

The neutral stability curve exhibits a critical point at $Re = 1.54$, $\alpha = 0.160$. The corresponding eigenfunction is shown in fig. 5, and the disturbance velocity components are shown in fig. 6.

The lower branch of the neutral curve has not been traced by previous authors seemingly because they consider $\alpha\,Re = 0(\alpha)$.

Esch [7] considered:

$$\alpha/(\alpha\,Re)^{\frac{1}{2}} = \text{cost.} \qquad (20)$$

The value of the constant calculated by Esch is 0.38. It is close to the value corresponding to the critical point:

$$\alpha_{crit.}/(\alpha_{crit.}\,Re_{crit.})^{\frac{1}{2}} = 0.32 \qquad (21)$$

Tatsumi and Gotoh [16] developed a method that holds for $Re \rightarrow 0$. While our results show a minimum Re. The numerical method proposed by Betchov and Szewczyk [1] does not work well for very low α.

In figs. 7a to 7c the streamlines of the flow obtained by superimposing the critical disturbance upon the basic velocity are shown, for three different values of the far-field velocities.

A computer simulation was considered injecting markers in a mixing layer with a ratio between the far-field velocities equal to 0.38 and the critical disturbance superimposed upon the basic velocity. In fig. 8 the streak-lines obtained are plotted. The computer plot appears quite similar to the well-known coherent structures experimentally carried out in the mixing layer by Konrad [10].

CONCLUSIONS

For the mixing layer a continuous spectrum of eigenvalues exists formed by two half straight-lines. These modes are stable and represent disturbances coming from the external flow.

By means of the present numerical method the neutral stability curve, generated by an eigenvalue of the discrete spectrum, is obtained. It shows a critical point in (α, Re) plane, i.e. a minimum Re, and a lower branch. In the critical condition the experimentally well-documented coherent structure is visualized by a computer simulation.

REFERENCES

[1] Betchov, R., Szewczyk, A., "Stability of a Shear Layer between Parallel Streams", The Physics of Fluids, Vol. 6, 1963.

[2] Browand, F.K., "An Experimental Investigation of the Instability of an Incompressible, Separated Shear Layer", J. Fluid Mechanics, Vol. 26, 1966.

[3] Breidenthal, R., "A Chemical Reacting, Turbulent Shear Layer", Cal. Ins. Tec., 1978.

[4] Corcos, G.M., "The Deterministic Description of the Coherent Structure of Free Shear Layer", Lecture Notes in Physics, No. 136, Springer Verlag, 1980.

[5] Di Carlo, A., Mele, P., Morganti, M., "Analisi Numerica della Stabilità Lineare di Flussi Paralleli su Domini Illimitati", VI Congresso AIMETA, Genova, 1982.

[6] Drazin, P.G., Reid, W.H., "Hydrodynamic Stability", Cambridge University Press, 1981.

[7] Esch, R.E., "The Instability of a Shear Layer between Two Parallel Streams", J. Fluid Mechanics, Vol. 3, 1957.

[8] Grosh, C.E., Salwen, H., "The Continuous Spectrum of the Orr-Sommerfeld Equation", J. Fluid Mechanics, Vol. 87, 1978.

[9] Jimenez, J., Martinez-Val, R., Rebollo, M., "The Spectrum of Large Scale Structures in a Mixing Layer", Proceedings Second Symposium on Turbulent Shear Flows, Imperial College, London, 1979.

[10] Konrad, J.H., "An Experimental Investigation of Mixing in Two-Dimensional Turbulent Shear Flows with Applications to Diffusion-Limited Chemical Reactions", Cal. Ins. Tec., 1976.

[11] Laufer, J., "New Trends in Experimental Turbulence Research", Annual Review of Fluid Mechanics, Vol. 7, 1975.

[12] Mele, P., Morganti, M., Di Carlo, A., Tatone, A., "Laminar to Turbulent Flow Study by Means of F.E.M.", Second International Conference on Numerical Methods in Laminar and Turbulent Flow, Venice, 1981.

[13] Mele, P., Morganti, M., Di Carlo, A., "Linear Stability of Free Jets", Fourth International Symposium on F.E.M. in Flow Problems, Tokyo, 1982.

[14] Michalke, A., "On the Inviscid Instability of the Hyperbolic-Tangent Velocity Profile", J. Fluid Mechanics, Vol. 19, 1964.

[15] Roshko, A., "Structure of Turbulent Shear Flows: A New Look", AIAA Journal, Vol. 14, 1976.

[16] Tatsumi, T., Gotoh, K., "The Stability of Free Boundary Layers between Two Uniform Streams", J. Fluid Mechanics, Vol. 7, 1960.

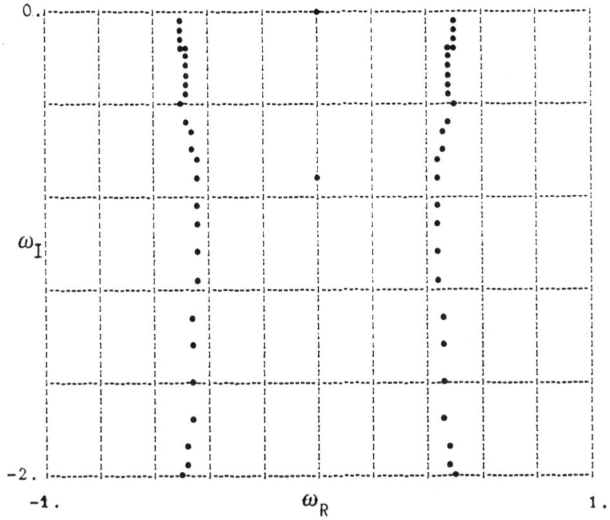

Fig. 1 Eigenvalues numerically obtained for Re = 5.02 and $\alpha = 0.494$, with $y_\infty = 40$. The eigenvalues corresponding to the continuous spectrum lie along the half straight-lines. $\omega \equiv \alpha c = 0.494 - i(1+k^2)0.0486$, $\omega \equiv \alpha c = -0.494 - i(1+k^2)0.0486$.

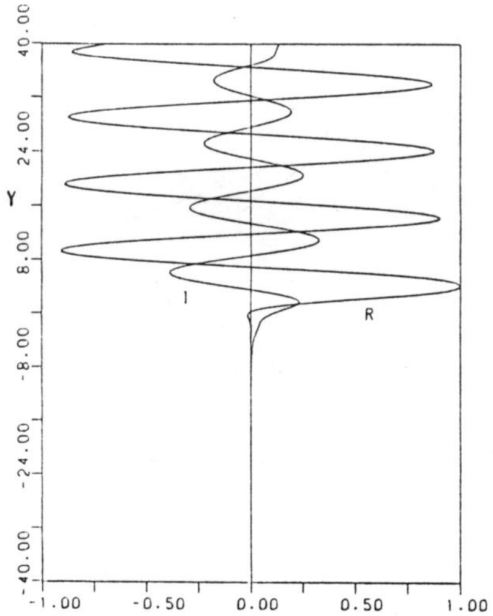

Fig. 2 Eigenfunction corresponding to the eigenvalue $c = 0.99 - i\,0.26$, belonging to the continuous spectrum. R = real part; I = imaginary part.

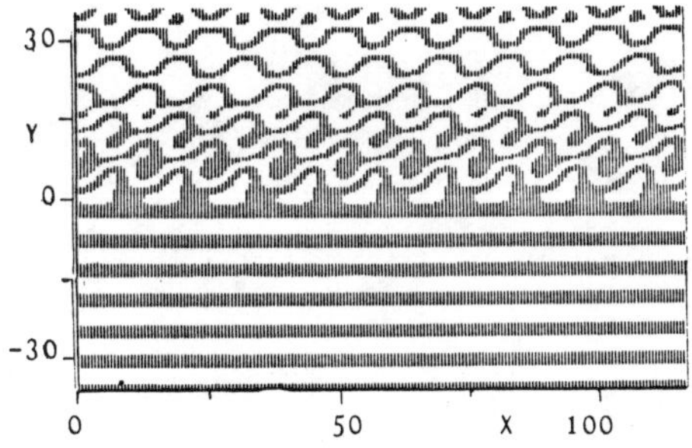

Fig. 3 Streamlines of the flow obtained by superimposing the disturbance coming from the external region upon the basic velocity: $U_{-\infty} = -1$ and $U_\infty = 1$

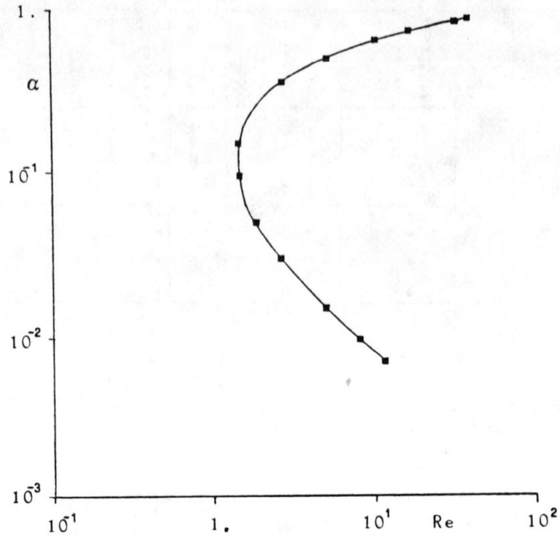

Fig. 4 Neutral stability curve in the (α,Re) plane

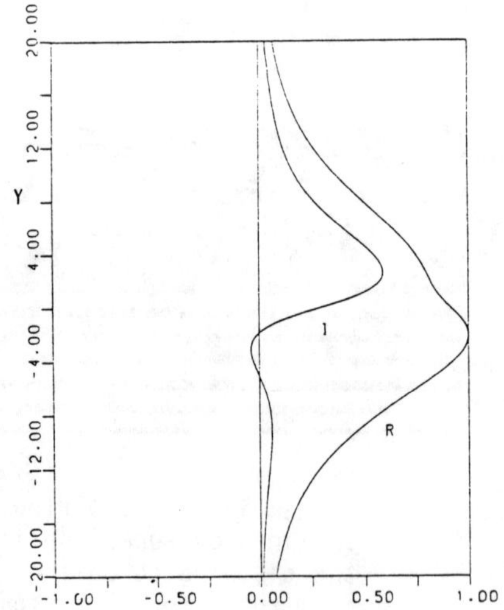

Fig. 5 Eigenfunction corresponding to the critical point for Re = 1.54 and α = 0.16. R = real part; I = imaginary part.

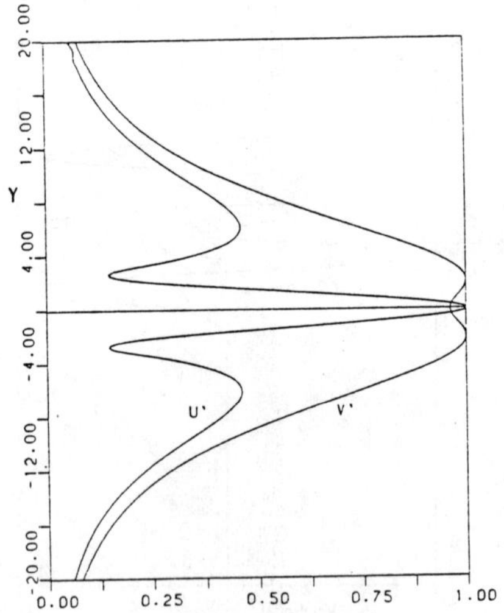

Fig. 6 Disturbance velocity components corresponding to the critical conditions: Re = 1.54 and α = 0.16.

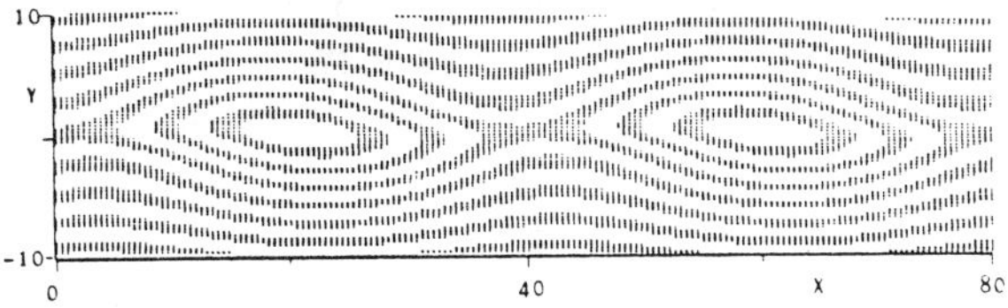

(a)

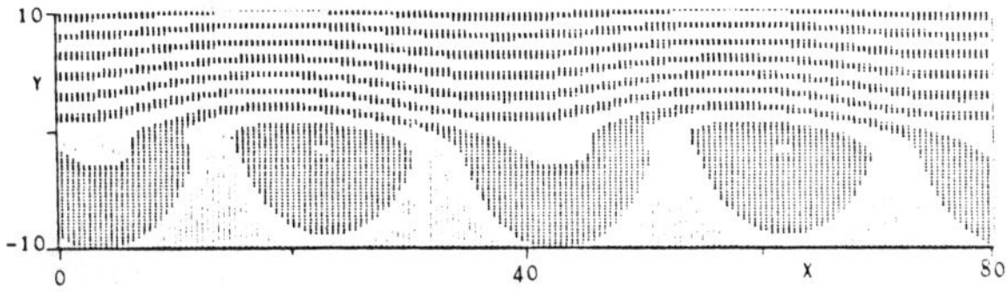

(b)

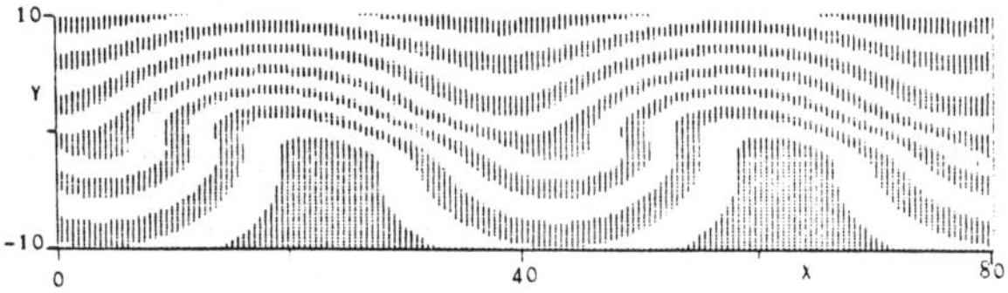

(c)

Fig. 7 Streamlines of the flow obtained by superimposing the critical disturbance upon the basic velocity.
(a) $U_{-\infty} = -1$ and $U_{\infty} = 1$, R.M.S. may 0.05;
(b) $U_{-\infty} = 0$ and $U_{\infty} = 2$, R.M.S. max 0.005;
(c) $U_{-\infty} = 1$ and $U_{\infty} = 3$, R.M.S. max 0.01.

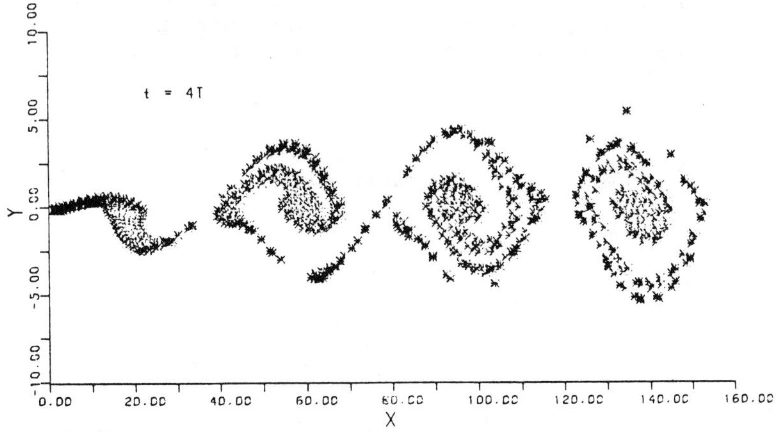

Fig. 8 Computer simulated streaklines

INFLUENCE OF COMPATIBILITY CONDITIONS IN NUMERICAL SIMULATION OF INHOMOGENEOUS INCOMPRESSIBLE FLOWS

F. Montigny-Rannou
Office National d'Etudes et de Recherches Aérospatiales (ONERA)
92320 Châtillon (France)

SUMMARY

The compatibility conditions in numerical simulation of inhomogeneous incompressible flows are examined for a two-dimensional Stokes problem defined on a slab geometry. A pseudo-spectral space-time method is used by means of Chebyshev approximations in space and in time. It is shown that this method associated with a Poisson solver for the pressure, gives less accurate results than the same method applied on a staggered grid to the direct treatment of the continuity equation.

1. INTRODUCTION

For numerical simulation of inhomogeneous incompressible flows, it is not always easy to obtain a compatible initial velocity field. Mathematically speaking, the initial velocity field must satisfy both the continuity equation and the boundary conditions. Moreover, some consistency equations appear. Our purpose is to show the influence of the initial velocity field and of the pressure treatment on the decay-rate coefficient of the solution of a simple 2-D Stokes problem. A very accurate time-scheme and a precise space discretization (space-time pseudo-spectral methods) are used.

First, compatibility conditions are given for Navier-Stokes equations. In section 3, the 2-D Stokes problem is presented. It was posed in a GAMM workshop held in Louvain-la-Neuve in October 1980 and was suggested by M. Deville. Next, the numerical methods are explained : a pseudo-spectral space-time method is used with a Poisson equation for the pressure. The same method is then employed with the continuity equation. In section 6, results are discussed. It appears that whatever the method may be, a compatible initial velocity field gives more quickly, good results. The first numerical technique with a Poisson solver strongly depends on the pressure initialization. The second technique with the direct resolution of the continuity equation is more satisfactory.

2. COMPATIBILITY RELATIONS

The unsteady Navier-Stokes equations for an inhomogeneous incompressible flow are the following ones :

$$\begin{cases} \dfrac{\partial \underline{U}}{\partial t} = -\underline{\nabla} P - (\underline{U}.\underline{\nabla})\underline{U} + \nu \Delta \underline{U}, & \text{in domain } \Omega, \quad (2.1) \\ \underline{\nabla}.\underline{U} = 0, & (2.2) \end{cases}$$

with no slip boundary conditions

$$\underline{U} = 0 \quad \text{on boundary } \partial\Omega . \qquad (2.3)$$

Generally, for each time level and particularly for the initial time, the pressure is obtained by solving a Poisson equation (divergence of momentum equation (2.1)) :

$$\Delta P = -\underline{\nabla}.(\underline{U}.\underline{\nabla})\underline{U} \quad \text{in domain } \Omega . \qquad (2.4)$$

From $\underline{\nabla} P = \nu \Delta \underline{U}$ on boundary $\partial\Omega$, (2.5)

we obtain two boundary conditions for the pressure : a Dirichlet condition

$$[(2.5).\underline{\tau}] \qquad \frac{\partial P}{\partial \tau} = \nu \underline{\tau}.\Delta \underline{U} , \qquad (2.6)$$

and a Neumann condition

$$[\,(2\text{-}5)\,.\,\underline{n}\,] \qquad \frac{\partial P}{\partial n} = \nu\, \underline{n}.\Delta \underline{U}\ . \tag{2.7}$$

But such conditions cannot generally be satisfied simultaneously.

O.A. Ladyzhenskaya [1] gives a regularity condition at the initial time for which the expressions of the pressure obtained by the two relations (2.6) and (2.7) may be the same. Although mathematically speaking, this condition is not necessary for the existence of the solution, it would appear that numerically this regularity condition on the initial velocity field is important. Severe oscillations appear near the boundary when that condition is not verified. An initial velocity field is **compatible** when Eqs. (2.6) and (2.7) give the same value for the pressure. But as it seems difficult to obtain regular compatible initial velocity fields, the numerical methods may have some damping properties to cope with the instabilities that appear at the beginning of the computation. J. Heywood and R. Rannacher [2] have examined the finite-element approximation. P. Moin [3] has analysed a finite difference scheme in space and has made a correction on the initial velocity field near the boundary. For a space-time spectral method, it is necessary to have a compatible initial velocity field in the whole space domain, because the derivatives at one point are influenced by all the other points.

3. EXAMPLE : 2-D STOKES PROBLEM

A simpler problem has been chosen to illustrate our purposes. The 2-D Stokes problem is defined on the infinite slab in the y direction with $|x| \leq 1$. The equations are :

$$\begin{cases} \dfrac{\partial \underline{U}}{\partial t} = -\underline{\nabla} P + \nu \Delta \underline{U} & (3.1) \\ \underline{\nabla}.\underline{U} = 0 & (3.2) \end{cases}$$

with no-slip boundary conditions : $\underline{U} = 0$ at $x = \pm 1$ and with initial conditions : $\underline{U}\ (\underline{x},\ t = 0)$. Periodicity in y direction is assumed in such a manner that the solution of Eqs. (3.1, 3.2) may be decomposed into Fourier modes :

$$\underline{U} = \underline{\tilde{U}}(x,t)\, e^{jky}\ ,\quad P = \tilde{P}(x,t)\, e^{jky}\ . \tag{3.3}$$

($j^2 = -1$). By insertion of Eq.(3.3) in Eqs.(3.1) and (3.2) and with the change of variables $U_1 = \tilde{U}_1$, $U_2 = j\tilde{U}_2$ $P = \tilde{P}$, where $\tilde{U}_1$, $\tilde{U}_2$ and the x -and y- components of $\underline{\tilde{U}}$, the following relations (E) are obtained :

$$(E)\quad \begin{cases} \dfrac{\partial U_1}{\partial t} = -\dfrac{\partial P}{\partial x} + \nu\left(\dfrac{\partial^2 U_1}{\partial x^2} - k^2 U_1\right), & (3.4) \\ \dfrac{\partial U_2}{\partial t} = kP + \nu\left(\dfrac{\partial^2 U_2}{\partial x^2} - k^2 U_2\right), & (3.5) \\ \dfrac{\partial U_1}{\partial x} + kU_2 = 0\ . & (3.6) \end{cases}$$

For problem (Π) that set of equations (E) is associated to no-slip boundary conditions

$$U_1 = 0\,,\ U_2 = 0 \ \text{ for } x = \pm 1, \tag{3.7}$$

and to the initial conditions defined below.

4. INITIAL VELOCITY FIELDS

Four initial conditions are considered. The first velocity field is not compatible and gives the two first initial conditions, the pressure satisfying Eq. (2.6) or Eq. (2.7). The third condition is obtained by a modification of the first field and is compatible. The fourth condition is the first eigenfunction of the problem.

First U_1 (x) is chosen as

$$U_1(x) = -k\,(1-x^2)^2, \tag{4.1}$$

and then U_2 (x) is easily calculated by means of the continuity equation (3.6). Because the numerical methods described below use an initial pressure field for the initialization of the iterative process, the pressure P (x,0) is sought out by means of a Poisson equation. The boundary conditions are obtained with one of the two first equations of the problem (Π). In fact, two values of the pressure are calculated as following :

1rst initial field

Boundary conditions on P (x,o) are obtained with the first equation of (Π) (Neumann condition), and using the no-slip condition $\frac{\partial U_1}{\partial t}=0$ at x = ± 1.

$$\begin{cases} U_1(x,0)=-k(1-x^2)^2, \\ U_2(x,0)=4x(x^2-1), \\ \text{Pressure for initialization : } P(x,0)=-8\nu\dfrac{\sinh kx}{\cosh k}. \end{cases} \tag{4.2}$$

2nd initial field

Boundary conditions on P (x,o) are obtained with the second equation of (Π) (Dirichlet condition), and using the no-slip condition $\frac{\partial U_2}{\partial t}=0$ at x = ± 1.

$$\begin{cases} U_1(x,0)=-k(1-x^2)^2, \\ U_2(x,0)=4x(x^2-1), \\ \text{Pressure for initialization: } P(x,0)=\dfrac{-24\nu}{k}\dfrac{\sinh kx}{\sinh k}. \end{cases} \tag{4.3}$$

3rd initial field

For consistency between initial and boundary conditions, the initial field U_2 (x,o) is modified by means of a parameter λ and takes the form :

$$U_2(x,0)=4x(x^2-1)+\lambda x^3(1-x^2). \tag{4.4}$$

U_1 (x,o) is then calculated by means of Eq. (3.6). The parameter λ is calculated in such a way that the Dirichlet condition (2.6) and the Neumann condition (2.7) are simultaneously satisfied. Then the third initial field is obtained :

$$\begin{cases} U_1(x,0)=-k(x^2-1)^2-\frac{2}{3}k\lambda(x^2-1)(x^4-\frac{1}{2}x^2-\frac{1}{2}), \\ U_2(x,0)=4x(x^2-1)(1+\lambda x^2), \\ \text{Pressure for initialization : } P(x,0)=3\nu\dfrac{\sinh kx}{\cosh k(k\tanh k-7)} \end{cases} \tag{4.5}$$

with $\lambda=\dfrac{3-k\tanh k}{k\tanh k-7}$.

4th initial field

A solution of the equations (E) can be expanded as follows :

$$\begin{aligned} U_1(x,t)&=-k\sum_{i=1}^{\infty}f_i(x).g_i(t), \\ U_2(x,t)&=\sum_{i=1}^{\infty}\frac{df_i}{dx}(x).g_i(t), \end{aligned} \tag{4.6}$$

where $f_i(x)$ is the ith eigensolution and ν_i is the ith value of problem (Π), with

$$\begin{aligned} f_i(x)&=-\frac{\cos\nu_i}{\cosh k}\cosh kx+\cos\nu_i x, \\ g_i(t)&=\exp(-\nu\sigma_i t), \end{aligned} \tag{4.7}$$

and where σ_i satisfies $\sigma_i=\nu_i^2+k^2$.

Then, the first eigensolution is used as a fourth initial velocity field.

$$\begin{cases} U_1(x,0)=-\dfrac{\cos\nu_1}{\cosh k}\cosh kx+\cos\nu_1 x, \\ U_2(x,0)=\dfrac{\cos\nu_1}{\cosh k}\sinh kx+\dfrac{\nu_1}{k}\sin\nu_1 x, \\ \text{Pressure for initialization : } P(x,0)=\dfrac{-\cos\nu_1}{k\cosh k}\nu\sigma_1\sinh kx, \end{cases} \tag{4.8}$$

with $\sigma_1 = 9.313739854$.

5. NUMERICAL SCHEMES

As time t and space x are quite similar variables for such equations, the discretizations used along time and space axis are similar. The solution of the problem (Π) will be obtained as the limit of a sequence of approximations :

$$\begin{array}{lll} U_1^0,\ U_1^1 \ldots U_1^l \ldots & & U_1^0(x,t) = U_1(x,0)\ , \\ U_2^0,\ U_2^1 \ldots U_2^l \ldots & \text{with} & U_2^0(x,t) = U_2(x,0)\ , \\ P^0,\ P^1 \ldots P^l \ldots & & P^0(x,t) = P(x,0)\ , \end{array} \qquad (5.1)$$

where each (U_1^l, U_2^l) satisfies, the initial and boundary conditions of (Π) and is defined on the whole space-time domain. Every quantity is changed according to the following relations :

$$\begin{cases} U_1^{l+1}(x,t) = U_1^l(x,t) + \eta\, u_1^l(x,t), \\ U_2^{l+1}(x,t) = U_2^l(x,t) + \eta\, u_2^l(x,t), \\ P^{l+1}(x,t) = P^l(x,t) + \eta\, p^l(x,t), \end{cases} \qquad (5.2)$$

where η is an under-relaxation coefficient and where the variations u_1^l, u_2^l, p^l come from a Newton method and satisfy homogeneous initial and boundary conditions [4]-[5]. That recursive process lets implicit computations that are difficult with pure spectral methods. So the velocity-pressure variations are calculated by means of finite-difference approximations used as preconditionning of the spectral operators.

The variation problem is then written at the lth iteration :

$$\begin{cases} \dfrac{\partial u_1^l}{\partial t} = -R_1^l - \dfrac{\partial p^l}{\partial x} + \nu\left(\dfrac{\partial^2 u_1^l}{\partial x^2} - k^2 u_1^l\right), & (5.3) \\ \dfrac{\partial u_2^l}{\partial t} = -R_2^l + k p^l + \nu\left(\dfrac{\partial^2 u_2^l}{\partial x^2} - k^2 u_2^l\right), & (5.4) \\ \dfrac{\partial u_1^l}{\partial x} + k u_2^l = -R_3^l\ . & (5.5) \end{cases}$$

In the right-hand sides of the equations (5.3 - 5.5) the residues R_1^l, R_2^l, R_3^l are given by the following expressions :

$$\begin{cases} R_1^l = \dfrac{\partial U_1^l}{\partial t} + \dfrac{\partial P^l}{\partial x} - \nu\left(\dfrac{\partial^2 U_1^l}{\partial x^2} - k^2 U_1^l\right), & (5.6) \\ R_2^l = \dfrac{\partial U_2^l}{\partial t} - k P^l - \nu\left(\dfrac{\partial^2 U_2^l}{\partial x^2} - k^2 U_2^l\right), & (5.7) \\ R_3^l = \dfrac{\partial U_1^l}{\partial x} + k U_2^l\ . & (5.8) \end{cases}$$

The residues R_1^l, R_2^l, R_3^l are approximated by **spectral method** in space and time, whereas the variations u_1^l, u_2^l and p^l are computed by means of **finite-difference** method with implicit schemes applied on the collocation points.

The programme structure is then :

	. initial conditions
iterative process	. residues computation [Eqs. (5.6- 5.8)]
	. variations computation [Eqs. (5.3-5.5)]
	. calculation of new approximations [Eqs. (5.2)]
	. convergence test

Two possibilities are proposed to compute the velocity-pressure variations.

The first technique is to solve a Poisson equation to obtain the pressure variation and after to compute the velocity variation.

The second possibility is to compute velocity-pressure variation using a staggered grid for solving the divergence equation (5.5) together with the momentum equations (5.3, 5.4).

5.1 1rst technique : Poisson solver for p

The pressure variations are given by solving a Poisson equation obtained by taking the divergence of Eqs. (5.3 - 5.4) and using Eq. (5.5)

$$\frac{\partial^2 p^l}{\partial x^2} - k^2 p^l = \frac{\partial R_3^l}{\partial t} + \nu k^2 R_3^l - \nu \frac{\partial^2 R_3^l}{\partial x^2} - k R_2^l - \frac{\partial R_1^l}{\partial x} \quad . \tag{5.9}$$

The computation is made on a classical time-space Chebyshev grid presented figure 1. The right-hand side of the Eq. (5.9) is calculated by formal derivation on a basis of Chebyshev polynomials. The equations (5.3), (5.4), (5.9) are solved by finite-difference schemes. The time derivatives in Eqs. (5.3) and (5.4) are approximated by an implicit Crank-Nicolson algorithm. Then a tri-diagonal matrix is obtained and solved.

5.1.1 Boundary conditions

The boundary conditions on the velocity variation are :

$$u_1^l = u_2^l = 0 \quad \text{at} \quad x = \pm 1 . \tag{5.10}$$

For the pressure variation solver, a backward iteration Neumann boundary condition is chosen. It is written as :

$$\frac{\partial p^l}{\partial x}(x,t) = - R_1^l(x,t) + \nu \frac{\partial^2 u_1^{l-1}}{\partial x^2}(x,t) \quad , \forall t \quad x = \pm 1 \quad . \tag{5.11}$$

$$\left(\text{instead of } \frac{\partial^2 u_1^l}{\partial x^2}(x,t)\right)$$

Condition (5.11) is approximated by a finite-difference scheme ; using the continuity equation (5.5), this condition is written for example at point x = 1 :

$$p^l(1,t) - p^l(2,t) = h_x(2) R_1^l(1,t) - h_x(2)\, \nu \frac{\partial R_3^l}{\partial x}(1,t) + \nu k u_2^{l-1}(2,t). \tag{5.12}$$

5.1.2 Remark

A backward iteration Dirichlet boundary condition has been also used with that pressure variation solver. It is written as :

$$p^l(x,t) = \frac{1}{k} R_2^l(x,t) - \frac{\nu}{k} \frac{\partial^2 u_2^{l-1}}{\partial x^2}(x,t) \quad , \quad \forall t \quad x = \pm 1 . \tag{5.13}$$

$$\left(\text{instead of } \frac{\partial^2 u_2^l}{\partial x^2}(x,t)\right)$$

The second derivative of u_2^l is approximated by a finite-difference form. The first technique associated at that boundary condition is not stable for viscosities greater than 10^{-2}, whatever the initial field is. It is for that reason that the Neumann boundary condition has been prefered.

5.2 2nd technique : direct treatment of the continuity equation

The variation equations (5.3 - 5.5) are solved simultaneously on a staggered grid presented on figure 2. The staggered points are Chebyshev collocation points of the form :

$$x_i = \cos \left[\frac{\pi (2i-1)}{2(N_x - 1)} \right] \tag{5.14}$$

For residues computation, Chebyshev extrapolation is used for the pressure term. The time derivatives in Eqs. (5.3) and (5.4) are also approximated by means of an implicit Crank-Nicolson scheme. The space derivatives are approximated by finite-difference forms. The matrix obtained so is not well-conditioned. A L.D.U. factorisation is used for solving the problem. As it is easily shown, boundary conditions are not necessary for the pressure variation. The pressure variation has one degree of freedom less than the velocity variation. This technique illustrates the "filtered" modes introduced by Y. Morchoisne [6].

The pressure variation p^l is known at the staggered points. The velocity variation components u_1^l and u_2^l are known at the classical Chebyshev points. Eqs. (5.3) and (5.4) are written at classical points and linear extrapolation is used for p^l in Eq. (5.4). Eq. (5.5) is written at the staggered points and linear extrapolation is used for u_2^l.

The boundary conditions on the variations u_1^ℓ and u_2^ℓ are written exactly (Eq. (5.10)).

Remark

A second staggered technique can be developed. This technique comes from the 2-D staggered grid method. The pressure variation p^l and the velocity variation component u_2^ℓ are given at the staggered points and Eqs. (5.4) and (5.5) are written at these points. The velocity variation component u_1^ℓ is given at the classical points and Eq. (5.3) is written at these points (see figure 2).

As u_2 is not defined for $x = \pm 1$, a fixed point method is used for the boundary condition U_2 $(\pm 1,\ t) = 0$:

$$u_2^\ell(\pm 1, t) = -\frac{1}{\eta}\, U_2^\ell(\pm 1, t)\,. \qquad (5.15)$$

The value of $u_2^\ell(\pm 1,t)$ is now introduced in the finite difference expression of $\partial^2 u_2^\ell/\partial x^2$ that appears in Eq. (5.4), for the first and last staggered points of the space discretization. But some convergence problems appear at the boundaries $(x = \pm 1)$ and the no-slip condition is not well-respected for the U_2 component.

6. RESULTS

Therefore, the results of these two computational techniques are presented. The first technique is based on a Poisson solver and a backward iteration Neumann boundary condition for the pressure variation. The second technique is based on the use of staggered grid (figure 2) to solve the momentum equation and the continuity relation together. For initial fields 1, 2 and 3 and if the viscosity ν is sufficiently high $(\nu > 0.75)$ and after a certain lapse of time $(t > 0.75)$, the first mode becomes dominant (contribution of the other modes $< 10^{-14}$). We neglect the other terms of the expansions (Eq. 4.6). The comparison is done between a numerical value σ and the theoretical decrease coefficient $\sigma_t \neq \sigma_1$ = 9.313739854.

Numerically an instantaneous decay-rate coefficient is computed and is defined as :

$$\sigma = \frac{-1}{\nu\, U_1(x=0,t)} \left.\frac{\partial U_1}{\partial t}\right)_{x=0,t} \qquad (6.1)$$

The time derivative is computed by time Chebyshev expansion. A relative error is given by

$$\frac{\delta\sigma}{\sigma} = \frac{|\sigma - \sigma_t|}{\sigma_t}$$

The viscosity ν is taken equal to 1 as the comparison time. The other parameters are the following :

wave number k = 1, space discretization : 17 points, time discretization : 257 points. To reduce the number of harmonics in time discretization, the total time integration domain may be divided into subdomains. The iterative process is performed in each subdomain ; the initialization values of the next time subdomain is given by the values of velocity pressure obtained at the end of the previous subdomain. In that case we use 17 time harmonics in each time subdomain.

The figure 3 shows that the variation of the Chebyshev harmonics is very small after 40 iterations (convergence level equal to 10^{-9}). The under-relaxation coefficient η takes two values : 0.80 for the 1rst technique, and 0.50 for the 2nd technique.

6.1 1rst technique

Figures 4 and 5 show the velocity components U_1 and U_2 at time t = 1 and for each initial field. Oscillations near the boundary appear on figure 5 for the non compatible initial field 2. On figure 6, the pressure P is drawn for time t = 1 and for each initial field.

Initial Fields	1		2		3		4	
Time mono-domain $N_t=257$	$\sigma = 9.3135563$ $\frac{\delta\sigma}{\sigma} = 2.10^{-5}$		$\sigma = 22.2797990$ $\frac{\delta\sigma}{\sigma} = 1.39$		$\sigma = 9.3137333$ $\frac{\delta\sigma}{\sigma} = 7.10^{-7}$		$\sigma = 9.3137334$ $\frac{\delta\sigma}{\sigma} = 7.10^{-7}$	
Iterations Number	82		341		32		32	
Divergence	4.10^{-6}		4.10^{-3}		3.10^{-8}		4.10^{-10}	
10 time sub-domains $N_t=17$	$\sigma = 9.3137427$ $\frac{\delta\sigma}{\sigma} = 3.10^{-7}$		$\sigma = 1.4820058$ $\frac{\delta\sigma}{\sigma} = 0.84$		$\sigma = 9.3137526$ $\frac{\delta\sigma}{\sigma} = 10^{-6}$		$\sigma = 9.3137329$ $\frac{\delta\sigma}{\sigma} = 7.10^{-7}$	
Iterations Number	52	21	92	22	50	20	39	20
Divergence	3.10^{-3}	4.10^{-11}	6.10^{-2}	5.10^{-4}	3.10^{-7}	2.10^{-11}	3.10^{-10}	3.10^{-11}
	1rst s.d	last s.d	1rst s.d	last s.d	1rst s.d	last s.d	1rst s.d	last s.d

Table I –

The decay-rate coefficient σ is shown at time t = 1, on table 1.

This technique gives accurate results for initial compatible fields (fields 3 and 4). If the initial field is not compatible the Neumann pressure field 1 gives better results than the Dirichlet pressure field 2. The same difficulties with non compatible initial field, appear with other numerical methods as the Influence Matrix Method developed by L. Kleiser [7].

In our method, the time subdomain technique has smoothing properties, which avoid the oscillations that appear near the boundary.

6.2 **2nd technique**

Table 2 shows the results concerning the coefficient σ.

The second technique gives precise results even if the initial velocity is not compatible. The convergence of the iterative process is obtained with less iterations than for the first technique. The iterations number is greater for non compatible fields (fields 1 and 2) than for compatible fields (fields 3 and 4).

Table II –

Initial Fields	1		2		3		4	
Time mono-domain $N_t=257$	$\sigma = 9.3137387$ $\frac{\delta\sigma}{\sigma} = 10^{-7}$		$\sigma = 9.3137386$ $\frac{\delta\sigma}{\sigma} = 10^{-7}$		$\sigma = 9.3137277$ $\frac{\delta\sigma}{\sigma} = 10^{-6}$		$\sigma = 9.3137335$ $\frac{\delta\sigma}{\sigma} = 6.10^{-7}$	
Iterations Number	45		45		33		34	
Divergence	10^{-10}		10^{-10}		4.10^{-11}		4.10^{-11}	
10 time sub-domains $N_t=17$	$\sigma = 9.3137746$ $\frac{\delta\sigma}{\sigma} = 4.10^{-6}$		$\sigma = 9.3138325$ $\frac{\delta\sigma}{\sigma} = 8.10^{-6}$		$\sigma = 9.3137079$ $\frac{\delta\sigma}{\sigma} = 3.10^{-6}$		$\sigma = 9.3137285$ $\frac{\delta\sigma}{\sigma} = 10^{-6}$	
Iterations Number	80	27	80	45	52	20	38	21
Divergence	5.10^{-11}	2.10^{-11}	5.10^{-11}	2.10^{-11}	5.10^{-11}	2.10^{-11}	7.10^{-11}	2.10^{-11}
	1rst s.d	last s.d	1rst s.d	Last s.d	1rst s.d	last s.d	1rst s.d	last s.d

7. CONCLUSIONS

The pseudo-spectral space-time method seems to be one of the most accurate method for solving partial differential equations.

This method associated with a Poisson equation for the pressure requires a compatible initial velocity field. If the compatibility condition is not satisfied, a time subdomains method has sufficient damping properties to cope with the initial singularities and to give accurate results.

This method associated with a simultaneous treatment of momentum and continuity equations on a staggered grid, gives precise results even if the initial velocity field is not compatible.

For solving Navier-Stokes equations in the case of inhomogeneous incompressible flows, the second technique seems more accurate than the technique based on a Poisson solver for the pressure.

Acknowledgments
The author is grateful to Y. Morchoisne and to Y. Maday for helpful discussions.

REFERENCES

[1] Ladyzhenskaya, O.A., "The Mathematical Theory of Viscous Incompressible Flow". Gordon and Breach (1969) pp. 162-168.

[2] Heywood, J.G., Rannacher, R., "Finite Element Approximation of the Nonstationary Navier-Stokes Problem I" Siam Journal of Numerical Analysis Volume 19 (1982) pp. 275-311.

[3] Moin, P., Reynolds, W.C., Ferziger, J.H. "Large Eddy simulation of Incompressible Turbulent Channel Flow" (1978) Dept. Mech. Engng, Stanford Univ. Rep TF-12.

[4] Morchoisne, Y., "Inhomogeneous Flow Calculations by Spectral Methods Monodomain and Multi-domain Techniques" Symposium on Spectral Methods for Partial Differential Equations" August 16-18, 1982. ICASE, NASA Langley Research Center Hampton, Va., U.S.A. Proceedings to be published by SIAM, Philadelphia, Pa.

[5] Deville, M., Haldenwang, P., Labrosse, G. "Comparison of Time Integration (finite difference and spectral) for the non Linear Burgers Equation", Proceedings of the Fourth GAMM Conference on Numerical Methods in Fluid Mechanics (H. Viviand, Editor) Vol. 5, Vieweg (1982) pp. 64-76.

[6] Huberson, S., Morchoisne, Y. "Large Eddy Simulation by Spectral Method or by Multi Level Particle Method". 6th AIAA Computational Fluid Dynamics Conference, DANVERS, U.S.A. July 13-15 (1983).

[7] Deville, M., Kleiser, L., Montigny-Rannou, F., "Pressure and time treatment for Chebyshev Spectral Solution of a Stokes problem" (1983) International Journal for Numerical Methods in Fluids (to be published).

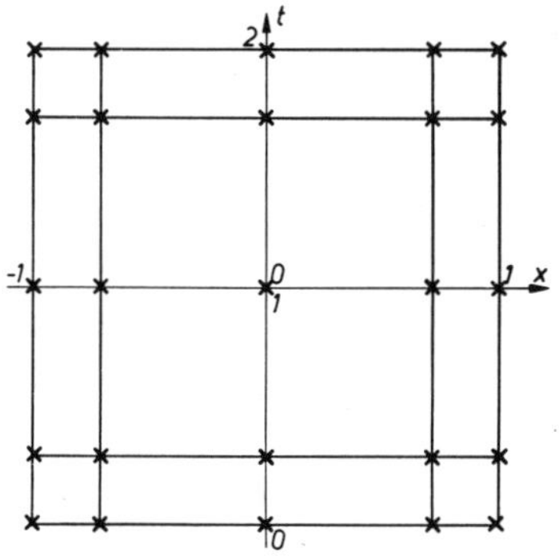

Fig. 1 — Time-space grid. 1rst technique.
x : U_1, U_2, P.

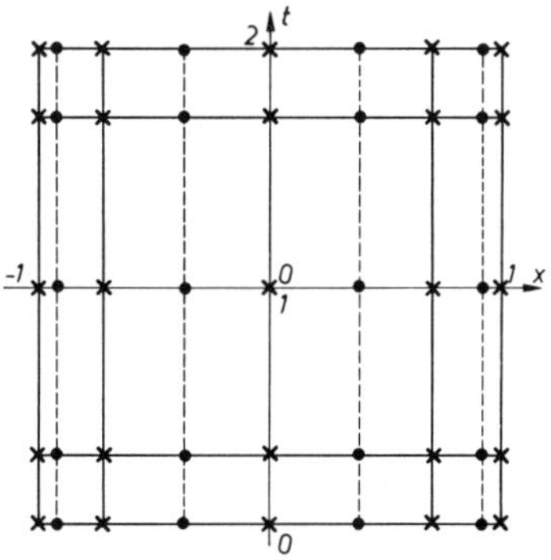

Fig. 2 — Time-space grid. 2nd technique.
x : U_1, U_2
•: P

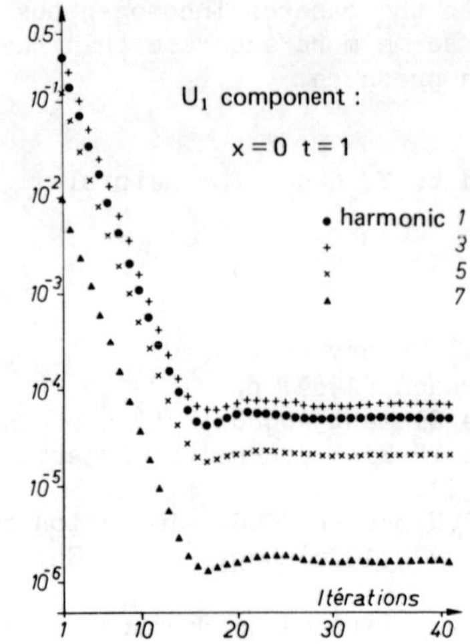

Fig. 3 – Chebyshev harmonics.

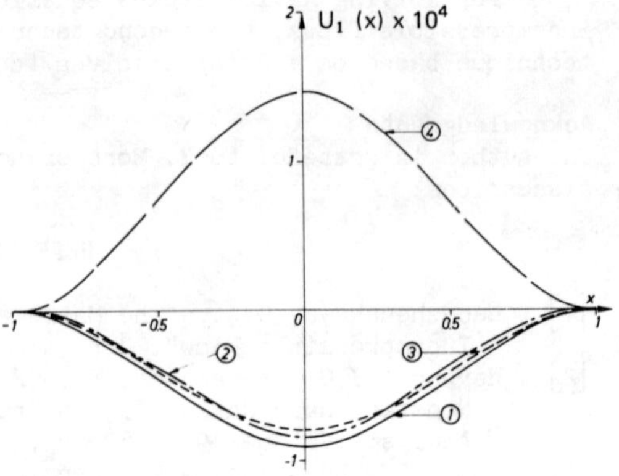

Fig. 4 – U_1 component at t = 1. 1rst technique

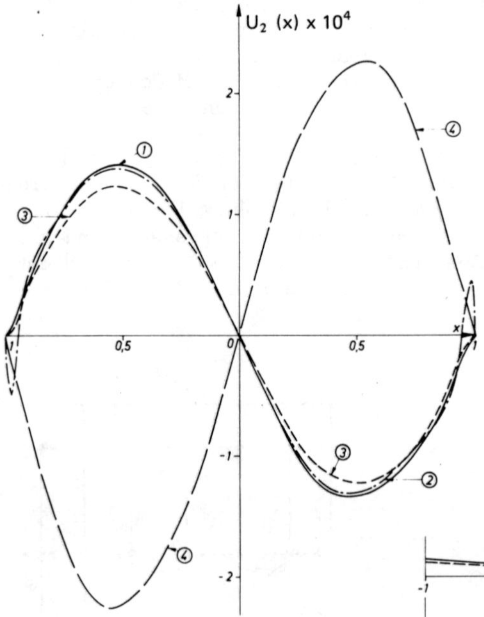

Fig. 5 – U_2 component at t = 1. 1rst technique.

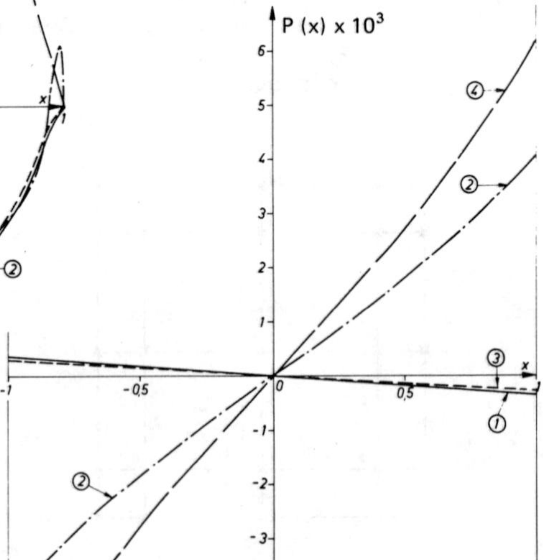

Fig. 6 – Pressure at t = 1rst technique.

CHARACTERISTIC GALERKIN METHODS FOR HYPERBOLIC PROBLEMS

K. W. Morton
University of Reading
Whiteknights, Reading, U.K.

SUMMARY

In this paper we present methods, which have been used by a number of authors, based on combining the following of characteristics with the Galerkin projection. We show how the idea of "information recovery" can be used to greatly extend the range of these methods; and for the linear advection equation we analyse key properties and relationships with other well-known methods. The main algorithm is for a scalar conservation law: extensions for the Euler equations are derived based on flux-difference splitting; and applications to shock modelling using shock recovery are given.

ECG SCHEMES FOR A SCALAR CONSERVATION LAW

The use of characteristics together with interpolation on a fixed mesh has a long history - see e.g. Courant, Isaacson & Rees [1] or Ansorge [2]. Linear interpolation used in a finite difference framework, however, gives only first order accuracy and excessive numerical damping. On the other hand, coupling it with the Galerkin approach yields dramatic improvements - a fact which has been observed and exploited by many authors.

Consider the scalar conservation law for $u(\underline{x},t)$ in the whole of real space

$$\partial_t u + \underline{\nabla}\cdot\underline{f}(u) = 0 \tag{1a}$$

or

$$\partial_t u + \underline{a}(u)\cdot\underline{\nabla} u = 0 \ , \tag{1b}$$

where $\underline{f}(u)$ is a flux vector and $\underline{a}(u) = \partial\underline{f}/\partial u$ the corresponding characteristic velocity vector. Then u is constant along the characteristics $d\underline{x}/dt = \underline{a}(u)$, which are therefore straight, and if we write $u^n(\underline{x})$ for $u(\underline{x},n\Delta t)$ at time-level n and use a similar notation for $\underline{f}$ and $\underline{a}$, we have for smooth flows

$$u^{n+1}(\underline{y}) = u^n(\underline{x}) \quad \text{where} \quad \underline{y} = \underline{x} + \underline{a}^n(\underline{x})\Delta t. \tag{2}$$

Now suppose $u^n(\underline{x})$ is approximated by the finite element expansion

$$U^n(\underline{x}) = \textstyle\sum_{(j)} U_j^n \phi_j(\underline{x}), \tag{3}$$

where we particularly have in mind piecewise linear approximations in one dimension and either piecewise linears over triangles or piecewise bilinears over rectangles in two dimensions. Then if the L^2 projection is used in a Galerkin formulation, (2.2) may be approximated by

$$\langle U^{n+1},\phi_i\rangle = \int U^n(\underline{x})\phi_i(\underline{y})d\underline{y}, \quad \underline{y} = \underline{x} + \underline{a}(U^n)\Delta t. \tag{4}$$

Here $\langle\cdot,\cdot\rangle$ denotes the usual L^2 inner product and we shall later use $\|\cdot\|$ to denote the corresponding norm. This form, or the related implicit form using $\underline{a}(U^{n+1})$, is that used for the linear advection terms in transport problems by Douglas & Russell [3], Bercovier et al. [4], Benqué &

Ronat [5] and for the Navier-Stokes equations by Brebbia & Smith [6] and Benqué et al. [7]. For the non-linear conservation law (1a,b) we prefer to start from the equivalent form, introduced in Morton [8,9],

$$\langle U^{n+1} - U^n, \phi_i \rangle + \Delta t \langle \underline{\nabla} \cdot \underline{f}(U^n),\ \Phi_i^n \rangle = 0, \tag{5}$$

where

$$\Phi_i^n(\underline{x}) = \frac{1}{|\underline{a}\,(U^n)|\Delta t} \int_{\underline{x}}^{\underline{x}+\underline{a}(U^n)\Delta t} \phi_i(\underline{z})\,d\ell. \tag{6}$$

The equivalence of the two forms follows from an integration-by-parts for ∂_a, the derivative in the direction from $\underline{x}$ to $\underline{y}$,

$$\int v(\underline{x})[\phi_i(\underline{y})d\underline{y} - \phi_i(\underline{x})d\underline{x}] = \int v(\underline{x})d\left[\int_{\underline{x}}^{\underline{y}} \phi_i(\underline{z})d\ell\right]$$
$$= -\int \partial_a v(\underline{x}) \left[\int_{\underline{x}}^{\underline{y}} \phi_i(\underline{z})d\ell\right] d\underline{x} = -\Delta t \int \left[\underline{a}(U^n)\cdot\underline{\nabla} v\right] \Phi_i^n d\underline{x}. \tag{7}$$

The form (5), (6) suggests a number of related simplified methods obtained by approximating the test functions $\Phi_i^n(\underline{x})$, or the integrals involving them, (see [8]) and enables relationships with other generalised Galerkin methods (see below) to be explored.

More importantly, (5) and (6) display the dependence on the flux vector and so indicate how improved methods for its estimation may be made use of. We call (5) [or (4)] the basic ECG (Euler-Characteristic-Galerkin) method. It gives for U^{n+1} the best L^2 fit to $E_\Delta U^n$ from the space spanned by $\{\phi_i\}$, where E_Δ denotes evolution through one time-step. Thus it leads naturally to the objective of carrying the best L^2 fit to the true solution u^n. However, even if U^n is the best fit to u^n, $\underline{f}(U^n)$ and $\underline{a}(U^n)$ may not be very good fits to $\underline{f}(u^n)$ and $\underline{a}(u^n)$: and it is these that appear in (5) and (6) when these equations relate the best fit to u^n to the best fit to u^{n+1}. Thus suppose we can find a recovery function $\tilde{u}^n$ such that $\underline{f}(\tilde{u}^n)$ and $\underline{a}(\tilde{u}^n)$ model better the key features of $\underline{f}(u^n)$ and $\underline{a}(u^n)$: then we can replace (5) by the more general form

$$\langle U^{n+1} - U^n, \phi_i \rangle + \Delta t \langle \underline{\nabla} \cdot \underline{f}(\tilde{u}^n), \tilde{\Phi}_i^n \rangle = 0, \tag{8}$$

where $\tilde{\Phi}_i^n$ is given by (6) with $\underline{a}(\tilde{u}^n)$ replacing $\underline{a}(U^n)$. To obtain $\tilde{u}^n$ we make use of any a priori knowledge that is available from the data and the equation - such as smoothness, monotonicity or positivity - and to retain the best fit property, set

$$\langle U^n - \tilde{u}^n, \phi_i \rangle = 0 \qquad \forall i. \tag{9}$$

Examples will be given below. It is usual for the basis functions to be such that the sum of ϕ_i over i gives the unit constant: then it follows automatically that (8) always ensures conservation of $\int U^n$.

Norms other than the L^2 norm may be used and we shall make some reference to such ECG schemes in the next section. Finally, other time-stepping procedures may be used to generate Characteristic-Galerkin methods which will give higher accuracy when the characteristics are curved.

PROPERTIES FOR LINEAR ADVECTION IN ONE DIMENSION

Suppose we use piecewise linear elements on a uniform mesh for the model problem

$$\partial_t u + a\partial_x u = 0, \tag{10}$$

where $a>0$ and we denote the CFL number $a\Delta t/\Delta x$ by μ. Then the basic ECG method (5) or (4) can be written out as a finite difference scheme for the nodal parameters U_j^n: the inner products $\langle\phi_j,\phi_i\rangle$ give rise to the familiar tridiagonal mass matrix with coefficients $(1/6)(1,4,1)$; but the inner products on the right of (4) will generally involve four terms depending on the size of μ. Suppose p is an integer such that

$$\mu = p + \hat{\mu} \quad \text{where} \quad p < \mu \leqq p+1. \tag{11}$$

Then the scheme becomes

$$(1 + \frac{1}{6}\delta^2)U_i^{n+1} = [(1 + \frac{1}{6}\delta^2) - \hat{\mu}(\Delta_0 - \frac{1}{2}\hat{\mu}\delta^2 + \frac{1}{6}\hat{\mu}^2\delta^2\Delta_-)]U_{i-p}^n, \tag{12}$$

where, in the usual notation, $\Delta_- U_i := U_i - U_{i-1}$, $\Delta_0 U_i := \frac{1}{2}(U_{i+1}-U_{i-1})$ and $\delta^2 U_i := U_{i+1}-2U_i + U_{i-1}$.

This is a compact, unconditionally stable, highly accurate scheme whose properties are typical of Characteristic-Galerkin methods. It is to be contrasted with the result of direct linear interpolation for U_i^{n+1} as in [1]: this gives only the familiar first order upwind scheme

$$U_i^{n+1} = (1-\hat{\mu})U_{i-p}^n + \hat{\mu}U_{i-p-1}^n. \tag{13}$$

In fact it is easy to see that (13) corresponds to the basic ECG method obtained with the simplest possible elements, namely piecewise constants on cells with boundaries at $(j+\frac{1}{2})\Delta x$, which was developed in [9] for shock modelling.

The usual Taylor series analysis of (12) would show that it is third order accurate but it is worth investigating the character of the error more thoroughly. The truncation error of the basic ECG method (5) can be written symbolically as

$$\text{T.E.} := (1/\Delta t)\,(P_2E_\Delta - P_2E_\Delta P_2)u, \tag{14}$$

where P_2 denotes the L^2 projection onto the approximation space, as in (9). This expresses the fact that the only error in (5) is due to the projection which is carried out before the (exact) evolution through time-step Δt. Suppose we write $\phi_j(x) = \phi(x/\Delta x-j)$, introduce the convolution function

$$\sigma(t) := \int \phi(s)\,\phi(s-t)ds \tag{15}$$

and denote by u_j^n the nodal parameters of P_2u^n. Then the key factor in (14) can be written more explicitly as

$$\sum_{(m)} u_{i+m}^{n+1}\,\sigma(m) - \sum_{(m)} u_{i+m}^n\sigma(m+\mu), \tag{16}$$

that is, as the difference between changing the coefficients in the expansion and changing the argument. For piecewise linears, ϕ is the familiar "hat" function which is the linear B-spline: then σ is the cubic B-spline. Thus the error is one of cubic spline approximation.

Substituting a Fourier mode e^{ikx} and introducing $\xi = k\Delta x$, we obtain the amplification factor $\lambda(\xi,\mu)$ for (12) in the form

$$\lambda(\xi,\mu) = \sum_{(m)} e^{im\xi}\sigma(m+\hat{\mu}) / \sum_{(m)} e^{im\xi}\sigma(m). \qquad (17)$$

Comparing with the exact result, we have

$$e^{-i\mu\xi} - \lambda(\xi,\mu) \sim \frac{1}{24}\hat{\mu}^2(1-\hat{\mu})^2\xi^4 \quad \text{as} \quad \xi \to 0. \qquad (18)$$

Note, that over the operative range $0 < \hat{\mu} \leqq 1$, the coefficient in this leading term is zero at the two ends and at the central maximum is only 1/384. For stability, we find that

$$|\lambda|^2 = 1 - (4/3)r^2[1 + (4/3)r]/q^2 \qquad (19)$$

where $r = \hat{\mu}(1-\hat{\mu})s^2$, $q = 1-(2/3)s^2$, $s = \sin \frac{1}{2}\xi.^2$
Thus (12) has the very useful property that it is stable well beyond the range in which it is to be used, viz. it is stable for $-1/2 \leqq \mu \leqq 3/2$.

A distinctive feature of (12) is its implicitness, through the mass matrix represented by $1 + (1/6)\delta^2$. Because of its diagonal dominance, this is easily inverted in practice to adequate accuracy. However, one could by use of a mixed norm projection obtain corresponding explicit formulae. Suppose we define

$$\langle u,v\rangle_\gamma := \langle u,v\rangle + \gamma^2\langle\partial_x u,\partial_x v\rangle . \qquad (20)$$

Then for $\gamma^2 = (\Delta x)^2/6$ it is easy to see that $\langle\phi_j,\phi_i\rangle_\gamma = \Delta x\delta_{ij}$. The scheme corresponding to (12) becomes

$$U_i^{n+1} = [1-\hat{\mu}\Delta_0 + \frac{1}{2}\hat{\mu}^2 \delta^2 + \frac{1}{6}\hat{\mu}(1-\hat{\mu}^2)\delta^2\Delta_-]U_{i-p}^n , \qquad (21)$$

which will be recognised as a third order accurate difference scheme due to Warming et al. [10]. The penalty is that now

$$e^{-\mu\xi} - \lambda(\xi,\mu) \sim \frac{1}{24}\hat{\mu}(1-\hat{\mu}^2)(2-\hat{\mu})\xi^4 \quad \text{as} \quad \xi \to 0, \qquad (22)$$

which is nine times larger than (18) at the maximum point $\hat{\mu} = \frac{1}{2}$; and stability holds only in the range $0 \leqq \hat{\mu} \leqq 1$.

For the general method (5) the main practical difficulty is the evaluation of the second inner product, or that on the right of (4). Various quadrature rules may be used, but it is often helpful to think in terms of approximating Φ_i^n by $\phi_i, \phi_{i-1}, \phi_i'$ etc. In [8] several schemes were given which all reproduce (12) in the case of linear constant-coefficient advection. A particularly suitable one if product integration is used, gives the approximation

$$\begin{aligned}\langle\phi_j',\Phi_i\rangle \simeq (1-\hat{\mu})^2\langle\phi_j',\phi_i\rangle - \hat{\mu}(1-\hat{\mu})\langle\phi_j',\phi_{i-1}\rangle \\ + \hat{\mu}(3-2\hat{\mu})\langle\phi_j,\phi_i-\phi_{i-1}\rangle .\end{aligned} \qquad (23)$$

This involves the evaluation of only those inner products which would be needed if one used the Galerkin method, in which Φ_i is replaced by its limit ϕ_i as $\Delta t \to 0$.

Use of the mixed norm may also be helpful in this approximation process. If one uses (20) with $\gamma^2 = (a\Delta t)^2/6$, one finds that the corresponding special test function Φ_i has zero integral outside the support of ϕ_{i-p} and that a suitable approximation is

$$\Phi_i \approx \phi_{i-p} + \tfrac{1}{2}a\Delta t\phi'_{i-p} \,. \tag{24}$$

Instead of (12) one then obtains the difference scheme

$$[1 + \frac{1}{6}(1-\hat{\mu}^2)\delta^2](U_i^{n+1} - U_{i-p}^n) + (\hat{\mu}\Delta_0 - \frac{1}{2}\hat{\mu}^2\delta^2)\hat{U}_{i-p}^n = 0. \tag{25}$$

This is exactly the EPG II scheme developed by Morton & Parrott [11] in their study of Petrov-Galerkin methods or, equivalently, the Taylor-Galerkin method based on Euler time-stepping given by Donea [12]: in both papers methods based on alternative time-stepping schemes are given and the approximate ECG schemes derived as above coincide with those of [12], which are not always the same as those of [11]. (See also [13] for similar schemes.) Corresponding to (18), scheme (25) gives

$$e^{-\mu\xi} - \lambda(\xi,\mu) \sim \frac{1}{24}\hat{\mu}^2(1-\hat{\mu}^2)\xi^4 \qquad \text{as} \quad \xi \to 0. \tag{26}$$

This has a maximum which is four times larger than that of (18) (though smaller than that of (22)) but, reflecting the fact that for $|\hat{\mu}| \leqq 1$ (25) is directionally independent, the maximum occurs at $\hat{\mu}^2 = \frac{1}{2}$ and stability holds for $-1 \leqq \hat{\mu} \leqq 1$. Note that as $|\hat{\mu}| \to 1$, (25) becomes more nearly explicit, eventually coinciding with (21).

Relationships with many other schemes may be explored by choice of norm and approximation of the corresponding Φ_i. Forms such as (24) are of course particularly reminiscent of the streamline diffusion schemes first introduced by Hughes & Brooks [14] for diffusion-convection problems.

All of the above was concerned with the basic ECG method, with no recovery. Now let us consider how (13) may be improved upon by a recovery process. Suppose the piecewise constant $U^n(x)$ is recovered by a quadratic spline. On a uniform mesh all the B-splines can be defined by convolution with the characteristic function $\chi^{(0)}(t)$ of the interval $(-\frac{1}{2},\frac{1}{2})$: thus

$$\chi^{(m+1)}(t) = (\chi^{(m)} * \chi^{(0)})(t) = \int \chi^{(m)}(s)\chi^{(0)}(s-t)ds; \tag{27}$$

$\chi^{(0)}(t)$ generates the piecewise constant basis, $\chi^{(1)}(t)$ the linears and so on. Thus let the recovery function $\tilde{u}^n(x)$ be given by

$$\tilde{u}^n(x) = \sum \tilde{u}_j^n \chi^{(2)}(x/\Delta x - j). \tag{28}$$

Then because inner products between $\chi^{(2)}$ and $\chi^{(0)}$ will give the same results as those between $\chi^{(1)}$ and $\chi^{(1)}$, the recovery operation (9) reduces to

$$(1 + \frac{1}{6}\delta^2)\tilde{u}_i^n = U_i^n \,. \tag{29}$$

Similarly (8) becomes

$$\begin{aligned} \langle U^{n+1},\phi_i\rangle &= \int \tilde{u}^n(y-a\Delta t)\phi_i(y)dy \\ &= \Delta x \sum \tilde{u}_j^n \int \chi^{(2)}(s-\mu)\chi^{(0)}(s-i+j)ds \\ &= \Delta x[(1+\frac{1}{6}\delta^2) - \hat{\mu}(\Delta_0 - \frac{1}{2}\hat{\mu}\delta^2 + \frac{1}{6}\hat{\mu}^2\delta^2\Delta_-)]\tilde{u}_{i-p}^n \end{aligned} \tag{30}$$

just as in (12). Operating throughout with $1 + (1/6)\delta^2$ we exactly reproduce (12). This is an example of a general result: for linear constant-coefficient advection on a uniform mesh, an ECG scheme based on splines of degree m together with recovery by splines of degree $m + 2p$ is equivalent to the basic ECG scheme based on splines of degree $m + p$.

SHOCK MODELLING WITH PIECEWISE CONSTANTS

The use of recovery techniques to obtain high accuracy for smooth flows combined with their natural aptness for shocked flows makes the piecewise constant ECG methods extremely powerful. However, neither (4) nor (5) have obvious meaning when U^n is discontinuous and $\underline{a}$ is not constant: at a "rarefaction" the characteristic mapping is undefined and at "compressions" it is multivalued. One solution in the one dimensional case is to resolve the discontinuities by solving the Riemann problem at each cell interface as was done by Godunov [15]: but this is time consuming and takes no account of the fact that these discontinuities are merely a consequence of the projection we are using and have no physical meaning. In [9] Morton introduced a recovery procedure on a uniform mesh using piecewise linears to spread the discontinuities $\frac{1}{2}\theta\Delta x$ either side of the cell boundaries. Taking the limit $\theta \to 0$, one obtains an update algorithm for use at each cell boundary which, if $f(u)$ has a single sonic point $\bar{u}$ and all CFL numbers are less than unity, has the form:-

$$\text{for } a^n_{i-1}, a^n_i \geq 0 \quad \text{add } (\Delta t/\Delta x)\Delta_- f^n_i \text{ to } U^{n+1}_i , \tag{31a}$$

$$\text{for } a^n_{i-1}, a^n_i \leq 0 \quad \text{add } (\Delta t/\Delta x)\Delta_- f^n_i \text{ to } U^{n+1}_{i-1} , \tag{31b}$$

$$\text{for } a^n_{i-1}\, a^n_i < 0 \quad \left.\begin{array}{l}\text{add } (\Delta t/\Delta x)(f^n_i - f(\bar{u})) \text{ to } U^{n+1}_i \\ \text{and } (\Delta t/\Delta x)(f(\bar{u}) - f^n_{i-1}) \text{ to } U^{n+1}_{i-1}\end{array}\right\} . \tag{31c}$$

In this case, the procedure exactly reproduces the Engquist-Osher scheme [16]. A further procedure was given for shock recovery: the presence of a shock in a cell is recognised by the crossing of the characteristics from the cells on either side, the position and strength recovered by the general formula (9), and the updates based on the recovered shock. The result is a shock motion satisfying the Rankine-Hugoniot conditions and an algorithm which was very effective for the inviscid Burger equation.

For $\theta > 0$ but sufficiently small, the recovery procedure gives for the constant sections in each cell values $\tilde{u}^n_i$ satisfying (cf. (29))

$$\tilde{u}^n_i + (\theta/8)\delta^2\tilde{u}^n_i = U^n_i \quad . \tag{32}$$

If we then approximate by

$$\tilde{u}^n_i \approx U^n_i - (\theta/8)\delta^2 U^n_i , \quad f(\tilde{u}^n_i) \approx f(U^n_i) - a(U^n_i)(\theta/8)\delta^2 U^n_i \tag{33}$$

we obtain when $a(U^n_i)$, $a(U^n_{i-1}) \geq 0$

$$\langle \partial_x f(\tilde{u}^n), \tilde{\Phi}^n_i \rangle \approx \Delta_- f(U^n_i) + (\theta\Delta x/8\Delta t)[(1-\mu_i)\delta^2 U^n_i + \mu_i \delta^2 U^n_{i-1}], \tag{34}$$

where $\mu_j = a(U^n_j)\Delta t/\Delta x$. For linear advection this is second order accurate if $\theta = 4\mu(1-\mu)$, when it is third order accurate for $\mu = 0, \frac{1}{2}, 1$. Thus θ

acts like a "flux limiter" - see Sweby [17] for a survey and rationalisation: in smooth parts of the flow it may be chosen by various criteria - for example, if the equation is monotonicity-preserving then ensuring that $\tilde{u}_i^n$ is monotone will ensure that U_i^{n+1} is also.

The shock modelling algorithm has recently been developed and programmed for the Euler equations in one dimension, including shock recovery, but otherwise for the limit $\theta \to 0$. In [9] the extension of (3.1) to a system of equations was outlined using a matrix of test function Φ_i^n but the present implementation is based on the flux-difference splitting of Roe [18]. Let $\underline{w}$ be the vector of conserved quantities, density ρ, momentum m and total specific energy e, and $\underline{f}$ the vector of fluxes. Then at each interface, between states $\underline{w}_R$ on the right and $\underline{w}_L$ on the left, averaged values of the velocity $\bar{u}$, the sound speed $\bar{a}$, the density $\bar{\rho}$ and the enthalpy $\bar{H}$ are defined so as to construct approximate eigenvectors $\underline{e}^{(k)}$ and eigenvalues $s^{(k)}$ (the characteristic speeds) to the Jacobian matrix $A = \partial\underline{f}/\partial\underline{w}$:-

$$s^{(1,2,3)} = [\bar{u}-\bar{a},\ \bar{u},\ \bar{u}+\bar{a}] \tag{35a}$$

$$\underline{e}^{(1,2,3)} = \begin{bmatrix} 1 & 1 & 1 \\ \bar{u}-\bar{a} & \bar{u} & \bar{u}+\bar{a} \\ \bar{H}-\overline{ua} & \tfrac{1}{2}\bar{u}^2 & \bar{H}+\overline{ua} \end{bmatrix} . \tag{36a}$$

These are such that coefficients $\alpha^{(k)}$ can be introduced to satisfy

$$\underline{w}_R-\underline{w}_L = \sum_{(k)}\alpha^{(k)}\underline{e}^{(k)}, \quad \underline{f}_R-\underline{f}_L = \sum_{(k)}s^{(k)}\alpha^{(k)}\underline{e}^{(k)} . \tag{37}$$

Then (31) is generalised by applying it to the split flux differences $s^{(k)}\alpha^{(k)}\underline{e}^{(k)}$ for $k = 1,2,3$ using the characteristic speeds $u-a, u, u+a$ in the R and L cells. The only difficult case is (31c), where we have $\underline{f} = \underline{0}$ at a sonic point, and $\underline{f}_R$ and $\underline{f}_L$ have each to be split.

Several criteria are used to recognise the presence of a shock or contact discontinuity in the ith cell:-

<u>right-facing shock</u>: $-2(\Delta t/\Delta x)\Delta_0(u_i+a_i) > \text{tol.}$ (≈ 1),

$$\rho_{i-1} > \rho_i > \rho_{i+1} \quad \text{and} \quad \Delta_0 p_i/\Delta_0\rho_i > \text{tol.}_{CD} \ (<< 1); \tag{38a}$$

<u>left-facing shock</u>: $-2(\Delta t/\Delta x)\Delta_0(u_i-a_i) > \text{tol.}$

$$\rho_{i-1} < \rho_i < \rho_{i+1} \quad \text{and} \quad \Delta_0 p_i/\Delta_0\rho_i > \text{tol.}_{CD} \ ; \tag{38b}$$

<u>contact discontinuity</u>: $\Delta_0 p_i/\Delta_0\rho_i \leq \text{tol.}_{CD}$

$$(\Delta_+\rho_i)(\Delta_-\rho_i) > 0 \quad \text{and} \quad |\Delta_0 u_i/(u_{i+1}+u_{i-1})| < \text{tol.}_{CD} . \tag{38c}$$

Shocks in contiguous cells cannot be dealt with so if (38a,b or c) is satisfied in a set of contiguous cells, that with the largest pressure jump is selected for the recovery procedure.

The recovery of the shock position is based on the density, i.e. the shock is taken to be at $(i+\eta-\frac{1}{2})\Delta x$, where

$$\eta = \Delta_+\rho_i/2\Delta_0\rho_i \ ; \tag{39}$$

and to maintain conservation of all quantities the states $\underline{\tilde{w}}_L$, between $(i-3/2)\Delta x$ and the shock, and $\underline{\tilde{w}}_R$, between the shock and $(i+3/2)\Delta x$, are given by

$$3\tilde{\underline{w}}_L = [(3-\eta)\underline{w}_{i-1} + \underline{w}_i - (1-\eta)\underline{w}_{i+1}]$$
$$3\tilde{\underline{w}}_R = [-\eta\underline{w}_{i-1} + \underline{w}_i + (2+\eta)\underline{w}_{i+1}]. \qquad (40)$$

Finally the flux difference $\tilde{\underline{f}}_R - \tilde{\underline{f}}_L$ has to be split as in (37) and the updates allocated to the three cells $i-1$, i and $i+1$ according to how $\tilde{s}^{(k)}\Delta t/\Delta x$ compares with $-\eta$ and $1-\eta$, the positions of the cell boundaries relative to the shock position. Extensive testing of this algorithm is now under way.

REFERENCES

[1] Courant, R., Isaacson E. & Rees, M. Comm. Pure & Appl. Math. 5 p243 (1952).

[2] Ansorge, R., Numer. Math. 5, p443 (1963).

[3] Douglas Jr.J., & Russell, T.F. SIAM J. Numer. Anal. 19, p871 (1982).

[4] Bercovier, M., Pironneau, O., Hasbani, Y. & Livne, E. Proc. MAFELAP 1981 Conf. (ed. J.R. Whiteman), Academic Press (London)471-478 (1982).

[5] Benqué, J.P. & Ronat, J. 5th Intl. Symp. on Comp. Meth. in Appld. Sc. & Engng., INRIA)1981).

[6] Brebbia, C.A. & Smith, S., Proc. 1st Int. Conf. on Finite Elements in Water Resources, Pentech Press, 4.205-4.230 (1977).

[7] Benqué, J.P., Ibler, B., Keramsi, A. & Labadie, G., Proc. 3rd Int. Conf. on Finite Elements in Flow Problems, Banff, Alberta, Canada (1980).

[8] Morton, K.W. Proc. IMA Conf. on Num. Meth. for Fluid Dynamics (eds. K.W. Morton & M.J. Baines), Academic Press 1-32 (1982).

[9] Morton, K.W. Proc. 8th Int. Conf. on Numerical Methods in Fluid Dynamics, Aachen 1982 (ed. E. Krause), Lect. Notes in Physics 170, Springer-Verlag, Berlin, 77-93 (1982).

[10] Warming, R.F., Kutler, P. & Lomax, H., AIAA Jnl. 11,189 (1973).

[11] Morton, K.W. & Parrott, A.K., J. Comp. Phys. 36, 249-270 (1980).

[12] Donea, J. Int. J. Num. Meth. in Engng. (to appear).

[13] Griffiths, D.F. The Mathematics of Finite Elements and Applications Proc. MAFELAP 1981 (ed. J.R. Whiteman), Academic Press, London 411-420 (1982).

[14] Hughes, T.J.R. & Brooks, A. Finite Element Methods for Convection Dominated Flows (ed. T.J.R. Hughes), AMD Vol. 34, Am. Soc. Mech. Eng. (New York), 19-35 (1979).

[15] Godunov, S.K. Mat. Sb. 47 (1959), 271-290.

[16] Engquist, B. & Osher, S. Math. Comp. 36, 321-352 (1981).

[17] Sweby, P., Proc. AMS-SIAM Conf. Large Scale Fluid Calculations, La Jolla June/July 1983.

[18] Roe, P.L. J. Comp. Phys., 43, 357-372 (1981).

APPLICATION OF A VARIABLE NODE FINITE-ELEMENT METHOD TO THE GAS DYNAMICS EQUATIONS

M. Calvin Mosher

David Taylor Naval Ship Research and Development Center
Bethesda, Maryland, U.S.A. 20084

SUMMARY

The variable node finite element method [1] which has both a variable space and time step has been applied to the gas dynamics equations in one space dimension in conservation form with a numerical viscosity. The method was implemented in the program VFEGD and tested successfully on a shock tube problem. Since the variable node method resolves regions of large gradients with high accuracy, it makes possible automatic numerical detection of contact discontinuities and shocks. Once they have been detected, they can be sharpened and consequently their diffusion controlled. The program VFEGD with modifications to detect and sharpen contacts and shocks was also successfully tested on a shock tube problem.

INTRODUCTION

This paper describes the application of a variable node method to the gas dynamics equations. This method was previously derived by the author and applied to single partial differential equations (PDE's) [1]. It has since been extended to cover systems of equations in one space dimension and in particular the gas dynamics equations [2].

The variable node method uses the finite element method and linear splines but allows the nodal points to move with time. Algebraic equations are derived which determine the position of nodes. They place many nodes where the gradient or curvature of the finite element splines is large and few nodes elsewhere. The derivatives of the algebraic equations are taken with respect to time and are equal in number to the number of nodes used in the finite element splines. The resulting new ordinary differential equations (ODE's) are integrated simultaneously with the equations of the finite element method. As a result, the node positions are determined simultaneously with the solution and nodes are concentrated where the solution has a large gradient or curvature. The concentration of nodes in regions where small internodal spacing is needed allows very efficient integration of equations with shock-like solutions and results in considerable savings in computer time and memory.

DERIVATION OF THE METHOD

Derivation of the method for systems of equations proceeds just as it does for a single equation [1] and is described in detail in [2]. The solution of a system of PDE's in the form $dU/dt = L(U)$ where L is a differential operator and U is a vector function on the space [0,1] is approximated by linear splines $v(j,t) = a(1,j,t)\alpha(1)+\ldots a(N,j,t)\alpha(N)$ for $j=1,\ldots,NE$ where NE is the number of PDE's. The coefficients $a(i,j,t)$ are the values $v(j,s(i),t)$ of the function at the nodes $s(1),\ldots,s(N)$ which are dependent on time. The $\alpha(i,x)$ equal $(x-s(i-1))/(s(i)-s(i-1))$ for $s(i-1) \leqslant x \leqslant s(i)$, $(s(i+1)-x)/(s(i+1)-s(i))$ for $s(i) \leqslant x \leqslant s(i+1)$ and zero otherwise. The L^2-norm of the residual of the PDE's, defined with the piecewise linear approximation $V = (v(1),\ldots,v(N))$, is formally minimized with respect to the da(i,j)/dt by requiring that the ODE's

$$d||dV/dt-L(V)||_2/d(da(i,j)/dt) = 0$$

hold for i = 1,...,N and j=1,...,NE.

Using the equations $e = \rho\varepsilon + 0.5\ \rho u^2$ and the equation $\varepsilon = p/\rho(\gamma-1)$ for a polytropic gas where γ is a constant greater than one, the gas dynamics equations can be written in conservation form with a numerical viscosity ν to smear out the discontinuities $U_t + F(U)_x = \nu U_{xx}$ where $U = (\rho,m,e)$ and $F(U) = (m,\ (\gamma-1)e+0.5(3-\gamma)m^2/\rho,\ \gamma me/\rho+.5(1-\gamma)m^3/\rho^2)$. Here ρ is the density, u is the velocity, m is the momentum ρu, ε is the internal energy per unit mass, p is the pressure, and e is the total energy. Replacing ρ, m and e by linear splines with coefficients a(i,1), a(i,2) and a(i,3) respectively the ODE's of our method for the gas dynamics equations take the following form for i=1,...,N

$$\sum_{k=-1}^{1} [(da(i+k,1)/dt)< \alpha(i+k), \alpha(i)> + (ds(i+k)/dt)< \beta(i+k,1), \alpha(i)>] =$$
$$- [-(s(i)-s(i-1))^{-1} \int_{s(i-1)}^{s(i)} m dx + (s(i+1)-s(i))^{-1} \int_{s(i)}^{s(i+1)} m dx]$$
$$+ \nu[-(s(i)-s(i-1))^{-1} \int_{s(i-1)}^{s(i)} \rho_x\, dx + (s(i+1)-s(i))^{-1} \int_{s(i)}^{s(i+1)} \rho_x\, dx] \qquad (1.a)$$

$$\sum_{k=-1}^{1} [(da(i+k,2)/dt) < \alpha(i+k), \alpha(i)> + (ds(i+k)/dt) < \beta(i+k,2), \alpha(i)>] =$$
$$- (\gamma-1)[-(s(i)-s(i-1))^{-1} \int_{s(i-1)}^{s(i)} e\, dx + (s(i+1)-s(i))^{-1} \int_{s(i)}^{s(i+1)} e\, dx]$$
$$-.5(3-\gamma)[-(s(i)-s(i-1))^{-1} \int_{s(i-1)}^{s(i)} (m^2/\rho)dx+(s(i+1)-s(i))^{-1} \int_{s(i)}^{s(i+1)} (m^2/\rho)dx]$$
$$+ \nu[-s((i)-s(i-1))^{-1} \int_{s(i-1)}^{s(i)} m_x dx + (s(i+1)-s(i))^{-1} \int_{s(i)}^{s(i+1)} m_x dx] \qquad (1.b)$$

$$\sum_{k=-1}^{1} [(da(i+k,3)/dt) < \alpha(i+k), \alpha(i)> + (ds(i+k)/dt) < \beta(i+k,3), \alpha(i)>] =$$
$$- \gamma[-(s(i)-s(i-1))^{-1} \int_{s(i-1)}^{s(i)} (me/\rho)dx + (s(i+1)-s(i))^{-1} \int_{s(i)}^{s(i+1)} (me/\rho)dx]$$
$$-.5(1-\gamma)[-(s(i)-s(i-1))^{-1} \int_{s(i-1)}^{s(i)} (m^3/\rho^2)dx+(s(i+1)-s(i))^{-1} \int_{s(i)}^{s(i+1)} (m^3/\rho^2)dx]$$
$$+ \nu[-(s(i)-s(i-1))^{-1} \int_{s(i-1)}^{s(i)} e_x dx + (s(i+1)-s(i))^{-1} \int_{s(i)}^{s(i+1)} e_x dx] \qquad (1.c)$$

The $\beta(i,j)$ for i=1,...,N and j=1,...,NE result from differentiating the $\alpha(i)$ with respect to time and equal -m(i-1,j)(x-s(i-1))/(s(i)-s(i-1)) for s(i-1) $\leq$ x $\leq$ s(i), -m(i,j)(s(i+1)-x)/(s(i+1)-s(i)) for s(i) < x $\leq$ s(i+1) and zero otherwise. The slopes m(i,j) of the splines are (a(i+1,j)-a(i,j))/(s(i+1)-s(i)) for i=1,...,N-1 and j=1,...,NE. The <f,g> are defined as $\int_0^1 f(x)g(x)dx$. The integrals $\int me/\rho dx$, $\int m^2/\rho dx$ and $\int m^3/\rho^2 dx$ can be computed by numerical integration or by exact series representations of the integrands. With regard to the series representations, if the difference a(i+1,1)-a(i,1) of the density is large enough on an interval [s(i), s(i+1)], the integrals can be reduced to algebraic expressions with the log function by dividing the numerator by the denominator and integrating; otherwise, the integral can be expressed as a series by expanding the denominator in a series, multiplying by the numerator and integrating the series term by term. The series representations for integrals are described in detail for one space dimension in [2] and for two space dimensions in [3].

There are 3N equations in (1) but 4N unknowns because the nodes s(1),...,s(N) are functions of time. The remaining N equations needed to complete the system of equations are derived by differentiating a system of N algebraic equations which distribute the nodes. The algebraic equations have the form

$$g(1) = 0,\dots,\ g(N) = 0 \qquad (2)$$

where g(1) = s(1), g(i) = f(i-1)-f(i) for 2 $\leq$ i $\leq$ N-1 and g(N) = s(N)-1 and

$$f(i,s(i-1),\ s(i),\ s(i+1),\ s(i+2)) = \Big(\sum_{j=1}^{NE} m(i,j)^2 + \varepsilon 1\Big)^{1/2} (s(i+1)-s(i))$$

$$+\ B1\ \Big(\sum_{j=1}^{NE} (m(i,j)-m(i-1,j))^2 + \varepsilon 2\Big)^{1/2}(s(i+1)-s(i-1))$$

$$+\ B2\ \Big(\sum_{j=1}^{NE} (m(i+1,j)-m(i,j))^2 + \varepsilon 2\Big)^{1/2}(s(i+2)-s(i))$$

$$= \text{constant} \qquad (3)$$

Here B1 is zero if i=1 and B2 is zero if i=N-1. Eqns. (2) require each interval [s(i-1),s(i)] to have the same total amount of gradient and curvature. If the gradient or curvature has larger values in the interval [s(i-1), s(i)] than in others, then that interval will be smaller. The equations g(1)=0 and g(N)=0 define the boundary points.

At time zero the N nonlinear equations g(1)=0,...,g(N)=0 must be solved for s(1),...,s(N). The method for solving the equations for a scalar initial solution U(x,0) implemented in the program GIV described in [1] can be modified to solve the new equations as indicated in [2].

IMPLEMENTATION OF THE METHOD

The method was implemented in the program VFEGD which uses the Gear-Hindmarsh program GEARIB [4] to integrate the system of 4N ODE's of the method. The implicit stiff equation solvers [5] (which use a finite difference version of Newton's method) and the variable time step employed by GEARIB are needed because variables a(i,j) and s(i) change rapidly in the shock region. Although we do not have a numerical method for computing values for the parameters $\varepsilon 1$, $\varepsilon 2$, B and the number of nodes N in Eqns. (2), we can use the heuristic procedure described in [1] where in computing $\varepsilon 1$, $|d\phi/dt|^2$ is replaced by $|d\phi(1)/dt|^2+\ldots+|d\phi(NE)/dt|^2$.

APPLICATION OF THE METHOD TO A SHOCK TUBE PROBLEM

The program VFEGD was applied to the test problem used by Sod [6]. In this problem a diaphragm divides a shock tube with closed ends in half with the gas having discontinuities in pressure and density at the diaphragm. At time zero the diaphragm is broken producing a rarefaction wave which moves to the left of the midpoint of the shock tube and a contact discontinuity and shock which move to the right. The initial conditions are for a shock tube of length one: $\rho 1=1.0$, $p1=1.0$, $u1=0.0$, $\rho 2=0.125$, $p2=0.1$, and $u2=0.0$. A ramp was used in the interval 0.5 to 0.51 as a transition region between the first three values and the second three. We let $\varepsilon 1=10$, $\varepsilon 2=10^{-6}$, $B=10^{-2}$, $\nu=10^{-3}$ and N=63 in Eqns. (2). The parameters EPS and MF in GEARIB were set to 10^{-5} and 22 and the series representations of the integrals were used. To measure the diffusion of the shock and contact we define the <u>interior width</u> as the distance between the first two nodes interior to the left and right endpoints of the shock or contact and the <u>exterior width</u> as the distance between the two endpoints. In Figure 1 the interior width of this contact is the distance between nodes 64 and 67 and the exterior width is the distance between nodes 63 and 68, which are the endpoints. Integrating for 0.14 time units on the CDC CYBER74 computer (at the Naval Ship Research and Development Center) with FTN5 and OPT=2, VFEGD computed the shock with an interior width less than 0.01 during the entire integration and the contact discontinuity with an interior width less than or equal to 0.05. The rarefaction wave was smeared at the endpoints because of the large numerical viscosity. The ratio of the time step to the minimum space step in the shock region was many times larger than that allowed by the Courant-Friedrichs-Lewy condition for the shock wave for an explicit method. Increasing the number of nodes to 100, we ran the program again keeping the other parameter values the same and present the plots of the density, pressure, velocity and internal energy versus distance at time 0.14 in Figure 2a. (In Figure 2 the circles represent values computed at nodal points and the solid lines represent exact solutions.) The CPU time was nearly doubled while the interior widths of the shock and contact were approximately the same.

The exterior widths of the shock and contact computed by VFEGD are somewhat larger than the interior widths. At time 0.14 for example the exterior widths of the shock and contact are 0.03 and 0.1 respectively for 63 nodes and 0.024 and 0.0825 for 100 nodes. These larger exterior widths are due to the large size of the intervals between the endpoints and the first interior nodes in the shock or contact which in Figure 1 are the intervals between nodes 63 and 64 and between nodes 67 and 68. However, these large <u>edge intervals</u> are not produced by diffusion but by Eqns. (2) which determine the positions of the nodes. Eqns. (2) make these edge intervals large because they make the transition between the region where the density, momentum and total energy are constant and the steep gradient region in the interior of the shock and contact. Nevertheless, since it is often sufficient to represent the contacts and shocks as discontinuities, we need to know only their locations and the values of the density, momentum and energy at these locations; the large edge values apparently do not interfere with this information in the test problem as seen in Figure 2a.

A disadvantage of the VFEGD program is that it has no mechanism for controlling the amount of diffusion of the contacts and shocks other than by reducing the numerical viscosity. For example, although it will compute much thinner contact discontinuities if the viscosity is reduced from 10^{-3} to 10^{-4}, it will use somewhat more CPU time. Modifications of VFEGD which allow control of the diffusion of the shock and contact are described in the next section.

DETECTION AND SHARPENING OF CONTACT DISCONTINUITIES AND SHOCKS

Since the variable node method concentrates nodes in monotone regions, i.e., where a variable of the solution is strictly increasing or decreasing, it can resolve them with high accuracy and therefore should be very reliable in detecting contact discontinuties and shocks. Once the discontinuties have been detected, they can be sharpened and consequently their diffusion controlled.

A primary consideration in selecting a sharpening strategy is that the ODE solvers in VFEGD are more efficient if a large numerical viscosity is used to make the solution appear smooth in regions of large curvature, i.e., in regions of smeared discontinuities, although the viscosity should be small enough to prevent excessive diffusion in regions of small curvature such as a rarefaction wave. Hence, the strategy for sharpening should be:

1. apply program VFEGD to the gas dynamics equations with a large numerical viscosity for (at least) a specific number of time steps which is called MINSTEP
2. after MINSTEP time steps, detect and sharpen the shocks and contact discontinuities
3. repeat steps (1) and (2).

For simplicity the detection and sharpening of contact discontinuities but not shocks will be discussed. The detection of contact discontinuities is handled by subroutine FIND. Since the density has a jump across a contact, but the velocity and pressure do not, a contact discontinuity is defined as a region where the density is monotonic with a maximum gradient larger than some minimal value and the velocity and pressure are constant with respect to some tolerance. We may add to this definition that the internal energy be monotonic with a maximum gradient larger than some minimal value and of a different sign from the maximum gradient of density. More specifically, the version of FIND used on the test problem in this paper defines a contact discontinuity as a region containing at least four nodes where all the slopes m(i,1) of the density spline have the same sign, the absolute values of the m(i,1) are larger than some minimal value $\tau 2$ and the maximum absolute value SLMAX of the m(i,1) must be larger than a minimal value $\tau 12$ in order to prevent a monotone region from being sharpened into a contact discontinuity prematurely but less than $\tau 13$ to prevent the contact from being sharpened too often. In addition the absolute value of the difference between the values of the velocity as well as the pressure at adjacent nodes must be less than some maximum value $\tau 3$.

The subroutine SHARP reduces the widths of and sharpens regions considered to be smeared contacts and shocks. We assume that if the monotone region between points x1 and x2 marked with a solid line in Figure 1 for example is sharpened into a stationary shock with area being conserved, it will look like the broken line in Figure 1 where $(SP-x1)(\rho 1-\rho 2)$ equals the area of the monotone region between values $\rho 1$ and $\rho 2$ and x1 and x2. Here SP is the location of the shock which has amplitude $\rho 1-\rho 2$. Given s(M1),..., s(M2), values a(M1),...,a(M2) and slopes m(M1),...,m(M2) in the monotone region with M1 and M2 being the numbers of the first and last nodes in the region, we can sharpen the smeared out shock to an arbitrary width without changing the SP or the shape, i.e., the ratios of the slopes m(i)/m(i+1) will remain the same. We let XL=s(M2)-s(M1) and select a desired width for the shock which we call WSH. Let

new s(M1)=SP-(SP-s(M1))WSH/XL
new s(M1+1)=new s(M1)+(s(M1+1)-s(M1))WSH/XL
new s(M1+2)=new s(M1+1)+(s(M1+2)-s(M1+1))WSH/XL and so on.

The result is a new linear spline having the same values a(M1),...,a(M2), the same shape and the same SP as the original but with the desired width WSH. The total area under the new spline will be exactly that under the original.

The parameter WSH can be replaced by XL(SLMAX)/SLMXNEW where SLMAX is the maximum absolute value of the slopes m(M1),...,m(M2) of the spline in the monotone region and SLMXNEW is the desired SLMAX. The value of SLMXNEW is designated $\tau 5$.

Modifications must be made to the ODE solver GEARIB in order to use FIND and SHARP. If SHARP is called, the values of the solution at previous time steps cannot be used to predict the solution at the new time step. Consequently, GEARIB which employs multi-step methods must use a first order explicit predictor. Since the equations of the variable node method have the form A(y)dy/dt=G(y), the inverse $A(y)^{-1}$ must be computed; the equations $dy/dt=A(y)^{-1}G(y)$ can then be solved by Euler's method. In GEARIB the subroutine DIFFUN computes $A(y)^{-1}G(y)$, i.e., it computes dy/dt. Since the first order predictor-corrector method in GEARIB which employs DIFFUN is not efficient, FIND is called only after a minimum number of time steps MINSTEP have elapsed from the last time it was called so that GEARIB will automatically change to a second order method and thereby achieve a large time step. Program GEARIB is allowed to use just a first and second order method.

Since SHARP modifies the output of GEARIB, the positions of the nodes must be determined to satisfy equations (2). We do not use GIV to satisfy these equations but a new method GIV2 described in [2] which is more efficient.

The program VFEGD with FIND, SHARP, GIV2, and the modifications to GEARIB is called VFGDSHARP. This program has two sets of parameters. The first set consists of parameters from the program VFEGD and are determined just as they were for that program. The second set consists of parameters in FIND and SHARP. To determine values for the second set we run the program VFEGD on the problem at hand and observe the widths of the smeared discontinuities and the rate at which they diffuse [2].

The program VFGDSHARP was applied to Sod's test problem with the same values for the VFEGD parameters used in the application of VFEGD with 63 nodes. Using the output of VFEGD on the problem, we selected the following values for the parameters for detecting and sharpening the contact in FIND and SHARP: MINSTEP=11, $\tau 2=0.4$, $\tau 3=0.02$, $\tau 5=25$, $\tau 12=5$, and $\tau 13=10$. Integrating for time 0.14 on the CDC CYBER74 with FTN5 and OPT=2, VFGDSHARP computed both the shock and contact with interior widths of approximately 0.01 at time 0.14. The interior width of the shock was less than 0.018 during the entire integration, whereas the interior width of the contact was 0.02 when first detected at time 0.0535 and was gradually reduced to 0.011 at time 0.14. The exterior widths for the shock and contact were less than 0.056 and 0.065 respectively during the integration and 0.045 and 0.051 at time 0.14, while the location SP of the contact computed at time 0.14 by FIND was in error by 0.008. The rarefaction wave was integrated with the same accuracy as the VFEGD program. The ratio of the average time step to the minimum space step in the shock wave occurring during the integration was larger than that allowed by the Courant-Friedrichs-Lewy condition for the shock wave for an explicit method. However, the program used 25% more CPU time than VFEGD for this test problem. Increasing the number of nodes to 100, we ran the program again keeping the other parameter values the same. Subroutine FIND was applied to the output of VFGDSHARP at time 0.14 and plots of the resulting values for the density, pressure, velocity and internal energy versus distance are presented in Figure 2b. The program used 70% more CPU time than with 63 nodes, detected the contact initially at time 0.0193 and halved the error in the location SP of the contact computed by FIND to 0.004. The interior widths of the shock and contact were approximately the same, while the exterior widths were less than or equal to 0.047 and 0.05 respectively during the integration and 0.031 and 0.043 at time 0.14.

REFERENCES

[1] M.C. Mosher, A Variable Node Finite Element Method, to be published in J. Comput. Physics.
[2] M.C. Mosher, "A Variable Node Method Applied to the Gas Dynamics Equations," to appear as DTNSRDC report.
[3] M.C. Mosher, "A Finite Element Method for the Gas Dynamics Equations in Two Dimensions and in Conservation Form," to appear as a DTNSRDC report.
[4] A.C. Hindmarsh, "GEARIB: Solution of Implicit Systems of Ordinary Differential Equations with Banded Jacobian", Lawrence Livermore Laboratory, 1976, UCID-30130.
[5] C.W. Gear, "Numerical Initial Value Problems in Ordinary Differential Equations", Prentice-Hall, Inc., Englewood Cliffs, New Jersey, 1971.
[6] G.A. Sod, J. Comput. Physics 27 (1978), 1-31.

ACKNOWLEDGEMENT

I want to thank Kathy Donaldson for her very capable typing of this paper.

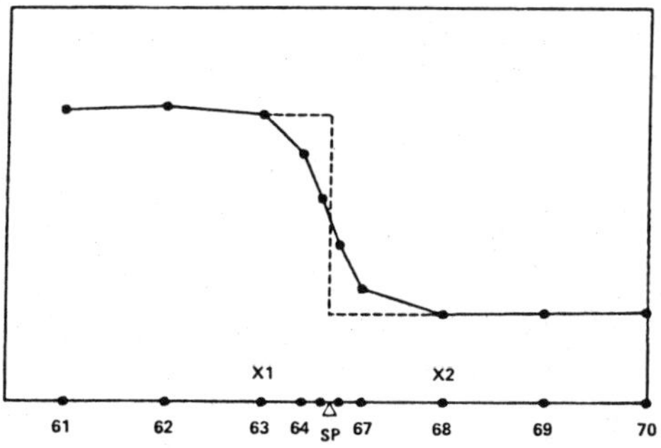

Fig. 1 Density versus distance computed by VFGDSHARP at t = 0.14 with N = 100 before application of SHARP.

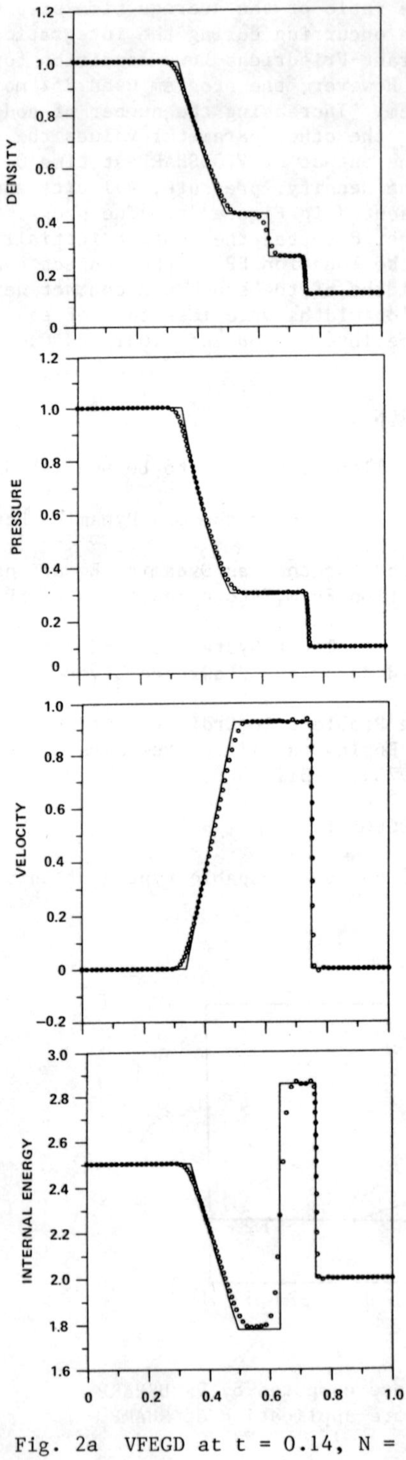

Fig. 2a VFEGD at t = 0.14, N = 100

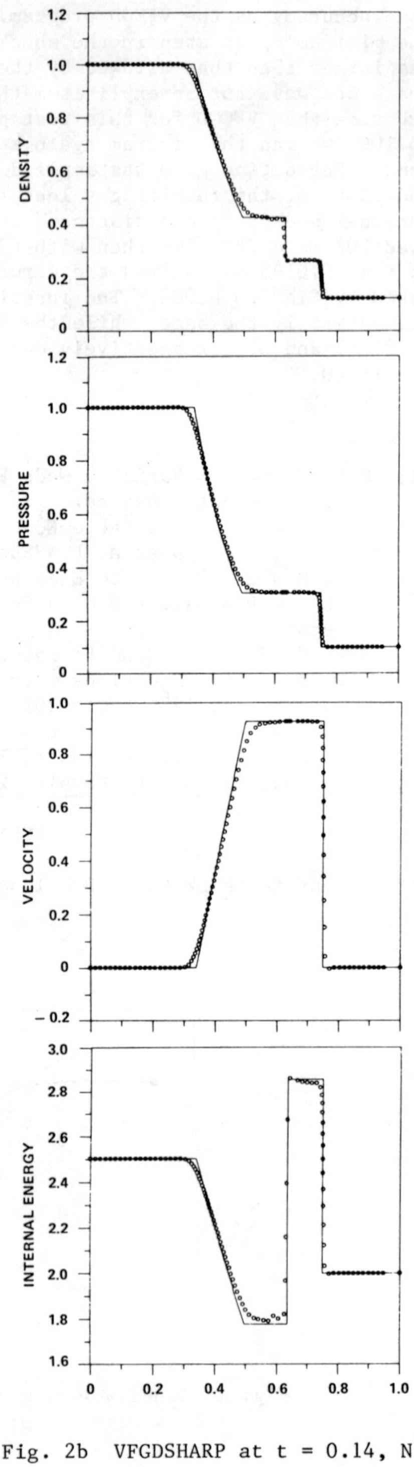

Fig. 2b VFGDSHARP at t = 0.14, N = 100

THREE-DIMENSIONAL IMPLICIT LAMBDA METHODS*

M. Napolitano and A. Dadone
Istituto di Macchine, via Re David 200, 70125 Bari, Italy

SUMMARY

This paper provides various block-explicit and block-implicit methods for solving the three-dimensional lambda-formulation equations numerically. Three model problems, characterized by subsonic, supersonic and transonic flow conditions, are used to assess the reliability and compare the efficiency of the proposed methods.

INTRODUCTION

Among the many theoretical models employed in the numerical simulation of compressible inviscid flows the so-called lambda-formulation has received considerable interest (see, e.g., [1-8]): the time-dependent Euler equations are recast into compatibility conditions of bicharacteristic variables along the corresponding bicharacteristic lines and discretized using windward differences, in order to account for the direction of wave propagation phenomena, correctly. Such an approach has many nice properties: it provides very accurate numerical results, even with rather coarse meshes (see, e.g., [2], [3], [6]); it requires only the physical boundary conditions, so that there is no need for any additional numerical boundary treatments, which are frequently the cause of numerical instability [7]; it handles in a most automatic and physically-sound way mixed supersonic-subsonic flow fields; and finally, it has a well documented, although controversial, capability of capturing shocks without any additional dissipation [2-6]. For these reasons, in spite of the fact that the "captured shocks" are only isentropic approximations to correct weak solutions of the Euler equations and do not correctly move within the flow field - unless properly fitted [4], the lambda-formulation is considered to be a very useful and reliable tool for predicting compressible flow fields and, therefore, very worthy of further studies and improvements; and in fact, in the last two years, for the cases of quasi-one dimensional and two dimensional flows, the development of various kinds of implicit integration schemes [5-8] has removed the only major limitation of previous lambda methods, namely, the CFL stability restriction associated with their explicit integration procedures.

It now appears very timely and worthwhile to develop efficient numerical methods, based on the lambda-formulation, for three-dimensional flows, as done in the present paper: the governing equations are discretized and linearized in time using a delta approach and various block-explicit as well as block-implicit numerical techniques are proposed to solve the resulting discrete equations approximately, at every time step; all of the proposed methods are then applied to solve three model problems, characterized by subsonic, supersonic and transonic flow conditions, respectively, in order to assess their reliability and efficiency.

---

*Research supported in part by the Consiglio Nazionale delle Ricerche and in part under NASA Contract No. NAS1-17070 while the first author was in residence at ICASE, NASA Langley Research Center, Hampton, Va 23665.

GOVERNING EQUATIONS AND NUMERICAL METHODS

The nondimensional lambda-formulation equations for three-dimensional homentropic flows of a perfect gas are derived for a general curvilinear orthogonal coordinate system in Ref. 10. Here, they are given for Cartesian coordinates [3] as:

$$C_t + D_t + (u+a)\,C_x + (u-a)D_x + v\left(C_y + D_y\right) + w\left(C_z + D_z\right) = 0 \tag{1}$$

$$E_t + F_t + u\left(E_x + F_x\right) + (v+a)E_y + (v-a)F_y + w\left(E_z + F_z\right) = 0 \tag{2}$$

$$G_t + H_t + u\left(G_x + H_x\right) + v\left(G_y + H_y\right) + (w+a)G_z + (w-a)H_z = 0 \tag{3}$$

$$\frac{1}{3}\left\{C_t - D_t + E_t - F_t + G_t - H_t\right\} + (u+a)C_x - (u-a)D_x + (v+a)E_y - (v-a)\,F_y + (w+a)G_z - (w-a)H_z = 0 \tag{4}$$

$$C - D + E - F = 0 \tag{5}$$

$$C - D + G - H = 0 \tag{6}$$

where u, v and w are the velocity components in the three Cartesian directions x, y and z, a is the speed of sound, t is the time and subscripts indicate partial derivatives; C, D, E, F, G and H are the six bicharacteristic variables, given as

$$C = u + \delta a \qquad D = u - \delta a \qquad E = v + \delta a \tag{7a,b,c}$$

$$F = v - \delta a \qquad G = w + \delta a \qquad H = w - \delta a \tag{7d,e,f}$$

where $\delta = 2/(\gamma-1)$, γ being the specific heats ratio.

It is noteworthy that all derivatives in eqns. (1-4) are associated with the convection of the bicharacteristic waves along their bicharacteristic lines (see [10], for details). Equations [1-6] are discretized and linearized in time using the delta form [11,5,6], to give

$$\frac{\Delta C}{\Delta t} + \frac{\Delta D}{\Delta t} + (u+a)^n\,\Delta C_x + (u-a)^n\,\Delta D_x + v^n\left(\Delta C_y + \Delta D_y\right) + w^n\left(\Delta C_z + \Delta D_z\right)$$
$$= -\,(u+a)^n\,C_x^n - (u-a)^n\,D_x^n - v^n(C_y + D_y)^n - w^n(C_z + D_z)^n \tag{8}$$

$$\frac{\Delta E}{\Delta t} + \frac{\Delta F}{\Delta t} + u^n\left(\Delta E_x + \Delta F_x\right) + (v+a)^n\,\Delta E_y + (v-a)^n\,\Delta F_y + w^n\left(\Delta E_z + \Delta F_z\right)$$
$$= -u^n(E_x + F_x)^n - (v+a)^n\,E_y^n - (v-a)^n\,F_y^n - w^n(E_z + F_z)^n \tag{9}$$

$$\frac{\Delta G}{\Delta t} + \frac{\Delta H}{\Delta t} + u^n(\Delta G_x + \Delta H_x) + v^n\left(\Delta G_y + \Delta H_y\right) + (w+a)^n\,\Delta G_z + (w-a)^n\,\Delta H_z$$
$$= -u^n\left(G_x + H_x\right)^n - v^n\left(G_y + H_y\right)^n - (w+a)^n\,G_z^n - (w-a)^n\,H_z^n \tag{10}$$

$$\frac{1}{3}\left\{\frac{\Delta C}{\Delta t} - \frac{\Delta D}{\Delta t} + \frac{\Delta E}{\Delta t} - \frac{\Delta F}{\Delta t} + \frac{\Delta G}{\Delta t} - \frac{\Delta H}{\Delta t}\right\} + (u+a)^n \Delta C_x - (u-a)^n \Delta D_x + (v+a)^n \Delta E_y$$

$$- (v-a)^n \Delta F_y + (w+a)^n \Delta G_z - (w-a)^n \Delta H_z = -(u+a)^n C_x^n + (u-a)^n D_x^n$$

$$- (v+a)^n E_y^n + (v-a)^n F_y^n - (w+a)^n G_z^n + (w-a)^n H_z^n \quad (11)$$

$$\Delta F = -\Delta C + \Delta D + \Delta E \quad (12)$$

$$\Delta H = -\Delta C + \Delta F + \Delta G, \quad (13)$$

where, Δt is the time step, $\Delta C = C^{n+1} - C^n$ (the superscripts n + 1 and n indicating the new and old time levels $t^{n+1} = t^n + \Delta t$ and t^n) etc. Equations (8-13) constitute a first-order-accurate implicit time discretization of the corresponding differential problem; eqns. (8-11) are then discretized in space using windward differences to properly account for the direction of wave propagation and the ΔF and ΔH unknowns are eliminated in favor of ΔC, ΔD, ΔE and ΔG by means of eqns. (12) and (13) to produce, together with appropriate boundary conditions, a large 4×4 block-sparse linear system of the type

$$\mathbf{A}\ \mathbf{f} = \mathbf{b}. \quad (14)$$

For the case of a cubic integration domain having N gridpoints in every spatial direction, **A** is a square matrix of order N^3 having only seven nonzero diagonals of 4×4-matrix-elements, **f** is the unknown vector having N^3 four-element-vector components and **b** is the known coefficient vector. It is noteworthy that in previous works [5,6] a second-order-accurate time linearization was employed. However, due to the use of a backward Euler time discretization, eqn. (14) is only first-order-accurate in time anyway. Moreover, the present linearization, coupled with windward difference approximations for the left-hand sides of eqns. (8-11), leads to a diagonally dominant matrix **A** and has been verified to increase the stability of all the implicit methods later proposed in this study. It is also noteworthy that second-order-accurate, three-point windward differences can be used to approximate the right-hand sides of eqns. (8-11) so that, if the flow reaches a steady state, the final solution is second-order-accurate [5,6].

The main reason to employ an implicit method is to remove the CFL stability restriction, thus improving the efficiency of the calculations. However, a direct solution of problem (14), even if feasible, is certainly impractical. Therefore the matrix **A** will be replaced by a matrix **B** which is easily invertible and is a first-order-accurate approximation (in time) of **A**.

A Block-Explicit Method

The simplest first-order-accurate approximation to **A** can be obtained by dropping all but the time-derivative terms in the left-hand sides of eqns. (8-11). The resulting matrix **B** is diagonal and a simple 4×4 linear system needs to be solved at every gridpoint to provide the local ΔC, ΔD, ΔE, and ΔG values. Furthermore, eqns. (8-11) can be rearranged to give

$$\frac{2\Delta C}{\Delta t} = RHS(8) + RHS(11) \tag{15}$$

$$\frac{\Delta C}{\Delta t} + \frac{\Delta D}{\Delta t} = RHS(8) \tag{16}$$

$$-\frac{\Delta C}{\Delta t} + \frac{\Delta D}{\Delta t} + \frac{2\Delta E}{\Delta t} = RHS(9) \tag{17}$$

$$-\frac{\Delta C}{\Delta t} + \frac{\Delta D}{\Delta t} + \frac{2\Delta G}{\Delta t} = RHS(10) \tag{18}$$

(where RHS(8) is a shorthand notation for the right-hand side of eqn. (8), etc.) so that every element of **B** is a lower triangular matrix which can be inverted directly. The present BE method has been developed mainly for assessing the efficiency of various implicit methods; however, due to its extreme coding simplicity, it could very well be a useful tool by itself, especially if implemented on a vector computer.

A Block-Alternating-Direction-Implicit (BADI) Method

An ADI technique has been developed, which is the direct extension to three-dimensional problems of the method of Refs. [5] and [6]. A three-sweep ADI process is used to solve problem (14) approximately. At the first sweep the t and x derivatives in the left-hand sides of eqns. (8-11) are evaluated implicitly, whereas the y and z derivatives are evaluated explicitly. At the second and third sweeps the t and y and the t and z derivatives are evaluated implicitly so that **A** is approximated by the product of three 4×4 block-tridiagonal matrices. In practice, at every sweep of the BADI method a 4×4 block-tridiagonal system of order N has to be solved along each line of the computational grid, so that $3N^2$ such systems need to be solved at every time step (i.e., to solve eqn. (14) approximately). With respect to two-dimensional flow problems, the present ADI method is less competitive as compared to a standard explicit method, for two reasons: the block size of the tridiagonal systems increases from three to four and, more importantly, the number of tridiagonal systems to be solved at every time step grows from 2N to $3N^2$. Actually, for the simple problems later considered in this study the computer time per step for an 11^3 mesh was found to be about 30 times greater than that required by the BE method. More efficient implicit methods need therefore to be devised for the three-dimensional lambda-formulation equations.

A Block-Line-Gauss-Seidel (BLGS) Method

Classical relaxation methods have been recently employed with considerable success in connection with "upwind schemes" for the one- and two-dimensional Euler equations [7,8,12]. Here an obvious choice, leading to a reduction of the computer time per step to about one third, is to employ a single step 4×4 block-line-Gauss-Seidel method: all of the time and x derivatives in the left-hand sides of eqns. (8-11) are evaluated implicitly together with the diagonal contributions of the y and z derivatives, so that only N^2 4×4 block-tridiagonal systems (of order N) are to be solved at every time level. By accounting for the previously evaluated nontridiagonal entries, explicitly, the matrix **A** is effectively replaced by its three main diagonals plus its two additional nonzero lower diagonals. Furthermore, the ordering of the solution process is changed at every time step so as to account for the two additional nonzero upper or lower diagonals, alternately.

A Block-Point-Gauss-Seidel (BPGS) Method

By taking to its extreme the logic behind the previous method, an obvious choice presents itself; that is, to replace the matrix **A** with its lower or upper triangular part. In eqns. (8-11) the diagonal contributions are accounted for implicitly and the previously evaluated off-diagonal contributions are brought to the right-hand sides of the equations and accounted for explicitly. At every gridpoint location a 4×4 linear system needs to be solved as in the BE method; however, due to its variable coefficients, the local 4×4 matrix cannot be triangularized and a complete Gauss-Jordan elimination procedure, using diagonal pivot strategy, has been employed (here, as well as to solve the local linear systems within a general block-tridiagonal inversion routine in all of the present implicit methods).

A Simplified-Line-Gauss-Seidel (SLGS) Method

From their very definitions (eqns. (7)) as well as from their compatibility conditions (see, e.g., [10] or from the governing eqns. (8)) it appears that the waves associated with the bicharacteristic variables C and D mainly propagate in the x direction, whereas the E and F waves and the G and H waves mainly propagate in the y and z directions, respectively. Therefore, it would seem appropriate to devise a numerical method exploiting such a property of the bicharacteristic variables as done in [7,8] for the case of one- and two-dimensional flows. Therefore the following simplified line-Gauss-Seidel method is proposed here: Equations (8) and (11) are solved coupled together for the ΔC and ΔD variables by means of a line-Gauss-Seidel method, implicit in the x direction, so that a 2×2 block-tridiagonal system of order N has to be solved at every y_j and z_k gridpoint location. Equations (9) and (10) are then solved by means of line-Gauss-Seidel methods implicit in the y and z direction, respectively, so that $2N^2$, additional scalar tridiagonal systems need to be solved. Obviously equations (12) and (13) are used to eliminate ΔF and ΔH from eqns. (8-11) and all of the $\Delta C,\cdots,\Delta H$ terms already evaluated at any level of the computation process are accounted for in the right-hand sides of the equations. Furthermore, since the pressure eqn. (11) does not have a main direction of propagation, it is coupled to eqn. (9), to evaluate ΔE and ΔF implicitly in the y direction, and to eqn. (10), to evaluate ΔG and ΔH implicitly in the z direction, at successive time steps.

It is noteworthy that, in general, the matrix **A** contains all the boundary conditions which are therefore accounted for, with the level of implicitness typical of every single method. However, for simplicity, in all of the present applications the exact solution of the continuum problem has been enforced at all boundaries to provide homogeneous boundary conditions for all of the incremental bicharacteristic variables. More general boundary conditions can be implemented as suggested in [6] and are not expected to cause any difficulty.

RESULTS

In order to test the proposed methods, a simple steady one-dimensional spherical source flow of air ($\gamma = 1.4$) has been considered; for such a flow field the continuity and energy equations are given as

$$a^5 \, v_r \, r^2 = c_1 \tag{19}$$

$$0.2 \, v_r^2 + a^2 = c_2 \tag{20}$$

v_r being the radial velocity component and r the radial distance from the origin. All of the calculations have been performed using a Cartesian coordinate system inside the unit cube such that: $2 \leq x \leq 3$; $-.5 \leq y \leq .5$; $-.5 \leq z \leq .5$. Three flow conditions have been considered: the subsonic flow corresponding to $c_1 = 3.2$ and $c_2 = 1.128$ and the supersonic and transonic flows corresponding to $c_1 = 4.2$ and $c_2 = 1.2205$. In the last case, an isentropic shock at $r = 2.15$ separates a supersonic region (for $r < 2.15$) from a subsonic one (for $r > 2.15$). The exact solution for the bicharacteristic variables has been imposed at all boundaries (the six sides of the computational cube) and a flow field having the exact values for u and a and zero v and w has been used as a suitable initial condition. The solution was advanced in time by means of any of the proposed methods using a constant (in time) and uniform (in space) value of Δt, until the average absolute value of ΔC at all interior points was less than 10^{-6}. Due to the use of the delta approach, the final steady solution is the same for all of the methods. The computed Mach number distribution along the x axis is plotted in Fig. 1 for the three flow cases versus the exact solution, for $\Delta x = \Delta y = \Delta z = .1$.

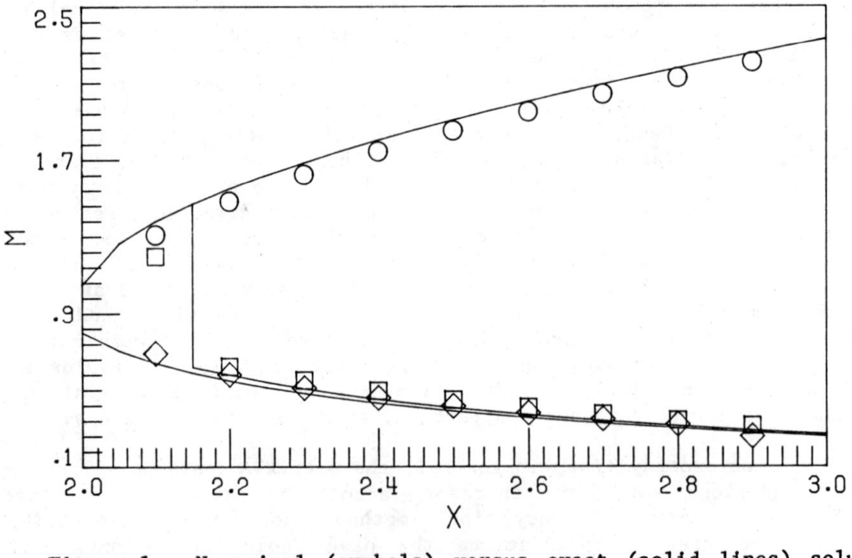

Figure 1. Numerical (symbols) versus exact (solid lines) solutions for spherical source-flows.

The solution is fairly good for the subsonic and supersonic case and qualitatively correct for the transonic one. In particular, the shock is captured in the correct mesh interval and no wiggles are present in spite of the absence of any additional dissipation. However, for shocks as strong as that given in Fig. 1, a shock-fitting procedure is warranted.

The computations were performed on a CDC Cyber 175 computer using two-point windward differences for all of the spatial derivatives.

The main purpose of this paper was to devise "efficient" implicit methods for the three-dimensional lambda-formulation equations. Therefore the performance of all of the present methods are given in Table I as the values of the Δt leading to the fastest convergence and the corresponding number of time steps (K) and CPU seconds.

TABLE I

	Subsonic Flow			Supersonic Flow			Transonic Flow		
Method	Δt	K	CPU	Δt	K	CPU	Δt	K	CPU
BE	.03	178	57	.03	62	21	.03	403	128
BADI	.3	35	361	.3	38	398	.3	86	887
BLGS	⩾5	18	75	⩾2	11	46	⩾10	71	290
BPGS	⩾5	29	49	⩾2	11	20	⩾10	76	126
SLGS	2.	22	34	.4	27	41	⩾10	99	147

From Table I the following conclusions can be drawn. For the supersonic flow case the BE and BPGS methods are clearly superior; this is obvious insofar as there is no upstream propagation in the x direction and thus the implicit methods use most of the CPU time accounting for zero entries; also, the performance of the BLGS and BPGS methods are identical as they should (all of the x derivatives being approximated with backward differences). For the more relevant transonic and subsonic flow cases the BLGS method always requires the smallest number of iterations to converge; however, the BPGS, SLGS and BE methods are the most efficient ones, whereas the BADI method is consistently the least competitive one. It is noteworthy that all of the Gauss-Seidel methods are very robust insofar as they maintain a quasi-optimal convergence rate over a wide range of Δt values. Among the three "best methods," the BPGS and the BE methods are considerably simpler to code and require less computer memory, a very critical resource when dealing with three-dimensional problems. Therefore, preliminary studies have been conducted to assess the influence on their convergence rate of the mesh size and of second-order-accurate discretizations for the nonincremental terms of the governing equations. The two methods converge in a number of iterations which is roughly inversely proportional to the step size (e.g., for the subsonic flow problem, convergence is reached after 261 and 52 iterations for a 17^3 mesh and after 309 and 59 iterations for a 21^3 mesh, for the BE and the BPGS methods, respectively). However, it is noteworthy that for these calculations (performed on a VAX 11/750 computer) the BE method required about 2.5 more CPU time than the BPGS method. This indicates that the solution routine for the local 4×4 linear systems works very inefficiently on the Cyber computer and that the superiority of the BPGS method over the BE one is greater than it actually appears from Table I. Also, the use of second-order-accurate differencing seems to deteriorate the convergence rate of the BPGS method less than that of the BE method. Finally, the superiority of the BPGS method (with respect to the BE method) is expected to increase even further by using a variable Δt [5,6,12] and when more general boundary conditions are employed; this, because the additional work will be relatively greater for the simpler BE method.

In conclusion, the BPGS method appears to be the most promising technique for solving three-dimensional compressible flow problems, by itself, or as a robust smoother within a more general multigrid procedure. However, both the BLGS and SLGS methods proposed in this study appear to be very promising alternatives to the ADI method of Refs. 5, 6 for solving two-dimensional steady flows, for which they are likely to outperform even the present BPGS method.

REFERENCES

[1] Butler, D. S., "The Numerical Solution of Hyperbolic System of Partial Differential Equations in Three Independent Variables," Proc. of the Royal Society of London, Series A, pp. 232-252, April 1960.

[2] Moretti, G., "The λ-scheme," Computers and Fluids, Vol. 7, pp. 191-205, 1979.

[3] Zannetti, L. and Colasurdo, G., "Unsteady Compressible Flow: A Computational Method Consistent with the Physical Phenomena," AIAA J., Vol. 19, pp. 851-856, July 1981.

[4] Chakravarti, S. R., Anderson, D. A. and Salas, M. D., "The Split Coefficient Matrix Method for Hyperbolic Systems of Gasdynamic Equations," AIAA Paper 80-0268.

[5] Dadone, A. and Napolitano, M., "An Implicit Lambda Scheme," AIAA Paper 82-0972, June 1982.

[6] Dadone, A. and Napolitano, M., "Efficient Transonic Flow Solutions to the Euler Equations," AIAA Paper 83-0258, January 1983.

[7] Moretti, G., "Fast Euler Solver for Steady One-Dimensional Flows," NASA CR-3689, June 1983.

[8] Moretti, G., "A Fast Euler Solver for Steady Flows," AIAA Paper 83-1940, AIAA 6th CFD Conference, Denvers, MA, July 13-15, 1983.

[9] Moretti, G., "A Physical Approach to the Numerical Treatment of Boundaries in Gas Dynamics," in Numerical Boundary Condition Procedures, NASA CP-2201, pp. 73-95.

[10] Napolitano, M. and Dadone, A., "Three-dimensional Implicit Lambda Methods," ICASE Report, 1983, to appear.

[11] Beam, R. M. and Warming, R. F., "An Implicit Factored Scheme for the Compressible Navier-Stokes Equations," AIAA. J., Vol. 16, pp. 393-402, April 1978.

[12] Van Leer, B. and Mulder, W. A., "Relaxation Methods for hyperbolic Equations, T. H. D. Report, Delf University of Technology, 1983.

ACKNOWLEDGMENTS

The authors are grateful to Drs. D. Dwoyer and B. Van Leer for their precious suggestions and discussions during most of this project. The ICASE environment was vital to the success of this effort.

NUMERICAL SOLUTION OF AN IMPINGING JET FLOW PROBLEM

Samuel Ohring

David Taylor Naval Ship Research and Development Center
Bethesda, Maryland, U.S.A. 20084

SUMMARY

Unsteady flow-surface interactions at leading and trailing edges are important because they are primary sources of noise in ship technology. The present paper is concerned with the two-dimensional computer simulation of impinging incompressible shear-layer flows involving leading edge/vortex interaction for an edgetone configuration. This configuration consists of a jet, emanating from a channel, impinging on a wedge-shaped body. Reynolds numbers, based on channel width and average velocity, of 250 and 650 are considered. To the author's knowledge such computer simulations have not been reported previously. The flow for this Reynolds number range is experimentally observed to be two-dimensional. The basic features of jet impingement flow are numerically simulated and the numerical results, which are in agreement with experimental results, indicate the presence of one major frequency in the flow for Re = 250 and two major frequencies for Re = 650.

INTRODUCTION

Significant advancement in understanding the edgetone type problem has resulted from the theoretical investigations of Powell [1], [2] and the more recent investigations summarized by Rockwell [3]. However details of the feedback mechanism and of the flow structure in the leading-edge region remain unclarified. To help clarify this structure, the fully nonlinear, viscous flow problem is considered in this paper which discusses a numerical integration of the Navier-Stokes equations.

MATHEMATICAL FORMULATION AND NUMERICAL METHOD

For the solution of fluid flow problems, the use of elliptic numerical transformations that map boundary-fitted coordinates (the body being a coordinate line) in physical space onto a Cartesian coordinate system of a computational space [4] has become quite commonplace [5]. Fig. 1 shows the physical space coordinate system used in this paper. This physical domain is geometrically identical, for the channel-wedge configuration, to the apparatus used at Lehigh University [6] for conducting physical experiments for the flow problem under consideration. The nondimensional length and height is shown. The coordinate origin is at the center of the channel nozzle. A Cartesian mesh (not shown) overlays the computational space. Physical space boundaries map onto computational space boundaries. The wedge-shaped body maps onto the slit in computational space. The mapping is such that geometrical corner points are excluded from the grid. The total number of grid points is 39,481.

A vorticity(ω)-stream function(ψ) formulation of the Navier-Stokes equations is employed. The vorticity-transport equation and the Poisson equation for the stream function ψ are written in computational space coordinates as

$$\omega_t = [\psi_\eta \omega_\xi - \psi_\xi \omega_\eta + (\alpha\omega_{\xi\xi} - 2\beta\omega_{\xi\eta} + \gamma\omega_{\eta\eta} + \sigma\omega_\eta + \tau\omega_\xi)/(\mathrm{Re}\ J)]J^{-1} \quad (1)$$

$$\alpha\psi_{\xi\xi} - 2\beta\psi_{\xi\eta} + \gamma\psi_{\eta\eta} + \sigma\psi_\eta + \tau\psi_\xi = J^2\omega \quad (2)$$

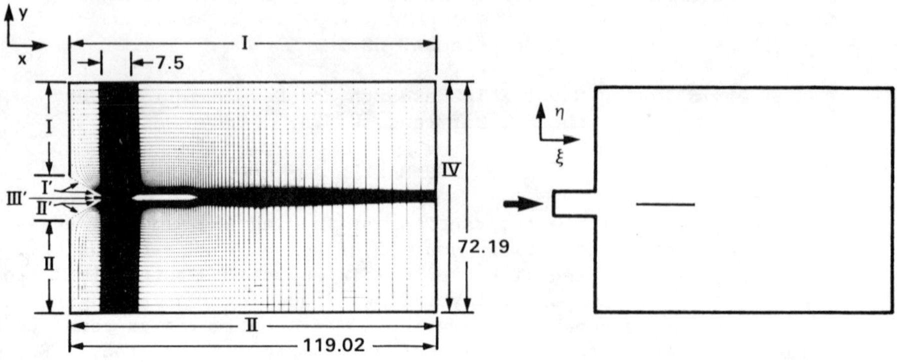

Physical Space | Computational Space

FIG. 1 NUMERICAL COORDINATE TRANSFORMATION

$$\text{where}\quad \alpha = x_\eta^2 + y_\eta^2,\quad \beta = x_\xi x_\eta + y_\xi y_\eta,\quad \gamma = x_\xi^2 + y_\xi^2,\quad J = x_\xi y_\eta - x_\eta y_\xi \tag{3}$$

$$\text{with}\quad \sigma = J^2 Q,\ \tau = J^2 P,\ P(\xi) = -x_{\xi\xi}/(x_\xi)^3,\ Q(\eta) = -y_{\eta\eta}/(y_\eta)^3 \tag{4}$$

Derivatives for P and Q are evaluated only at the outer boundaries to produce orthogonal grids at the outer boundaries of the physical space [7].

The following scaling has been used:

$$\omega' = (U/\delta)\omega;\ \psi' = \delta U\psi;\ t' = (\delta/U)t;\ p' = \rho U^2 p;\ Re = U\delta/\nu;\ St = f\delta/U \tag{5}$$

Primed quantities are dimensional. All velocities and lengths have been scaled by U and δ, respectively. The constants U, δ, ρ, ν are, respectively, the average velocity of the Couette flow in the channel; the width of the channel; density; and kinematic viscosity. (t is time and p is pressure.) The nondimensional parameters Re and St are, respectively, the Reynolds number and Strouhal number (nondimensional frequency) with f a dimensional frequency to be defined later.

The boundary conditions employed are (referring to Fig. 1): $\psi = -1/2$, $\omega = 0$ at boundary I; $\psi = -1/2$, ω generated at I′; $\psi = 1/2$, $\omega = 0$ at II; $\psi = 1/2$, ω generated at II′; $\psi = -6(y/4 - y^3/3)$, $\omega = 12y$ (Couette flow) at III′; and $\psi = \psi_b(t)$, ω generated at the wedge shaped body. The function $\psi_b(t)$ is part of the solution. Convective outflow conditions for ω and ψ_ξ with convection by the local horizontal velocity [8], are used at IV.

All boundaries are walls except III′ and IV (the inflow and outflow boundaries, respectively). The condition $\omega = 0$ at a wall is equivalent to imposing perfect slip [9]. The large distance of the outer walls from the jet, in combination with this boundary condition, results in the jet being unaffected by the outer walls. Vorticity generated at no slip walls is given by $\omega = v_x - u_y$ with application of no-slip conditions. (u and v are, respectively, the horizontal and vertical velocity.) The value ψ_b (t) at a time instant is computed by a procedure given in [10] that keeps the pressure single valued in the flow field.

The initial-boundary value problem is completed with the initial conditions of Couette flow in the channel, potential flow elsewhere and vorticity sheets at no slip walls I′, II′ and at the wedge shaped body.

An implicit, second order (in time and space) Crank-Nicholson finite-difference procedure is used to advance the solution in time. At each time step several iterations were necessary to satisfy all the following convergence criteria. The % change from iterate to iterate in ψ_b and wall generated vorticity (all no slip walls) had to be less than 1% with the tests not applying if $\psi_b < .001$ or $\omega_{wall} < .05$. In addition the global L_2 error norm for ω from Eq. (1) and for ψ from Eq. (2) each had to be less than .001 for convergence. Each iteration consisted of a sweep of Eq. (1) followed by a sweep of Eq. (2) using a four-color overrelaxation scheme for each sweep with an overrelaxation factor of 1.6. These iterations were fully vectorized on the Cray 1-S upon which the numerical computation was performed. Initial starting guesses for ψ, ω at a new time step were obtained by extrapolation from several previous time steps. Time step sizes of .008 and .006 were used for Re = 250 and 650, respectively, after much smaller time steps are used near t = 0. Approximately five hours of computer time were used to obtain the results of this paper.

RESULTS

Fig. 2 shows computer drawn equi-vorticity lines and streamlines at selected times for the numerical solution for Re = 250. (Throughout this paper equi-vorticity line values are ± 1, ± 2,.... and streamline values are ....-.2, -.1, 0, .1, .2 ....). Negative equi-vorticity lines are dashed; positive are solid. In areas of high grid point density, the dashed lines will appear solid. At t = 0, due to the initial conditions, the jet flow separates at the corners of the channel orifice resulting in starting vortices which are then convected downstream. This is seen at t = 11.4. The effect of the starting vortices is very similar to the cast-off vortex from the trailing edge of a wing abruptly started from rest [11]. The flow is still symmetric at t = 11.4. In nature the jet is unstable and therefore an asymmetric disturbance is abruptly introduced into the numerical scheme at t = 11.4. The disturbance consists of applying a "moving belt" condition in the lower channel wall for .8 nondimensional time units followed by applying the same condition in the upper channel wall for .8 time units.

Because of lingering transient effects from the asymmetric disturbance and limited availability of computer time, a complete cycle for Re = 250 could not be obtained. In the laboratory many cycles are required before essentially periodic quasi-steady-state cycles can be obtained for Re = 250 which is believed to be very close to the lower limit at which Strouhal numbers can be easily established. Recent research [6], [12] found St = .04 for Re = 250. However, the numerical results do show a long jet stem, the dominant feature for Re = 250 as seen in the laboratory. The jet stem is first, below the wedge (t = 63.4) (and induces a vortex of opposite sign (see t = 71.4)). Then it swings up and impinges directly on the wedge (t = 75.4). Finally the stem is above the wedge (t = 87.4). This constitutes a half-cycle of the jet stem oscillation. This half cycle still contains start-up effects.

Fig. 3 shows a time sequence of streamlines and equi-vorticity lines for the numerical solution for Re = 650 which is abruptly started from the Re = 250 case at t = 87.4. For Re = 650 there are two major frequencies. Let β be the frequency of vortex formation in the jet shear layers. Vortices form in these layers every 6 to 9 nondimensional time units which gives $\beta \approx 1/7.5 = .133$. See t = 96.4, 102.4, 111.4, 120.4. The other major frequency is that at which the stem of the jet oscillates (or that at which the configuration of three vortices occurs (as at t = 114.4) on the same side of the wedge). This occurs every 18 time units. See t = 96.4, 114.4 for the upper side of the wedge and t = 105.4 and t = 123.4 for the lower side of the wedge. This jet stem oscillation frequency is $1/18 = .055 \simeq 5\beta/12$. Additional numerical results discussed shortly indicate this is a rough approxi-

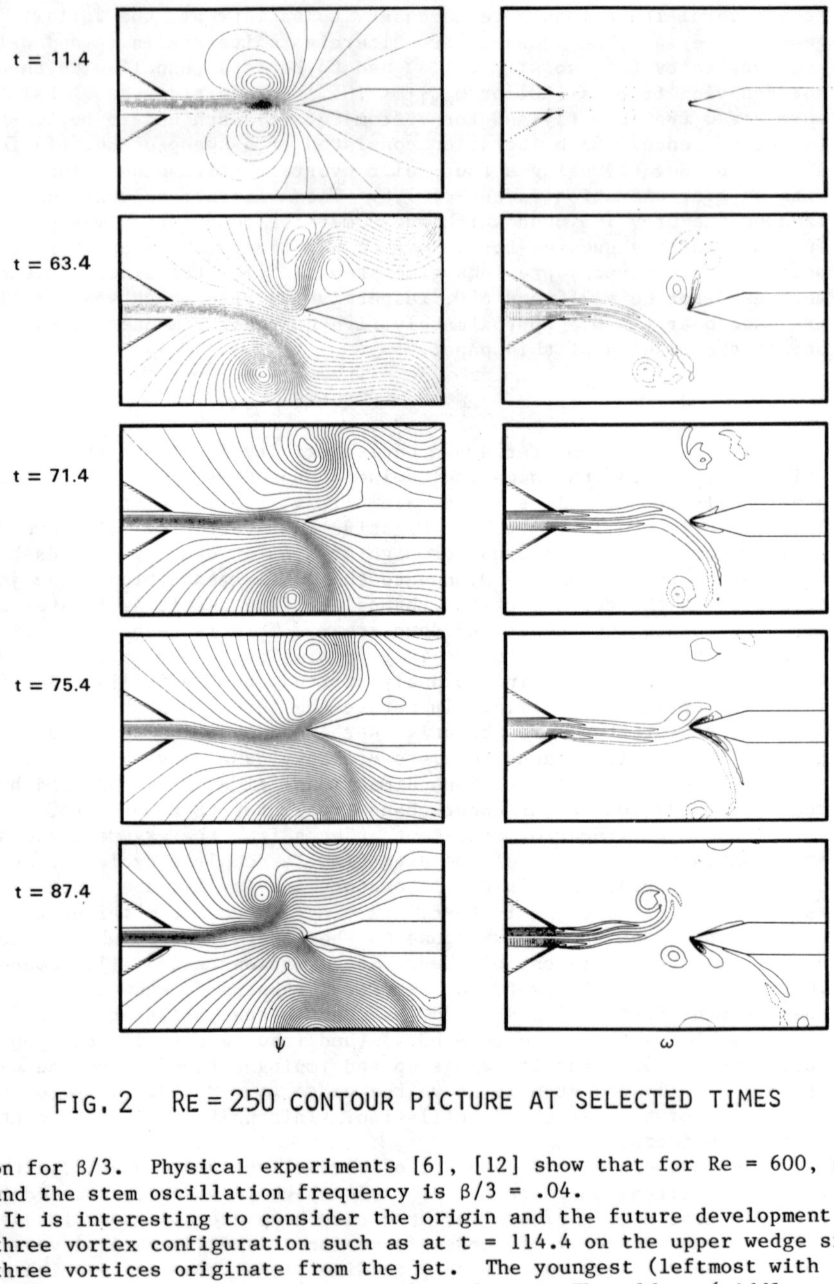

FIG. 2 RE = 250 CONTOUR PICTURE AT SELECTED TIMES

mation for β/3. Physical experiments [6], [12] show that for Re = 600, β = .12 and the stem oscillation frequency is β/3 = .04.

It is interesting to consider the origin and the future development of the three vortex configuration such as at t = 114.4 on the upper wedge side. All three vortices originate from the jet. The youngest (leftmost with arrow) vortex is formed from the jet shear layer. The oldest (middle one above) and the rightmost vortices are formed from the jet vorticity tongues being intercepted by the wedge. The middle vortex wraps around the leftmost vortex from t = 117.4 to 120.4. They then appear to fuse at t = 123.4. This process has been observed experimentally for Re = 600. [6], [12]. The Re = 250 time sequence (Fig. 2) does not exhibit the three vortex configuration primarily due to greater diffusion effects.

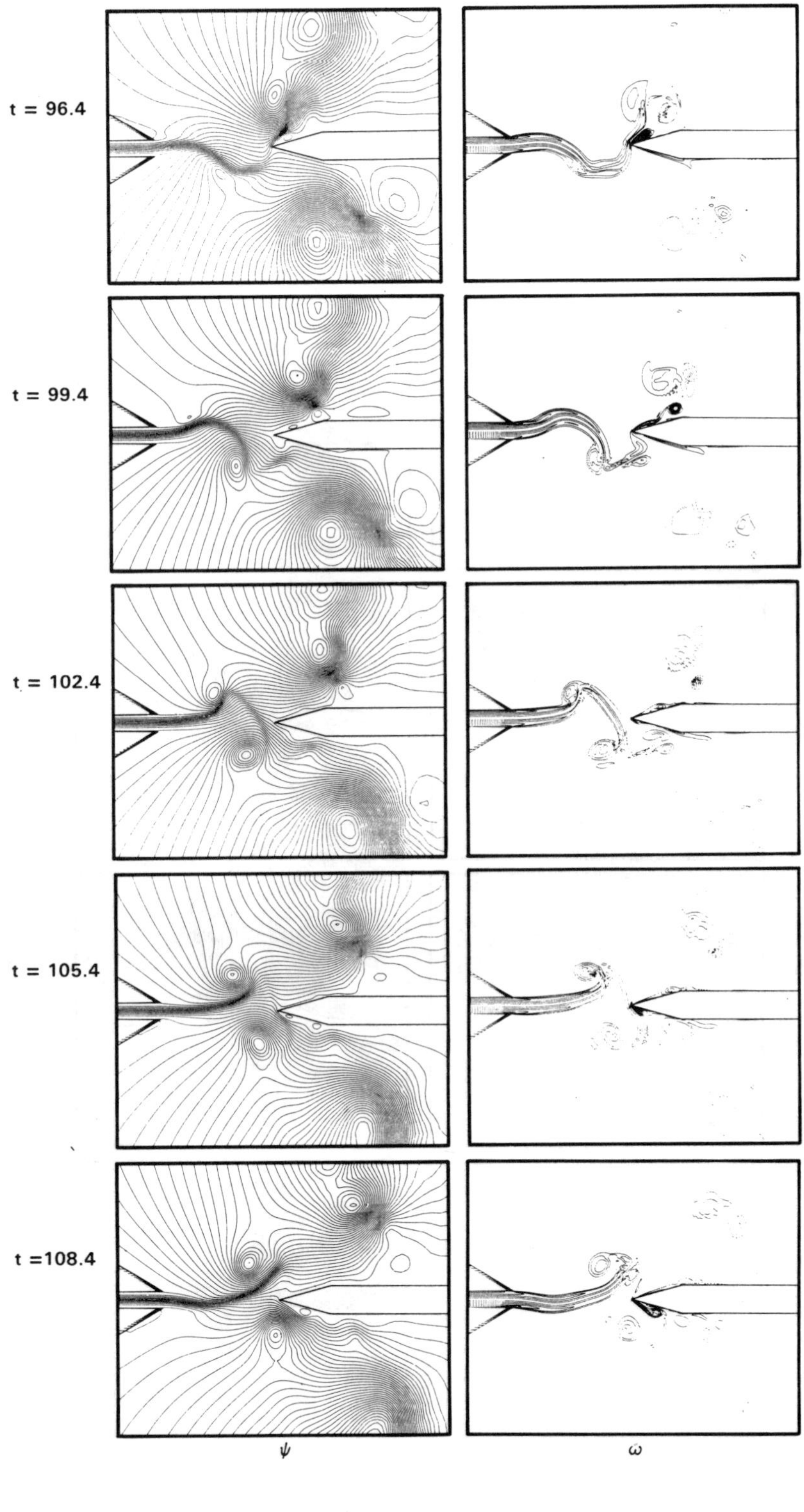

FIG. 3 RE = 650 TIME SEQUENCE OF CONTOUR PICTURES

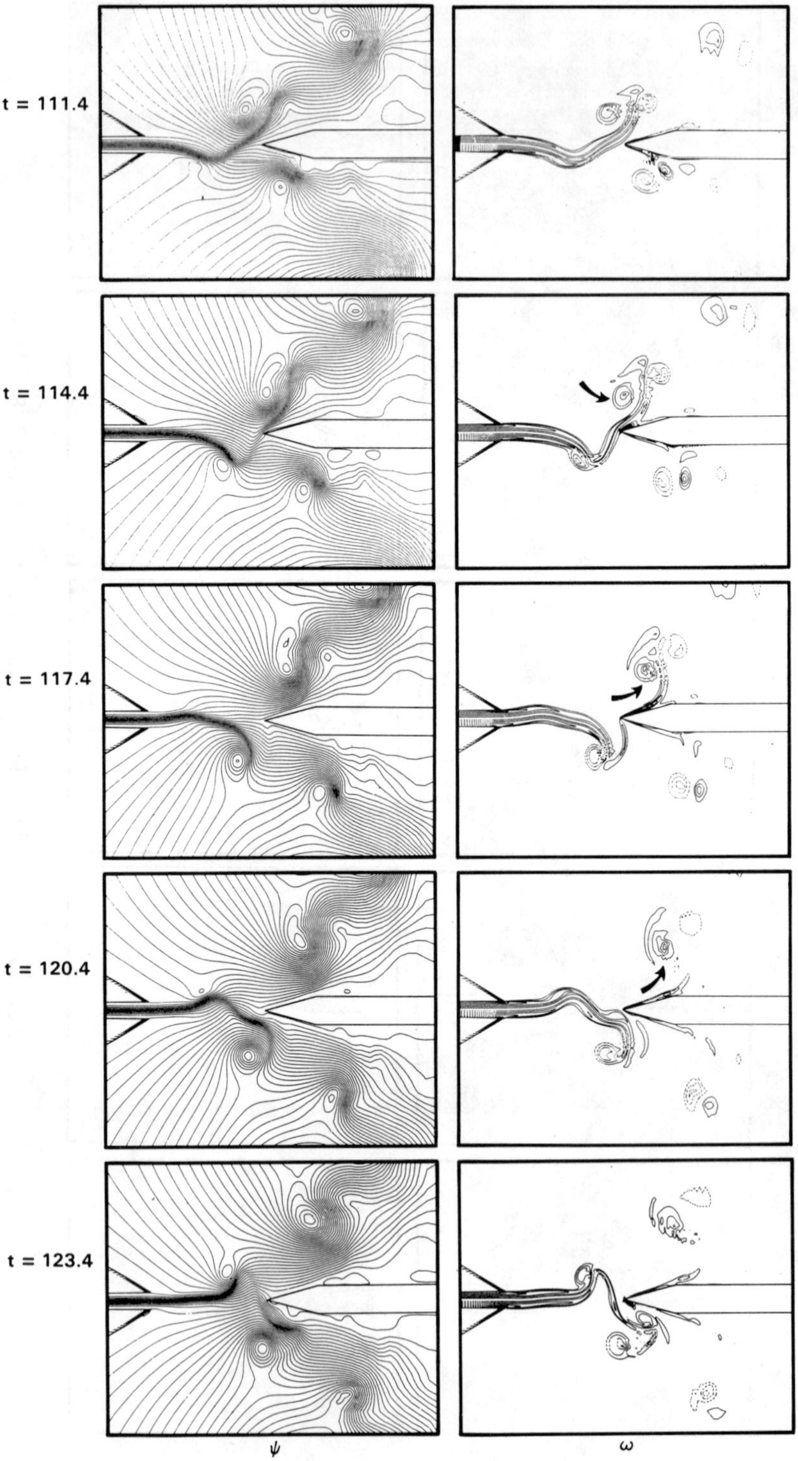

FIG. 3A RE = 650 TIME SEQUENCE OF CONTOUR PICTURES

It is worth noting that the flow from the downstream end of the body to the outflow boundary IV (see Fig. 1) remains essentially potential flow.

Fig. 4 shows entire time histories of the vertical (transverse) velocity at the jet axis (center) at three different locations: near the wedge (98% of the nondimensional distance L = 7.5 from orifice to wedge); halfway between the wedge and orifice (56% of L); and near the orifice (7% of L). Consider the Re = 650 case. As expected, the jet's energy (in the transverse direction) increases in going from the orifice to the wedge as seen from the increasing amplitude. By the orifice the spectral composition consists of essentially one component, whereas the two major frequencies β, $\beta/3$ can be seen both halfway between the wedge and orifice and near the wedge. These

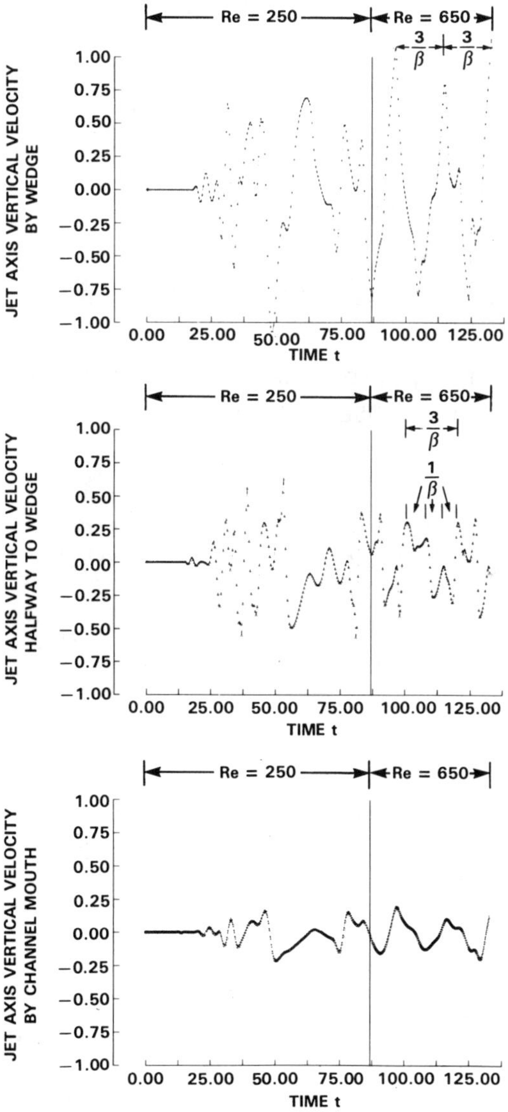

FIG. 4 TIME HISTORIES OF TRANSVERSE VELOCITIES AT JET CENTER

curves for Re = 650 are qualitatively and quantitatively quite similar to experimental results [6], [12] except that the amplitude is somewhat higher for the numerical curve at the wedge.

Fig. 5 shows pressure plotted vs. time at corresponding top and bottom (partially drawn and dashed) wedge surface points very close to tip (at a distance of .14 in the x-direction from wedge tip). The pressure is set to zero at the upstream end of the channel with a negative pressure gradient in the essentially Couette flow of the channel. This explains the mostly negative values of pressure in Fig. 5. The β frequency is clearly evident. A subharmonic frequency β/4 is also found.

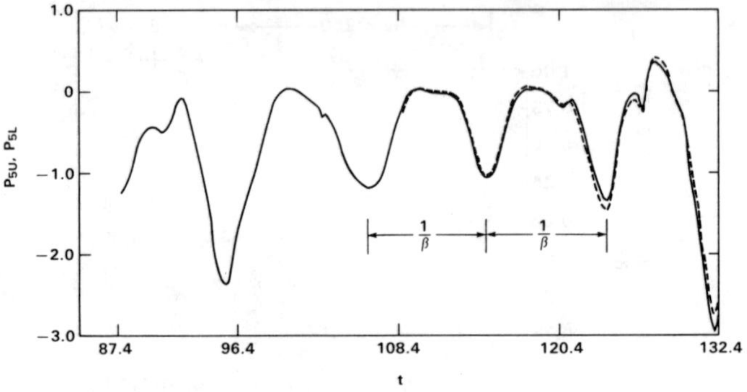

FIG. 5 TIME HISTORY OF COMPUTED PRESSURE AT WEDGE VERY CLOSE TO TIP

ACKNOWLEDGEMENT

The author would like to thank Professor D. Rockwell and Mr. M. Lucas of Lehigh University, Bethlehem, Pa. and Drs. H.J. Lugt and H.J. Haussling of the David Taylor Naval Ship R&D Center, Bethesda, Md. for valuable discussions and comments.

REFERENCES

[1] Powell, A., J. of Acoustical Society of America, vol. 33, no. 4 (1961) pp. 359-409.
[2] Powell, A., "Advances in Aeroacoustics", Rapports du 5^{e} Congres International d'Acoustique, vol. 11 Conferences Generales, Liege (1965).
[3] Rockwell, D., AIAA J., vol. 21, no. 5 (1983) pp. 645-664.
[4] Thompson, J.F., Thames, F.C., Mastin, C.W., and Shanks, S.P., in "Proc. of the AIAA 2nd Comp. Fluid Dynamics Conf.", Hartford, Conn., 1975.
[5] Thompson, J.F., Warsi, Z.A.U., and Mastin, C.W., J. of Comp. Physics, vol. 47 (1982) pp. 1-108.
[6] Rockwell, D., Lehigh U., Bethlehem, Pa., U.S.A., Private Comm. (1983).
[7] Plant, T.J. Technical Report AFFDL-TR-77-116, Air Force Flight Dynamics Laboratory, Wright-Patterson Air Force Base, Ohio, (1977).
[8] Mehta, U.B. and Lavan, Z., J. of Fluid Mech., vol. 67, (1975) p. 227.
[9] Lugt, H.J. and Ohring, S., Phys. of Fluids 18 (1975), pp. 1-8.
[10] Sood, D.R. and Elrod Jr., H.G., AIAA J. 12, (1974) pp. 636-641.
[11] Goldstein, S. "Modern Developments in Fluid Dynamics: Vol. 1" (1938).
[12] Lucas, M. and Rockwell, D. "Self-excited Jet: Low Frequency Modulation and Multiple Frequencies" to be published.

A PSEUDO-SPECTRAL SOLUTION OF BINARY GAS MIXTURE FLOWS

J. Ouazzani and R. Peyret
Département de Mathématiques
Université de Nice
06034 Nice Cedex, France

SUMMARY: This paper presents a pseudo-spectral method for the calculation of steady binary gas mixture flows with application to vapor crystal growth. The method uses a pseudo-unsteady approach. The scheme is implicit for the diffusion terms and mainly explicit for the other terms. The spatial approximation makes use of Chebyshev polynomials expansions associated with a collocation technique. This reduces the computational effort to matrix-vector products and Fast-Fourier-Transforms. These operations are efficiently performed on a vector computer (CRAY 1).

1. INTRODUCTION

Flows of binary mixtures are governed by the Navier-Stokes equations associated with an advection-diffusion-type equation for the species conservation. The flows considered here are assumed to be isothermal : the variations of density are due to variations of concentration and pressure. The governing equations are similar to the Navier-Stokes equations for compressible fluids.

In the case of a very slow flow, the density variations are due mainly to changes in concentration, so that the density is no longer connected to the pressure, in a first approximation. In such a case, the equations of motion are of the same nature as the incompressible Navier-Stokes equations.

This paper is devoted to the presentation of a collocation pseudo-spectral method based on Chebyshev polynomials approximations to solve the two-dimensional steady equations in both situations exposed above. The method is applied to the computation of internal plane flows related to the transport of species occuring in vapor crystal growth process[1].

2. THE EQUATIONS OF BINARY GAS MIXTURE FLOWS

We consider isothermal flows of a binary gas mixture in a rectangular domain Ω $(0\leqslant x\leqslant \mathcal{L}\ ,\ 0\leqslant y\leqslant \ell)$. The governing partial differential equations are :

$$\frac{\partial \rho}{\partial t}+\nabla\cdot\rho\vec{V}=0 \tag{1}$$

$$\rho\frac{\partial \vec{V}}{\partial t}+\rho(\vec{V}\cdot\nabla)\vec{V}+\nabla p=\mu\ \Delta\vec{V}+\frac{\mu}{3}\nabla(\nabla\cdot\vec{V}) \tag{2}$$

$$\rho\frac{\partial W_B}{\partial t}+\rho\vec{V}\cdot\nabla W_B=\nabla\cdot(\rho D_{AB}\nabla W_B) \tag{3}$$

where $\vec{V}=(u,v)$ is the mean velocity vector such that

$$\rho\vec{V}=\rho_A\vec{V}_A+\rho_B\vec{V}_B \tag{4}$$

$$\rho=\rho_A+\rho_B \tag{5}$$

where the subscripts A and B refer to each component of the mixture; ρ is the total density and

$$p=p_A+p_B \tag{6}$$

is the total pressure. The equation (3) describes the behaviour of the weight fraction $W_B=\rho_B/\rho$ of the component B. An identical equation is satisfied by $W_A=\rho_A/\rho$ since $D_{BA}=D_{AB}$ and $W_A=1-W_B$ from Eq.(5). In the above equations, the viscosity coefficient of the mixture μ is assumed to be a constant. Each component of the mixture follows the equation of state of perfect gas, so that

$$p=\left[1+\left(\frac{M_A}{M_B}-1\right)W_B\right]\frac{RT}{M_A}\rho\equiv g(\rho,W_B) \tag{7}$$

where M_A and M_B are the molecular weights. The temperature T is assumed to be constant. Finally, the coefficient of diffusion D_{AB} is expressed by

$$D_{AB} = D^{\circ}_{AB} \frac{p_\circ}{p} \left(\frac{T}{T_\circ}\right)^2 \equiv D(p) \qquad (8)$$

where D°_{AB} is the value at standard conditions $p_\circ, T_\circ$ (see [2]).

Dimensionless variables are defined by using the following characteristic values : ℓ (length), ρ^* (density), W_B^* (weight fraction), $D^* = D[g(\rho^*, W_B^*)]$ (diffusion), $D^*/\mathcal{L}$ (velocity), $\rho^*(D^*/\mathcal{L})^2$ (pressure) and $\ell\mathcal{L}/D^*$ (time). The choice of ρ^* and W_B^* depends on the problem under consideration and will be specified in Section 4 when boundary conditions are described.
When keeping the same symbols for dimensionless variables, the equations (1)-(3) can be written in the vector form

$$\mathscr{A}\frac{\partial f}{\partial t} - \Lambda f + F = 0 \qquad (9)$$

where $f = (\rho, W_B, u, v)^T$, $\Lambda = \Lambda_x + \Lambda_y$ is a differential operator such that

$$\Lambda_x = \mathscr{B}\frac{\partial^2}{\partial x^2}, \quad \Lambda_y = \mathscr{C}\frac{\partial^2}{\partial y^2} \qquad (10)$$

$\mathscr{A}$, $\mathscr{B}$ and $\mathscr{C}$ are diagonal matrices with the elements :

$\mathscr{A}$: $1, \rho(1+HW_B), \rho, \rho$

$\mathscr{B}$: $0, \gamma(1+H), 4\gamma\, Sc/3, \gamma\, Sc$

$\mathscr{C}$: $0, \gamma(1+H), \gamma\, Sc, 4\gamma\, Sc/3$

and $F = F_1 + F_2 + F_3$ with

$$F_1 = \begin{bmatrix} \nabla\cdot\rho\vec{V} \\ \rho(1+H W_B)\vec{V}\cdot\nabla W_B + \gamma H \dfrac{1+H}{1+H W_B}(\nabla W_B)^2 \\ \rho\vec{V}\cdot\nabla u \\ \rho\vec{V}\cdot\nabla v \end{bmatrix}$$

$$F_2 = \begin{bmatrix} 0 \\ 0 \\ \dfrac{\partial p}{\partial x} \\ \dfrac{\partial p}{\partial y} \end{bmatrix}, \qquad F_3 = \begin{bmatrix} 0 \\ 0 \\ -\gamma\dfrac{Sc}{3}\dfrac{\partial^2 v}{\partial x\,\partial y} \\ -\gamma\dfrac{Sc}{3}\dfrac{\partial^2 u}{\partial x\,\partial y} \end{bmatrix}$$

The equation of state (7) is now written as :

$$p = \frac{1}{M^2}(1+H W_B)\rho \equiv g(\rho, W_B). \qquad (11)$$

In the above equations, the following dimensionless parameters appear :

$$\gamma = \mathcal{L}/\ell, \qquad Sc = \mu/\rho^* D^*$$

$$H = \left(\frac{M_A}{M_B} - 1\right)W_B^*, \qquad M = \sqrt{\frac{M_A}{RT}}\,\frac{D^*}{\mathcal{L}},$$

γ is the aspect ratio, Sc is the Schmidt number, H characterizes, in particular, the difference between the molecular weights of the two components of the mixture, M is analogous to a Mach number: it is the ratio between the characteristic velocity and the sound speed in the component A.

3. THE NUMERICAL PSEUDO-SPECTRAL METHOD

The numerical method is an orthogonal collocation pseudo-spectral method based on Chebyshev polynomials expansions in a way analogous to [3]. The time discretisation of Eq.(9) is implicit for the term Λf and mainly explicit for the term F. However, the fact that $\mathscr{A}$ is variable in time should necessitate matrix inversions at each time-step. This is avoided by replacing the matrix $\mathscr{A}$ in Eq.(9) by the identity matrix I as it was done in [4]. This change in the equations is made possible here because we are only interested in the steady-state solution: the resulting numerical method is a "pseudo-unsteady" method [5],[6]. Therefore, the time-discretization (characterized by the superscript n, $t = n\Delta t$) is

$$(I - \alpha \Delta t \Lambda)(f^{n+1} - f^{n}) = \Delta t \mathscr{F} \quad (12a)$$

with

$$\mathscr{F} = \Lambda f^{n} - \tfrac{1}{2}(3F_1^{n} - F_1^{n-1}) + F_2^{n+1} + F_3^{n} \quad (12b)$$

where α is a constant ($0 \leqslant \alpha \leqslant 1$). The equation (12) is solved by means of the generalized ADI procedure

$$(I - \alpha \Delta t \Lambda_x)\psi^{*} = \Delta t \mathscr{F} \quad (13a)$$

$$(I - \alpha \Delta t \Lambda_y)\psi^{**} = \psi^{*} \quad (13b)$$

$$f^{n+1} = f^{n} + \psi^{**} \quad (13c)$$

Note that : (1) the calculation of ρ^{n+1} (the first component of f) is explicit and the resulting new value $p^{n+1} = g(\rho^{n+1}, W_B^{n+1})$ is used in the momentum equations through F_2^{n+1}; (2) the calculation of W_B, u and v is uncoupled and leads to the inversion of constant matrices resulting from the approximation of $\partial^2/\partial x^2$ and $\partial^2/\partial y^2$ in the left hand-side of Eqs.(13a) and (13b), respectively;
(3) the value $\alpha = \frac{1}{2}$ has been used in the calculations reported below.

The spatial approximation makes use of Chebyshev polynomials expansions, let, for any dependent variable $\varphi(X,Y,t)$:

$$\varphi(X,Y,t) = \sum_{k=0}^{K} \sum_{\ell=0}^{L} \hat{\varphi}_{k,\ell}(t)\, T_k(X)\, T_\ell(Y) \quad (14)$$

where X and Y refer to the transformed plane $X = 2x/\gamma - 1$, $Y = 2y - 1$, ($-1 \leqslant X,\ Y \leqslant 1$).

The orthogonal collocation technique [7] consists to express the derivatives at one collocation point in terms of the values of the function at all points. For example, the X-derivative is expressed by

$$\frac{\partial\varphi}{\partial X}(X_i, Y_j, t) = \sum_{m=0}^{K} d_{im}\, \varphi(X_m, Y_j, t) \quad (15)$$

at collocation points ;

$$X_i = \cos(\pi i/K)\ ,\ i=0,\dots,K\ ,\quad Y_j = \cos(\pi j/L)\ ,\ j=0,\dots,L \quad (16)$$

The coefficients d_{im} (see Appendix) are obtained by using the property of orthogonality of the Chebyshev polynomials and usual trigonometrical formulas. In vector notation, $\Phi = [\varphi(X_0, Y_j, t), \dots, \varphi(X_K, Y_j, t)]^T$, Eq.(15) writes as

$$\frac{\partial\Phi}{\partial X} = \mathscr{D}\Phi\ ,\quad \mathscr{D} = [\,d_{im}\,] \quad (17)$$

then $\partial^2\Phi/\partial X^2 = \mathscr{D}^2\Phi$. When using approximations of this type in the left-hand-side of Eqs. (13a,b), one obtains algebraic systems of the form

$$\mathscr{M}\Phi = \Psi \quad (18)$$

The boundary conditions are easily included in the system: they replace the first and the last equation in Eq.(18). The inversion of the resulting $[(K+1)\times(K+1)$ or $(L+1)\times(L+1)]$ matrices $\mathscr{M}$ are made once, at the beginning of the calculation; then, only matrix vector products have to be performed at each time-step, so that the method is well adapted to a vector computer.

The right-hand-side of Eqs.(13a,b) is evaluated by the standard pseudo-spectral technique[8], which is more efficient than the use of formulas of type (17). The derivatives are evaluated in the spectral space through the calculation of the coefficients $\hat{\varphi}_{k,\ell}$ and the products are performed in the physical space. The two spaces are connected by a Fast-Fourier-Transform algorithm[9].

No theoretical study of the stability of the numerical scheme has been made. For the heat equation the scheme is unconditionnally stable. However, the explicit treatment of F, particularly for the ρ-component, introduces a restriction on the time-step Δt. In fact, two time-steps have been simultaneously used: one (Δt) for W_B , u and v and the other ($\Delta t_\rho = \Delta t/q$, $q > 1$) for ρ. The solution for ρ is advanced q times during ont time-cycle for the other variables. The overall property of convergence of this procedure has been found generally better that the one which would consist to advance together (with different time-steps) all the equations.

4. NUMERICAL APPLICATIONS

The applications concern steady diffusive flows occuring in crystal growth process ("Physical Vapor Transport"[1]). Axisymmetric finite-difference solutions have been obtained in [10]-[11] and in [12] for the double diffusive case. This latter case in a plane configuration is calculated in [13] by means of a pseudo-spectral method applied to the vorticity-stream function formulation. The flows considered here take place in the domain Ω ($0\le x\le\gamma$, $0\le y\le 1$, Fig.1). The component A (solute) diffuses from the source to the crystal through the component B (solvent). The boundary conditions are :

- At the source (x=0) :

$$\rho=1,\ W_B=1,\ \rho u=\frac{\gamma(1+H)}{(1+HW_B)\,W_B}\frac{\partial W_B}{\partial x}h\ ,\qquad v=0 \qquad (19)$$

- At the crystal ($x=\gamma$) :

$$W_B=W_\gamma>1,\ \rho u=\frac{\gamma(1+H)}{(1+HW_B)\,W_B}\frac{\partial W_B}{\partial x}h\ ,\qquad v=0 \qquad (20)$$

- At the walls (y=0 and y=1) :

$$\partial W_B/\partial y=0,\qquad u=0,\qquad v=0 \qquad (21)$$

These conditions are slightly different that those considered in [11]; their choice is discussed in [14].

The above boundary conditions define the characteristic quantities ρ^* and W_B^* as their values at the source $x=0$.

Due to the symmetry, the solution is effectively calculated in a half-domain ($0\le x\le\gamma$, $\frac{1}{2}\le y\le 1$) with symmetry conditions at $y=\frac{1}{2}$:

$$\partial\rho/\partial y=0,\quad \partial W_B/\partial y=0,\quad \partial u/\partial y=0\quad ,\quad v=0 \qquad (22)$$

The condition on ρu in Eqs.(19) and (20) derives from Fick's Law[2]. It expresses the relationship between the diffusive mass flux $\rho_A u_A \equiv \rho u$ (because $u_B=0$ at the interfaces) of the A-component, the weight fraction W_A ($=1-W_B$) and its gradient. The function h(y), satisfying to $h(0)=h(1)=0, h(\frac{1}{2})=1$ is artificially introduced here in order to regularize the solution : otherwise [i.e. if $h(y)\equiv 1$], the velocity u would be discontinuous at the corners [11].

At $x=\gamma$, the density is calculated from the continuity equation. At the wall y=1, it has been found better to calculate the pressure by using the normal momentum equation (density and pressure are connected through the equation of state). Therefore, the gradient $\partial p/\partial y$ is deduced from the normal momentum equation evaluated at the wall. The knowledge of this value $\partial p/\partial y$, associated with the knowledge of the inner values of p, allows us to calculate p at the wall using a formula analogous to Eq.(17). The same technique is used at the axis of symmetry where $\partial p/\partial y=0$.

4.1 Compressible flow

The first application concerns a situation in which the effect of pressure variations on density are important. Such a situation (not realistic in Physical Vapor Transport process) occurs when M is not small. More precisely, we choose here M=1 and Sc=1, H=0, $W_\gamma=1.284$, $\gamma=1$, $h(y)=4y(1-y)$.

Initial conditions were derived from the one-dimensional solution obtained in [11]; they are :

$$\rho=\frac{1+H}{1+HW_B}\ ,\quad W_B=\frac{\exp(Px/\gamma)}{1+H[1-\exp(Px/\gamma)]}\ ,\quad u=\frac{1}{\gamma W_B}\frac{\partial W_B}{\partial x}h(y),\quad v=0 \qquad (23)$$

where $P=\ln[(1+H)W_\gamma/(1+HW_\gamma)]$.

A first solution is calculated in a 9×9 mesh (K=L=8) with $\Delta t=0.1$, $\Delta t_\rho=0.01$.

Then, this solution is refined in a 17×17 mesh (K=L=16) with $\Delta t=0.025$, $\Delta t_\rho=0.0025$. A difficulty was experimented in the convergence of the continuity equation. This was due to a lack of verification of the steady total flux condition

$$\int_\Gamma \rho \vec{V}\cdot\vec{n}\, d\Gamma = 0 , \qquad \Gamma=\partial\Omega \tag{24}$$

when W_B, hence $\rho\vec{V}$ at the boundary Γ, has converged. This difficulty was surmounted by slightly modifying $h(y)$ at $x=\gamma$, such that $h=4C\,y(1-y)$. The constant C is determined in order to satisfy (24); we found $C=1.069$. The number N of time-cycles needed to reach convergence ($|\nabla\cdot\rho\vec{V}|\leqslant 10^{-5}$) is N=1105 in the 9×9 mesh and N=2100 in the 17×17 mesh. The CPU time (CRAY 1) is 0.04s/time-cycle in the coarse mesh and 0.15 s in the finest mesh. The differences between results in the two cases are of order 10^{-3}.

The fig.2 shows the u-velocity profiles at various sections x=0. It can be remarked that the evolution of u with respect to x is not monotonous. This is better seen in fig.3. Finally, the iso-density lines (identical here with the isobar lines) are shown in fig.4.

4.2 Quasi-incompressible flow

We denote here by "quasi-incompressible" a flow in which the variations of the density are mainly due to changes in concentration and practically not in pressure. This situation occurs when $M\ll 1$ and it is the case in P.V.T. process for which $M=10^{-3}-10^{-5}$. In this case we can use an approximate equation of state. This equation is obtained by writing the equation (11) in the form

$$\rho = \frac{1+H}{1+HW_B} + M^2\,\frac{p-p_s}{1+HW_B} \tag{25}$$

where p_s is the value of p at x=0. Then, we neglect the M^2-term in Eq.(25): the density depends only on the weight fraction W_B. The resulting equations of motion are analogous to the incompressible Navier-Stokes equations.

The solution is calculated from the equations (12) with $f=(p,W_B,u,v)^T$. That is to say, the steady-state constraint $\nabla\cdot\rho\vec{V}=0$ is obtained through the equation

$$\frac{\partial p}{\partial t} + \nabla\cdot\rho\vec{V} = 0 \tag{26}$$

The pseudo-unsteady method constructed in this way is nothing else than the artificial compressibility method[6]. The density is deduced from (25)where M=0. The boundary conditions are given by (19)-(21); the condition on ρ being automatically satisfied. As in purely incompressible flow, there is no boundary condition for the pressure: it is calculated either from the continuity equation (26) (at the crystal) or from the normal momentum equations (at the other parts of the boundary).

Figures 5 to 10 show results obtained in the case $\gamma=1$, Sc=0.103, H=0.248, $W_\gamma=4.262$ and $h=1-(2y-1)^8$. The initial conditions are given by (23) and p=1.
In the 9×9 mesh: $\Delta t=0.1$, $\Delta t_\rho=0.05$, N=925, CPU time = 0.03 s/time-cycle.
In the 17×17 mesh: $\Delta t=0.08$, $\Delta t_\rho=0.01$, N=997, CPU time = 0.11 s/time-cycle.
The constant C was found to be 1.077.
The fig. 9 shows the mass average velocity field and the velocity field of the component B. This latter is obtained from

$$\vec{V}_B = \vec{V} - (\nabla W_B)/(\rho\, W_B) \tag{27}$$

It can be remarked that the component B has a recirculating motion; this feature was already found in [11]. Finally, the evolution of the velocity component u on the axis of symmetry for various values of the parameters is shown in fig. 11 .

The authors are grateful to J.M.Lacroix for his help during this research.

The calculations have been performed on the CRAY 1 computer of **Centre de Calcul Vectoriel pour la Recherche.**

• REFERENCES

[1] Rosenberger,F., *Fluid Dynamics in crystal growth from vapors,* Physico-Chemical Hydrodynamics,1, 3-26 (1980)

[2] Rosenberger,F., *Fundamental of Crystal Growth I.,* Springer Verlag (1979)

[3] Patera,A.T. & S.A. Orszag, *Finite amplitude stability of axisymmetric pipe flow,* J.Fluid Mech.,112, 467-474 (1981)

[4] Peyret,R. & H. Viviand, *Calcul de l'écoulement d'un fluide visqueux compressible autour d'un obstacle de forme parabolique,* Lect.Notes in Physics, vol.19, 222-229, Springer-Verlag (1973)

[5] Peyret,R. & H. Viviand, *Pseudo-unsteady methods for inviscid or viscous flow computation,* to be published in Recent Advances in Aerospace Science

[6] Peyret,R. & T.D. Taylor, *Computational Methods for Fluid Flow,* Springer-Verlag (1983)

[7] Finlayson,B.A., *The method of Weighted Residuals and Variational Principles,* Academic Press (1972)

[8] Gottlieb,D., & S.A. Orszag, *Numerical Analysis of Spectral methods,* Monograph n°26, SIAM (1977)

[9] Lhomme,B., J.Morgenstern, & P.Quandalle, *Description de nouveaux programmes de transformations de Fourier rapides, à une ou deux dimensions,* ESA Contrat 4277/80/GP-1, Depart.Math., Univ. Nice, (Dec.1981)

[10] Markham,B.L. & F. Rosenberger, *Velocity and concentration distribution in a Stephan diffusion tube,* Chem.Eng.Commun.,5, 287-298 (1980)

[11] Greenwell,D.W., B.L.Markham & F.Rosenberger, *Numerical modeling of diffusive physical vapor transport in cylindrical ampoules,* J.of Crystal Growth, 51, 413-425 (1981)

[12] Markham,B.L., D.W.Greenwell & F.Rosenberger, *Numerical modeling of diffusive-convective physical vapor transport in cylindrical vertical ampoules,* J. of Crystal Growth, 51, 426-437 (1981)

[13] Elie,F., A.Chikhaoui, P.Bontoux, & B.Roux, *Spectral approximation for Boussinesq double diffusion,* 5th GAMM-Conf. Numerical Methods in Fluids, Rome (October 5-7, 1983)

[14] Ouazzani,J., *Méthode pseudo-spectrale pour la résolution des équations d'un mélange binaire. Application à la croissance cristalline,* Thèse de 3°cycle, Mécanique des Fluides, Univ. Nice (to be published).

♦

APPENDIX : Coefficients d_{im} [Eq.(15)]

$$X_k = \cos(\pi k/K),\ k=0,\dots,K\ ;\quad c_o = c_K = 2\ ;\ c_k = 1,\ k=1,\dots,K-1$$

$$d_{oo} = (2K^2+1)/6\ ,\quad d_{om} = \frac{2(-1)^m}{c_m(1-X_m)}\ ,\ m=1,K$$

$$d_{io} = \frac{(-1)^{i+1}c_i}{2(1-X_i)}\ ,\ i=1,\dots,K-1;\quad d_{iK} = \frac{(-1)^{i+K}c_i}{2(1+X_i)}\ ,\ i=1,\dots,K-1$$

$$d_{ii} = -\frac{X_i}{2(1-X_i^2)}\ ,\quad i=1,\dots,K-1\ ;\quad d_{im} = \frac{(-1)^{i+m}}{(X_i-X_m)},\ i=1,\dots,K-1; m=1,\dots,K-1;\ i\neq m$$

$$d_{Km} = \frac{2(-1)^{m+K+1}}{c_m(1+X_m)}\ ,\ m=0,\dots,K-1\ ;\ d_{KK} = -(2K^2+1)/6.$$

•

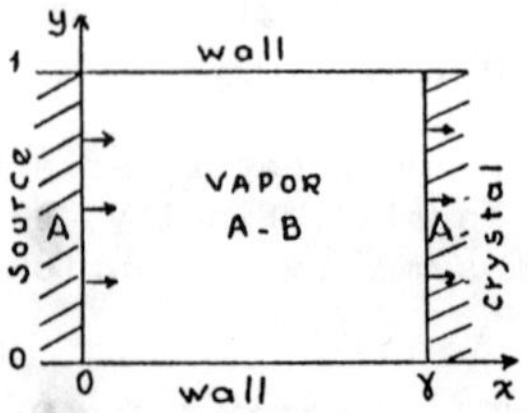

Fig 1: *Vapor crystal growth model*

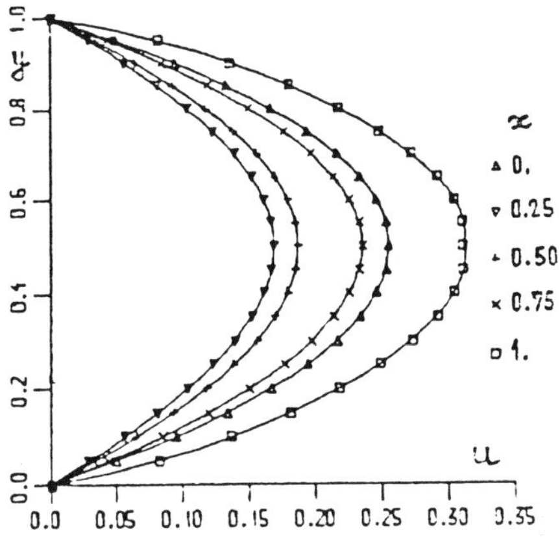

Fig. 2 u *versus* y *at various sections* x; M=1, Sc=1, H=0, W_γ = 1.284.

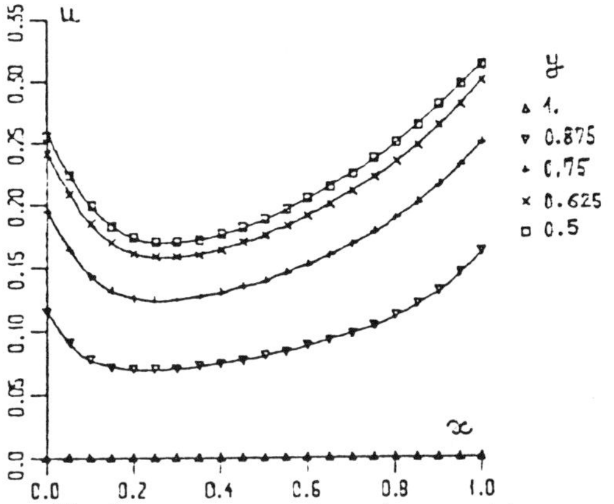

Fig. 3 u *versus* x *at various sections* y; M=1, Sc=1, H=0, W_γ =1.284.

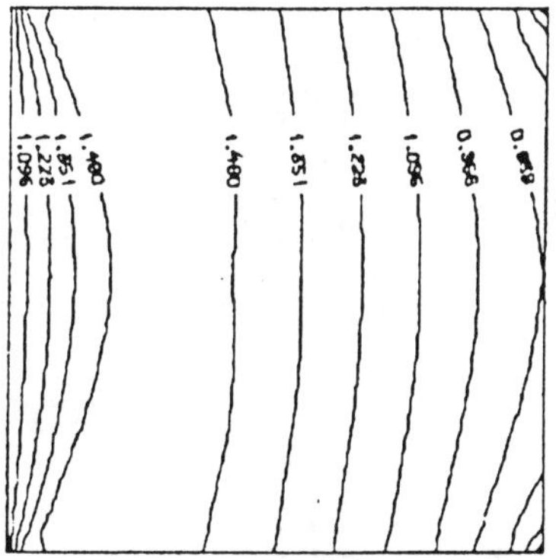

Fig. 4 *Iso-density lines;* M=1, Sc=1, H=0, W_γ = 1.284.

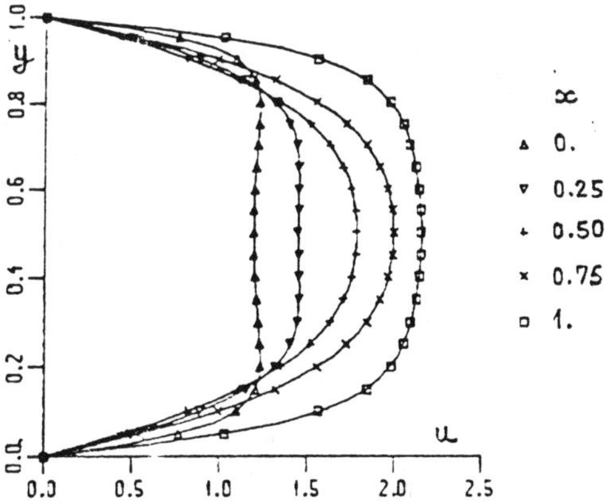

Fig. 5 u *versus* y *at various sections* x; M=0, Sc=0.103, H=0.248, W_γ=4.262.

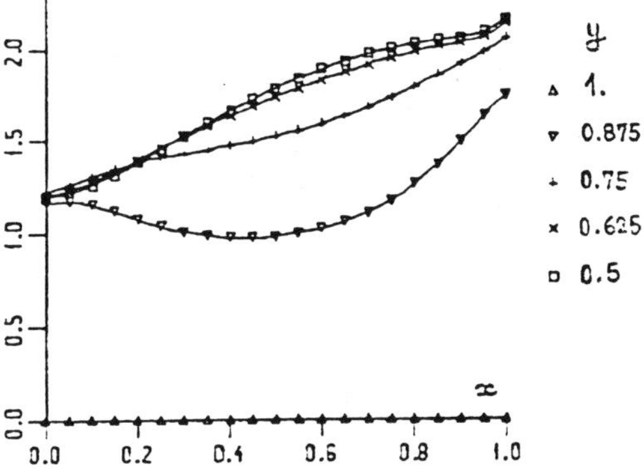

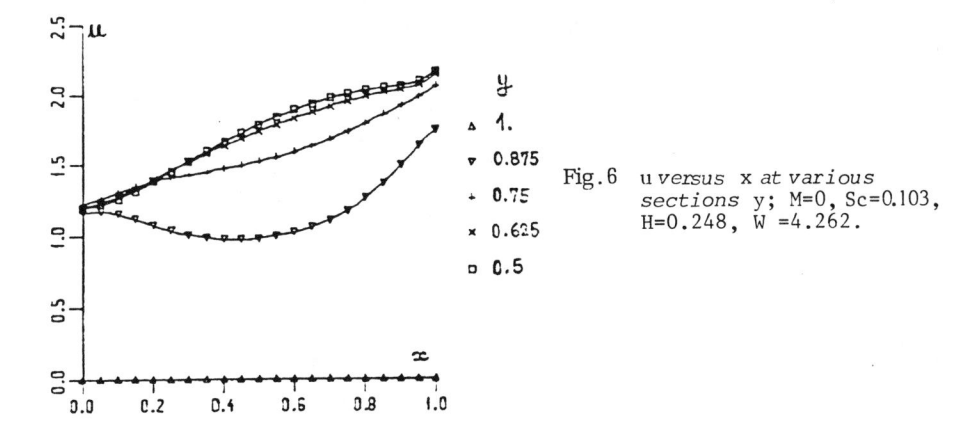

Fig. 6 u *versus* x *at various sections* y; M=0, Sc=0.103, H=0.248, W =4.262.

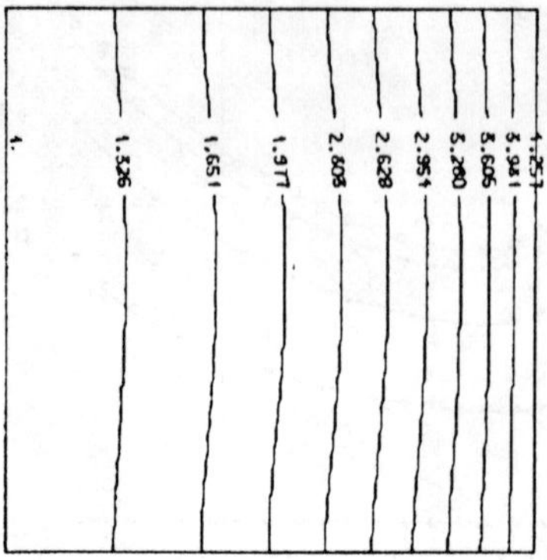

Fig. 7 *Iso-*W_B *lines;* M=0, Sc=0.103, H=0.248, W_γ=4.262.

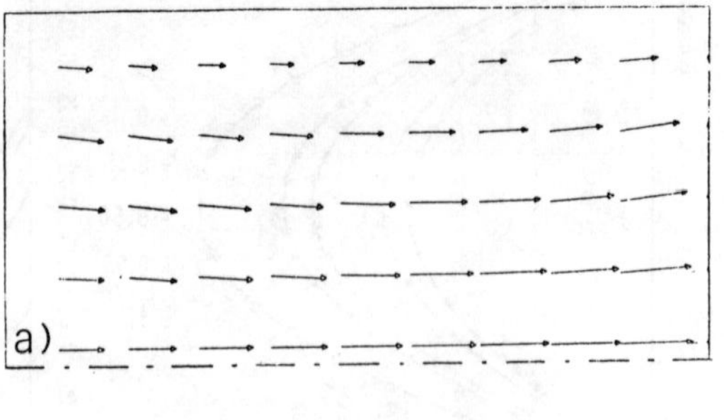

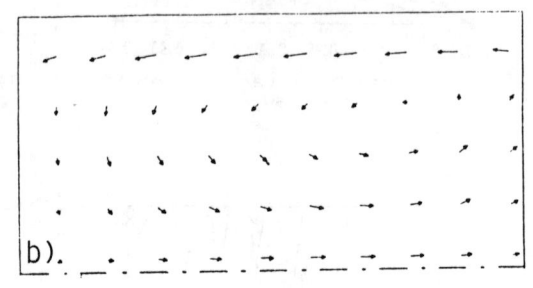

Fig. 9 *Velocity field (upper half domain);* M=0, Sc=0.103, H=0.248, W_γ=4.262; *(a): Mass average velocity,* *(b):* B-*component velocity.*

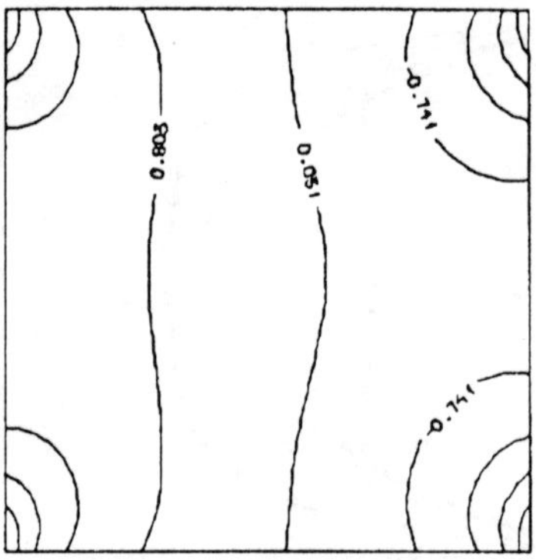

Fig. 8 *Isobar lines;* M=0, Sc=0.103, H=0.248, W_γ=4.262.

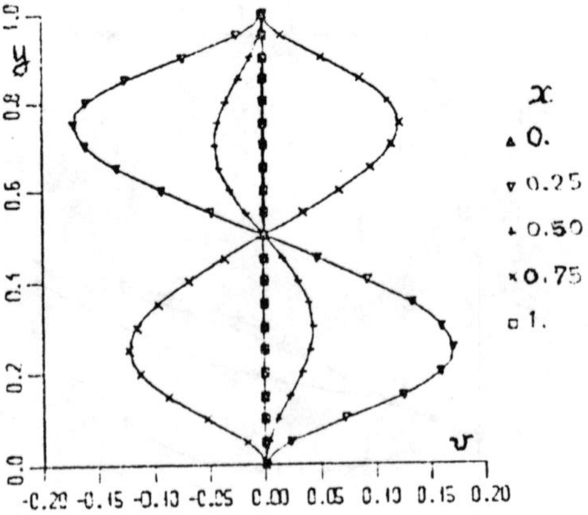

Fig. 10 v *versus* y *at various sections* x; M=0, Sc=0.103, H=0.248, W_γ = 4.262.

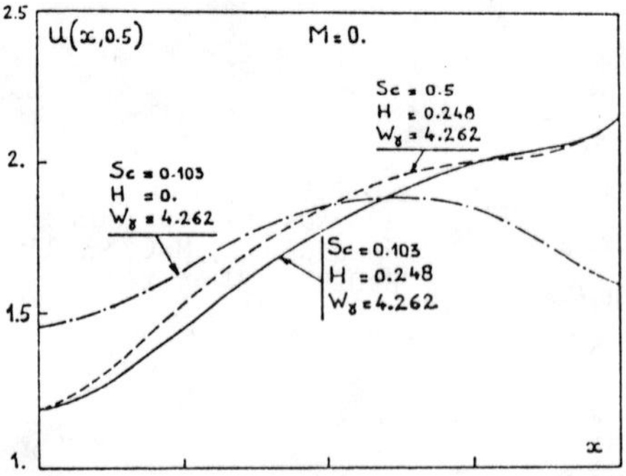

Fig. 11 u *on the axis of symmetry* y=0.5 *for various parameters.*

ASYMPTOTIC BEHAVIOUR OF SOME NON-LINEAR SCHEMES FOR LINEAR ADVECTION

P L Roe, Royal Aircraft Establishment, Bedford, UK
M J Baines, Department of Mathematics, University of Reading, UK

SUMMARY

In this paper we discuss numerical methods for solving the linear advection equation, which in one space dimension is

$$u_t + au_x = 0 \tag{1}$$

and in two space dimensions is

$$u_t + au_x + bu_y = 0 \tag{2}$$

Here u is a scalar variable, x,y are cartesian coordinates, t is time, and a,b are constants. These studies are motivated by the belief that the results can be applied to practical problems (in which u is a vector, x,y are curvilinear, and a,b are matrices) by mechanisms discussed elsewhere [1-5]. These discussions are not repeated here, but results of practical applications are shown.

INTRODUCTION

As in a paper presented to the previous GAMM conference [3], we consider only explicit algorithms for solving (1), (2), but distinguish between linear and homogeneous algorithms. A linear explicit algorithm for solving (2) has the general form

$$u^{n+1}_{i,j} = \sum_{p,q} C_{p,q}\, u^n_{i+p,j+q} \tag{3}$$

In this equation, $u^n_{i,j}$ is some approximation to the value of u at $x = i\Delta x$, $y = j\Delta y$, $t = n\Delta t$, and $\{C_{p,q}\}$ is a set of constant coefficients whose non-zero members define the stencil of the algorithm (ie the numerical domain of dependence). A more general algorithm may use coefficients which are data-dependent, but can still be thought of as a mapping, from those data values $u^n_{i+p,j+q}$ picked out by the stencil, onto the desired value $u^{n+1}_{i,j}$, ie

$$\begin{aligned} u^{n+1}_{i,j} &= H(u^n_{i+p1,j+q1},\ u^n_{i+p2,j+q2},\ \text{etc}) \\ &= H(h_1, h_2, \text{etc}) = H(\underline{h}) \end{aligned} \tag{4}$$

Homogeneous algorithms are those which meet the restriction that for any scalar multiplier k, $H(k\underline{h})=kH(\underline{h})$. That is, the algorithm is invariant with respect to units of measurement. Linear algorithms result from the further restriction that for any two distinct data sets $\underline{h}_1$, $\underline{h}_2$, then $H(\underline{h}_1 + \underline{h}_2) = H(\underline{h}_1) + H(\underline{h}_2)$. It is well-known that this second restriction hinders the algorithm in achieving certain desirable properties, such as monotonicity, monotonicity preservation, and total variation diminution (TVD), all of which terms are defined and discussed by Harten [6]. For linear problems, such as (1), (2), and linear algorithms (3), all these properties amount to requiring that none of the coefficients $C_{p,q}$ are negative. It is easy to show that another equivalent property is one which we define as follows. We require that for all i,j, if the subset of data values picked out by the stencil possesses a least upper bound and a greatest lower bound, then the predicted value $u_{i,j}^{n+1}$ lies between these same bounds. This local bounding (LB) property cannot be achieved by any linear algorithm which is better than first-order accurate [6]. No such limitation exists for homogenous algorithms, and LB homogeneous algorithms which are for all practical purposes second- or third-order accurate were constructed in [3,7]. Rather similar methods have been proposed elsewhere [6,8,9], but we expect to establish certain additional merits of simplicity for our work, especially with regard to systems of equations.

In this paper we concentrate on two significant developments. In Section 2 we describe how the use of certain averaging functions creates one-dimensional algorithms with rather surprising asymptotic properties, which are both of theoretical interest and practical use. In Section 3 we report the discovery of two-dimensional methods which combine the LB property with a more natural treatment of conservation than in [3].

ONE-DIMENSIONAL SCHEMES AND NON-LINEAR AVERAGING

We recapitulate the class of homogeneous one-dimensional algorithms studied in [3]. It has proved convenient (to us) to change the notation slightly. For each interval x_i, x_{i+1}, we carry out the following steps:

1 Evaluate the fluctuation $\phi_{i+\frac{1}{2}}$ and the total signal $\Phi_{i+\frac{1}{2}}$

$$\phi_{i+\frac{1}{2}} = a(u_{i+1} - u_i) \tag{5}$$

$$\Phi_{i+\frac{1}{2}} = \frac{-\Delta t}{\Delta x} \phi_{i+\frac{1}{2}} \tag{6}$$

2 Let $\sigma = \text{sgn}\, a = \pm 1$. Add $\Phi_{i+\frac{1}{2}}$ to $u_{i+\frac{1}{2}+\frac{1}{2}+\sigma}$. This completes an upwind, first-order, conservative, LB scheme.

3 Let $\nu = a\Delta t/\Delta x$ = Courant number. Let $b_{i+\frac{1}{2}} = \frac{1}{2}(1-|\nu|)\Phi_{i+\frac{1}{2}}$. Transfer $b_{i+\frac{1}{2}}$ from $u_{i+\frac{1}{2}+\frac{1}{2}\sigma}$ to $u_{i+\frac{1}{2}-\frac{1}{2}\sigma}$. This completes a second-order scheme (in fact the Lax-Wendroff scheme) which maintains conservation but is not LB. This stage resembles the anti-diffusion stage in FCT algorithms [11,12].

To obtain an LB property, we modify the third stage, so that the transferred quantity is $B_{i+\frac{1}{2}}$ where

$$B_{i+\frac{1}{2}} = B(b_{i+\frac{1}{2}}, b_{i+\frac{1}{2}-\sigma}) \tag{7}$$

and B is some homogeneous averaging function, representable as a function of one variable through the identity $B(b_1, b_2) = b_2\, B(r, 1)$ where $r = b_1/b_2$. Conditions to be satisfied by B are discussed in [3, 4, 10]. Natural conditions which simplify the analysis are (a) $B(r, 1) = 0$ if $r \leq 0$; (b) $\min(r, 1) \leq B(r,1) \leq \max(r,1)$. Note that $B(r,1) = r$ recovers the Lax-Wendroff algorithm, and $B(r,1) = 1$ recovers the second-order upwind scheme. Neither choice meets the LB condition, which is [3, 4].

$$B(r,1) \leqslant 2/|\nu| \; ; \; B(r,1) \leqslant 2r\,(1-|\nu|) \qquad (8a,b)$$

These conditions restrict B within the shaded region of the diagram in sketch (i). A simple averaging technique which yields an LB algorithm is $B(r,1) = \min\,(r,1)$. A sophisticated form of averaging [4], obtained by demanding that the algorithm returns an exact result for any exponentially varying data $u_i^o = A + Be^{\lambda i}$, is the hypergeometric function

$$B(r,1) = r\; {}_2F_1\,(1,\; 1+\nu;\; 3;\; 1-r) \qquad (9)$$

These two functions are shown as M and H in sketch (i). Each is biassed toward the smaller input; such bias has become, in various contexts, a standard device for producing well-behaved high-order schemes. The LB conditions do permit, however, over a limited range of r, averages which are biassed the other way, and we have recently experimented with functions which exploit this possibility. Of particular interest is the function which takes, for all r, its largest permitted value; also the function which takes the largest permitted value which does not depend on ν. These are respectively

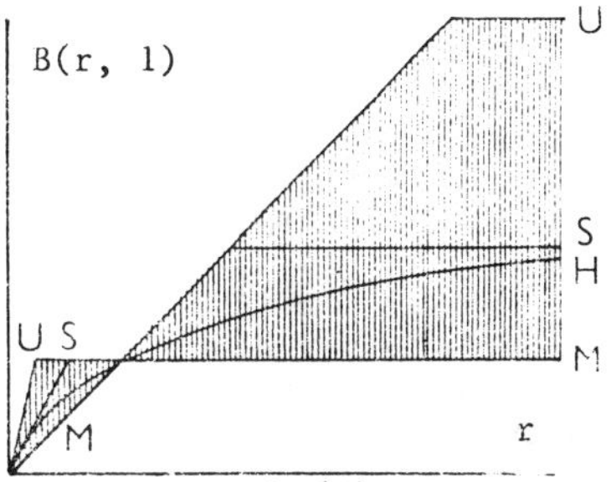

Sketch (i)

$$B(r,\,1) = 2r/(1-|\nu|) \text{ if } r < (1-|\nu|)/2;\; 2/|\nu| \text{ if } r > |\nu|/2;$$
$$= \max\,(r,\,1) \text{ otherwise} \qquad (10)$$

$$B(r,\,1) = 2r \text{ if } r < \tfrac{1}{2};\; 2 \text{ if } r > 2$$
$$= \max\,(r,\,1) \text{ otherwise} \qquad (11)$$

For mnemonic purposes, we refer to (9) as Hyperbee, (10) as Ultrabee and (11) as Superbee.

It is interesting to know how the choice of averaging will affect the operation of the algorithm over many time steps. Particularly we wish to know whether sharp steps are preserved, whether the extreme values of peaks are maintained, and whether smooth data becomes distorted. Firstly we study the advection of steps, and recall a known result [13] for linear schemes applied to step data ($u_i^o = U_L$ if $i \leqslant 0$; $u_i^o = U_R$ otherwise). For sufficiently large time, the results become asymptotically self similar;

$$u_i^n \simeq \text{fn}\,\frac{x - at}{t^{1/(p+1)}} = \text{fn}\,\frac{i - \frac{1}{2} - \nu n}{n^{1/(p+1)}} \qquad (12)$$

where p is the order of accuracy of the method. In words, the values u_i^n generated by the algorithm become samples drawn from a <u>continuous</u> wave whose width grows with time (eg like the cube root of time for a second-order method). We confirmed by experiment [14] that the same result appears to hold for certain algorithms containing logical elements similar to those under study. One plausible measure for the transition width is T_ε^n, the number of points at time level n for which u_i^n does not lie in either of the intervals $[-\varepsilon,\varepsilon]$, $[1-\varepsilon, 1+\varepsilon]$. Fig 1 shows, on a log-log plot, the way that T_ε^n changes with n for various averaging functions, and various Courant numbers. The value of ε was taken to be 0.005, so we are counting the number of points which cover 99% of the transition. The three parts of the figure show results for Courant numbers $\nu = 0.25, 0.5, 0.75$. For simple first-order upwind differencing(F), the results are well fitted by a trend line with slope $\frac{1}{2}$, which is what we expect for a first-order linear scheme, (see eqn (12)). For the homogeneous schemes employing the minimum average (M, which gives a second-order method) and Hyperbee (which gives a third-order method [4]), convincing trend lines with slopes 1/3, 1/4 respectively indicate that the spreading rate predicted by eqn (12) still holds good. Since all our methods have four-point stencils, and since third-order accuracy is the best that can be achieved, these results would appear to be optimal, but in fact the calculations using Superbees and Ultrabees fairly quickly take on transition widths that remain constant, and, indeed, quite small. Detailed examination of the numerical results leads us to conjecture that if ν is a rational fraction m/n, then the solution enters a limit cycle with period n which displaces the data through m grid points. However, there is no indication that the quality of the results depends strongly on ν, or that simple fractional values are in any practical sense unrepresentative.

Secondly, we studied the behaviour of test data comprising four cycles of a sine wave, spread over 60 mesh points. Algorithms which perform well on step data often do badly on oscillatory data; the LB property ensures that local extrema are always diminished, never restored. Ultrabee is in one respect an exception to this rule. Because the LB conditions are only just met, the amplitude is actually preserved. In all other cases, we measured a mean amplitude of the waves, excluding the first and last peaks, which were not always typical. We found that this mean amplitude decayed exponentially (Fig 2a), and that in all cases the phase error remained remarkably small. The waveform, however, was very slowly distorted toward a series of square pulses (Fig 2b). In tests involving oscillations at lower frequencies, this clipping was also observed, but took place even more slowly.

From these studies of simple model problems, it becomes apparent that we have available schemes in which the diffusive effects of numerical error (in other words, the artificial viscosity) can be kept extremely small. We next present some evidence that these enable better calculations of inviscid gasdynamic flows. Our first example is a shock-tube problem studied by Woodward and Colella [15]. The flow of an ideal gas with $\gamma = 1.4$ is assumed to obey the unsteady Euler equations; the techniques needed to extend our methods to these and similar hyperbolic systems are discussed in [1-5]. The initial data is taken to be $\rho \equiv 1$, $u \equiv 0$, with $p = 10^3$ $(0 < x < 0.1)$, $p = 10^{-2}$ $(0.1 < x < 0.9)$, $p = 10^2$ $(0.9 < x < 1.0)$, and perfectly reflecting boundaries are placed at $x = 0, 1.0$. The shockwaves which are generated by the initial discontinuities move inward and collide at about $t = 0.28$. The computed solution on a fine mesh of 1200 intervals is shown in Fig 3a, b at $t = 0.34$, by which time several more interactions have taken place. Shock waves and contacts are both present; they are easily distinguished because the former cause discontinuities in both density and velocity, whereas the latter appear only in the density distribution. In Fig 3c, d we show the density distribution only, on a mesh of 400 intervals. On this somewhat coarser grid it is possible to distinguish more clearly between the different methods. In Fig 3c the LB property has been enforced through the minimum mean, and in 3d through Superbee. It is apparent that Superbee greatly increases the resolution of contacts, and that in consequence the local extrema of density are much better represented. Actually 3b is very near the exact solution [15], and 3d reproduces all its main features. Further one-dimensional calculations, with interesting comments and comparisons, have been made by Sweby [10].

Fig 4 shows some results for a two-dimensional problem, also proposed by Woodward and Colella [15]. This is to find the flow due to a step of unit height created impulsively in a channel of height five units containing uniform flow at M = 3.0. The results shown are density contours after the uniform flow has travelled three units. They were computed by Glaister [16] using a time-split version of the present method incorporating Superbee, on a 40 x 120 grid. All features of the flow are nicely resolved, including the slip line which is forming behind the Mach stem on the upper wall. Finally we exhibit Fig 5, which is due to Lytton [5], showing the steady flow at M = 0.60 past a circular cylinder, computed by a pseudo-time-marching algorithm in which, again, Superbee has been incorporated. Due to limitations of space we show only the streamline pattern, but further details, and applications to transonic aerofoil problems, will be found in [5].

Despite these successes, however, the application to multidimensional problems still offers some difficulties, of which the most significant is perhaps a rather slow convergence to steady states, attended by "numerical noise" generated near slowly-moving oblique shockwaves. These difficulties, which seem [15] to be common to other methods which attempt to compute highly-resolved multidimensional shocks by appealing in part to one-dimensional concepts, prompted us to renew our efforts at creating a two-dimensional algorithm, and these are reported in the next section.

TWO-DIMENSIONAL ALGORITHMS

One of the main difficulties in designing algorithms for multi-dimensional problems is the increased multiplicity of possible schemes, even if attention is restricted to one type of strategy. As in [3], we retain the fluctuation-signal approach, and attempt to use the insights gained from our experience in one dimension. At first it was expected that the natural definition of a two dimensional fluctuation would be

$$\iint_{\text{cell}} (au_x + bu_y)\,dx\,dy \qquad (13).$$

If this expression is transformed by the Divergence Theorem and approximated by trapezium rule quadrature to give (dropping the superscript n)

$$\begin{aligned}\phi_{i+\frac{1}{2},j+\frac{1}{2}} &= \tfrac{1}{2}a\left[(u_{i+1,j+1} + u_{i+1,j}) - (u_{i,j+1} + u_{i,j})\right]\Delta y \\ &\quad + \tfrac{1}{2}b\left[(u_{i,j+1} + u_{i+1,j+1}) - (u_{i,j} + u_{i+1,j})\right]\Delta x \qquad (14)\end{aligned}$$

then a corresponding signal

$$\Phi_{i+\frac{1}{2},j+\frac{1}{2}} = \frac{-\Delta t}{\Delta x \Delta y}\,\phi_{i+\frac{1}{2},j+\frac{1}{2}} \qquad (15)$$

could be used, without splitting, to update u at adjacent nodes. It is easy to find linear algorithms which achieve second-order accuracy in this way. One such, equivalent to Lax-Wendroff, appears in the work of Ni [17]. By modifying such a scheme along the lines of the homogeneous algorithms described in Section 2, we hoped to obtain an LB scheme. However, it can be shown that this is not in fact possible. The technique adopted in [3, eqn (18)] which defines ϕ in a way that depends on the signs of a, b, is unsatisfactory in more general problems such that a or b may change sign; there are difficulties with maintaining conservation. We have therefore adopted a new approach which parallels more closely the one-dimensional theory, but does not completely reduce to operator splitting. Note first that the sum of the fluctuations (14) taken over all rectangles, is equal to the sum over all sides of rectangles, of fluctuations

$$\phi_{i+\frac{1}{2},j} = a(u_{i+1,j} - u_{i,j})\Delta y; \quad \psi_{i,j+\frac{1}{2}} = b(u_{i,j+1} - u_{i,j})\Delta x \tag{16}$$

Now proceed as follows:

1. Let each such fluctuation generate, as in one dimension, a signal

$$\Phi_{i+\frac{1}{2},j} = \frac{-\Delta t}{\Delta x \Delta y}\,\phi_{i+\frac{1}{2},j}; \quad \Psi_{i,j+\frac{1}{2}} = \frac{-\Delta t}{\Delta x \Delta y}\,\psi_{i,j+\frac{1}{2}} \tag{17}$$

2. Let $\sigma 1 = \operatorname{sgn} a = \pm 1$, $\sigma 2 = \operatorname{sgn} b = \pm 1$. Add $\Phi_{i+\frac{1}{2},j}$ to $u_{i+\frac{1}{2}+\frac{1}{2}\sigma 1,j}$ and $\Psi_{i,j+\frac{1}{2}}$ to $u_{i,j+\frac{1}{2}+\frac{1}{2}\sigma 2}$, which completes an upwind first-order, conservative, LB scheme. Note that in a splitting scheme Φ would be added before evaluating ψ.

3. Let $\nu_1 = a\Delta t/\Delta x$, $\nu_2 = b\Delta t/\Delta y$. Let

$$b_{i+\frac{1}{2},j} = \tfrac{1}{2}(1-|\nu_1|)\Phi_{i+\frac{1}{2},j}; \quad b_{i,j+\frac{1}{2}} = \tfrac{1}{2}(1-|\nu_2|)\Psi_{i,j+\frac{1}{2}} \tag{18}$$

As in one dimension, transfer $b_{i+\frac{1}{2},j}$ from $u_{i+\frac{1}{2}+\frac{1}{2}\sigma 1,j}$ to $u_{i+\frac{1}{2}-\frac{1}{2}\sigma 1,j}$, and $b_{i,j+\frac{1}{2}}$ from $u_{i,j+\frac{1}{2}+\frac{1}{2}\sigma 2}$ to $u_{i,j+\frac{1}{2}-\frac{1}{2}\sigma 2}$. This is not yet a second-order scheme; there is an error of order $\Delta x \Delta y\, u_{xy}$. This error is removed by the following stage, which has no parallel in a time-split scheme.

4. Let θ be an arbitrary parameter. Let

$$c_{i+\frac{1}{2},j} = \nu_2 \cos^2\theta\, \Phi_{i+\frac{1}{2},j}; \quad c_{i,j+\frac{1}{2}} = \nu_1 \sin^2\theta\, \Psi_{i,j+\frac{1}{2}} \tag{19}$$

Transfer $c_{i+\frac{1}{2},j}$ from $u_{i+\frac{1}{2}+\frac{1}{2}\sigma 1,j}$ to $u_{i+\frac{1}{2}+\frac{1}{2}\sigma\ ,j+\sigma 2}$, and $c_{i,j+\frac{1}{2}}$ from $u_{i,j+\frac{1}{2}+\frac{1}{2}\sigma 2}$ to $u_{i+\sigma 1,j+\frac{1}{2}+\frac{1}{2}\sigma 2}$. The simplest choice for θ is $\pi/4$, but $\tan^{-1}(b/a)$ is attractive also.

The scheme 1-4 above is not LB, but is second-order, conservative, and has an upwind bias. If a, b are both positive, then its stencil is as in sketch (ii), and the coefficients $\{C_{p,q}\}$ are as marked. It may be shown analytically [7] that the scheme is stable for Courant numbers

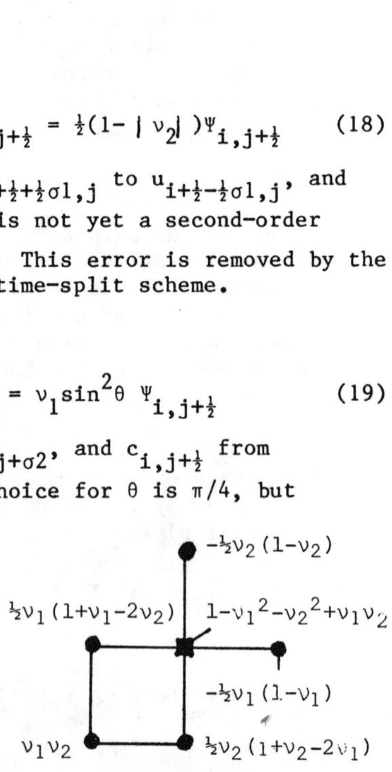

Sketch (ii)

ν_1, ν_2 sufficiently small, and numerical evidence indicates stability for all $\nu_1, \nu_2 \in [0,1] \times [0,1]$. If a, b are not both positive, the stencil changes appropriately. Stability and conservation are both preserved.

To obtain an LB property we modify stages 3 and 4 above, so that the transferred quantities are

$$B_{i+\frac{1}{2},j} = B(b_{i+\frac{1}{2},j},\ b_{i+\frac{1}{2}-\sigma 1,j});\quad B_{i,j+\frac{1}{2}} = B(b_{i,j+\frac{1}{2}}, b_{i,j+\frac{1}{2}-\sigma 2}) \qquad (20)$$

$$C_{i+\frac{1}{2},j} = C(c_{i+\frac{1}{2},j},\ c_{i+\frac{1}{2},j+\sigma 2});\quad C_{i,j+\frac{1}{2}} = C(c_{i,j+\frac{1}{2}}, c_{i+\sigma 1,j+\frac{1}{2}})$$

where B, C are homogeneous averaging functions as used in the one-dimensional theory. In particular, if B, C are each taken to be the minimum function, then a sufficient condition [7] for the algorithm to be LB is that $|\nu_1| + |\nu_2| \leqslant 2/3$.

We have as yet rather little experience to report with this algorithm on practical problems. Glaister [16] has found acceptable results for the problem shown in Fig 4, but so far only for coarse grids and small times. More detailed evaluation is needed before a practical advantage can be claimed for the unsplit scheme.

REFERENCES

1. P.L. Roe, Proc. 7th Int. Conf. Num. Meth. Fl. Dyn. Springer 1981.
2. P.L. Roe, J. Comp. Phys., 43 p 357 1981.
3. P.L. Roe, M.J. Baines, Proc. 4th GAMM Conf. Num. Meth. Fl. Mech. Vieweg 1982.
4. P.L. Roe, Proc. AMS/SIAM Seminar (San Diego 1982) to appear.
5. C.C. Lytton, RAE TR in preparation.
6. A. Harten, J. Comp. Phys., 49, p 357, 1983.
7. M.J. Baines, Univ. of Reading, Num. Anal. Rept. 1/83.
8. B. van Leer, J. Comp Phys., 32, p 101, 1979.
9. P. Colella, P.M. Woodward, Univ. of California preprint 1982.
10. P.K. Sweby, Proc. AMS/SIAM Seminar (San Diego 1983) to appear.
11. S.T. Zalesak, J. Comp. Phys., 31, p 335, 1979.
12. A.Y. LeRoux, R.A.I.R.O. Num. Anal., 15, p 151, 1981.
13. A. Harten, C.P.A.M., XXX p 611, 1977.
14. P.L. Roe, RAE TR 81047, 1981.
15. P.M. Woodward, P. Colella, Univ. of California preprint 1982.
16. P. Glaister, Univ. of Reading, unpublished.
17. R.H. Ni, AIAA Jnl, 20, p 1565, 1982.

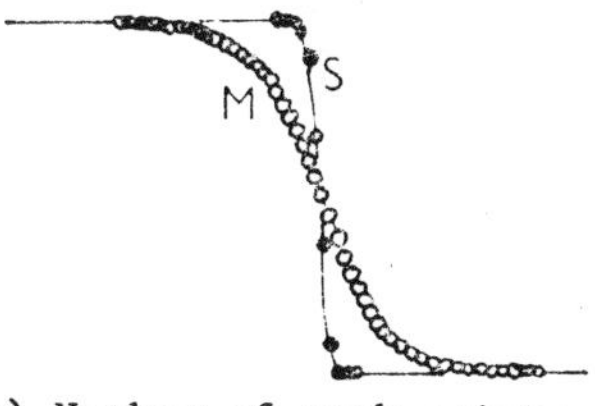

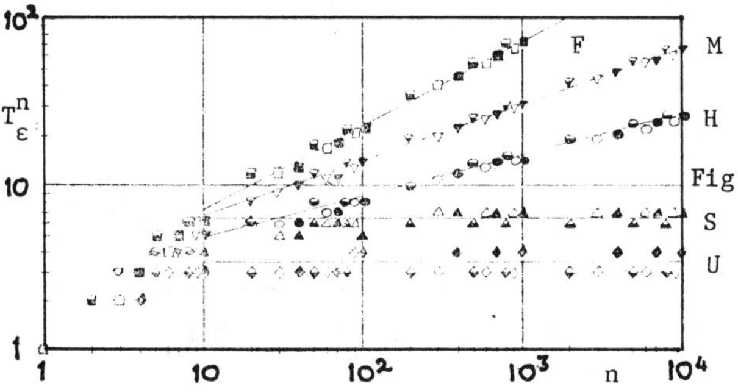

Fig 1(a) Number of mesh points in transition vs number of time steps.
(b) Two transition profiles, (M) and (S) after 10,000 steps at $\nu = 0.1$.

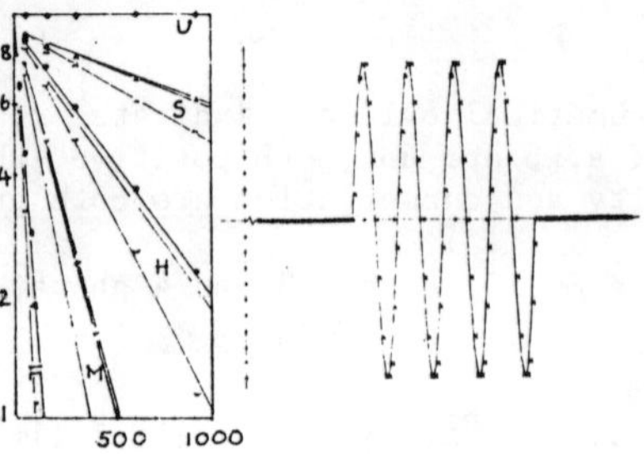

Fig 2(a) Mean amplitude of waves vs number of time steps.
(b) Results from (U) after 915 steps at $\nu = 0.5$.

In 1(a), 2(a) Solid symbols are $\nu = 0.25$, half-open are $\nu = 0.50$, open are $\nu = 0.75$.

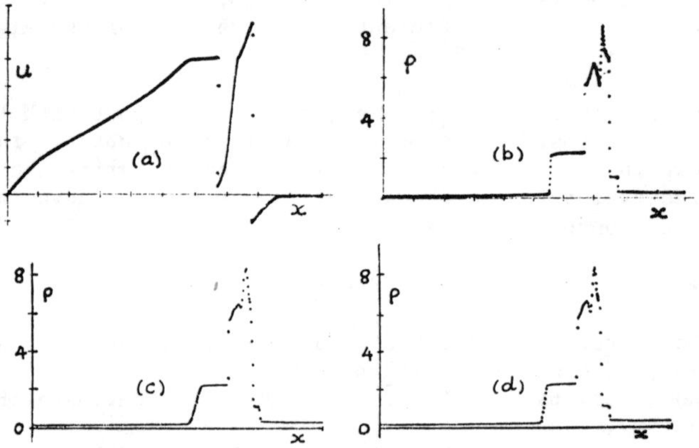

Fig 3 Results for shock-tube problem.

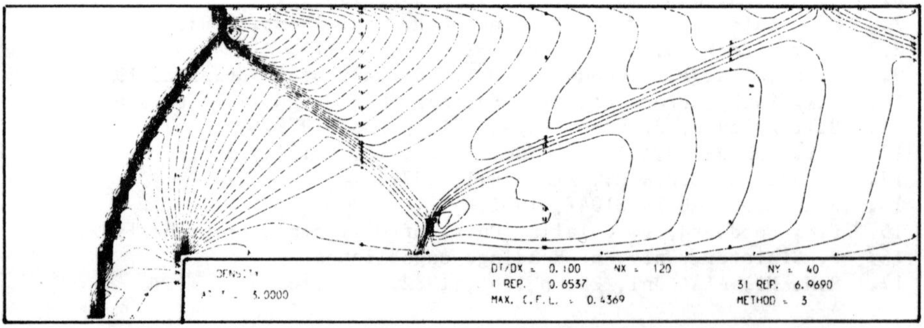

Fig 4 Density contours in the unsteady flow over a step at M = 3.0.

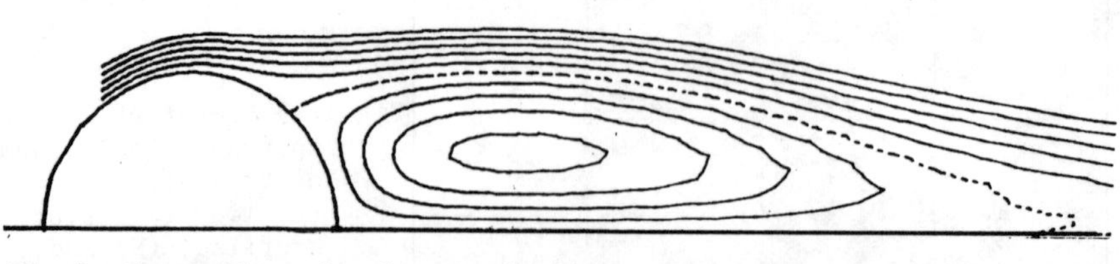

Fig 5 Streamlines in the steady flow over a cylinder at M = 0.6.

UNCONDITIONALLY STABLE EXPLICIT METHOD FOR THE NUMERICAL SOLUTIONS OF THE COMPRESSIBLE NAVIER-STOKES EQUATIONS

Nobuyuki Satofuka
Department of Mechanical Engineering
Kyoto Technical University
Matsugasaki, Sakyo-ku
Kyoto 606, Japan

SUMMARY

An unconditionally stable explicit method is devised for solving the Navier-Stokes equations for unsteady compressible flow. The spatial derivatives of the time-dependent equations are discretized by using the modified differential quadrature (MDQ) method. The resulting system of ordinary differential equations in time are then integrated by the rational Runge-Kutta (RRK) scheme. The method is fully explicit, requires no matrix inversion, and is stable at much larger time step than the usual explicit methods. Applications are presented for one- and two-dimensional flow test problems to demonstrate the accuracy and efficiency of the proposed algorithm. The present method is found to be very accurate and efficient when results are compared with other numerical solutions.

1. INTRODUCTION

Time-dependent approach is sometimes used to solve steady-state flow problems described by the Navier-Stokes equations. Such calculations, however, usually take a large number of time steps to reach the desired time-limit. One reason for this is that stiffness of the problem by virtue of disparate characteristic speed and/or length scale severely restricts the length of time step at least for explicit method. By far the most common way to avoid any form of stiffness is to use implicit algorithms. The existing implicit methods, such as Beam-Warming [1], requires costly inversions of large block-tridiagonal matrices. A method recently developed by MacCormack [2] eliminates this disadvantage. However, programming complexity, as well as computer time per time step, of this method is still larger than that of the explicit method. If one can devise a new method that possesses the conceptual simplicity of an explicit method as well as the property of unconditional stability, the methods will have immense practical importance. The purpose of this paper is to present such a development and to demonstrate its application to some relevant problems.

The new method consists of the modified differential quadrature (MDQ) method combined with a rational Runge-Kutta (RRK) time integration scheme [3]. The MDQ method is an extension of the differential quadrature (DQ) method proposed originally by Bellman et al. [4]. In the present method, spatial derivatives are approximated by a weighted sum of the values of an unknown function at properly chosen neighboring points to generate a set of ordinary differential equations (ODEs) in time, whereas in the original DQ method these are approximated by using values at all mesh points in the computational domain. As a result, computational efficiency is significantly improved. The resulting set of ODEs is then solved by using an explicit A_0 and/or $A(\alpha)$-stable RRK time integration scheme. In this sence, the present method is taken to be a method of lines. It should be noted that the present method is computationally explicit and yet unconditionally stable for some class of parabolic partial differential equations

(PDEs).

The present paper is organized in the following way. The numerical procedure of the present method for a nonlinear scalar model equation, that is, the Burgers equation, is described briefly in Sec. 2; application to the Burgers equation and the compressible form of the Navier-Stokes equations and some numerical results for one and two-dimensional flow problems are presented in Sec. 3 to demonstrate the accuracy and efficiency of the present method.

2. DESCRIPTION OF THE METHOD

In this section we will describe the basic elements of the present method when it is applied to the following nonlinear scalar model equation, viz., Burgers equation,

$$\frac{\partial u}{\partial t} = -u\frac{\partial u}{\partial x} + \nu\frac{\partial^2 u}{\partial x^2} , \qquad (1)$$

with the initial condition

$$u(x,0) = f(x) , \qquad (2)$$

and boundary conditions

$$B_j u(x,t) = 0 , \quad j = 1,2. \qquad (3)$$

(a) Spatial Discretization

If the function u satisfying Eq. (1) is sufficiently smooth, we can write the approximate relation

$$\frac{\partial u_i(t)}{\partial x} \cong \sum_{j=1}^{N} a_{ij}u_j(t) , \quad i = 1,2,\cdots,N , \qquad (4)$$

where we adopt the notation $u_i(t) = u(x_i,t)$. Viewing Eq. (4) as a linear transformation of u, we see that the second-order derivatives can be approximated by

$$\frac{\partial^2 u_i(t)}{\partial x^2} \cong \sum_{j=1}^{N} b_{ij}u_j(t) , \qquad (5)$$

where $b_{ij} = \sum_{k=1}^{N} a_{ik}a_{kj}$. In this paper we have modified the approximate relations, Eqs. (4) and (5), to use the values of u at the nearest M mesh points centered around x_i, instead of using those at all mesh points in the computational domain, as is the case in the original DQ method. By using these values of u, the number of arithmetic operations to be performed for every mesh point is significantly reduced; moreover, in the case of a uniform mesh, the weighting coefficients a_{ij} become independent of index i. Therefore, the approximate relations, Eqs. (4) and (5), can be rewritten as

$$\frac{\partial u_i(t)}{\partial x} \cong \sum_{j=-m}^{m} a_j u_{i+j}(t) \equiv D_M(u_i) , \qquad (6)$$

$$\frac{\partial^2 u_i(t)}{\partial x^2} \cong \sum_{j=-m}^{m} b_j u_{i+j}(t) \equiv D_M^2(u_i) , \qquad (7)$$

where $a_j = a_{mj}$, $b_j = b_{mj}$, and $m = (M - 1)/2$.

(b) Determination of Weighting Coefficients

There are many ways of determining the coefficients a_{ij}. In the DQ method, Bellman et al. determined a_{ij} explicitly, choosing x_i to be the root of shifted Legendre polynomial of degree of N, $P_N^*(x)$. In this paper we have determined a_{ij} numerically, similarly to Lagrangian interpolation and choose the test function in the following form,

$$\Phi_j(x) = \Pi(x)/[(x - x_{i+j})\Pi'(x_{i+j})] , \tag{8}$$

where $\Pi(x)$ is a polynomial of degree M,

$$\Pi(x) = (x - x_{i-m})\cdots(x - x_i)\cdots(x - x_{i+m}) . \tag{9}$$

If follows that $\Phi_j(x)$ is a polynomial of degree M - 1 such that $\Phi_j(x_{i+k}) = \delta_{jk}$ and $\Pi(x_{i+k}) = 0$, where δ_{jk} denotes the classical Kronecker delta. If the values of u(x) are known at M points, $x = x_{i-m},\cdots,x_i,\cdots,x_{i+m}$, a polynomial of degree M - 1, $\tilde{u}(x)$, which coincides with u(x) at these collocation points, can be written as

$$\tilde{u}(x) = \sum_{j=-m}^{m} \Phi_j(x)u(x_{i+j}) . \tag{10}$$

By differentiating Eq. (10) with respect to x, we have the relation,

$$\tilde{u}'(x) = \sum_{j=-m}^{m} \Phi_j'(x)u(x_{i+j}) . \tag{11}$$

Using the fact that such a relation as Eq. (4) is to be exact for $u(x) = \Phi_j(x)$, we see that

$$a_{ij} = \Pi'(x_i)/[(x_i - x_{i+j})\Pi'(x_{i+j})] , \; j \neq 0 . \tag{12}$$

For the case j = 0, use of l'Hospital's rule gives

$$a_{i0} = \Pi''(x_i)/[2\Pi'(x_i)] . \tag{13}$$

Using Eqs. (12) and (13), the coefficients b_{ij} in Eq. (5) can be calculated. Then the coefficients a_j and b_j in Eqs. (6) and (7) are determined, using the relations $a_j = a_{mj}$ and $b_j = b_{mj}$ in which $m = (M - 1)/2$. The coefficients a_j and b_j are computed once and for all at the beginning of the calculation and stored.

(c) Time Integration

Substitution of the approximate relations, Eqs. (6) and (7), into Eq. (1) yields the set of N ODEs in time,

$$u_i'(t) = -u_i(t)D_M(u_i) + \nu D_M^2(u_i) , \tag{14}$$

or in matrix form

$$\vec{U}' = \vec{F}(\vec{U}) , \tag{15}$$

where $U = (u_1,u_2,\cdots,u_N)^T$ and the prime denotes differentiation with respect to time. The numerical solution of such a system, Eq. (15), is a simple task, using the rational Runge-Kutta (RRK) time-integration scheme. As applied to Eq. (15), the scheme can be written in the following two-stage form,

$$\vec{g}_1 = \Delta t\vec{F}(\vec{U}) , \quad \vec{g}_2 = \Delta t\vec{F}(\vec{U}^n + c_2\vec{g}_1) ,$$
$$\vec{U}^{n+1} = \vec{U}^n + [2\vec{g}_1(\vec{g}_1,\vec{g}_3) - \vec{g}_3(\vec{g}_1,\vec{g}_1)]/(\vec{g}_3,\vec{g}_3) . \tag{16}$$

where $\vec{g}_3 = b_1\vec{g}_1 + b_2\vec{g}_2$, $b_1 + b_2 = 1$ and $(\vec{d},\vec{e})$ denotes the scalar product of $\vec{d}$ and $\vec{e}$. The RRK scheme is generally of order 1 but is of order 2 if in addition, $b_2c_2 = -1/2$. The method is fully explicit and $A(\alpha)$-stable if $b_2c_2 < -1/2$. Therefore, the method is free from severe stability restriction to which most explicit methods are subject.

3. APPLICATION TO COMPRESSIBLE NAVIER-STOKES EQUATIONS

(a) Governing Equations

The compressible Navier-Stokes equations in Cartesian coordinates for two-dimensional or axisymmetric flow can be written in dimensionless, conservation-law form for a perfect gas without external forces in the following form [5]:

$$\frac{\partial\vec{U}}{\partial t} + \frac{\partial\vec{E}}{\partial x} + \frac{\partial\vec{G}}{\partial y} + \delta\,\frac{\vec{G} + \vec{H}}{y} = \frac{1}{Re}\left(\frac{\partial\vec{R}}{\partial x} + \frac{\partial\vec{S}}{\partial y} + \delta\,\frac{\vec{S} + \vec{V}}{y}\right) \tag{17}$$

where $\vec{U}$, $\vec{E}$, $\vec{G}$, $\vec{H}$, $\vec{R}$, $\vec{S}$, and $\vec{V}$ are the following four-components vector:

$$\vec{U} = \begin{bmatrix}\rho\\ \rho u\\ \rho v\\ e\end{bmatrix} , \quad \vec{E} = \begin{bmatrix}\rho u\\ p + \rho u^2\\ \rho uv\\ (e + p)u\end{bmatrix} , \quad \vec{G} = \begin{bmatrix}\rho v\\ \rho uv\\ p + \rho v^2\\ (e + p)v\end{bmatrix} ,$$
$$\vec{H} = \begin{bmatrix}0\\ 0\\ -p\\ 0\end{bmatrix} , \quad \vec{R} = \begin{bmatrix}0\\ \tau_{xx}\\ \tau_{xy}\\ R_4\end{bmatrix} , \quad \vec{S} = \begin{bmatrix}0\\ \tau_{xy}\\ \tau_{yy}\\ S_4\end{bmatrix} , \quad \vec{V} = \begin{bmatrix}0\\ 0\\ -\tau_{\theta\theta}\\ 0\end{bmatrix} . \tag{18}$$

with

$$\tau_{xx} = (\lambda + 2\mu)\frac{\partial u}{\partial x} + \lambda\frac{\partial v}{\partial y} + \delta\lambda\frac{v}{y} ,$$
$$\tau_{xy} = \mu\left(\frac{\partial u}{\partial y} + \frac{\partial v}{\partial x}\right) ,$$
$$\tau_{yy} = (\lambda + 2\mu)\frac{\partial v}{\partial y} + \lambda\frac{\partial u}{\partial x} + \delta\lambda\frac{v}{y} ,$$
$$\tau_{\theta\theta} = \delta(\lambda + 2\mu)\frac{v}{y} + \lambda\left(\frac{\partial u}{\partial x} + \frac{\partial v}{\partial y}\right) , \tag{19}$$
$$R_4 = u\tau_{xx} + v\tau_{xy} + \frac{\gamma}{\gamma - 1}\,\frac{\kappa}{Pr}\,\frac{\partial T}{\partial x} ,$$
$$S_4 = u\tau_{xy} + v\tau_{yy} + \frac{\gamma}{\gamma - 1}\,\frac{\kappa}{Pr}\,\frac{\partial T}{\partial y} .$$

In the conservative variables of Eq. (18), the pressure p is nondimensionalized by p_0, the density ρ by ρ_0, and the velocity components u and v in the x and y directions by $a_0/\sqrt{\gamma}$. Re represents the reference Reynolds number defined as $Re = c_0\rho_0 L/(\sqrt{\gamma}\,\mu_0)$, Pr the Prandtl number, κ the coefficient of thermal conductivity, μ the viscosity coefficient, and γ the ratio of specific heats. The coefficient of thermal conductivity and the viscosity coefficient are nondimensionalized with respect to their reference values. In Eq. (17), $\delta = 0$ for two-dimensional flow, $\delta = 1$ for axisymmetric flow (y becomes cylindrical radius). The pressure, density and velocity components are related to the total energy per unit volume e by the following equation for an ideal gas

$$e = p(\gamma - 1) + \rho(u^2 + v^2)/2. \tag{20}$$

To transform Eq. (17) for mapping complicated physical regions into rectangular computational planes, the following independent variable transformation is chosen,

$$\xi = \xi(x,y) , \quad \eta = \eta(x,y) . \tag{21}$$

Subject to the transformation, Eq. (21), the governing equations, Eq. (17), can be written as

$$\frac{\partial \hat{U}}{\partial t} + \frac{\partial \hat{E}}{\partial \xi} + \frac{\partial \hat{G}}{\partial \eta} + \delta \hat{H} = \frac{1}{Re}\left\{ \frac{\partial \hat{R}}{\partial \xi} + \frac{\partial \hat{S}}{\partial \eta} + \delta \hat{T} \right\} , \tag{22}$$

where

$$\hat{U} = \vec{U}/J , \quad \hat{E} = \left(\frac{\partial \xi}{\partial x}\vec{E} + \frac{\partial \xi}{\partial y}\vec{G}\right)/J , \quad \hat{G} = \left(\frac{\partial \eta}{\partial x}\vec{E} + \frac{\partial \eta}{\partial y}\vec{G}\right)/J ,$$

$$\hat{R} = \left(\frac{\partial \xi}{\partial x}\vec{R} + \frac{\partial \xi}{\partial y}\vec{S}\right)/J , \quad \hat{S} = \left(\frac{\partial \eta}{\partial x}\vec{R} + \frac{\partial \eta}{\partial y}\vec{S}\right)/J , \tag{23}$$

$$\hat{H} = (\vec{G} + \vec{H})/(yJ) , \quad \hat{V} = (\vec{S} + \vec{V})/(yJ) .$$

and derivatives such as $\partial u/\partial x$ are expanded by chain rule,

$$\partial u/\partial x = (\partial \xi/\partial x)(\partial u/\partial \xi) + (\partial \eta/\partial x)(\partial u/\partial \eta).$$

Here J is the transformation Jacobian

$$J = \frac{\partial \xi}{\partial x}\frac{\partial \eta}{\partial y} - \frac{\partial \xi}{\partial y}\frac{\partial \eta}{\partial x} . \tag{24}$$

The geometrical factors (metrics) in Eq. (22) resulting from the coordinate transformation are defined as follows in terms of the derivative of the Cartesian coordinates of the grid points;

$$\frac{\partial \xi}{\partial x} = J\frac{\partial y}{\partial \eta} , \quad \frac{\partial \xi}{\partial y} = -J\frac{\partial x}{\partial \eta} , \quad \frac{\partial \eta}{\partial x} = -J\frac{\partial y}{\partial \xi} , \quad \frac{\partial \eta}{\partial y} = J\frac{\partial x}{\partial \xi} . \tag{25}$$

In general, the metrics of Eq. (25) are not known analytically and must be determined numerically at the beginning of the calculation and stored.

The second order MDQ method ($M = 3$) combined with the RRK time-integration scheme, Eq. (16), is used to solve the conservation-law form of the Navier-Stokes equations, Eq. (22), together with the initial and boundary conditions.

(b) Stability Properties of the Method

The RRK time-integration scheme described in Eq. (16) is computationally explicit and yet unconditionally stable for some class of parabolic PDEs. Hairer [6] has proved that for a system of equations, the scheme, Eq. (16), is A_0-stable if $b_2c_2 \leq -1/2$ and $A(\alpha)$-stable if $b_2c_2 \leq -1/[2\cos\alpha(2-\cos\alpha)]$. According to his stability definition, a method is said to be A_0-stable if $\{x \in R \mid x \leq 0\}$ is the stability region and $A(\alpha)$-stable if

$\{z \mid |\pi - \arg z| \leq \alpha\}$ is the stability region. The relation between α and value of parameter, b_2c_2, is shown in Fig. 1. Figure 1 shows that with $b_2c_2 = -1$, the scheme is A(72.97°)-stable and with $b_2c_2 = -10$, A(88.55°)-stable. The scheme is also I-stable if $b_2c_2 \leq -1/2$.

Before application to the Navier-Stokes equations, the present method was applied to the Burgers equation, Eq. (1), written in terms of a transformed coordinate, $x' = x - t/2$. For the sake of convenience, we will hereinafter drop the prime for the transformed coordinate. The initial condition was specified such that

$$f(x) = 0 \quad x>0, \quad f(x) = 1/2 \quad x=0, \quad f(x) = 1 \quad x<0 . \tag{26}$$

with the boundary conditions

$$u \to 1 \text{ as } x \to -\infty \quad ; \quad u \to 0 \text{ as } x \to \infty . \tag{27}$$

A steady-state solution of the Burgers equation subject to these boundary conditions is known to be

$$u(x) = [1 - \tanh(x/4\nu)]/2 , \tag{28}$$

which gives a measure of the accuracy of the numerical results. The numerical solutions at 50 time steps are compared in Fig. 2 with that of the implicit MacCormack scheme. The solutions are identified by different symbols (Δ for the present and $\square$ for the MacCormack scheme) and the exact steady-state solution is shown by a solid line. The numerical solutions were obtained for a uniform mesh with $\Delta x = 1.0$, which contained 17 points in the computational domain, $-8 \leq x \leq 8$. The boundary conditions at these finite locations were assigned values predicted from the exact solution. Convergence history of the present method is compared in Fig. 3 with that of the implicit MacCormack method. The CFL number is defined as

$$CFL = \{|u_{max} - 0.5| + \frac{2\nu}{\Delta x}\} \frac{\Delta t}{\Delta x} . \tag{29}$$

In Fig. 3, the abscissa is the number of time steps and the ordinate is the L_2 error defined as

$$L_2 \text{ error} = \left[\sum_{i=1}^{N-1} (\delta u_i)^2/(N - 1) \right]^{1/2} \tag{30}$$

where the δu_i denotes the difference between the analytic and numerical solution. The present method is more accurate and does not show any steady-state solution dependency on Δt, while in the MacCormack method, the steady -state solution depends on Δt.

The present method was then applied to a series of viscous quasi-one-dimensional nozzle flow test problems. The results for the steady-state pressure distribution for the case with $p_e/p_0 = 0.8$ and $Re = 100$ are compared in Fig. 4 with that obtained by using the MacCormack method. The values of parameters in the RRK scheme were chosen as $b_2c_2 = -1/2$, $c_2 = 1/2$. The numerical solutions were obtained for a uniform mesh $\Delta x = 2^{-6}$, which contained 65 grid points in the computational domain, $0 \leq x \leq 1$. The analytic steady-state solution for a inviscid flow is shown by a solid line, and the numericals are identified by a solid line with symbols. Convergence history of the present method for the case with $p_e/p_0 = 0.8$, $Re = 100$ is compared in Fig. 5 with that of the implicit MacCormack method. As is the case for the Burgers equation, the present method is more accurate and does not show any steady-state solution dependency on Δt. Stability properties of the RRK time-integration scheme, Eq. (16), were investigated for the same problem as shown in Fig. 5 by repeating the calculation for series of fixed CFL number. The results of numerical stability investigations for the first-order ($b_2c_2 = -1$ and -10) and the second-order ($b_2c_2 = -1/2$) RRK schemes with various values of c_2 are summarized in Table 1. It is remarkable that although the present method is explicit, it gives stable solution with CFL number up to $O(10^4)$.

(c) Viscous Blunt Body Problem

The new explicit method was then extended to two-dimensional and axisymmetric arbitrary non-orthogonal mesh systems generated by using elliptic systems and applied to compute viscous hypersonic blunt body problems. A sphere or a elliptic body of revolution is located in a hypersonic flow at a free stream Mach number $M_\infty = 1.5$. The free stream Reynolds number Re_∞ is set equal to 10 and the Prandtl number equal to 0.75. The 30X42 computational mesh was used to compute the flow field around the sphere. The computed density contours a shown in Fig. 6. The calculation ran for 500 steps at which time the flow was near steady-state and required about 1.5 minutes of that CPU time. The result was computed with much larger time step than that corresponds to the CFL number = 1.0.

4. CONCLUSIONS

A new method has been presented for solving the equation of compressible viscous flows. For many applications this method is more efficient than other methods in use today.

The present method has the following features:

1) fully explicit and requires no matrix inversion,
2) stable at much larger time step than the usual explicit methods,
3) first or second-order accurate in time,
4) arbitrary order accurate in space,
5) simple and straightforward to program, and
6) easier to adapt to vector and parallel computer architectures.

5. REFERENCES

[1] Beam, R.M. and Warming, R.F.: An Implicit Factored Scheme for the Compressible Navier-Stokes Equations, AIAA Journal, Vol 16, No. 4, 1978, 393-402.

[2] MacCormack, R.W.: A Numerical Method for Solving the Equations of Compressible Viscous Flow, AIAA paper 81-110, 1981.

[3] Wambecq, A.: Rational Runge-Kutta Methods for Solving Systems of Ordinary Differential Equations, Computing, 20, 1978, 333-342.

[4] Bellman, R., Kashef, B.G., and Casti, J.: Differential Quadrature: A Technique for the Rapid Solution of Nonlinear Partial Differential Equations, Journal of Computational Physics, 10, 1972, 40-52.

[5] Kutler, P., Chakravarthy, S.R., and Lombard, C.K.: Supersonic Flow Over Ablated Nosetips Using an Unsteady Implicit Numerical Procedure, AIAA paper 78-213, 1978.

[6] Hairer, E.: Unconditionally Stable Explicit Methods for Parabolic Equations, Numerische Mathematik, 35, 1980, 57-68.

TABLE 1. EXPERIMENTAL MAXIMUM STABLE CFL NUMBER FOR THE NOZZLE FLOW PROBLEM

b_2c_2	-0.5		-1.0			-10.0		
c_2	0.5	1.0	1/128	0.5	1.0	1/128	0.5	1.0
λ_{max}	3.2	3.2	4096.	64.	32.	4096.	64.	32.
CFL	7.5	7.5	9600.	150.	75.	9600.	150.	75.

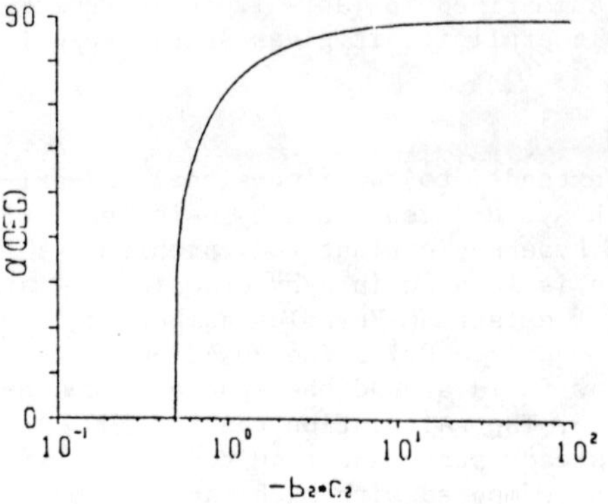

Fig. 1 Relation between α and $-b_2c_2$ for A(α)-stability.

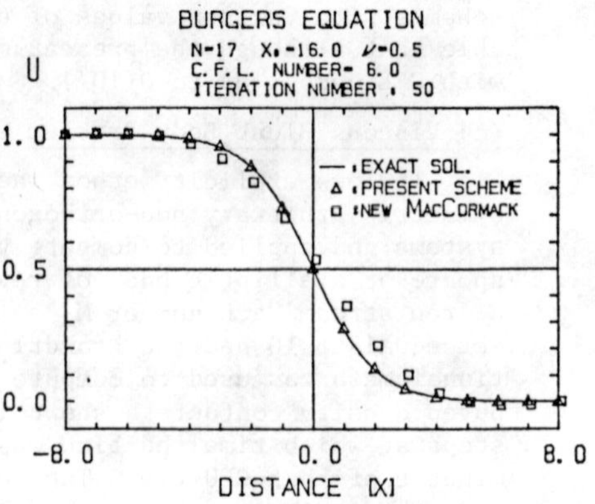

Fig.2 Comparison of the results for the Burgers equation.

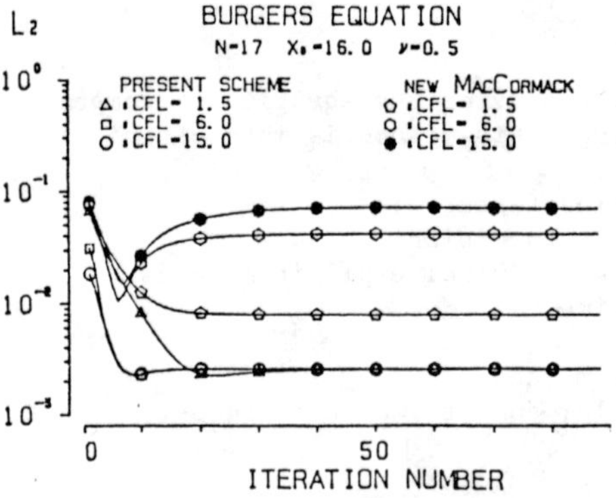

Fig. 3 Convergence history for the Burgers equation

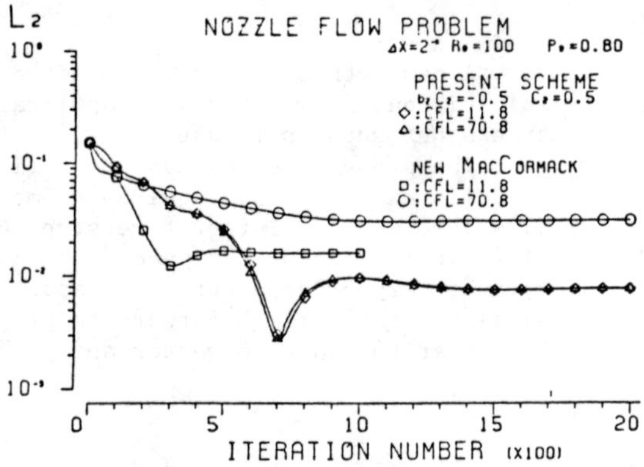

Fig. 5 Convergence history for quasi-one-dimensional nozzle flow problem.

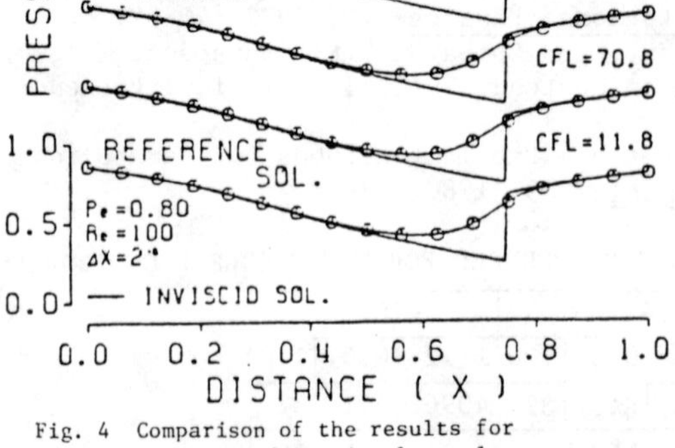

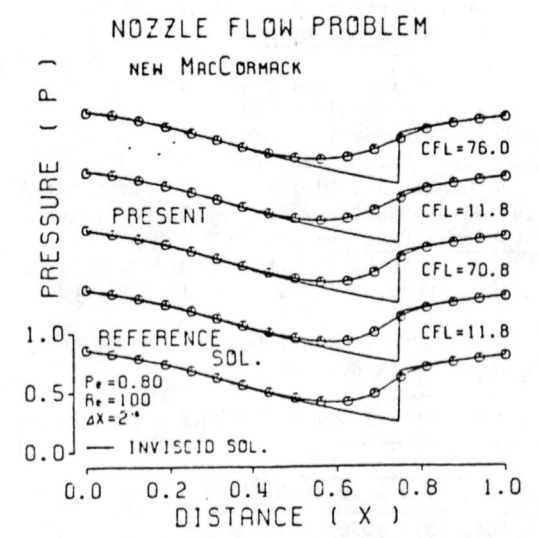

Fig. 4 Comparison of the results for quasi-one-dimensional nozzle flow problem.

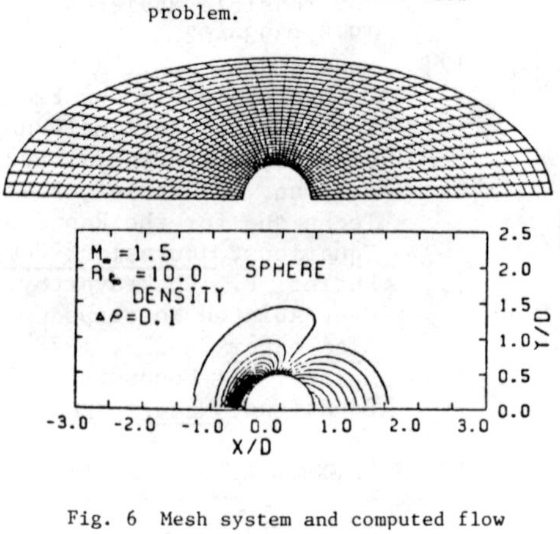

Fig. 6 Mesh system and computed flow field for viscous hypersonic flow around a sphere.

LARGE-EDDY SIMULATION OF TURBULENT BOUNDARY-LAYER FLOW

L.Schmitt and R.Friedrich
Lehrstuhl für Strömungsmechanik
Technische Universität München
Arcisstraße 21, 8000 München 2
Federal Republic of Germany

SUMMARY

A numerical method for large-eddy simulation is described. The method is an extended version of Schumann's 'volume balance procedure' [1]. It consists in formulating the transport equations as integral conservation equations for each grid volume and decomposing the flow quantities into grid scale (GS) and subgrid scale (SGS) components. In this way 'finite difference' formulae for the GS components are generated in a natural manner. The action of the SGS components is reflected in additional SGS fluxes which appear as grid surface mean values. Depending on the importance of these SGS fluxes one distinguishes between 'direct' and 'large-eddy' simulations. In the latter case modeling of these fluxes is necessary. For statistically steady and two-dimensional high Reynolds number turbulent flat plate boundary layer flow preliminary results are presented.

INTRODUCTION

Large-eddy simulation (LES) of turbulent flows is based on two experimental observations. First, the large-scale turbulence structure providing most of the important turbulent transport varies greatly from flow to flow and is consequently difficult, if not impossible, to model in a general way. Secondly, the small-scale turbulence is nearly isotropic, universal in character, and hence much more amenable to general modeling. Therefore in LES one advances as follows:

- derive dynamical equations for quantities describing only the large-scale field
- account for the action of the small-scale field by simple models
- calculate the large-scale motion in a time-dependent, three-dimensional computation
- extract the necessary statistical informations by proper ensemble averaging.

In the past mainly two techniques - the 'filter technique' [2] and the 'volume balance procedure' [1] - have been applied to derive the basic equations. In both cases the small-scale turbulence was modeled assuming an eddy-viscosity concept [3,1,4]. For wall bounded shear flows the theory has been applied successfully to fully developed turbulent channel and annular flows where periodicity boundary conditions at the free edges of the computational domain could be taken [3,1,5,4].

Our aim is to simulate developing turbulent flows bounded

by flat or curved walls which makes it necessary to use more general boundary conditions. In this paper we describe a 'volume balance procedure' in curvilinear orthogonal co-ordinates and a computer program allowing us to study wall curvature and nonperiodical boundary conditions. A first application to flat plate boundary layer flow is discussed.

BASIC EQUATIONS FOR FINITE VOLUMES AND TIME INTERVALS

Assuming a Newtonian fluid with constant density and viscosity the dimensionless integral conservation equations for mass and momentum can be written for a grid volume ΔV defined in a curvilinear orthogonal co-ordinate system ξ_α with unit vectors $\underline{\delta}_\alpha$ (see Fig.1) as

$$\sum_\alpha \left(\Delta A_\alpha^+ \; {}^{\Delta A_\alpha^+}\overline{v_\alpha} - \Delta A_\alpha^- \; {}^{\Delta A_\alpha^-}\overline{v_\alpha} \right) = 0 \tag{1}$$

$$\begin{aligned}
\sum_\alpha \Big\{ & \Delta V \left({}^{\Delta V}\overline{v_\alpha}(t^+) - {}^{\Delta V}\overline{v_\alpha}(t^-) \right) \\
& + \Delta t \sum_\beta \Big[\Delta A_\beta^+ \Big({}^{\Delta t \Delta A_\beta^+}\overline{v_\beta} \; {}^{\Delta t \Delta A_\beta^+}\overline{v_\alpha} + {}^{\Delta t \Delta A_\beta^+}\overline{v_\beta' v_\alpha'} + {}^{\Delta t \Delta A_\beta^+}\overline{p}\, \delta_{\alpha\beta} - {}^{\Delta t \Delta A_\beta^+}\overline{\tau_{\beta\alpha}} \Big) \\
& \qquad - \Delta A_\beta^- \Big({}^{\Delta t \Delta A_\beta^-}\overline{v_\beta} \; {}^{\Delta t \Delta A_\beta^-}\overline{v_\alpha} + {}^{\Delta t \Delta A_\beta^-}\overline{v_\beta' v_\alpha'} + {}^{\Delta t \Delta A_\beta^-}\overline{p}\, \delta_{\alpha\beta} - {}^{\Delta t \Delta A_\beta^-}\overline{\tau_{\beta\alpha}} \Big) \Big] \\
& - \Delta t \; {}^{\Delta t}\overline{(\mathrm{Term})}_\alpha \Big\} \; \underline{\delta}_\alpha^* = \underline{0}
\end{aligned} \tag{2}$$

where

$$\begin{aligned}
(\mathrm{Term})_\alpha = \sum_\beta \Big\{ & \Big[(\xi_\beta^+ - \xi_\beta^*)\, \Delta A_\beta^+ \left. \frac{\partial h_\beta}{h_\alpha \partial \xi_\alpha} \right|_{\xi_1^*,\xi_2^*,\xi_3^*} \Big({}^{\Delta A_\beta^+}\overline{v_\beta v_\beta} - {}^{\Delta A_\beta^+}\overline{\tau_{\beta\beta}} + {}^{\Delta A_\beta^+}\overline{p} \Big) \\
& - (\xi_\beta^- - \xi_\beta^*)\, \Delta A_\beta^- \left. \frac{\partial h_\beta}{h_\alpha \partial \xi_\alpha} \right|_{\xi_1^*,\xi_2^*,\xi_3^*} \Big({}^{\Delta A_\beta^-}\overline{v_\beta v_\beta} - {}^{\Delta A_\beta^-}\overline{\tau_{\beta\beta}} + {}^{\Delta A_\beta^-}\overline{p} \Big) \Big] \\
& - \Big[(\xi_\alpha^+ - \xi_\alpha^*)\, \Delta A_\alpha^+ \left. \frac{\partial h_\alpha}{h_\beta \partial \xi_\beta} \right|_{\xi_1^*,\xi_2^*,\xi_3^*} \Big({}^{\Delta A_\alpha^+}\overline{v_\alpha v_\beta} - {}^{\Delta A_\alpha^+}\overline{\tau_{\alpha\beta}} + {}^{\Delta A_\alpha^+}\overline{p}\, \delta_{\alpha\beta} \Big) \\
& - (\xi_\alpha^- - \xi_\alpha^*)\, \Delta A_\alpha^- \left. \frac{\partial h_\alpha}{h_\beta \partial \xi_\beta} \right|_{\xi_1^*,\xi_2^*,\xi_3^*} \Big({}^{\Delta A_\alpha^-}\overline{v_\alpha v_\beta} - {}^{\Delta A_\alpha^-}\overline{\tau_{\alpha\beta}} + {}^{\Delta A_\alpha^-}\overline{p}\, \delta_{\alpha\beta} \Big) \Big] \Big\}
\end{aligned} \tag{3}$$

$$\tau_{\alpha\beta} = \frac{1}{Re} D_{\alpha\beta} \tag{4}$$

$$D_{\alpha\beta}\, \underline{\delta}_\alpha \underline{\delta}_\beta = \left[\frac{h_\alpha}{h_\beta} \frac{\partial}{\partial \xi_\beta}\left(\frac{v_\alpha}{h_\alpha} \right) + \frac{h_\beta}{h_\alpha} \frac{\partial}{\partial \xi_\alpha}\left(\frac{v_\beta}{h_\beta} \right) + 2\, \delta_{\alpha\beta} \sum_\gamma \frac{v_\gamma}{h_\alpha h_\gamma} \frac{\partial h_\alpha}{\partial \xi_\gamma} \right] \underline{\delta}_\alpha \underline{\delta}_\beta \tag{5}$$

Curvature effects contained in $(\mathrm{Term})_\alpha$ follow from expanding the varying unit vectors in a finite volume in Taylor series relative to the center of the volume and neglecting second order terms. $D_{\alpha\beta}$ is the strain tensor and the h_α are the so-called scale factors [6]. GS mean values are defined as

$$ {}^{\Delta\varphi}\overline{\phi} = \frac{1}{\Delta\varphi} \int_{\Delta\varphi} \phi \, d\varphi \qquad (6) $$

and the SGS fluxes ${}^{\Delta t \Delta A_\alpha}\overline{V_\alpha' V_\beta'}$ arise from the nonlinear convection terms by introducing the decomposition

$$ V_\alpha = {}^{\Delta t \Delta A_\beta}\overline{V_\alpha} + V_\alpha' \qquad (7) $$

In equations (1) - (5) models are needed to determine SGS fluxes and GS quantities since their number is larger than the number of equations.

MODEL FOR THE SGS FLUXES

The SGS fluxes represent the action of the unresolved scales of motion on those that are resolved and are only important if the grid does not resolve the detailed structure of the flow field. For high Reynolds number turbulent flows this is the case and the SGS fluxes must be modeled. We adopt an eddy viscosity concept and relate the SGS fluxes to the GS strain tensor ${}^{\Delta t \Delta A_\alpha}\overline{D_{\alpha\beta}}$ in a way suggested by Schumann [1]:

$$ {}^{\Delta t \Delta A_\alpha}\overline{V_\alpha' V_\beta'} = - {}^{\alpha\beta}\mu_{iso} \left({}^{\Delta t \Delta A_\alpha}\overline{D_{\alpha\beta}} - \langle {}^{\Delta t \Delta A_\alpha}\overline{D_{\alpha\beta}} \rangle \right) - {}^{\alpha\beta}\mu_{inh} \langle {}^{\Delta t \Delta A_\alpha}\overline{D_{\alpha\beta}} \rangle \qquad (8) $$

where

$$ {}^{\alpha\beta}\mu_{iso} = c_2 \left(\Delta A_\alpha \, {}^{\alpha}c_5 \, {}^{\Delta V}\overline{E'_{iso}} \right)^{1/2} \, {}^{\alpha\beta}c \qquad (9) $$

$$ {}^{\Delta V}\overline{E'_{iso}} = \frac{c_2 c_{20}}{c_3} \Delta V^{1/3} \, \mathrm{Min}(c_{31} l_{mix}, \Delta V^{1/3}) \, {}^{\alpha\beta}c \left({}^{\Delta t \Delta A_\alpha}\overline{D_{\alpha\beta}} - \langle {}^{\Delta t \Delta A_\alpha}\overline{D_{\alpha\beta}} \rangle \right)^2 \qquad (10) $$

$$ {}^{\alpha\beta}\mu_{inh} = l_{mix}^2 \left| \langle {}^{\Delta t \Delta A_\alpha}\overline{D_{\alpha\beta}} \rangle \right| \cdot {}^{\alpha\beta}f \qquad (11) $$

$$ {}^{\alpha\beta}f = \delta_{\alpha 3} \delta_{\beta 1} \, \mathrm{Min}\left(1, c_{10} \left(\frac{\Delta A_3 \Delta A_1}{A_{10}^2} \right)^{1/4} \right) \qquad (12) $$

and $\langle (..) \rangle$ denotes an ensemble average. The model accounts for bad resolution, inhomogenities and grid volume anisotropies. The 'isotropic' eddy viscosity ${}^{\alpha\beta}\mu_{iso}$ is determined under assumption of locally isotropic turbulence and validity of the Kolmogorov spectrum [7] for the turbulent kinetic energy. Thus, all appearing coefficients in (9) and (10) can be evaluated [1]. While most of the coefficients (for definitions see [1]) account for geometrical details of the grid and are of order one the coefficient c_2 must be determined such that the energy dissipation due to the eddy viscosity in the

simulated large-scale motion is of the same magnitude as the viscous dissipation in reality. The expression (10) for the SGS energy $^{\Delta V}\overline{E'_{iso}}$ results from its transport equation (see [1]) by equating production and dissipation. The inclusion of the minimum function makes the model suited for large grid volumes, where the mixing length l_{mix} can become smaller than the grid scale $\Delta V^{1/3}$ [5]. This feature has been found important in our calculations, too.
The second part in (8) plays a predominant role in the near-wall region and in regions of the computational domain where only large grid volumes are taken to simulate the flow. Here the 'inhomogeneous' eddy viscosity $^{\alpha\beta}\mu_{inh}$ is derived from common mixing length models. Since for very large grid volumes all turbulent momentum is transported by the SGS motion the damping function $^{\alpha\beta}f$ - in (9) formulated according to Grötzbach [5] only for inhomogenities in the 3-direction - is designed such that in this limiting case relation (8) turns into a Reynolds stress model. The important parameter is A_{10}, the grid surface required to make the SGS fluxes equal to the total turbulent fluxes.

MODELS FOR THE GS QUANTITIES

We assume that the 'measurable' GS quantities are the grid surface mean values of the velocity components $^{\Delta A_\alpha}\overline{v}_\alpha$ appearing in the equation for the mass transport (1) and the pressure mean value $^{\Delta t \Delta V}\overline{p}$ defined for the same grid volume (see Fig.2). In the resulting 'staggered grid' system no approximation is necessary with respect to equation (1). Concerning the α-component of equation (2) a volume ΔV_α is used which is shifted so as to surround the position of the measurable $^{\Delta A_\alpha}\overline{v}_\alpha$ component. Then the unknown GS quantities are related to the measurable ones by physical reasoning. For a Cartesian equidistant grid this results in a 'centered' difference scheme in space and an explicit 'Euler-leapfrog' scheme in time.

COMPUTER PROGRAM

The theory is implemented in a FORTRAN 77 code named FLOWSI. What is FLOWSI able to handle: explicit time advancement of the full time-dependent and three-dimensional volume averaged Navier-Stokes-equations with/without SGS model, box-like flow region described either in Cartesian, cylindrical or natural co-ordinates (Fig.3), non-equidistant grid in one direction, with/without walls, various combinations of velocity, pressure or periodicity boundary conditions, solution of the Poisson-like pressure equation by combinations of Fast Fourier Transformation, Cyclic Reduction and Tridiagonal Matrix Algorithm depending on the boundary conditions, dynamic management of large data sets which allows to store those data that do not fit into the main core memory in an external direct access file.
Typical computing times to update all flow quantities (u,v,w, p,E) for one time step and grid point are less than 0.4 ms on a CDC Cyber 175.

APPLICATION TO TURBULENT FLAT PLATE BOUNDARY LAYER FLOW

We want to study the turbulent flow in a fixed space section of a flat plate boundary layer without pressure gradient. A main aspect hereby is the influence of the computational boundary conditions.
For the presented results the following boundary conditions have been used. While in transverse direction the geometry allows for periodic boundary conditions the ensemble averaged velocity components $\langle {}^{\Delta A\alpha}\overline{v}_\alpha \rangle$ develop in mean stream direction and prohibit a periodicity assumption for the GS components ${}^{\Delta A\alpha}\overline{v}_\alpha$. Since the fluctuating GS values ${}^{\Delta A\alpha}\overline{v}_\alpha'' = {}^{\Delta A\alpha}\overline{v}_\alpha - \langle {}^{\Delta A\alpha}\overline{v}_\alpha \rangle$ at the inflow edge are not known a priori we take computed values from a position far enough downstream, adjust them to the prescribed rms values and superimpose the mean velocity components $\langle {}^{\Delta A\alpha}\overline{v}_\alpha \rangle$. This procedure gives rise to a physically reasonable turbulence structure at the inflow edge and is permitted because of the weak inhomogenity of the mean flow in x-direction and the fact that the turbulent structures arise periodical and have a limited life-time [8]. At the outflow edge the GS quantities are extrapolated. The same conditions are applied to the SGS energy ${}^{\Delta V}\overline{E'_{iso}}$. Instead of resolving the steep gradients of the $\langle u \rangle$-profiles in the viscous sublayer we prescribe at the wall - in addition to the no-slip condition - the wall shear stress in a manner proposed by Schumann [1]. At the free stream edge we assume potential flow: ${}^{\Delta Ax}\overline{u} = \langle {}^{\Delta Ax}\overline{u} \rangle_\infty$, ${}^{\Delta Ay}\overline{v} = 0$, ${}^{\Delta V}\overline{E'_{iso}} = 0$. The normal velocity component ${}^{\Delta Az}\overline{w}$ is extrapolated. Note that for velocity components normal to bounding surfaces the continuity equation requires values on the new time level. The imposed ensemble mean values have been derived from the 1/7-power law [9] and the fluctuating components correspond to the rms values of Klebanoff [10].
The figures show the turbulent flow field after 2000 time steps and a dimensionless problem time of 4. The Reynolds-number $u_\tau \delta/\nu$ is 3240. The flow field has an extension of 4 units in x-direction and of 2 in y- and z-direction. An equally spaced grid of 16x8x16 volumes has been used. All quantities are made dimensionless by means of the boundary layer thickness δ and the friction velocity u_τ at the inflow edge. As initial conditions for the fluctuating velocity components we have taken random Gaussian deviates fulfilling only the rms values and the continuity equation.
The contour-lines and velocity vectors for the fluctuating velocity components show the known quasi-random behaviour (Fig.9,10). The build-up of spatial correlations is clearly seen in comparing figures 4a and 4b. The inclination of the velocity contour lines (Fig.9), the wavy interface (Fig.10) and the marker particle lines (Fig.12) correspond to experimental observations. The peak in the SGS energy near the wall (Fig.11) is a consequence of the high velocity fluctuations. The ensemble averaged turbulence structure agrees fairly well with Klebanoff's data [10] and show that most of the turbulent transport is due to the GS quantities (Fig.6,7,8).

CONCLUSIONS

We have presented a numerical method for LES in curvilinear co-ordinates. The results of a first application to developing flat plate boundary layer flow with a very coarse grid are encouraging. Our further aim is to improve ensemble averaging and boundary conditions, to enlarge the flow domain and to refine the grid. Then the validity of the model will be tested for boundary layers with pressure gradient, with surface roughness and curvature.

ACKNOWLEDGEMENTS

We thank Dr.U.Schumann, Dr.G.Grötzbach and the Kernforschungszentrum Karlsruhe for providing the code TURBIT from where we could extract useful information. This study was sponsored by the Deutsche Forschungsgemeinschaft.

REFERENCES

[1] Schumann,U.: Subgrid scale model for finite difference simulations of turbulent flows in plane channels and annuli. J.Comp.Phys. 18, 376-404, 1975.

[2] Leonard,A.: On the energy cascade in large-eddy simulations of turbulent fluid flows. Adv.Geophys. 18 A, 237-248, 1974.

[3] Deardorff,J.W.: A numerical study of three-dimensional turbulent flow at large Reynolds numbers. J.Fluid Mech. 41, 453-480, 1970.

[4] Moin,P.,Kim,J.: Numerical investigation of turbulent channel flow. J.Fluid Mech. 118, 341-377, 1982.

[5] Grötzbach,G.,Schumann,U.: Direct numerical simulation of turbulent velocity-, pressure-, and temperature-fields in channel flows. Proc. of the Symp. on Turbulent Shear Flows. Penn.State Univ., Apr. 18-20, 1977.

[6] Morse,P.,Feshbach,H.: Methods of Theoretical Physics. New York: McGraw-Hill, 1953.

[7] Hinze,J.O.: Turbulence. New York: McGraw-Hill, 1975.

[8] Cantwell,B.J.: Organized motion in turbulent flow. Ann. Rev. Fluid Mech. 13, 457-515, 1981.

[9] Schlichting,H.: Boundary-Layer Theory. New York: McGraw Hill, 1968.

[10] Klebanoff,P.S.: Characteristics of turbulence in a boundary layer with zero pressure gradient. NACA TR 1247, 1955.

FIGURES

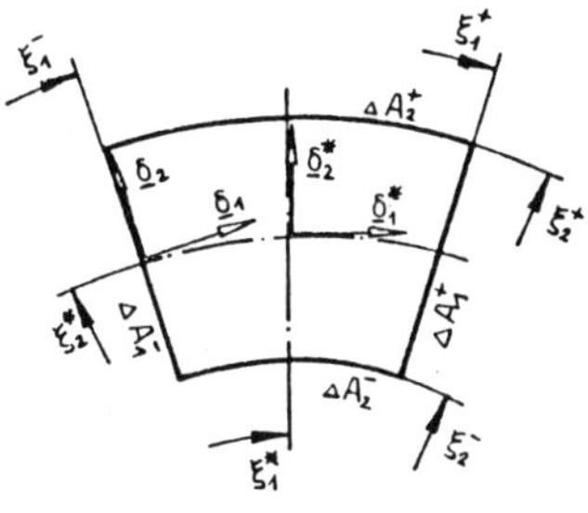

Fig.1
Grid volume

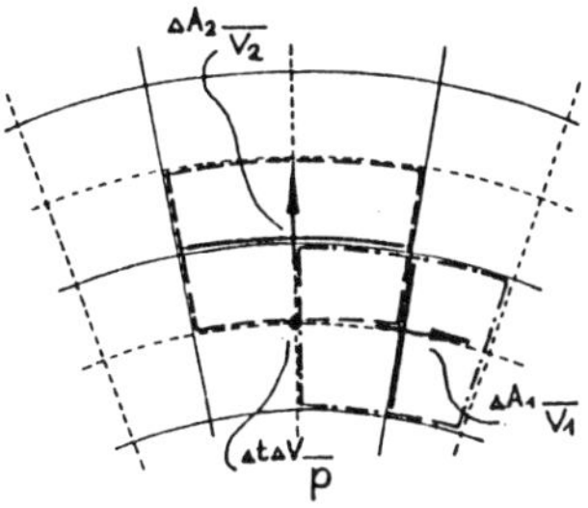

Fig.2
Staggered grid

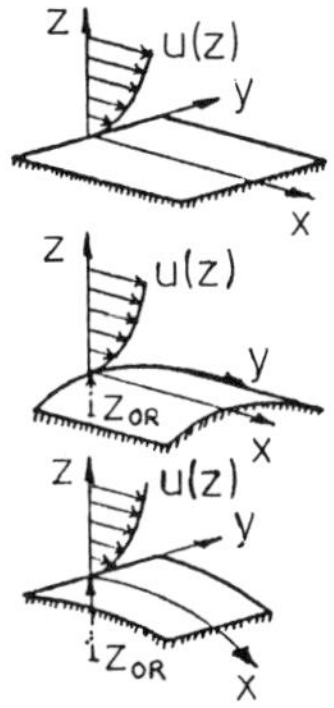

Fig.3
Geometry

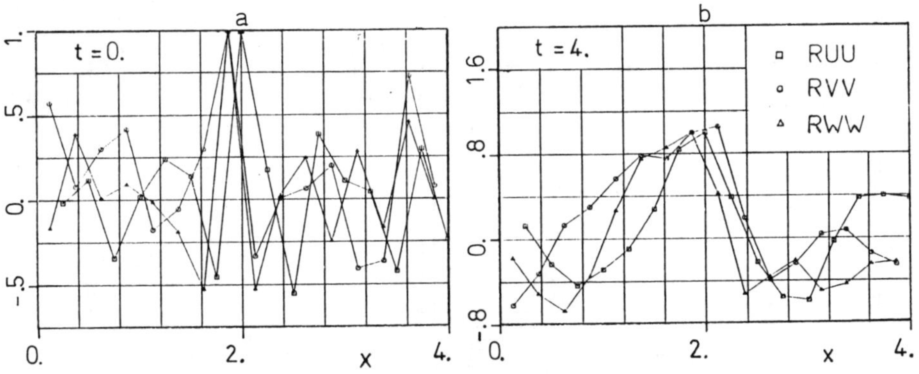

Fig.4 Two-point correlation functions ($x_0=2., z_0=.3$)

$$R V_\alpha V_\alpha = \langle {}^{\Delta A\alpha}\overline{V}''_\alpha (x_0,y,z_0)\ {}^{\Delta A\alpha}\overline{V}''_\alpha (x_0+r_x,y,z_0) \rangle / V^2_{\alpha\, rms}$$

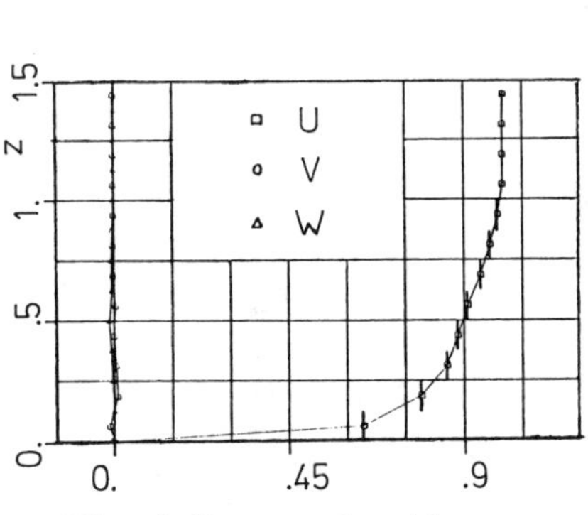

Fig.5 Mean velocity components (x=1.)

$$V_\alpha = \langle {}^{\Delta A\alpha}\overline{V_\alpha} \rangle / u_\infty$$

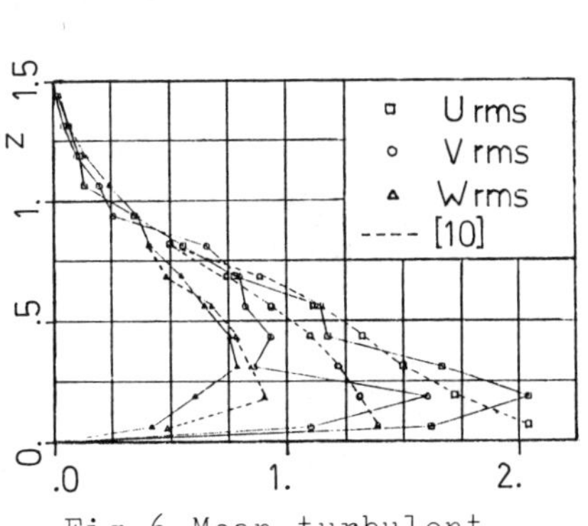

Fig.6 Mean turbulent rms velocity fluctuations (x=1.)

$$V_{\alpha\, rms} = \langle {}^{\Delta A\alpha}\overline{V}''_\alpha\ {}^{\Delta A\alpha}\overline{V}''_\alpha \rangle^{1/2}$$

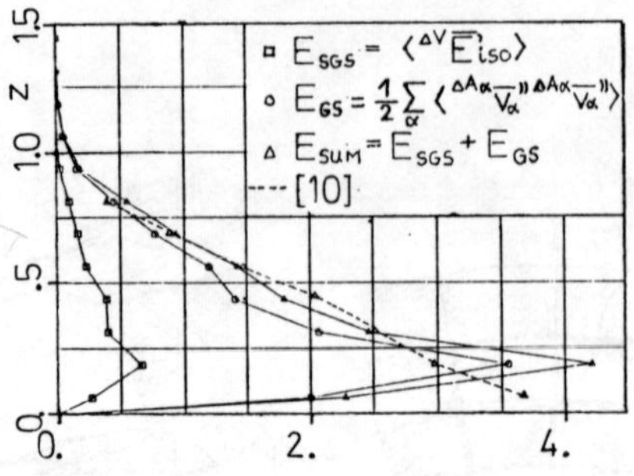

Fig.7 Mean turbulent kinetic energy (x=1.)

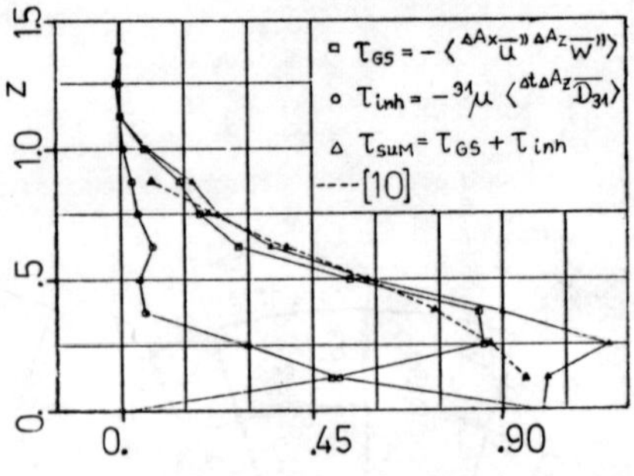

Fig.8 Mean turbulent shear stress (x=1.)

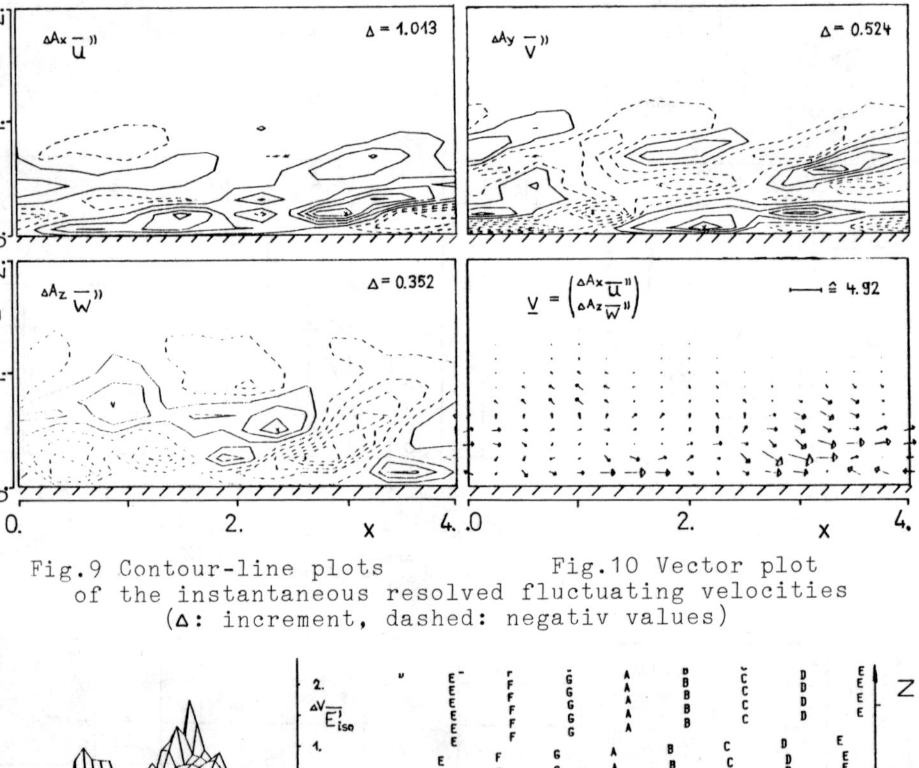

Fig.9 Contour-line plots Fig.10 Vector plot of the instantaneous resolved fluctuating velocities (Δ: increment, dashed: negativ values)

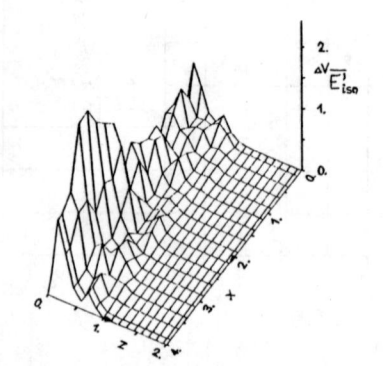

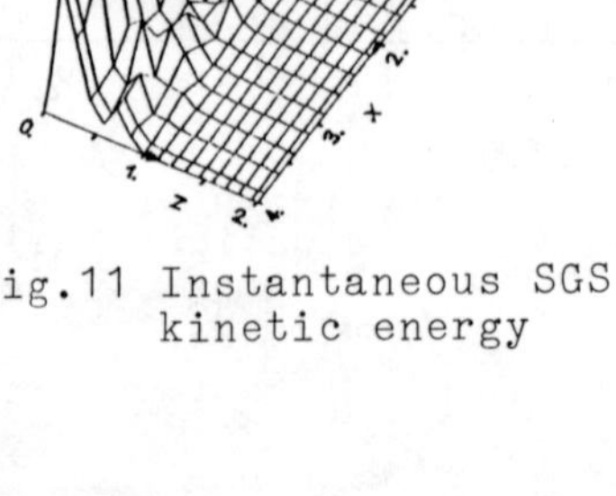

Fig.11 Instantaneous SGS kinetic energy

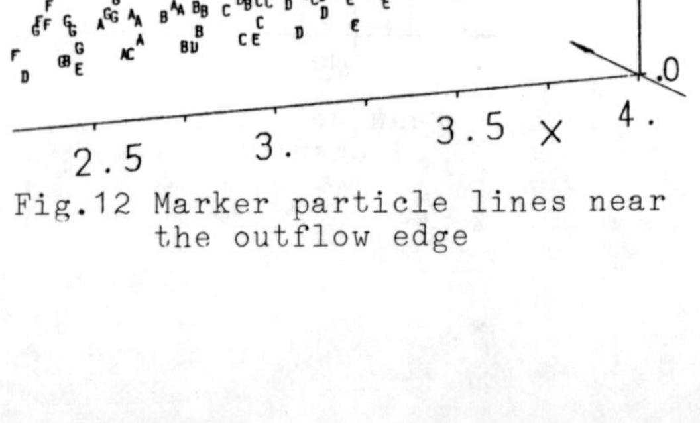

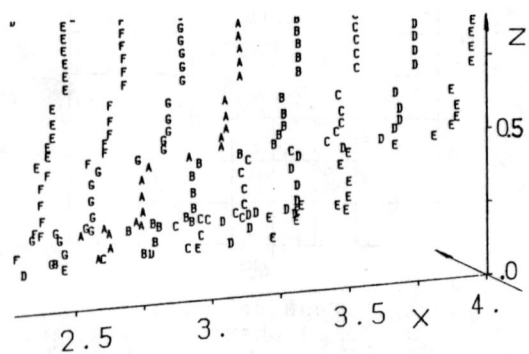

Fig.12 Marker particle lines near the outflow edge

THE CALCULATION OF STREAMLINE-POTENTIALLINE COORDINATES FOR CONFIGURATIONS WHICH ARE GIVEN ONLY BY A SET OF POINTS

W. Schönauer, K. Häfele, K. Raith
Rechenzentrum der Universität,
Postfach 6380, D-7500 Karlsruhe 1

SUMMARY

The 3-D configuration and the inviscid velocity potential ϕ are given by a set of points $P_i(\bar{x}_i,\bar{y}_i,\bar{z}_i,\phi_i)$ from which the streamline-potentialline coordinates are to be computed. From local approximations of the surface F and of the potential ϕ the vector grad ϕ of the streamlines is determined. The existence of dividing streamlines (attachement and detachement lines) forces to compute the streamlines <u>into</u> the front and rear stagnation point. For the determination of the metric coefficients of the streamline coordinates x,y, local approximations $\bar{x}(x,y)$, $\bar{y}(x,y)$, $\bar{z}(x,y)$ are determined. On the dividing streamlines the second metric coefficient h_2 becomes zero, causing a singularity of the transformation.

INTRODUCTION

Streamline-potentialline coordinates are for many 3-D fluid dynamical problems the most natural coordinates, e.g.for 3-D boundary layers or parabolized Navier Stokes equations. But it is not a trivial task to compute the streamlines with high accuracy. In 3-D inviscid flow there are <u>dividing streamlines</u>, see Fig.1, namely diverging or attachement lines on the windward side, see Hirschel, Kordulla [1], but also converging or detachement lines on the leeward side, where the fluid escapes in the normal direction, but there is no separation (not even in 3-D boundary layer calculations). Therefore the naive idea that the streamlines emanate like rays only from the stagnation point is wrong in 3-D. The consequence is that it is not possible to compute the streamlines out of the stagnation point, but that they must be computed <u>into</u> the stagnation point.

In the present paper only a brief sketch of the related problems can be presented. There are a lot of further minor details, but only all these details together make the problem treatable in a robust form. In [2] there is given the full report together with a commented description of the related computer programs, including the solution of the 3-D laminar boundary layer equations.

APPROXIMATION OF SURFACE F AND POTENTIAL ϕ

We assume that the 3-D surface F and the inviscid velocity potential ϕ are given by a set of points $P_i(\bar{x}_i,\bar{y}_i,\bar{z}_i,\phi_i)$, with $\bar{x},\bar{y},\bar{z}$ as basic cartesian coordinates. These points P_i are arranged on panel rings (with equal number of points) and panel lines, see Fig.2, thus we have a structured set. For a local approximation of F we use the quadric

$$F(\bar{x},\bar{y},\bar{z}) = a_1\bar{x}+a_2\bar{y}+a_3\bar{z}+a_4\bar{x}^2+a_5\bar{y}^2+a_6\bar{z}^2+ +a_7\overline{yz}+a_8\overline{xz}+a_9\overline{xy} = 1 \quad . \tag{1}$$

The 9 coefficients a_k are determined from m = 18, 30, 45 or 63 points P_i which are selected that equal information in $\bar{x},\bar{y},\bar{z}$ results. Distant points get lower weight. In Fig.3 the arrangement of the points with their weights is presented for m = 18 points, P_j is the "working point" of the streamline calculation. The resulting equations are transformed to normalized coordinates : $\bar{x}^* = \bar{x}/\bar{x}_{max}$, $\bar{y}^* = \bar{y}/\bar{y}_{max}$, $\bar{z}^* = \bar{z}/\bar{z}_{max}$, where max indicates maximal absolute value in the selected set, and then normalized coefficients, e.g. $a_4^* = a_4\bar{x}^2_{max}$, $a_9^* = a_9 x_{max} y_{max}$, are determined by the least squares method. The resulting linear system could only be solved by the Householder method. From (1) the normal to the surface, grad F, can easily be determined.

For the approximation of the inviscid potential ϕ the naive idea to use $\phi = \phi(\bar{x},\bar{y},\bar{z})$ is wrong, because on the surface $\bar{z} = \bar{z}(\bar{x},\bar{y})$, thus ϕ is only a function of two variables. Therefore we determine in the tangential plane of the working point P_o, see Fig.4, with the use of the projection P_1' of the nearest neighbour P_1, a local coordinate system ξ,η with origin in P_o and unit vectors $\mathfrak{w}_1, \mathfrak{w}_2$. Then we use the 5th order polynomial

$$\phi(\xi,\eta) = b_1+b_2\xi+b_3\xi^2+b_4\eta+b_5\eta^2+\ldots.+b_{21}\eta^5 \quad . \tag{2}$$

The coefficients b_k are computed from selected points which are projected onto the tangential plane, but excluding points of the "shadow side" of the surface. We use weighting, normalization of coordinates and least squares method as for the determination of F (1). Then the direction of the streamline is given (with $\xi = \eta = 0$ in P_o, and $\mathfrak{w}_1, \mathfrak{w}_2$ as vectors in $\bar{x},\bar{y},\bar{z}$)

$$\text{grad}\,\phi = b_2\mathfrak{w}_1+b_4\mathfrak{w}_2, \quad \mathfrak{w}_\psi = \text{grad}\,\phi/|\text{grad}\,\phi| \quad . \tag{3}$$

COMPUTATION OF THE STREAMLINES

We start the streamline calculation at the panel points of the central panel ring, computing into the front and rear stagnation point, see Fig.5. Because of surface curvature the streamline leaves the surface, see Fig.6, and the point A" must be projected back onto the surface, point A'. The curvature of the streamline causes a drift from one streamline to the other, see Fig.7. Therefore we use a second order method, following the mean vector $\mathfrak{w}_{\psi m}$, see Fig.7.

The determination of the step size Δs is an important part of the algorithm and is presented here only verbally, the formulae are given in [2]. A maximal step size Δs_M is determined by the maximum of the mean local distance of the panel points and a given fraction of the maximal diameter of the body, a minimal step size Δs_{min} is a given fraction of Δs_M. A maximal admitted length ℓ of the normal in Fig.6 and a maximal admitted angle α in Fig. 7 determine step sizes Δs_F and Δs_W. Then the new step size is

$$\Delta s = \max\,(\Delta s_{min},\ \min(\Delta s_M, \Delta s_F, \Delta s_W))\ , \tag{4}$$

thus $\Delta s_{min} \leq \Delta s \leq \Delta s_M$. If $\Delta s < \Delta s_{old}/2$ the old step is dropped and repeated with Δs.

The naive idea that the potential decreases in computing towards the front stagnation point may not hold if a streamline abruptly bends into a type of stagnation line at an airfoil-like configuration. Therefore a sophisticated check of the potential of A^* and A', see Fig.7, is made before accepting a point and before stopping near the stagnation point. Finally the two halfes of a streamline are stored as one sequence of points. As can be seen in Fig.5 many streamlines contract into a dividing streamline, leaving regions unconvered. Therefore by a complicated logic at four selected panel rings a check can be made if the streamlines are dense enough (compared to the density of the panel points) and further streamlines may be inserted. Finally the streamline coordinates x,y are determined by

$$\begin{aligned} x &= \phi - \phi_o\ , \quad \phi_o = \text{potential of front stagn.pt.}, \\ y &= 2\pi(\psi-1)/Z\ , \quad 0 \leq y \leq 2\pi\ , \end{aligned} \tag{5}$$

with ψ as actual number of the streamline and Z as total number of the streamlines. Thus the potential, normalized to zero in the front stagnation point, determines the x-scale and the numbering of the streamlines determines the y-scale.

It is worthwhile to mention here a suggestion by Hirschel [3], who proposes to present the primary information instead directly by $P_i(\overline{x}_i, \overline{y}_i, \overline{z}_i, \phi_i)$ now indirectly as functions of surface coordinates ξ, η, e.g. as $\overline{x}(\xi,\eta), \ldots$ or pointwise as $P_i(\overline{x}(\xi_i,\eta_i), \ldots)$ and to compute the streamlines in the ξ,η system. Then no projection back onto the surface F or the interpolation of ϕ in the tangential plane is needed which would save much computer time. But then there are the problems with the geometric singularities in ξ,η and possibly the representation of complicated geometries, and the method looses generality. But nevertheless it is an interesting alternative.

COMPUTATION OF THE METRIC COEFFICIENTS

Now the whole information about the configuration is stored in the streamlines, which constitute an ordered set of points $Q_i(x_i, y_i, \overline{x}_i, \overline{y}_i, \overline{z}_i)$. For the solution, e.g. of the 3-D boundary layer equations, the metric coefficients

$$h_1^2 = \overline{x}_x^2 + \overline{y}_x^2 + \overline{z}_x^2\ , \quad h_2^2 = \overline{x}_y^2 + \overline{y}_y^2 + \overline{z}_y^2 \tag{6}$$

and their derivatives are needed. Therefore we use the cubic ansatz

$$\overline{x}(x,y) = c_o + c_1 x + c_2 y + c_3 x^2 + \ldots + c_9 y^3 \tag{7}$$

and similar forms for $\overline{y}(x,y)$, $\overline{z}(x,y)$. But near the stagnation point there is a squareroot singularity and we use

$$\bar{x}(x,y) = \bar{x}_o + (f_{1,o}+f_{1,1}y+f_{1,2}y^2+f_{1,3}y^3)\sqrt{x} + + (f_{2,o}+..+f_{2,3}y^3)x + (f_{3,o}+...+f_{3,3}y^3)x^{3/2}, \quad (8)$$

with x_o as coordinate of the stagnation point, and similar expansions for $\bar{y},\bar{z}$. The coefficients of (7),(8) are computed for a selected number of the streamline points, using neighbouring points, with weighting etc as for the coefficients of (1). The final result then are these stored coefficients of the expansions (7),(8), from which (6) can be computed.

In a working point, e.g. of a 3-D boundary layer calculation, see Fig.8, one is tempted to use the nearest stored expansion, but this would result in a jump of the derivatives $\bar{x}_x$,... if the expansion changes. Therefore we recommend to take the values of the two nearest expansions and to interpolate linearly between the values, see lower part of Fig.8.

THE SINGULARITY OF THE DIVIDING STREAMLINE

The coincidence of several streamlines into the dividing streamline, see Fig.1, has the consequence that on the dividing streamline all y-derivatives vanish: $\bar{x}_y = \bar{y}_y = \bar{z}_y = 0$, similarly for the dependent variables. This induces for the second metric coefficient, see (6), $h_2 = 0$, which means that on the dividing streamlines the transformation degenerates. This singularity of the streamline coordinates is obviously not well known and represents the most serious drawback of the streamline coordinates. There e.g. the 3-D boundary layer equations degenerate to 2-D equations which are not explicitly known for general 3-D configurations.

As a useful criterion where there is a dividing streamline we use that $h_2 < 0.1\, h_{2,max}$, where $h_{2,max}$ is the maximum value over all streamlines at the same position x. All criteria based on a distance proved to be useless.

NUMERICAL RESULTS

The most serious problem in the development of a rather general program code was the problem of robustness.The program should work for quite different geometries. Therefore we used general ellipsoids of fuselage-like or airfoil-like shape, for which we generated "artificial" panel data from Lamb's solution [4]. Fig.9 and 10 show the computed streamlines for a prolate spheroid and an airfoil-like general ellipsoid. In Fig. 9 from each stagnation point a single dividing streamline emanates, while in Fig.10 from the stagnation points dividing streamlines emanate in both directions as a stagnation line.

ACKNOWLEDGEMENT

This research has been supported by the Stiftung Volkswagenwerk.

REFERENCES

[1] Hirschel,E.H., Kordulla,W., "Shear Flow in Surface Oriented Coordinates", Vieweg (1981).

[2] Schönauer,W., Glotz,G., Däubler,H.-G., Raith,K., Häfele,K., "Die Entwicklung eines selbststeuernden fehlerüberwachten Differenzenverfahrens zur Lösung der dreidimensionalen laminaren inkompressiblen Grenzschichtgleichungen bei punktweise numerisch gegebener Kontur", Interner Bericht Nr. 24/83 des Rechenzentrums der Universität Karlsruhe (1983).

[3] Hirschel,E.H., MBB Ottobrunn, private communication (1983).

[4] Lamb,H., "Lehrbuch der Hydrodynamik", Teubner (1931).

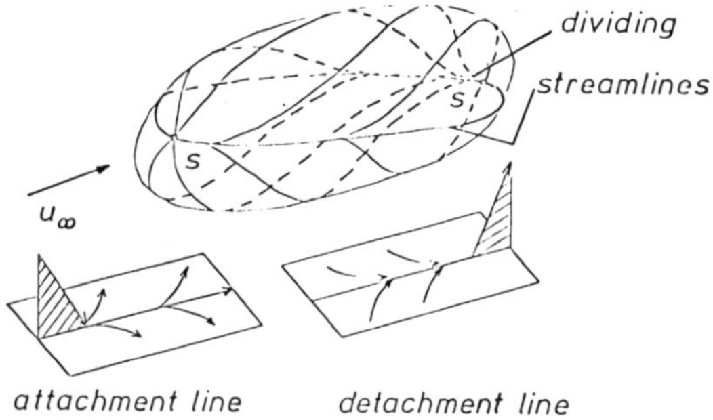

Fig.1 Definition of dividing streamlines.

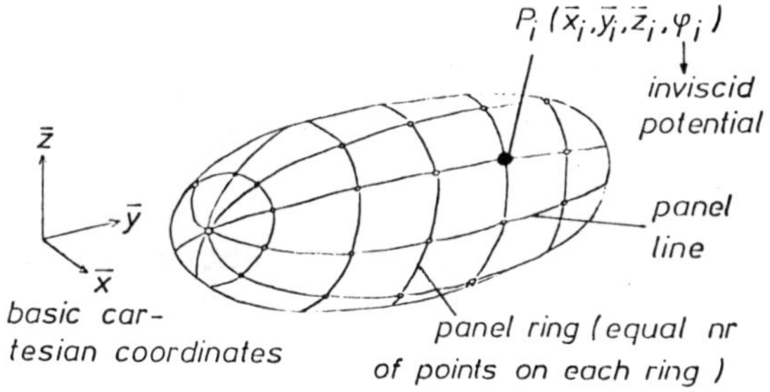

Fig.2 Structure of the given set of points P_i.

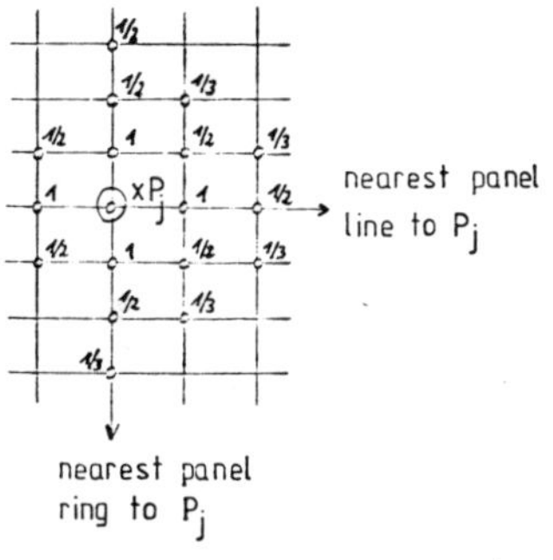

Fig.3 Distribution of points and weights for m = 18. $P_j(x)$ is the working point.

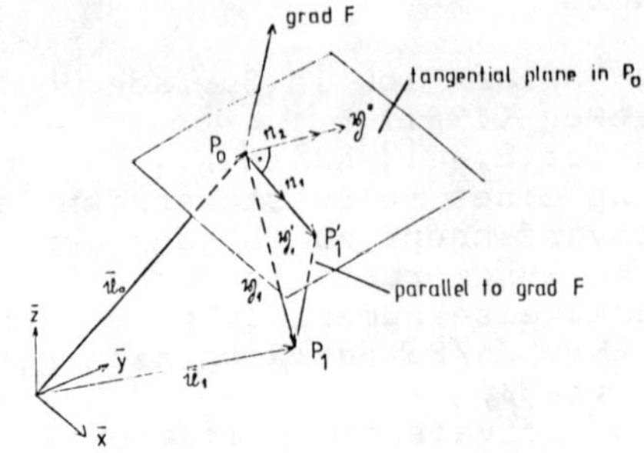

Fig.4 Local coordinates in tangential plane for P_o.

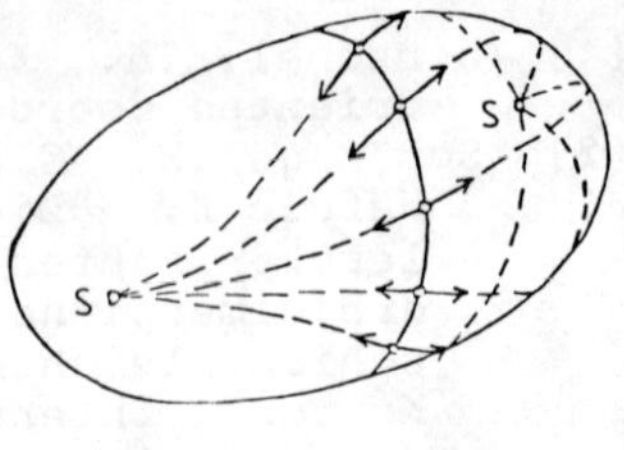

Fig.5 Starting points for the streamline calculation into front and rear stagnation point.

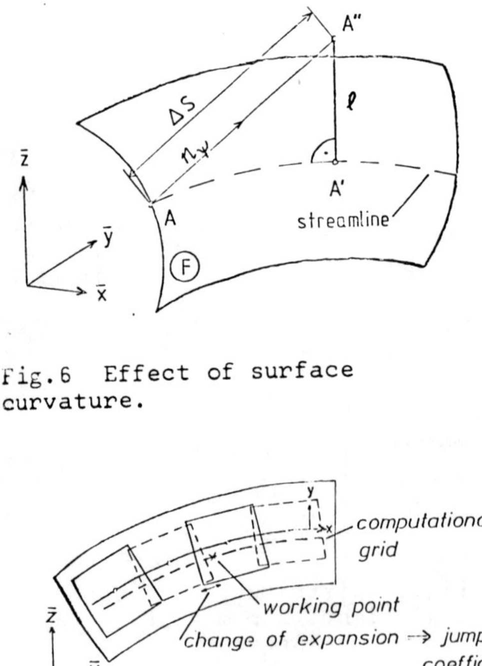

Fig.6 Effect of surface curvature.

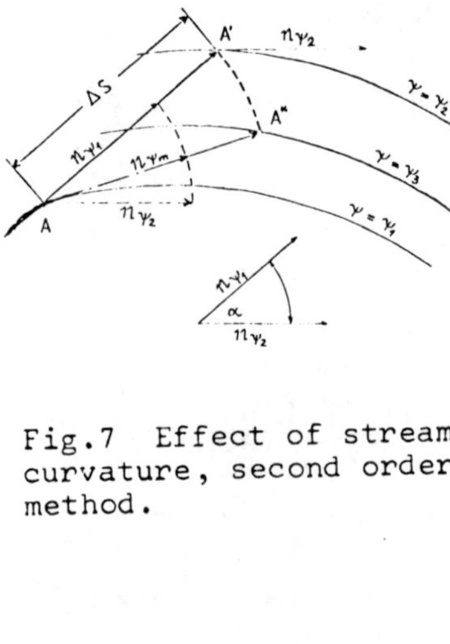

Fig.7 Effect of streamline curvature, second order method.

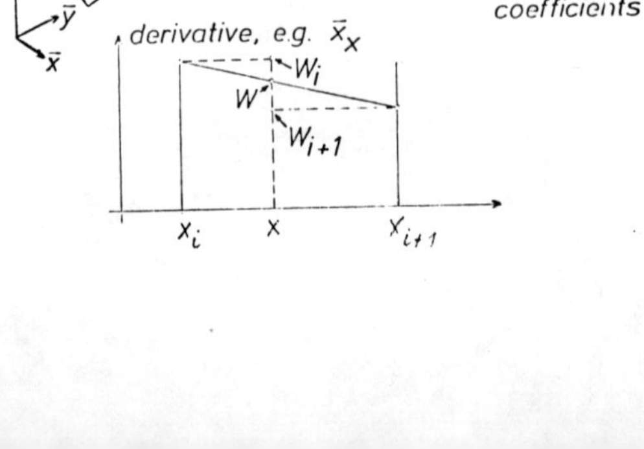

Fig.8 Generation of continuous derivatives by interpolation between neighbouring expansions.

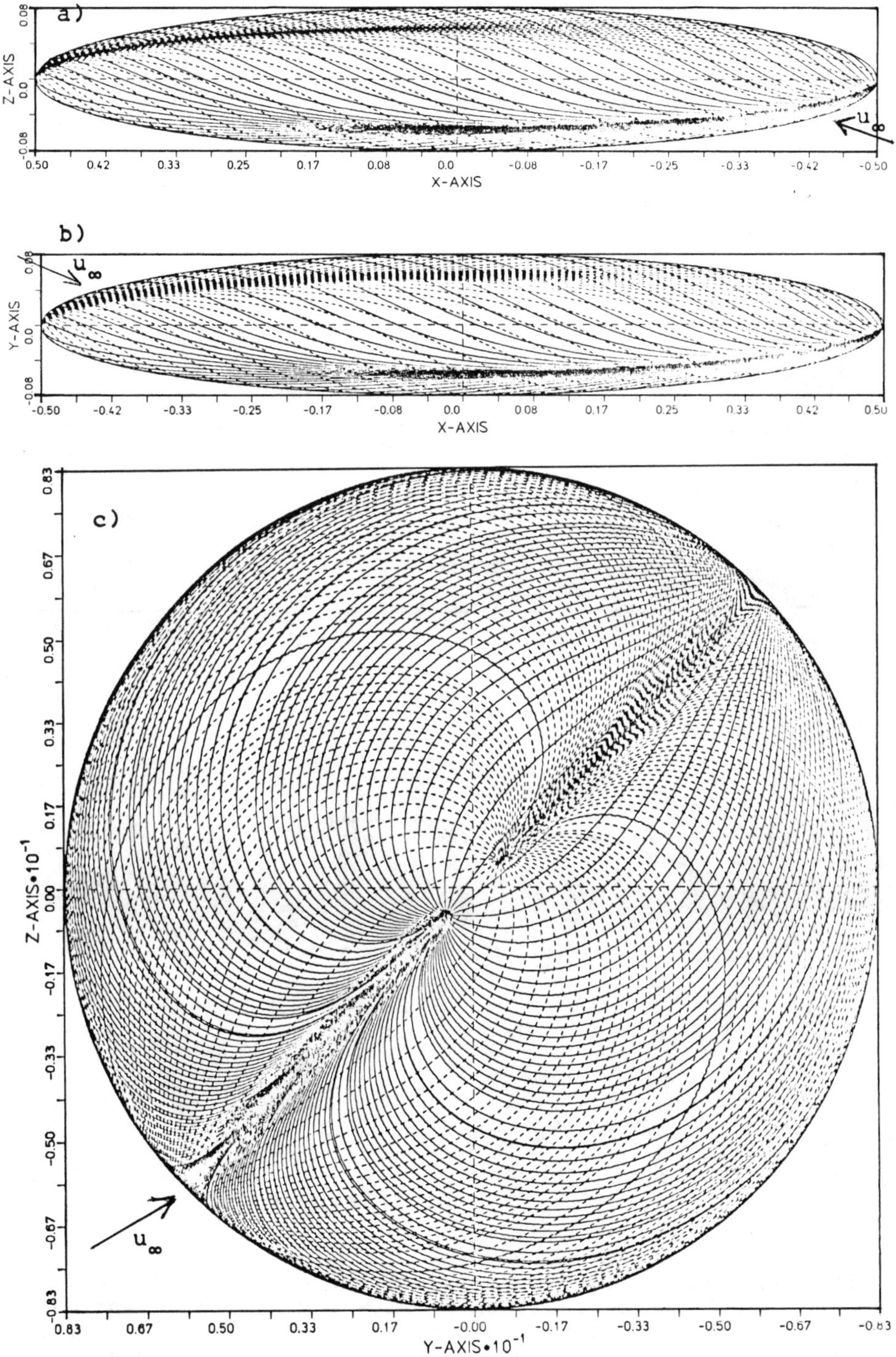

Fig.9 Inviscid streamlines (continued on next page)

of 6:1 prolate spheroid with 12.53° angle of attack in x,y- and x,z-plane and with 45° in y,z-plane (17.44° effective angle of attack). a) side view, b) top view (only 1/3 of the computed streamlines are plotted), c) blow-up of front view. The dividing streamlines are clearly visible.

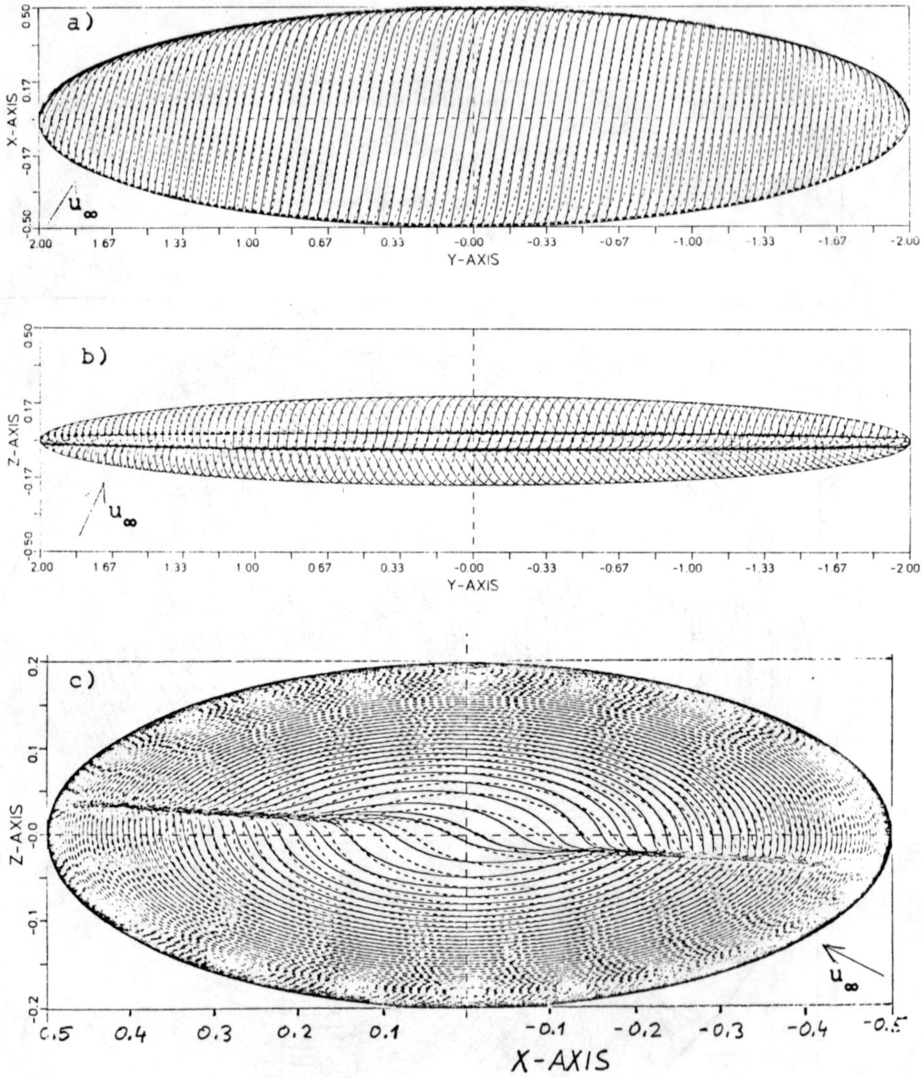

Fig.10 Inviscid streamlines for general 1/4:1:1/10 ellipsoid with 12.53° angle of attack in x,y- and x,z-plane and with 45° in y,z-plane. a) top view, b) front view, c) blow-up of side view. Dividing streamlines run around the whole airfoil-like shape.

BOUNDARY LAYERS ON WINGS

D. Schwamborn
DFVLR, Institute for Theoretical Fluid Mechanics
Bunsenstrasse 10, D-3400 Göttingen, West Germany

SUMMARY

A second-order accurate finite-difference method for the calculation of three-dimensional boundary layers on winglike bodies has been developed by the author such that the region of the attachment line is considered properly [1,2]. The method is now extended to compressible flow and applied to realistic wing configurations with pointwise given body and inviscid outer flow field. Initial data for the 3-D calculation are obtained by the "locally infinite swept wing" concept in a chordwise section near the root of the wing, and the accuracy of this concept is discussed. Furthermore, a new method is introduced to reduce oscillations in the numerical solution which are known to occur for Crank-Nicolson-like schemes if large gradients are present. This method markedly reduces the oscillations around the solution without loosing the second-order accuracy. Results for a sample calculation of the compressible laminar boundary layer on a realistic wing are presented.

INTRODUCTION

The initial data needed for the chordwise integration of the boundary layer equations for the flow past wings are often estimated at a position close to the leading edge. The error resulting from these initial data is well known to decay with the chordwise distance from the starting position, since the outer flow is accelerated in the region of the leading edge. There are some reasons, however, that make it desirable to have as accurate a solution as possible in the vicinity of the leading edge. We need such a solution e.g. in investigations regarding laminar flow control. Since there is a reduction in drag if the boundary layer is kept laminar just on a part of the wing surface only, the questions arise whether the boundary layer flow in the region of the attachment line and beyond is stable or not and where laminar separation takes place. Only if we can answer such questions there is a chance of controlling the boundary layer flow and for that purpose one needs accurate results.

On the other hand we need an exact method for comparison with methods that yield approximate solutions. Such an exact method for the calculation of the three-dimensional incompressible boundary layer in the region of the attachment line of winglike bodies has been developed by SCHWAMBORN [1,2]. This second-order finite difference method is extended to compressible flow and applied to realistic wing configurations where the body contour and the inviscid flow field may now be given pointwise only.

BOUNDARY LAYER EQUATION AND SOLUTION PROCEDURE

For the boundary layer calculation on a wing a surface-oriented coordinate system is used, consisting of lines of constant chord (x^1=const.) and lines of constant span (x^2=const.), with the third coordinate x^3 perpendicular to the surface. The boundary layer equations for steady compressible flow in contravariant formulation [3] then read:

Continuity equation:

$$(k_{01}\rho v^1)_{,1} + (k_{01}\rho v^2)_{,2} + (k_{01}\rho v^3)_{,3} = 0 \quad ,$$

momentum equation for the x^α direction ($\alpha = 1,2$):

$$\rho(v^1 v^\alpha_{,1} + v^2 v^\alpha_{,2} + v^3 v^\alpha_{,3} + k_{\alpha 1}(v^1)^2 + k_{\alpha 2} v^1 v^2 + k_{\alpha 3}(v^2)^2) =$$

$$k_{\alpha 4} p_{,1} + k_{\alpha 5} p_{,2} + (\mu v^\alpha_{,3})_{,3} \quad ,$$

energy equation

$$c_p \rho(v^1 T_{,1} + v^2 T_{,2} + v^3 T_{,3}) = Pr^{-1}_{ref}(kT_{,3})_{,3} +$$

$$+ E_{ref}(v^1 p_{,1} + v^2 p_{,2} + \mu(k_{41}(v^1_{,3})^2 + k_{42} v^1_{,3} v^2_{,3} + k_{43}(v^2_{,3})^2))$$

where indices after a comma denote partial differentiation with respect to the corresponding coordinate.

All quantities in these equations are nondimensionalized with appropriate free stream reference values. Note, that for the pressure $(\rho u^2)_\infty$ is used as reference value. The coordinate and the velocity component normal to the wall are stretched by multiplication with the square root of the reference Reynolds number. The contravariant velocities v^α ($\alpha = 1,2$) are related to the physical ones by $\overset{*}{v}^\alpha = \sqrt{a_{\alpha\alpha}} v^\alpha$, where $a_{\alpha\alpha}$ are the covariant elements of the metric tensor of the surface, which are also used in the computation of the metric properties k_{ij} [3].

The set of equations needed for the solution is complemented with the equation of state and the laws of viscocity and thermal conductivity. The boundary conditions at the wall are the no-slip condition, impermeability of the wall and given wall temperature or wall temperature gradient. The boundary conditions for v^1, v^2 and T at the outer edge of the boundary layer are taken from the inviscid flow. All metric properties needed for the calculation are obtained by a subroutine which also transforms the Cartesian velocity components of the inviscid outer flow to the surface-oriented coordinate system. The input for this routine is the pointwise given contour of the wing specified at several span stations as well as the velocity components in certain contour points, all in cartesian coordinates.

To start the calculation initial data are needed for a wing section of constant span close to the root of the wing. These data are obtained by a quasi-twodimensional solution using the so-called "locally infinite swept wing" concept, where it is assumed that the derivatives of the physical flow variables ($\overset{*}{v}^\alpha$, not v^α) in x^2-direction vanish locally along a line $x^2 = \text{constant}$ [3]. The spanwise variation of the metric, however, is taken into account. The "locally infinite swept wing" (LISW) concept is believed to give a relatively good approximation of the threedimensional boundary layer when the isobars nearly coincide with lines of constant chord.

The boundary layer equations for the 3D case as well as for the LISW-approximation are solved by second-order accurate finite difference methods based on [4,5]. Starting with the data of the LISW solution the solution then proceeds at first in spanwise direction in a relatively small domain along the attachment line. This guarantees that there is only outflow at both chordwise boundaries of the attachment line domain and a positive spanwise velocity v^2 as well, i.e. we have an appropriate domain of dependence such that marching in spanwise direction is possible [1,2]. After the solution in this domain is obtained the calculation is continued from this region in both chordwise directions until separation occurs.

A REMARK ON THE ACCURACY OF THE LOCALLY INFINITE SWEPT WING CONCEPT

As mentioned earlier an exact solution for the 3D boundary layer equations including the attachment line region can be used to judge the quality of approximate methods. In the present method we used such an approximation (the LISW solution) to compute initial data in a chordwise section of the wing. To justify this several 3D boundary layer calculations were made for the same wing, but each starting with the LISW solution at a different span station. Then a comparison of the solutions at several chord- and spanwise locations was made, with the calculation which started at 5 % halfspan.

The result was that there is only a little difference between the LISW-solution and the exact solution at most positions. The largest differences between LISW and exact solution were found close to the attachment line, where the difference in shear stress was 5 % and more. This is at least partially due to an incompatibility in the LISW solution at the attachment line [3]. Farther away from the attachment line the differences dropped below 1 %. Furthermore, a few spanwise steps with the exact method were sufficient to bring the differences between solutions started at different spanwise positions down to the order of the accuracy of the scheme.

From these results the LISW concept can be regarded as a fairly accurate approximation method, at least in the present case.

AN IMPROVEMENT FOR BOUNDARY LAYER SOLUTIONS

It is well known that the use of the Crank-Nicolson (C-N) scheme may lead to the so-called Crank-Nicolson noise [6], i.e. an oscillation of the numerical solution about the exact solution. This effect is also observed for all C-N-like schemes used in boundary layer calculations, i.e. all difference schemes which can be considered as 3-D extensions of the C-N-scheme, e.g. the rectangular scheme or the zig-zag scheme of KRAUSE [5]. As an example we present some results of 3-D boundary layer calculations for the winglike spheroid, already discussed in [1,2].

In Figure 1 the wall shear stress on the upper surface of the spheroid at 5 ° incidence is given for stepsizes $\Delta x^1 = \Delta x^2 = 0.03$ and $x^3 = 0.02$. Due to the oscillation the calculation breaks down at $x^1 = 0.9$ ($x^1 = 0$ is the leading edge; $x^1 = \pi$ the trailing edge) where the shear stress component τ^1 becomes zero. Reduction of the step size Δx^1 to 0.02 results in a smaller amplitude of the oscillations and breakdown occurs at $x^1 = 1.8$. The "separation" line for this case is shown dashed in figure 3, where the viscous and inviscid streamlines are presented. Since reducing the stepsize results in higher computation time, we look for another method to reduce the oscillations.

Consider a "boundary layer" like PDE

$$a u_x + b u_y - c u_{yy} = 0 \qquad \text{with } a,b,c = f(x,y,u) \geqq 0.$$

Discretisation with the C-N scheme and von Neumann stability analysis yield an amplification factor

$$\xi = \frac{1-\alpha+i\gamma}{1+\alpha+i\gamma}$$

with

$$\alpha = \frac{c\Delta x}{a\Delta y^2}\,(1-\cos\beta\Delta y)\quad(\geqq 0)$$

$$\gamma = \frac{b\Delta x}{a\Delta y}\,\sin\beta\Delta y \qquad ;\ \beta \text{ arbitrary, but real.}$$

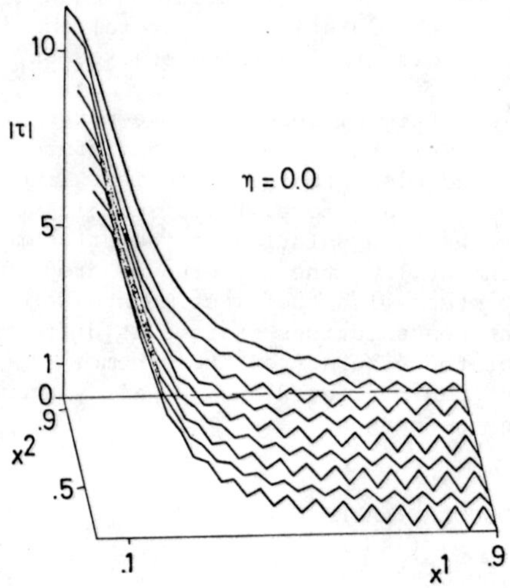

Fig. 1

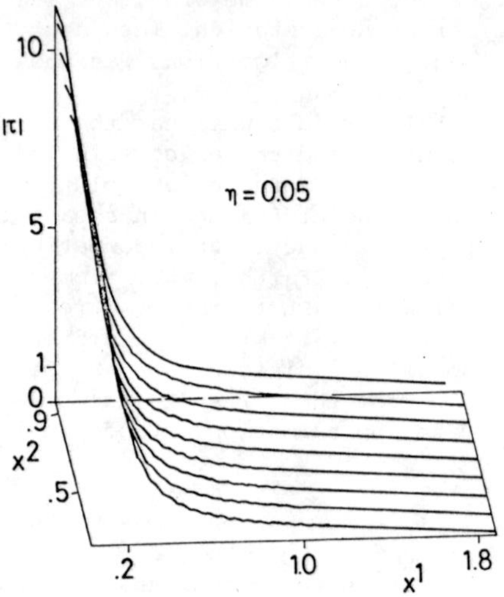

Fig. 2

The necessary stability criterion $|\xi| \leqq 1$ is fulfilled for all $\alpha(\geqq 0)$ and γ. If we consider, however, the case $\alpha \to \infty$ it follows $\xi \to -1$, i.e. disturbances of the solution are no longer damped out, but "conserved" as oscillations around the solution. Now α becomes arbitrary large only if a tends to zero, which corresponds to the velocity decreasing to zero at the wall in boundary layers. Large values of $\Delta x/\Delta y^2$, often used in C-N discretisation because of the second-order accuracy, support the tendency towards oscillations.

We therefore recommend a small modification in the C-N discretisation which only affects the treatment of the diffusion term by forward weighting $(\eta > 0)$:

$$u_{yy} = \frac{1}{2}((u_{yy})_{m+1}(1+\eta) + (u_{yy})_m(1-\eta)) .$$

and with $\alpha \to \infty$:

$$\xi \to \frac{1-\eta}{1+\eta}$$

so that marginal stability is avoided for $\eta > 0$.

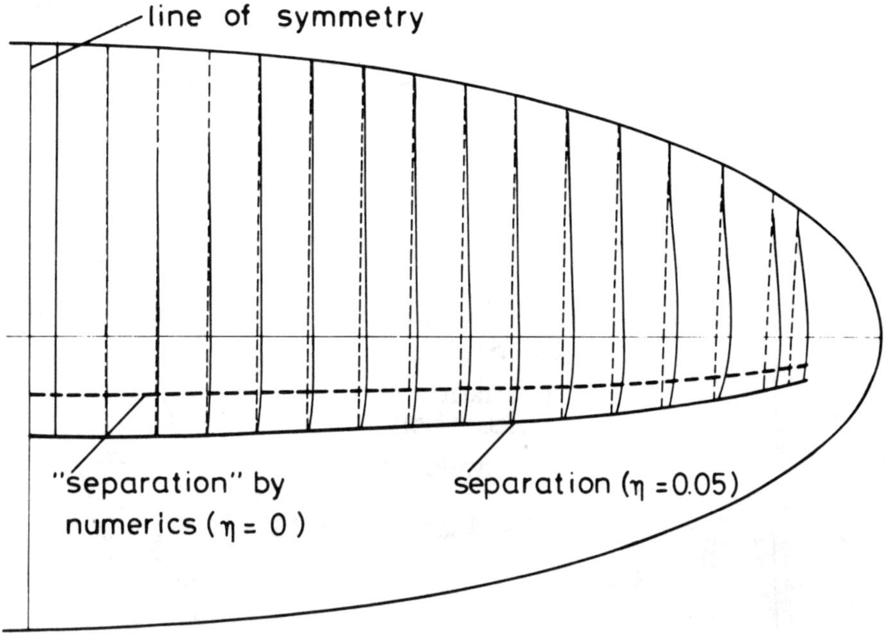

Fig. 3: Streamlines on upper surface of winglike spheroid

The weighting of the second derivative also influences the truncation error of the C-N scheme, which is now $O(\Delta x^2 + \Delta y^2 + \eta\Delta x)$, i.e. the scheme is still of the same order of accuracy, if we choose $\eta \approx \Delta x$. Different tests showed that an η of this order of magnitude is sufficient in most cases to get a solution with no or every few oscillations that otherwise would have been found only by reducing the stepsize to a third or less. In figure 2 a result for the same calculation as in figure 1 is shown, this time using a weighting with $\eta = 0.05$. The separation line for this case is also given in figure 3.

SAMPLE CALCULATION

In the following results for a calculation of the laminar boundary layer on a realistic wing configuration *) are discussed. The Machnumber is M=0.2 and the Reynolds number is about 20 million, i.e. in general, the flow will become turbulent at a very short distance from the leading edge. Although an appropiate turbulence model can be incorporated in the boundary layer code this has not been done yet, since the laminar solution, especially near the attachment line shall serve for future investigations regarding laminar flow control.

In figure 4 a view at the wing from the symmetry plane is given, showing the profiles at 0, 40 and 100 percent halfspan. The wing planform and a front view are presented in figure 5, where also the wall streamlines of the outer inviscid flow (dashed) and of the viscous flow (proper-

*) The data of the geometry and the inviscid outer flow have been friendly provided by Prof. Thiede, MBB.

Fig. 4

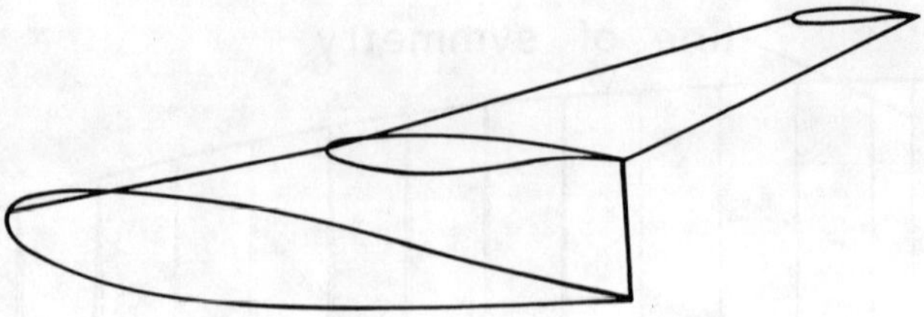

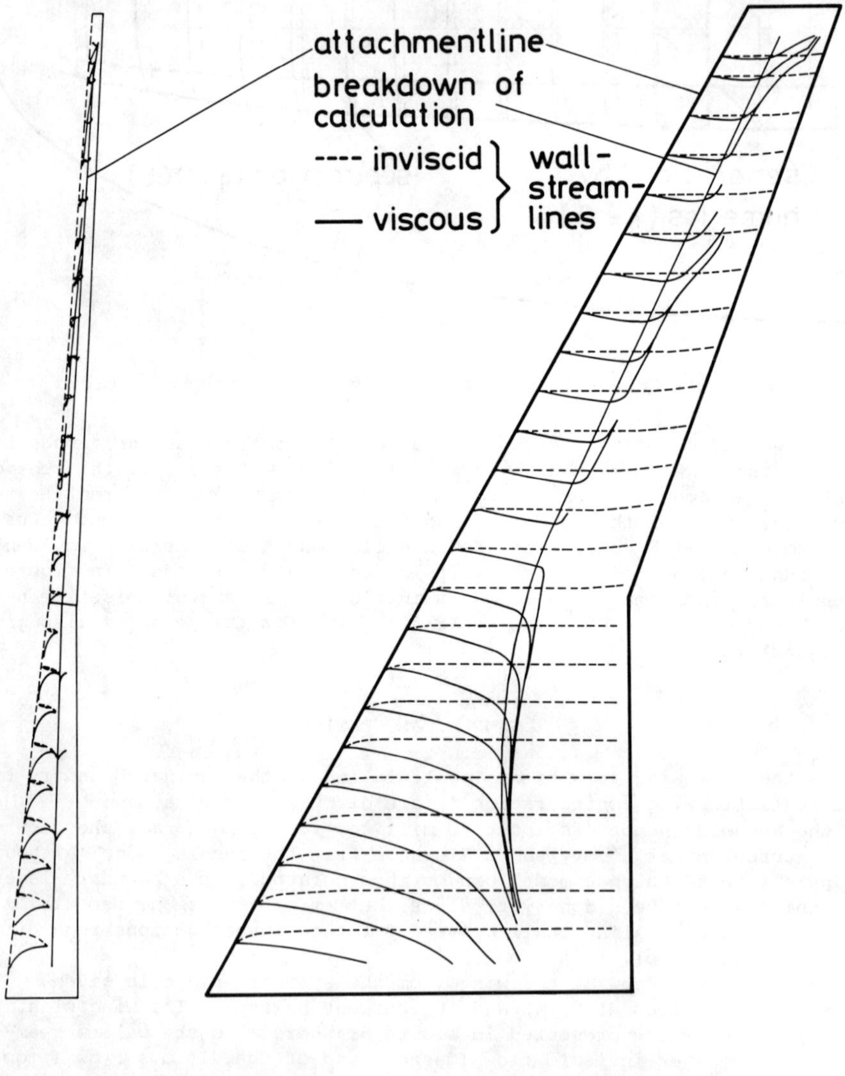

Fig. 5: Streamlines on the lower surface of the wing

ly: wall shear stress lines) at the lower surface of the wing are shown. The breakdown of the calculation, which is assumed whenever the solution does not convergence within 14 iterations, is indicated, too. This "breakdown" does not coincide with separation since the solution can be proceeded a few steps farther by allowing for more iterations and using smaller stepsize there. From an extrapolation of the shear stress field, however, it is possible to integrate the wall streamlines beyond the line of breakdown and it is obvious that laminar separation takes place shortly behind that line.

Results for the upper surface beyond the $x^1 = 0.01$ are not shown here, because the computation breaks down between $x^1 = 0.03$ and 0.05, due to the adverse pressure gradient behind the suction peak (fig. 6). Here a laminar separation bubble with turbulent reattachment could exist.

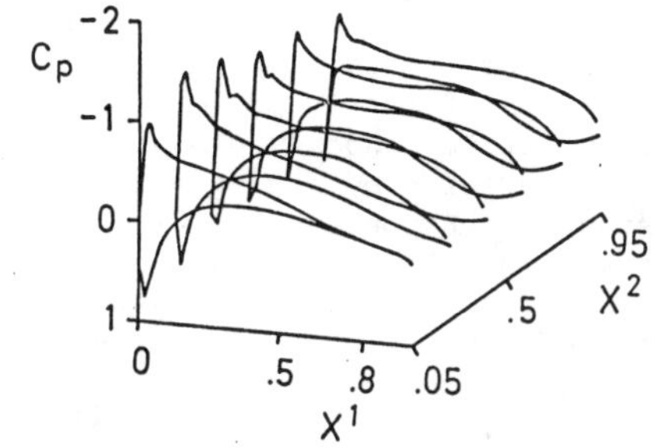

Fig. 6: Pressure distribution on the wing

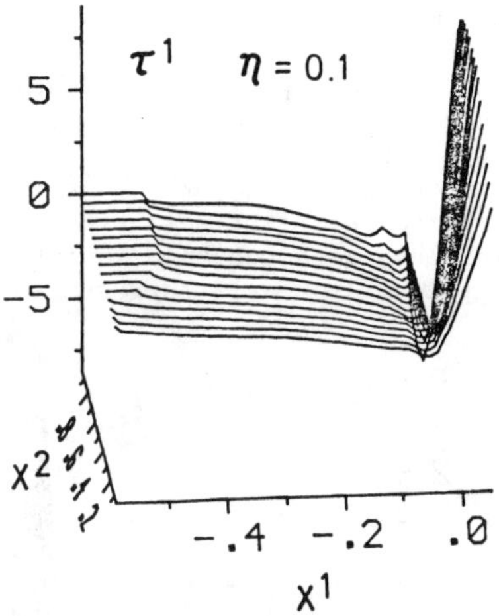

Fig. 7

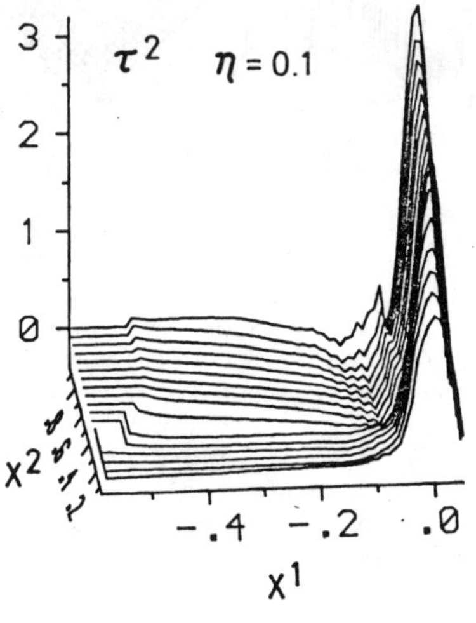

Fig. 8

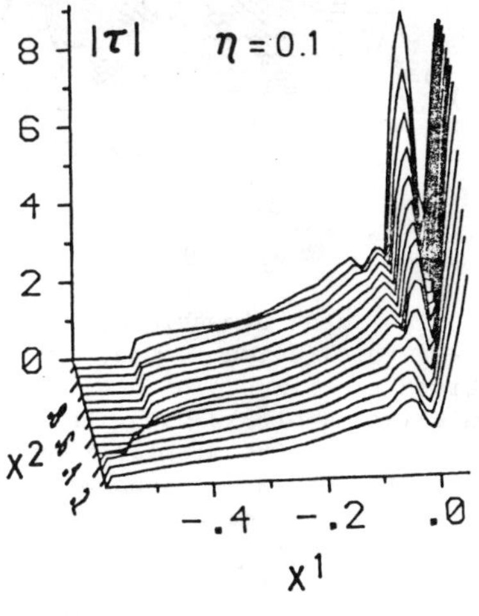

Fig. 9

In figures 7, 8 and 9 the components and the absolute value of the wall shear stress are shown, where the shear stress is non-dimensionalized by $(\rho u^2)_\infty$ and stretched by the quare root of the Reynoldsnumber. In all three diagrams the region of the attachment line (between $x = -0.03$ and $x = 0.01$) is stretched to double scale. Furthermore wall shear stress is set to zero behind the line of calculation breakdown and the component in x^1-direction is negative when the flow is to the trailing edge at the lower surface.

Since the flow is accelerated in spanwise direction along the attachment line a maximum of τ^2 results there (fig. 8), whereas maxima of the modulus of τ^1 exist at both sides of the attachment line (fig. 7) where the chordwise acceleration is largest (compare the pressure distribution in fig. 6).

The disturbances in the shear stress near $x^1 \approx -0.05$, especially for the outer portion of the wing result from an unsatisfactory representation of the metric properties in that region. Here a better subroutine for the interpolation of the geometry and the evaluation of metric properties and outer velocity has to be installed yet. The disturbances would usually result in longlasting C-N-noise, but this is prevented by a weighting factor of $\eta = 0.1$. Figure 10 gives an impression what happens to the results without weighting of the second derivatives ($\eta = 0$). This shows again the importance of the weighting in C-N-like schemes for boundary layer calculations, although this cannot prevent the errors in the region where the metric is not well represented.

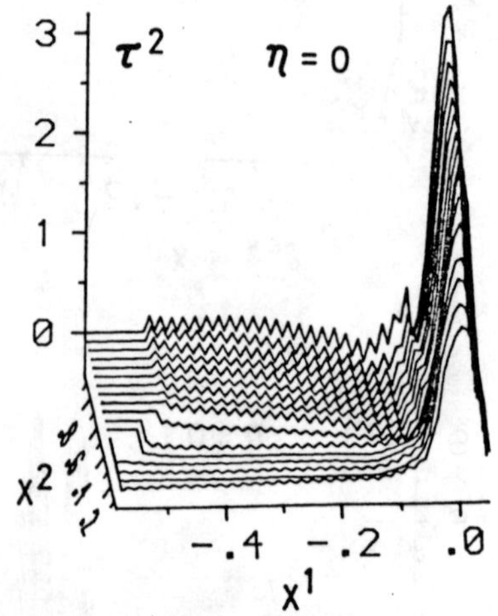

Fig. 10

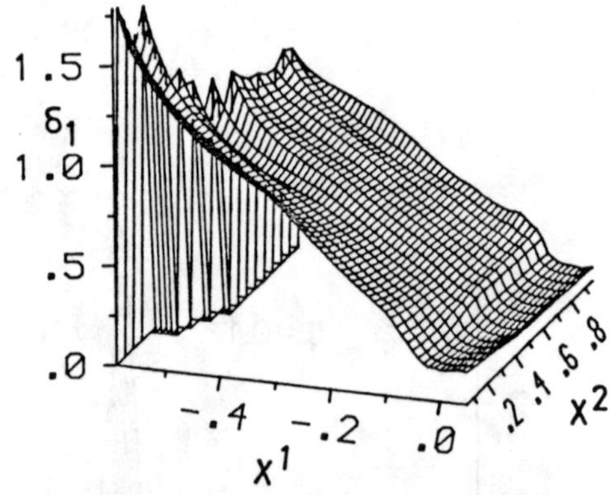

Fig. 11

The three dimensional displacement thickness is shown in figure 11, where the region of the attachment line ($-0.003 < x^1 < 0.01$) is again enlarged two times. The displacement thickness is also set to zero when the calculation breaks down. In the region of the attachment line the chordwise variation of the displacement thickness is very small. There is, however, a little decrease in spanwise direction due to the strong divergence of the flow in that region. In general, the displacement thickness for the same chordwise position is smaller near the tip then near the root of the wing, which mainly depends on the decrease of the physical chordlength in spanwise direction. An exception is the spanwise increase of δ_1 at $x^2 \approx 0.4$ where the outer and inner part of the wing meet, forming a curvature discontinuity.

CONCLUDING REMARKS

The method for the integration of thc compressible three dimensional boundary layer equation has been successfully applied to a realistic wing. The "locally infinite swept wing" concept used to obtain initial data near the root of the wing has been proven to be a good approximation. A method to improve the behaviour of the solution by weighting of the second-order derivatives has been discussed and successfully used.

REFERENCES

[1] Schwamborn, D., "Laminare Grenzschichten in der Nähe der Anlegelinie an Flügeln und flügelähnlichen Körpern mit Anstellung", DFVLR-FB 81-31, 1981; also: Laminar Boundary Layers in the Vicinity of the Attachment Line on Wings and Winglike Bodies at Incidence, ESA-TT-752, 1982.

[2] Schwamborn, D., "Boundary Layers on Finite Wings and Related Bodies with Consideration of the Attachment Line Region", in Proceedings of the 4th GAMM-Conference on Numerical Methods in Fluid Mechanics; Notes on Numerical Fluid Mechanics, Vol.5, Vieweg Verlag, Braunschweig, 1982.

[3] Hirschel, E.H., Kordulla, W., "Shear Flow in Surface-Oriented Coordinates", Notes on Numerical Fluid Mechanics, Vol.4, Vieweg Verlag, Braunschweig, 1981.

[4] Krause, E., "Numerical Solution of the Boundary Layer Equations", AIAA J.5, p.1231-1237, 1967.

[5] Krause, E., Hirschel, E.H., Bothmann, Th., "Die numerische Integration der Bewegungsgleichungen dreidimensionaler laminarer kompressibler Grenzschichten", Fachtagung Aerodynamik, Berlin 1968, DGLR Fachbuchreihe Band 3, Braunschweig, 1969.

[6] Mitchell, A.R., Griffiths, D.F., "The Finite Difference Method in Partial Differential Equations", John Wiley & Sons, New York, 1980.

NUMERICAL SIMULATION OF FLAME PROPAGATION IN A CLOSED VESSEL

J.A. Sethian

Department of Mathematics
and
Lawrence Berkeley Laboratory

University of California
Berkeley, California, 94720

SUMMARY

We present a numerical simulation of a flame propagating in a swirling, premixed, combustible fuel inside a closed vessel. The model we approximate is appropriate for viscous, turbulent combustion, and includes the effects of exothermic volume expansion along the flame front. The method, which is particularly suited for flow at high Reynolds number, uses random vortex element techniques coupled to a flame propagation algorithm based on Huyghen's principle. We analyze the various effects of pressure, boundary conditions, exothermicity and viscosity on the speed and shape on the burning flame.

INTRODUCTION

A particularly challenging problem in the study of turbulent combustion is the interaction between hydrodynamic turbulence and the propagation of a flame. At high Reynolds number, turbulent eddies and recirculation zones form, due to viscous effects, which affect the motion of a flame. Conversely, exothermic effects along the flame front influence the fluid motion. In many situations, this interaction is of great importance. For example, in the design of internal combustion engines, one might attempt to direct the flow in such a way that the largest amount of fuel is burned as quickly as possible, thus minimizing the amount of unburnt fuel expelled at the end of a stroke.

Questions of flame stability and the interaction between hydrodynamics and flame propagation have received considerable attention over the past few decades. Much of the analysis has concentrated on perturbation analysis of various models of combustion; for example, in [1], the effect of viscosity on the hydrodynamic stability of a plane flame front was examined. An excellent, though now outdated review of such techniques may be found in [2]; a more recent review may be found in [3].

Taking a different approach, in our work we are concentrating on numerical methods to analyze such problems. At the foundation of our investigations is the Random Vortex Method [4], a numerical technique that is specifically designed for high Reynolds number flow, and portrays in an accurate and natural manner the formation of turbulent eddies and coherent structures in the flow. This technique has been successfully applied to a variety of situations, for example, flow past a cylinder [5] and blood flow past heart valves [6], and was first applied to turbulent combustion over a backwards facing step in [7].

In [8], we used these techniques to model turbulent combustion in open and closed vessels, and in [9] analyzed the effect of viscosity on the rate of combustion. We showed that viscosity wrinkles the flame front, increasing the surface area of the flame and thus accelerating the combustion process. In these investigations, exothermic effects, that is, volume expansion along the flame front, were ignored, hence the fluid motion affected the flame, but there was no feedback mechanism by which combustion could influence fluid motion. Recently, a model for combustion in closed vessels was formulated in [10] which includes exothermic effects in confined chambers and allows the flame to influence the hydrodynamics. In this paper, we use the techniques described in [8] to approximate the solution of these equations. We investigate the role of pressure, boundary conditions, exothermicity and viscosity on the speed and shape of the burning flame.

THE MODEL: EQUATIONS OF MOTION

We consider two-dimensional, viscous flow inside the vessel. On solid walls, we require that the normal and tangential velocities be zero. In this model, combustion is characterized by a single step, irreversible chemical reaction; the fluid is a pre-mixed fuel in which each fluid particle exists in one of two states, burnt and unburnt. When the temperature of a particle becomes sufficiently high, it undergoes an instantaneous change in volume due to heating and becomes burnt. Thus we regard the interface between the burnt and unburnt regions as an infinitely thin flame front, acting as a source of specific volume and propagating in a direction normal to itself into the unburnt fluid. We assume that the Mach number M (the ratio of typical fluid velocities to typical sound speeds) is small, thus acoustic wave interactions are ignored.

The equations of motion for the above model applied to turbulent combustion in unconfined chambers may be found in [8]. In an unconfined vessel, the pressure remains constant, since expansion along the flame front merely pushes the fluid through the exit. Recently, Majda [10] has formulated a set of equations for low Mach number combustion in closed vessels that includes a time-dependent spatially uniform mean pressure term. Under the physically reasonable assumptions that the Mach number M is small, the initial pressure is spatially uniform within terms of order M^2, and the initial conditions for velocity, pressure and mass fraction are consistent within order M, formal asymptotic limits of the equations for fully compressible combustion can be taken to obtain a mathematically rigorous model of combustion that removes the detailed effects of acoustic waves, while retaining exothermic effects and spatial density variations. One can think of this model as existing "in between" constant density models, in which the fluid mechanics essentially decouples from the flame propagation, and the fully compressible combustion equations. In the case of combustion in unconfined chambers, the pressure remains constant and the equations reduce to those presented in [8]. We now describe the equations, for details, see [10].

Let $\vec{u} = (u,v)$ be the velocity of the fluid in the domain D. We let $\vec{u} = \vec{w} + \nabla\varphi$ where $\vec{w}$ is divergence-free (that is, $\nabla\cdot\vec{w} = 0$) and $\nabla\varphi$ is irrotational ($\nabla\times\nabla\varphi = 0$). Defining ξ to be the vorticity ($\xi = \nabla\times\vec{w}$), we take the curl of the momentum equation to produce the vorticity transport equation

$$\frac{D\xi}{Dt} = \frac{1}{R}\nabla^2\xi \tag{1}$$

where R is the Reynolds number and $\frac{D}{Dt}$ is the total derivative $\partial_t + (\vec{u}\cdot\nabla)$. Here, we have ignored the term $(\nabla\times\frac{\nabla P}{\rho})$ which corresponds to vorticity production across the flame front. The boundary conditions are that $\vec{u} = 0$ on ∂D.

With the assumption of an infinitely thin reaction zone, we view the flame front as a curve γ separating the burnt fluid from the unburnt fluid, where $\gamma(s,t)$ parameterizes by s, $0 \le s \le S$, the position of the front at time t. Thus, for each s and t, $\gamma(s,t)$ yields the coordinates (X_F, Y_F) of a fluid particle that is "on fire". The burning of the front in a direction normal to itself, plus the advection of the front by the fluid, can be described by the system of partial differential equations

$$\frac{\partial X_F}{\partial t} = k\,(Y_F)_s\left(\frac{1}{((X_F)_s^2 + (Y_F)_s^2)^{1/2}}\right) + u\,(X_F, Y_F) \tag{2}$$

$$\frac{\partial Y_F}{\partial t} = -k\,(X_F)_s\left(\frac{1}{((X_F)_s^2 + (Y_F)_s^2)^{1/2}}\right) + v\,(X_F, Y_F) \tag{3}$$

The burning speed k may be determined by examining the mass flux m across the flame front to obtain

$$k = \frac{m(\rho_u(t), P(t))}{\rho_u(t)} \tag{4}$$

where $\rho_u(t)$ and $P(t)$ are the density of the unburnt gas and mean pressure, at time t, respectively. A typical form for the mass flux (see [11]) is

$$m(\rho_u, P) = Q\rho_u^{1-a} P^a \tag{5}$$

where Q is the local laminar flame velocity and a is a constant. Under the assumption of a γ-gas law, the unburnt fluid density may be obtained from the pressure through the relation

$$\rho_u(t) = (P(t))^{1/\gamma} \rho_u(0) \tag{6}$$

where $\rho_u(0)$ is the density of the unburnt fuel initially. The mean pressure in the vessel changes as a result of the expansion of fluid particles along the flame front as they change from unburnt to burnt. The rate of change $\frac{\partial P}{\partial t}$, which clearly depends on the length of the flame front and the volume of the vessel, is of the form

$$\frac{dP}{dt} = \frac{q_0 \gamma m(\rho_u(t), P(t))}{Vol(D)} \int_0^S \gamma(s,t)ds \tag{7}$$

where q_0 is a constant corresponding to exothermic expansion, and $Vol(D)$ is the volume of the vessel.

Finally, we need to determine $\nabla\varphi$, the exothermic velocity field resulting from volume expansion along the flame front. The divergence of this velocity field corresponds to the amount of expansion (or compression) at any point. Since $\nabla\varphi \cdot \vec{n} = 0$ on ∂D (no flow through the walls), by the divergence theorem we must have that $\int_D \nabla^2\varphi = 0$. Using this fact, together with (7), yields the elliptic equation

$$\nabla^2\varphi = \frac{1}{\gamma P}\left(-\frac{dP}{dt} + q_0 \gamma m(\rho_u, P)\, \delta_F\right) \tag{8}$$

$$\nabla\varphi \cdot \vec{n} = 0$$

where δ_F is the surface Dirac measure concentrated on the flame front. Equations (1-8) form our model for combustion in closed vessels.

THE NUMERICAL APPROXIMATION

Numerical modeling of high Reynolds number flow is typically accomplished through the application of finite difference schemes to the above equations. Some of the problems inherent in these techniques are 1) the necessity of a fine grid in the boundary layer region near walls where sharp gradients exist 2) the introduction of numerical diffusion; the error term associated with the approximation equation looks like a diffusion term, and 3) the intrinsic smoothing of finite difference schemes which damps out physical instabilities. The random vortex element, introduced in [4], is specifically designed to deal with these problems. The equations of motion are written in vorticity form, and the motion of vorticity is followed by means of a collection of vorticity approximation elements. By avoiding the averaging and smoothing associated with finite difference formulations, this technique allows us to follow the development of large-scale coherent, turbulent structures within the flow. In [8], vortex methods were applied to problems in turbulent combustion in open vessels. Our application of these techniques to the equations for combustion in closed vessels is very similar, and is described below.

The vorticity ξ in (1) is approximated by a set of vortex "blobs", whose positions and strengths at any time yield the associated velocity field $\vec{w}$. The distribution of vorticity is updated in two stages. First, the vortex elements are moved under the flow field $\vec{w}$, corresponding to the advection of vorticity by the velocity field it induces. Second, viscous diffusion is simulated by a random walk imposed on the vortex motion. The normal boundary condition on $\vec{w}$ is met through the addition of a potential flow solution, and the tangential boundary "no-slip" condition is satisfied by a vorticity creation algorithm (vortex sheets).

To model the motion of the flame (Equations 2 and 3) , one is tempted to place marker particles along the boundary between the burnt and unburnt fluid and update their position and hence the location of the flame front in time. Because of the difficulty involved in determining the normal direction to the front (the direction in which the flame burns) from such an approximation, the flame front usually becomes unstable and develops wild oscillations (see [12]). We avoid this problem by imposing a grid on the domain and assigning each cell a number (a "volume fraction", see [13]) corresponding to the amount of burnt fluid in that cell at any given time. We allow each cell on the boundary of the burnt gas to ignite all its neighbors at the prescribed rate k; this is an approximation based on Huyghen's principle, which states that the envelope of all disks centered on the front corresponds to the front displaced in a direction normal to itself, (see [14]). The motion of the flame is broken up into two stages: first, we model burning by allowing the flame to propagate in a direction normal to itself at the prescribed speed and second, we advect the burned fluid by the yet to be determined velocity field $\vec{u}$. By updating these volume fractions according to the advection and burning processes, we may track the motion of the flame.

To determine the velocity field $\vec{u}$, we must solve for the exothermic velocity field $\nabla\varphi$ produced by volume expansion along the flame front. We calculate $\frac{dP}{dt}$ from (7) using the position of the flame as determined by the Huyghen's principle construction described earlier. This allows us to determine the right hand side of (8); a fast Poisson solver is used to solve the Neumann problem for φ. Straightforward finite differences on the fast solver grid provide $\nabla\varphi$ and hence $\vec{u}$. Again, the tangential boundary "no-slip" condition is satisfied by the creation of vortex sheets. The vortex elements are then advected under the field $\nabla\varphi$, and the flame is advected by the velocity field $\vec{u} = \vec{w} + \nabla\varphi$ to produce the new positions for the vortex blobs and flame. Finally, the non-linear ordinary differential equation (7) is solved to update the pressure, and (6) is used to update the density of the unburnt gas used in the mass flux calculation (5).

RESULTS

We performed a series of experiments to measure the effect of exothermicity on the propagation of a flame. We began by igniting a motionless, inviscid fluid at the center of a closed square. We chose a non-dimensional local laminar flame velocity $Q=.2$, with $a=.5$ (Equation 5). We took $P(0)=1.$ and $\rho_u(0)=1.$, and assumed that a fluid particle increased its volume by a factor of five upon burning, this corresponded to q_0=1.333 (see [9] for details). In Figure 1, the results of this experiment are shown. The black region corresponds to burnt fluid, and the velocity field is displayed on a 30×30 grid placed in the flow, where the magnitude of the vector at each point denotes the relative speed of the flow there. The fluid motion results entirely from expansion along the flame front. One can clearly see the mechanism by which the boundary shapes the front; although the front starts off circular, it soon becomes square-like in response to the boundary conditions on the exothermic velocity field $\nabla\varphi$, and thus burns into the corners. The final value ($t=1.55$) of the pressure in the vessel is 2.93 and the final value for k the propagation speed was .24 (compared with $k=.2$ at $t=0$).

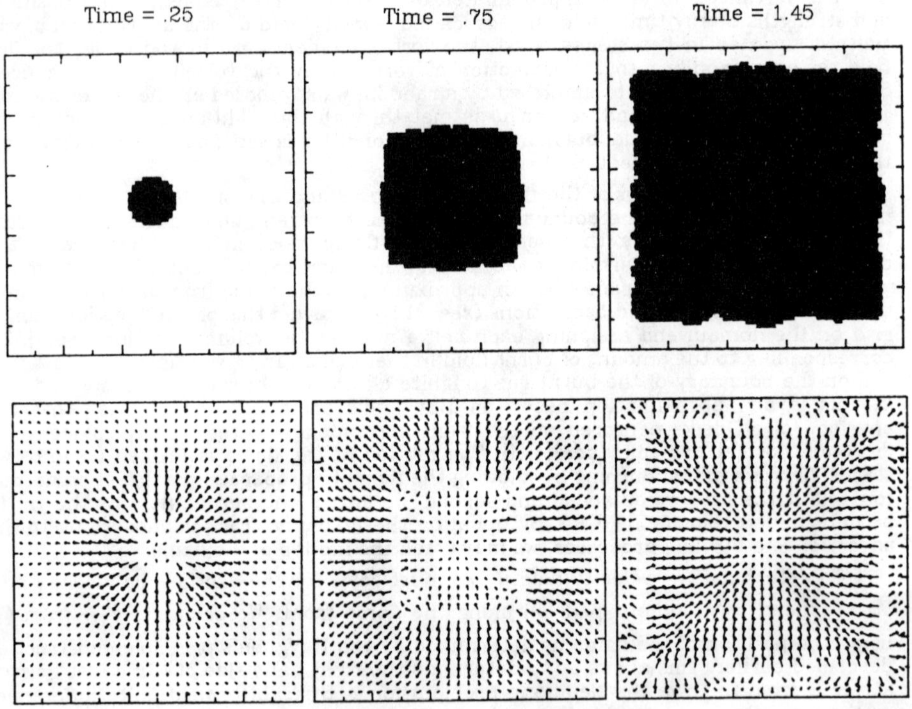

Figure 1: Inviscid, Motionless Fluid Ignited in Center, $q_0=1.333$

In the second set of experiments (Figure 2), we investigated the relative effects of viscosity and exothermicity on the rate at which combustion takes place in the vessel. In these experiments, fluid motion was generated by a vortex placed in the center of a square of sufficient strength so that the velocity tangential to each wall at its midpoint was 1. With $Q = .14$, we performed four different experiments. The top row corresponds to inviscid flow with $q_0=0$ (no exothermicity allowed) , the next row is inviscid flow with $q_0=1.333$ (factor of five expansion) , the next row is viscous flow with Reynolds number $R=1000$ and the bottom row corresponds to viscous flow, $R=1000.$, $q_0=1.333$. In the two viscous runs, the flow was started two seconds before ignition so that recirculation zones would have time to develop.

A. Inviscid Flow/Constant Density

B. Inviscid Flow/Volume Expansion

C. Viscous Flow/Constant Density

D. Viscous Flow/Volume Expansion

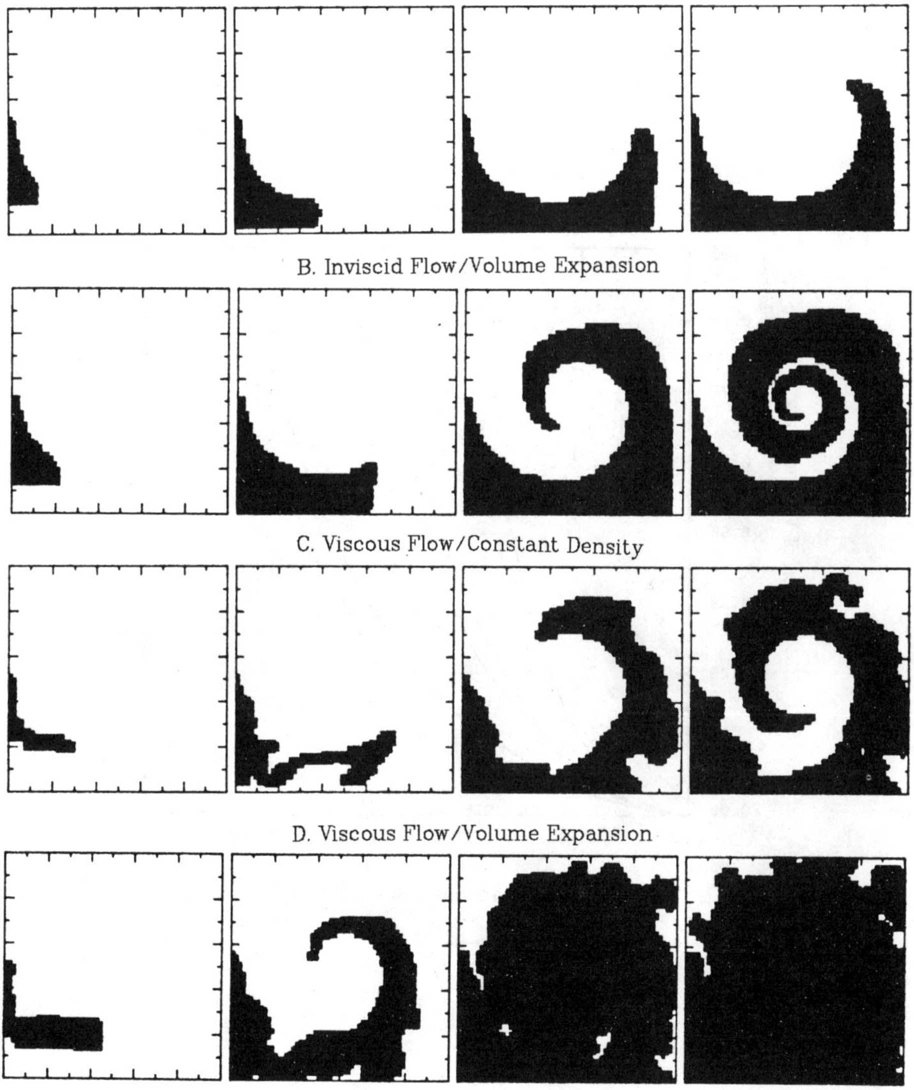

Time = .45 Time = .81 Time = 1.36 Time = 1.52

Figure 2: Swirling Fluid

The results may be summarized as follows. In the inviscid, constant density case, the flame wraps smoothly around the center, since the flow is smooth and there is no feedback mechanism from the flame to the hydrodynamics. When volume expansion effects are added, the resulting velocity field carries the flame around the center at a faster rate, in addition to the slightly higher propagation speed. In the viscous, constant density case, the flame motion is strongly influenced by the counterrotating eddies that grow in the corners as a result of vorticity production along solid walls; the flame is carried around each large eddy and then dragged backwards into the corner. These eddies are of prime importance in bringing the flame into contact with unburnt parts of the vessel. The front becomes jagged and wrinkled, increasing the surface area of the flame available for burning. In the viscous case with volume expansion, the flame is both wrinkled due to the turbulence of the flow and carried by the volume expansion velocity field, greatly decreasing the time required for complete conversion of reactants to products. In Figure 3, we illustrate these comments by plotting the percentage of the volume burnt as a function of time elapsed since ignition.

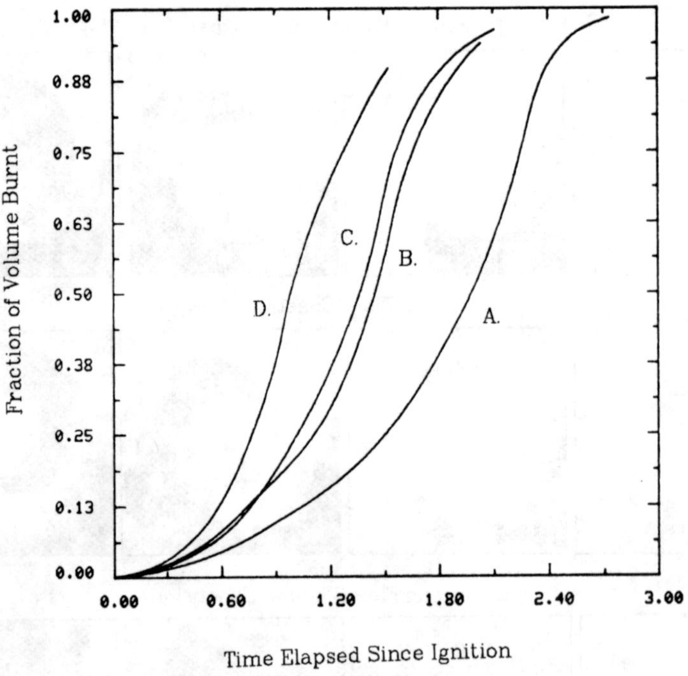

Figure 3

REFERENCES

1. Frankel, M.L., and Sivashinsky, G.I., "The Effect of Viscosity on Hydrodynamic Stability of a Plane Flame Front", Combustion and Science Technology, Vol. 29, pp.207-224.
2. Markstein, G.H., Nonsteady Flame Propagation, Pergammon Press, MacMillan and Company, New York, 1964.
3. Sivashinsky, G.I., "Instabilities, Pattern Formation, and Turbulence in Flames", to appear.
4. Chorin, A.J., "Numerical Studies of Slightly Viscous Flow", Journal of Fluid Mechanics, Vol. 57, 1973, pp. 785-796
5. Cheer, A.Y. "A Study of Incompressible 2-D Vortex Flow Past A Circular Cylinder" Lawrence Berkeley Laboratory, LBL-9950, 1979, to appear in Siam Journal of Scientific and Statistical Computing, 1983.
6. McCracken, M.F. and Peskin, C.S., "A Vortex Method For Blood Flow Through Heart Valves", Journal of Computational Physics, Vol.35, 1980, p.183-205.
7. Ghoniem, A.F., Chorin, A.J., Oppenheim, A.K. "Numerical Modeling of Turbulent Flow in A Combustion Tunnel", Philosophical Transactions of the Royal Society of London, Vol. 304, 1982, pp. 303-325.
8. Sethian, J.A. "Turbulent Combustion in Open and Closed Vessels", Lawrence Berkeley Laboratory, LBL-15744, 1983, to appear in the Journal of Computational Physics.
9. Sethian, J.A. "The Wrinkling of a Flame Due to Viscosity", Fire Dynamics and Heat Transfer, Proc. 21st Nat. Heat Transfer Conf., J.Quintiere, Ed.,p.29-32, 1983.
10. Majda, A., "Equations of Low Mach Number Combustion", Center for Pure and Applied Mathematics, University of Cal., Berkeley, 1982, to appear, SIAM Jour.of Appl. Math.
11. Kurylo, J., Dwyer,H.A., and Oppenheim, A.K., "Numerical Analysis of Flow Fields Generated by Accelerated Flames", AIAA J., Vol.18, 1980, pp.302-308.
12. Sethian, J.A. "An Analysis of Flame Propagation", Lawrence Berkeley Laboratory, LBL-14125, 1982
13 Noh, W.T. and Woodward, P., SLIC (Simple Line Interface Calculation). Proc. 5th Int. Conf. Numer. Math. Fluid Mechanics, Springer-Verlag, Berlin, 1976, pp.330-339.
14. Chorin, A.J., "Flame Advection and Propagation Algorithms", Journal of Computational Physics, Vol.35, 1980, pp.1-11

ON THE INVESTIGATION OF THE COMPLETELY CONSERVATIVE PROPERTY OF DIFFERENCE SCHEMES BY THE METHOD OF DIFFERENTIAL APPROXIMATION

Yu.I.Shokin, Z.I.Fedotova
Computing Center, Akademgorodok,
660036 KRASNOYARSK, 36, USSR

SUMMARY

In the present paper the method of the first differential approximation is applied to the investigation of the completely conservative property of difference schemes for the equations of gas dynamics in the case of Eulerian coordinates. The sixteen-parameter family of difference schemes was considered and among them the completely conservative difference schemes were found.

INTRODUCTION

Computational experience shows that the most important properties which a difference scheme for the equations of gas dynamics should possess are the properties of conservativity and completely conservativity [1, 2] . In the monograph [2] there is descriebed the two-parameter family of completely conservative difference schemes in Lagrangean coordinates which have proved themselves to be good by calculating different problems of gas dynamics using a crude mesh [2] . In [1,3] the connection between the completely conservative property of difference schemes in Lagrangean coordinates and properties of their first differential approximation is showed.

The first completely conservative difference scheme approximating gasdynamic equations in Eulerian coordinates is proposed in [4] . This difference scheme is three-layer. In [5] there is studied an eleven-parameter family of two-layer difference schemes and by the help of a special algorithm it is showed that completely conservative difference schemes are absent among them.

In this paper for the research of properties of conservativity and completely conservativity there is applied the algorithm based on the method of the first differential approximation. This investigation allowed to construct the new class of completely conservative schemes for gas dynamics in Eulerian coordinates.

PROBLEM

Consider the sixteen-parameter family of difference schemes of the first order of accuracy

$$\frac{\Delta_0}{\tau}\rho^{n-\frac{1}{2}}(x+\tfrac{h}{2})+\frac{\Delta_1}{h}f(x)\Big|_{\alpha_1}^{\sigma_1,\sigma_2}=0;$$

$$\frac{1}{2}\left\{\rho^{n+\frac{1}{2}}(x+\tfrac{h}{2})\frac{\Delta_0}{\tau}u^n(x)+\rho^{n-\frac{1}{2}}(x+\tfrac{h}{2})\frac{\Delta_0}{\tau}u^{n-1}(x)\right\}+$$

$$+\frac{1}{2}\left\{f(x+h)\Big|_{\alpha_2}^{\sigma_3,\sigma_4}\frac{\Delta_1}{h}u^n(x)+f(x)\Big|_{\alpha_3}^{\sigma_3,\sigma_4}\frac{\Delta_1}{h}u^n(x-h)\right\}+$$

$$+\frac{\Delta_1}{h}p^{\sigma_5}(x-\tfrac{h}{2})=0; \qquad (1)$$

$$\rho^{\sigma_6}(x+\tfrac{h}{2})\frac{\Delta_0}{\tau}\varepsilon^{n-\frac{1}{2}}(x+\tfrac{h}{2})+p^{\sigma_{10}}(x+\tfrac{h}{2})\frac{\Delta_1}{h}u^{\sigma_{11}}(x)+$$

$$+\alpha_3 f(x+h)\Big|_{\alpha_4}^{\sigma_7,\sigma_8}\frac{\Delta_1}{h}\varepsilon^{\sigma_9}(x+\tfrac{h}{2})+(1-\alpha_3)f(x)\Big|_{\alpha_5}^{\sigma_7,\sigma_8}\frac{\Delta_1}{h}\varepsilon^{\sigma_9}(x-\tfrac{h}{2})=0,$$

which approximate the system of equations of gas dynamics in Eulerian coordinates:

$$B\frac{\partial w}{\partial t}+A\frac{\partial w}{\partial x}=0\ ,$$

$$w=\begin{pmatrix}\rho\\ u\\ \varepsilon\end{pmatrix},\quad B=\begin{pmatrix}1&0&0\\0&\rho&0\\0&0&\rho\end{pmatrix},\ A=\begin{pmatrix}u&0&0\\p_\rho&f&p_\varepsilon\\0&p&f\end{pmatrix}.$$

Here $t=n\tau$ is the time coordinate, $x=ih$ is the Eulerian variable, u is the velocity, ρ is the density, $f=\rho u$, ε is the specific internal energy, $p=p(\varepsilon,\rho)$ is the pressure; τ, h are the grid steps, $\Delta_0=T_0-E$, $\Delta_1=T_1-E$, $E\varphi(x,t)=\varphi(x,t)$, $T_0\varphi(x,t)=\varphi(x,t+\tau)$, $T_1\varphi(x,t)=\varphi(x+h,t)$, $f(x)\Big|_{\alpha_1}^{\sigma_1,\sigma_2}=(\alpha_1\rho^{\sigma_1}(x+\frac{h}{2})+$ $+(1-\alpha_1)\rho^{\sigma_1}(x-\frac{h}{2}))\,u^{\sigma_2}(x)$, $\rho^{\sigma_1}(x+\frac{h}{2})=\sigma_1\rho^{n+\frac{1}{2}}(x+\frac{h}{2})+$ $+(1-\sigma_1)\rho^{n-\frac{1}{2}}(x+\frac{h}{2})$, $u^{\sigma_2}(x)=\sigma_2 u^{n+1}(x)+(1-\sigma_2)u^n(x)$,

the other notations are introduced by an analogous way; $\sigma_1,\dots,\sigma_{11}$; $\alpha_1,\dots,\alpha_5$ are parameters, $0\le\sigma_i\le 1$, $i=1,\dots,11$, $0\le\alpha_j\le 1$, $j=1,\dots,5$. The velocity is taken in the integer mesh points $(n\tau,ih)$, but all thermodynamical functions are connected with the half-integer mesh points $((n+\frac{1}{2})\tau,(i+\frac{1}{2})h)$.

In accordance with the definition of the completely conservative property there are considered the transformations of difference schemes defined by following matrices:

$$\tilde{\mathcal{R}}_1=\begin{pmatrix}1&0&0\\u^n(x)&1&0\\0&0&1\end{pmatrix},\quad \tilde{\mathcal{R}}_2=\begin{pmatrix}1&0&0\\0&1&0\\\varepsilon^{\gamma_1}(x+\frac{h}{2})&0&1\end{pmatrix},$$

$$\tilde{\mathcal{R}}_3=\begin{pmatrix}1&0&0\\\frac{1}{2}(u^n(x))^2&u^n(x)&0\\0&0&1\end{pmatrix},\ \tilde{\mathcal{R}}_4=\begin{pmatrix}1&0&0\\0&1&0\\-\dfrac{p^{\gamma_3}(x+\frac{h}{2})}{\rho^{\gamma_2}(x+\frac{h}{2})}&0&1\end{pmatrix},$$

$$\tilde{\mathcal{R}}_5 = \begin{pmatrix} 1 & 0 & 0 \\ u^n(x) & 1 & 0 \\ \frac{1}{2}(u^n(x))^2 + \varepsilon^{\gamma_1}(x+\frac{h}{2}) & u^n(x) & 1 \end{pmatrix}.$$

These matrices approximate the matrices $\mathcal{R}_i$ of the transformations by the help of which we can obtain the whole set of linear independent conservative laws and balance relations for the system of equations of gas dynamics.

Consider the Γ-form of the first differential approximation [1] of the difference scheme (1)

$$B \frac{\partial w}{\partial t} + A \frac{\partial w}{\partial x} = \Omega \tag{2}$$

and represent the matrices $\tilde{\mathcal{R}}_i$ by the following way: $\tilde{\mathcal{R}}_i = \mathcal{R}_i + \Xi_i^{(1)} + \Xi_i^{(2)}$, where elements of the matrix $\Xi_i^{(2)}$ have the order $O(\tau^2, h^2, \tau h)$.

It is easy to show that transforming the system of equations (2) by the help of matrices $\tilde{\tilde{\mathcal{R}}}_i = \mathcal{R}_i + \Xi_i^{(1)}$ one can obtain the set of necessary conditions for the difference scheme (1) of being completely conservative.

RESULT

It is shown that the considered multiparameter difference scheme is completely conservative if

$$\begin{gathered} \alpha_1 = \alpha_2 = \alpha_3, \ \alpha_4 = 1, \ \alpha_5 = 0, \ \sigma_2 = \sigma_4 = \sigma_8 = \sigma_{11} = 0, \\ \sigma_5 = \sigma_{10}, \quad \sigma_3 = \sigma_7 = \sigma_9 = 1 - \sigma_6 = \sigma_1 \text{ and } \sigma_1 = 0 \text{ or } 1. \end{gathered} \tag{3}$$

Besides it is obtained the conditions for the parameters of transformation matrices γ_i $(i = 1, 2, 3)$:

$$\gamma_1 = \gamma_2 = \sigma_1, \quad \gamma_3 = \sigma_5.$$

Applying to the difference scheme (1) with parameters (3) the transformation defined by the matrix $\tilde{\mathcal{R}}_5$ one can get the difference scheme approximating gasdynamic equations in a divergence form.

Write down the difference schemes which are obtained with the help of above-mentioned transformations:

1. The difference scheme of conservation of the momentum:

$$\frac{\Delta_0}{2\tau}\left[\rho^{n-\frac{1}{2}}\left(x+\frac{h}{2}\right)\left(u^n(x)+u^{n-1}(x)\right)\right]+$$

$$+\frac{\Delta_1}{h}\left[f(x)\Big|_{d_1}^{\sigma_1,0}\left(u^n(x)+u^n(x-h)\right)\right]+\frac{\Delta_1}{h}\,p^{\sigma_5}\left(x-\frac{h}{2}\right)=0\,.$$

2. The difference scheme for the balance of the internal energy

$$\frac{\Delta_0}{\tau}\left[\rho^{n-\frac{1}{2}}\left(x+\frac{h}{2}\right)\varepsilon^{n-\frac{1}{2}}\left(x+\frac{h}{2}\right)+p^{\sigma_5}\left(x+\frac{h}{2}\right)\frac{\Delta_1}{h}u^n(x)\right]+$$

$$+\frac{\Delta_1}{h}\left\{\left[d_1\rho^{\sigma_1}\left(x+\frac{h}{2}\right)\varepsilon^{\sigma_1}\left(x+\frac{h}{2}\right)+(1-d_1)\rho^{\sigma_1}\left(x-\frac{h}{2}\right)\varepsilon^{\sigma_1}\left(x-\frac{h}{2}\right)\right]u^n(x)\right\}=0.$$

3. The difference scheme for the balance of the kinetic energy

$$\frac{\Delta_0}{\tau}\left[\rho^{n-\frac{1}{2}}\left(x+\frac{h}{2}\right)u^n(x)\,u^{n-1}(x)\right]+$$

$$+\frac{\Delta_1}{2h}\left[f(x)\Big|_{d_1}^{\sigma_1,0}u^n(x)\,u^n(x-h)\right]+u^n(x)\,\frac{\Delta_1}{h}\,p^{\sigma_5}\left(x-\frac{h}{2}\right)=0\,.$$

4. The difference scheme for the balance of the entropy

$$\rho^{1-\sigma_1}\left(x+\frac{h}{2}\right)\left[\frac{\Delta_0}{\tau}\varepsilon^{n-\frac{1}{2}}\left(x+\frac{h}{2}\right)+p^{\sigma_5}\left(x+\frac{h}{2}\right)\frac{\Delta_0}{\tau}\left(1/\rho^{n-\frac{1}{2}}\left(x+\frac{h}{2}\right)\right)\right]+$$

$$+d_1u^n(x+h)\,\rho^{\sigma_1}\left(x+\frac{3}{2}h\right)\left[\frac{\Delta_1}{h}\varepsilon^{\sigma_1}\left(x+\frac{h}{2}\right)+p^{\sigma_5}\left(x+\frac{h}{2}\right)\frac{\Delta_1}{h}\left(1/\rho^{\sigma_1}\left(x+\frac{h}{2}\right)\right)\right]+$$

$$+(1-d_1)u^n(x)\,\rho^{\sigma_1}\left(x-\frac{h}{2}\right)\left[\frac{\Delta_1}{h}\varepsilon^{\sigma_1}\left(x-\frac{h}{2}\right)+p^{\sigma_5}\left(x-\frac{h}{2}\right)\frac{\Delta_1}{h}\left(1/\rho^{\sigma_1}\left(x-\frac{h}{2}\right)\right)\right]=0.$$

Adding the second difference scheme to the third one we get the difference scheme which approximate the equation of the total energy.

The given method of constructing of completely conservative schemes can be applied to another multiparameter families of difference schemes.

REFERENCES

[1] Shokin Yu.I., "The method of differential approximation", Novosibirsk, "Nauka", 124 p. (1979). English translation: "Springer-Verlag", Springer series in Computational Physics, 296 p. (1983).

[2] Samarskii A.A., Popov Yu.P., "Difference methods for the solution of problems of gas dynamics", M., "Nauka", 352 p. (1980).

[3] Shokin Yu.I., Fedotova Z.I., Marchuk A.G., "On the connection between the conservative property of difference schemes and properties of their first differential approximations", Soviet Math. Dokl., Vol. 19, NO. 5, pp. 1104-1108 (1978).

[4] Kuzmin A.V., Makarov V.L., Meladze G.V., "On one completely conservative scheme for equations of gas dynamics in Eulerian coordinates", Zh. vych. mat. i mat. fiziki, Vol. 20, NO. 1, pp. 171-181 (1980).

[5] Kuzmin A.V., Makarov V.L., "On one algorithm of constructing of the completely conservative difference schemes", Zh. vych. mat. i mat. fiziki, Vol. 22, NO. 1, pp. 124-132 (1982).

3D FINITE DIFFERENCE FOR NATURAL CONVECTION IN CYLINDERS

SMUTEK, C., ROUX, B., BONTOUX, P.
I.M.F.M., 13003 Marseille, FRANCE

and

DE VAHL DAVIS, G.
Univ. of N.S.W.,1348 Kensington,Australia

SUMMARY

The false transient method based on Samarskii-Andreyev ADI scheme, developed by the group of De Vahl Davis, is applied to the solution of three-dimensional Boussinesq flow inside cylindrical enclosures with different end temperatures. Comparaisons are proposed with asymptotical solution in long horizontal cylinder and with recent experiments at aspect ratio R/L = 0.10. The onset of instabilities in vertical Rayleigh-Benard situation is studied and the numerical solution are compared with the prediction of the linear stability analysis.

1. INTRODUCTION

The free convection in differentially heated enclosures is a topic of practical interest for many applications. Crystal growth by vapor process is very dependent on the type of the hydrodynamical patterns in the growth ampoules. The flow has been studied in vertical cylindrical models with heated circular end walls, by the stability analysis (Charlson and Sani, 1970, 1971, 1975, Behringer and Ahlers, 1982, Rosenblat, 1982), by experimental approach (Koschmieder and Pallas, 1982, Stork and Müller, 1975, Olson and Rosenberger, 1979, Kirchartz et al, 1981). Recent experiments in horizontal cylinders (Schiroky and Rosenberger, 1983) have shown the fully three-dimensional structure of the flow. At low Rayleigh numbers (core driven regime) the asymptotical solution (Bejan and Tien, 1978) is valid to predict the flow in the core region. At large Rayleigh number (boundary layer driven regime) the flow is no more parallel and the "improved" asymptotical solution are unable to predict the flow very satisfactorily (Schiroky and Rosenberger, 1983).

Three-dimensional numerical methods are available in rectangular boxes (Ozoe et al, 1976, 1977, Mallinson and De Vahl Davis, 1973, 1977, Mallinson et al, 1981, Upson et al, 1981, Mc Laughin and Orszag, 1982). Centered finite differences methods based on Samarskii-Andreyev ADI scheme have been developed in the group of Prof. De Vahl Davis for cylindrical geometries (Leong and De Vahl Davis, 1979, Leong, 1981, Leonardi, Reizes and De Vahl Davis, 1981). In the present paper, the direct simulation of the natural convection flow is studied in the vertical and horizontal cylinders. The programme is derived from De Vahl Davis's one, the boundary layers required the increase of the discretizing points in the end regions. Improvement of the efficiency of the method was then necessary to remain within moderate computational cost . The convergence of the numerical solution was studied with respect to analytical and experimental solutions.The three-dimensional structures are particularly emphasized when the Rayleigh number is varying.

2. PHYSICAL AND MATHEMATICAL MODELS

2.1. The cylindrical enclosure is defined by the radius, R_o, the length, L, and the inclination with respect to the gravitational acceleration, $\bar{g}$, (fig. 1). The aspect ratio is $A = R_o/L$. At a point $(\bar{r},\phi,\bar{z})$, the velocity is $\bar{U}$ $(\bar{u},\bar{v},\bar{w})$. The cavity is heated by the two circular end-walls, T_h $(\bar{z}=0) > T_c$ $(\bar{z} = L)$. The side walls are assumed perfectly conducting.

2.2. The mathematical model is given by the Navier-Stokes and energy equations with the simplified Boussinesq approximation (Joseph, 1976).

$$\nabla \cdot \bar{U} = 0 \tag{1}$$

$$\frac{D}{Dt} \bar{U} = - \frac{\nabla P}{\rho_o} - \beta (T-T_o) \bar{g} - \nu \nabla \times (\nabla \times \bar{U}) \tag{2}$$

$$\frac{D}{Dt} T = \kappa \nabla^2 T \tag{3}$$

where $T_o = \frac{1}{2} (T_h + T_c)$

Using R_o and κ/R_o as scaling factor for the length and the velocities respectively the system is written with the vorticity, $\zeta(\zeta_r,\zeta_\phi,\zeta_z)$, the velocity, $\bar{U}$, and the temperature, $\theta = 2\,(T-T_o)/\Delta T$, (Leong and De Vahl Davis, 1979, Leong, 1981) :

$$\frac{\partial \zeta}{\partial t} = \nabla \times (U \times \zeta) - \frac{Ra\ Pr}{2} \nabla \times (\theta \hat{g}) - Pr\ \nabla \times (\nabla \times \zeta) \tag{4}$$

$$\nabla^2 U = - \nabla \times \zeta \tag{5}$$

$$\frac{\partial \theta}{\partial t} = - \nabla \cdot (U\, \theta) + \nabla^2 \theta \tag{6}$$

where the Rayleigh and Prandtl numbers are $Ra = \beta\, g\ \Delta T\, R_o^3\, Pr/\nu^2$, $Pr = \nu/\kappa$, with $\bar{g} = g\,\hat{g}$, and $\Delta T = T_h - T_c$.

2.3. The thermal boundary conditions correspond to the isothermal hot and cold end walls, $\theta = \pm 1$, and the linear temperature profile along the side walls, $\theta = 1 - 2z/A$. The boundary conditions on the vorticity are derived from the natural conditions on the velocity $(u = v = w = 0)$:

- at the circular end walls $(z = 0,\ 1/A)$:

$$\zeta_r = - \frac{\partial v}{\partial z} \quad , \quad \zeta_\phi = \frac{\partial u}{\partial z} \quad , \quad \zeta_z = 0 \tag{7a}$$

- at the cylindrical side walls $(r = 1)$

$$\zeta_r = 0 \quad , \quad \zeta_\phi = - \frac{\partial w}{\partial r} \quad , \quad \zeta_z = \frac{\partial v}{\partial r} \tag{7b}$$

2.4. In three-dimensional problems, the vorticity-velocity formulation proposed by Fasel (1975) has simple boundary conditions for the three Poisson equations (5), compared to the primitive variables end vorticity - vector potential formulations and requires also less arrays than the vorticity-vector potential system.

3. NUMERICAL SCHEME

3.1. Centered finite difference are used. The mesh is uniform and composed of L x M x N discretizing points in r, ϕ, z directions. As proposed by De Vahl Davis (1979) no mesh points are located along the axis in order to avoid the problem of singularity (Fig. 2). The step size is $\Delta r = (L - 1/2)^{-1}$ and the first mesh is at $\Delta r/2$ from the axis where second order forward differences are used. The azimuth ϕ is refered from the half vertical plane below the axis.

3.2. The Poisson equations (5) with Dirichlet conditions are solved with the Fourier series direct method (Le Bail, 1972) and using the FFT algorithm developed by Cooley and Tukey (1965).

3.3. The advancement in time is made with the ADI scheme of Samarskii - Andreyev for equations (4) and (6) (Mallinson and De Vahl Davis, 1973, Leong, 1981).

$$(1 - \frac{\alpha_f \Delta t}{2} A_r)\, d^* = (A_r + A_\phi + A_z)\, f^n + S^n$$

$$(1 - \frac{\alpha_f \Delta t}{2} A_\phi)\, d^{**} = d^*$$

$$(1 - \frac{\alpha_f \Delta t}{2} A_z)\, d^{***} = d^{**}$$

$$f^{n+1} = f^n + \alpha_f \Delta t\, d^{***}$$

where A_r, A_ϕ and A_z are the point operators representing the finite difference approximation of the products and the derivatives of the variable f. The false transient factors, α_f, are used to enhance convergence. When $\alpha_\theta = \alpha_\zeta$ the solution is true transient. With $\Delta t = \frac{1}{2}$ Min $(\Delta r^2, \Delta\phi^2, \Delta z^2)$, the factor α_θ must be reduced from 20 to 1 when Ra is increased from about 70 to 20000. Similarly the factor α_ζ is varied as $\alpha_\theta/10$. The method requires the solution of tridiagonal and cyclic - tridiagonal systems. The boundary conditions on the vorticity are derived from the numerical approximation of conditions (7). Supplementary details on the method are given in Mallinson and De Vahl Davis (1973) and Leong (1981).

4. RESULTS AND DISCUSSION

4.1. Algorithm

The code was run on a vector-computer, CRAY 01-S. The efficiency of the method was improved in terms of computing time by vectorizing the highly serial algorithms. Most of the time is spent in the FFT and tri-diagonal algorithms, respectively 32 % and 28 % for the 9 x 32 x 65 mesh points-solution. Comparisons between vectorized and non-vectorized calculations have shown a reduction in computing time of about 78 % for the tridiagonal system and 72 % for the FFT. The resulting reduction for one iteration is about 70 %.

4.2. Convergence and accuracy

Asymptotical solution exists in long horizontal cylinders ($\gamma = 90°$)

when the longitudinal temperature is the driving force. The two - dimensional first order solution writes as follows in the core (Bejan and Tien, 1978).

$$w_I = - \frac{k_1}{8} (r^2 - 1)\, r \cos \phi \qquad (8)$$

where k_1 is the temperature gradient in the core (= R/L, for the core driven regime).

The solution is used as initial condition with a matching function in the third dimension. The convergence is shown on Fig. 3 for A = 0.10 and a 9 x 32 x 33 mesh. For the lower Ra, the convergence is obtained after about 60 iterations only. Up to Ra = 3580, the use of lower Ra solution (Ra = 660) as initial conditions does not improve strongly the convergence compared to the initialization with relation (8). After the transition from the core driven regime, Ra $\sim$ 3500 for A = 0.10, the S-profile solution (8) is no more a correct solution of the problem. The flow is very three-dimensional and non parallel. Then the initialization with lower Ra solution prevails on the initialization with (8) and very small false transient factor α_θ and α_ζ.

The drawback of the vorticity-velocity method compared to the vorticity-vector potential method resides in a lack of accuracy due to the non exact achievement of the discrete continuity equation. The accuracy of the solution is controlled by the discretized divergence equation (1). The values are plotted on Fig. 4 along the axis at ϕ = 0, r $\sim$ 2/3 with the maximum of relation (8) as scale factor. The maximum occurs in the end regions. When the step size is reduced the divergence is shown to tend toward zero everywhere in the cavity.

4.3. Horizontal cylinder A = 0.10 (γ = 90°)

The numerical solutions in the core were compared on Fig. 5 to experiments of Schiroky and Rosenberger (1983) for the core driven regime (Ra = 660), the transition (Ra = 3580) and the boundary layer regime (Ra = 18700). The computations corresponding to 11 x 16 x 17 and 9 x 32 x 33 mesh points fit very well with the experiments. The 9 x 32 x 33 solutions differs only of about 1 % with the 9 x 32 x 65 solution in the core. However the 9 x 32 x 65 calculation improves the accuracy of the solution (7 %) in the vertical boundary layer near the end walls, when secondary vorticies spread out (Ra = 18700).

The Fig.6 shows the variation of the maximum of the w velocity from experiments and 9 x 32 x 33 solution when Ra is varying. For low Ra (Ra < 3500), the numerical solution is in very good agreement with the experiments and the asymptotic core solution. When Ra > 3500 the driving force comes from the boundary layer near the end walls. The magnitude and the shape of the w profile are very different from the theoretical solution. On the other hand the numerical solution fits very well up to Ra $\sim$ 20000 with the experiments. Above the 9 x 32 x 65 mesh must be used to ensure the good approximation for the driving boundary layer flow.

The three-dimensional solution is plotted on Fig. 7 for different azimuth at Ra = 660, 3580 and 18700. At Ra = 660, the flow is parallel in the core for any ϕ. When Ra = 3580, the S-shape and the flow parallelism are already lost in the symmetry plane ϕ = 0-π. At Ra = 18700 the flow become more complex when very strong recirculations develop near the end walls.

The secondary flow is shown on Fig. 8 for several circular sections end Ra = 3580 and 18700. The spiral structures expand differently due to the onset of the vorticies in the symmetry plane (Fig. 7). In the middle of the cylinder, the circular structures are found in the four quadrants pretty similar to the ones predicted by the second order asymptotic solution (Bejan and Tien, 1978).

4.4. Vertical cylinder, $\gamma = 180°$

The linear stability analysis (Charlson and Sani, 1971) has given criterion for the onset of axisymmetric and antisymmetric convection in vertical cylinder heated from below.

For an aspect ratio A= 0.5,the first transition correspond to an antisymmetric state for Ra $\sim$ 1060. The onset of convection was studied numerically with the present code.When Ra = 800, it is verified that initial disturbances taken as 1/10 of the first order asymptotic solution (8) vanishes. For Ra = 1200 antisymmetric solution is obtained as predicted by the theory. For higher Ra, the antisymmetric structure still remains dominant (1200 < Ra < 3500).

For an aspect ratio A = 0.8, axisymmetry convection is obtained at Ra = 1000 in agreement with stability analysis.

References

1. Behringer and Ahlers, J. Fluid. Mech., 125, 219-258, 1982.
2. Bejan and Tien, Int. J. Heat Mass Transfer, 21, 701-708, 1978.
3. Charlson and Sani, Int.J. Heat Mass Transfer, 13, 1479-1496, 1970.
4. Charlson and Sani, Int. J. Heat Mass Transfer, 14, 2157-2160, 1971.
5. Charlson and Sani, J. Fluid Mech., 71, 209-229, 1975.
6. Cooley and Tukey, Math. of Comp., 19, 297-301, 1965.
7. Fasel, LN in Physics, Springer-Verlag, 35, 151-160, 1975.
8. Joseph, Springer Tracts in Natural Philosophy, 28, 1976.
9. Kirchartz, Müller, Oertel and Zierep, Acta Mech.,40,181-194, 1981.
10. Koschmieder and Pallas, Int. J. Heat Mass Transfer, 17, 991-1002,1974.
11. Le Bail, J. Comp. Phys., 9, 440-465,1972.
12. Leonardi, Reizes and De Vahl Davis,Num. Meth.Lam.Turb.Flow, Pineridge Press, 995-1006, 1981.
13. Leong and De Vahl Davis, Num.Meth. Thermal Prob., Pineridge Press, 287-296, 1979.
14. Leong, PHD Thesis, Univ. of New South Wales, Australia, 1981.
15. Mallinson and De Vahl Davis, J. Comp. Phys.,12, 435-461, 1973.
16. Mallinson and De Vahl Davis, J. Fluid Mech., 83,1,1-31, 1977.
17. Mallinson, Graham and De Vahl Davis, J. Fluid Mech.,109,259-275,1981.
18. Mc Laughin and Orszag, J. Fluid Mech.,122, 123-142, 1982.
19. Olson and Rosenberger, J. Fluid Mech.,92, 609-629, 1979.
20. Ozoe, Yamamoto, Churchill and Sayama, J.Heat Trans.,ASME,202-207,1976.
21. Ozoe, Yamamoto, Sayama and Churchill,Int.J.Heat Mass Trans., 20, 123-139,1977.
22. Rosenblat, J. Fluid Mech., 122, 395,410, 1982.
23. Schiroky and Rosenberger, Int. J. Heat Mass Trans.,to appear, 1983.
24. Stork and Müller, J. Fluid Mech., 71, 2, 231-240, 1975.
25. Upson, Grèsho, Sani, Chan and Lee, Livermore N.Lab.,UCRL-85555,1983.
26. De Vahl Davis, Num. Heat Transfer, 2, 261-266, 1979.

Acknowledgments

The Research was supported by the Centre National d'Etude Spatiale (CNES) and the Centre National de Recherche Scientifique (CNRS). Most of the computations were run on the CRAY01-S of the GC_2VR (CNRS). The authors would like to acknowledge S.S. Leong, R. Sani, R. Pèyret, B. Lhomme, J.M. Lacroix and F. Rosenberger for fruitfull discussion and helpfull correspondence.

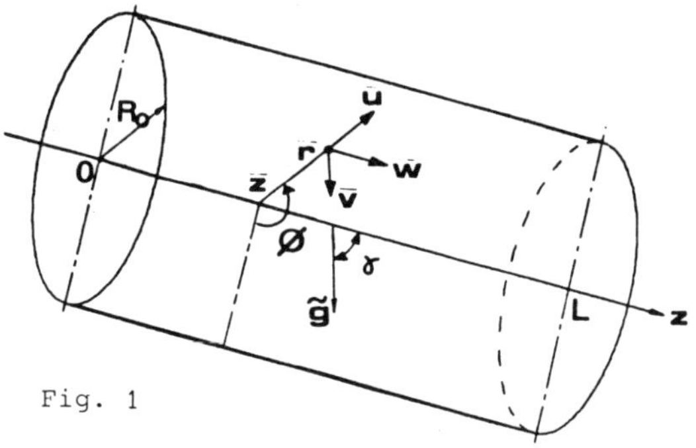

Fig. 1

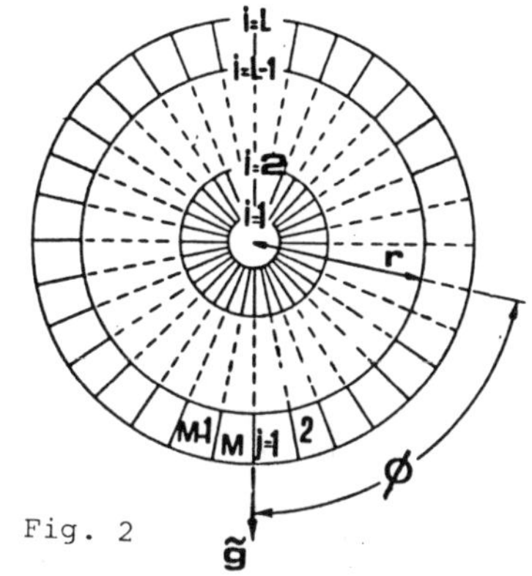

Fig. 2

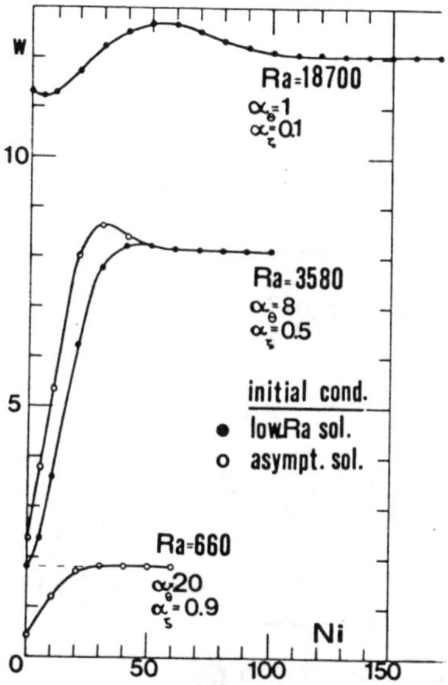

Fig. 3

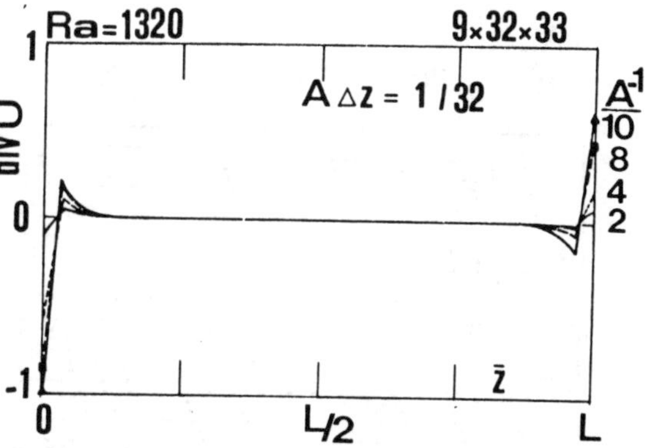

Fig. 4

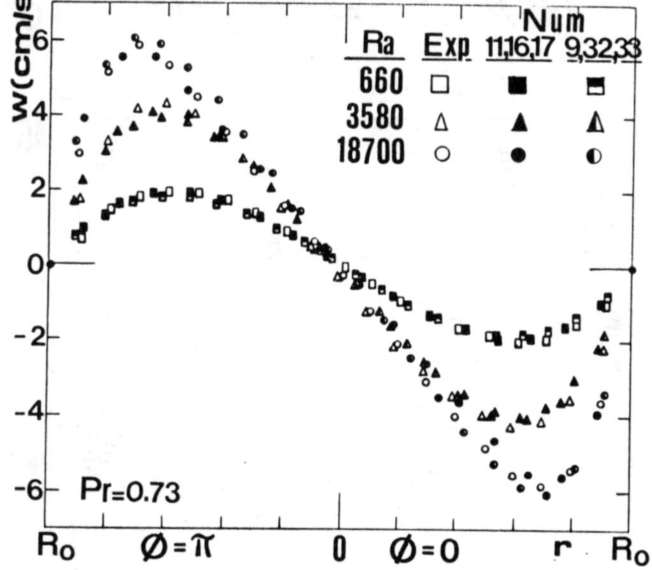

Fig. 5

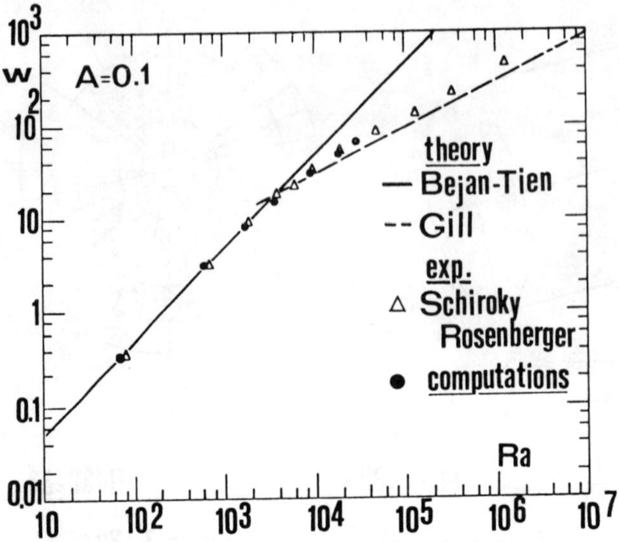
w
A=0.1
theory
— Bejan-Tien
-- Gill
exp.
△ Schiroky Rosenberger
● computations
Ra
0.01
0.1
1
10
10
10
10
10
10^2
10^3
10^4
10^5
10^6
10^7

Fig. 6

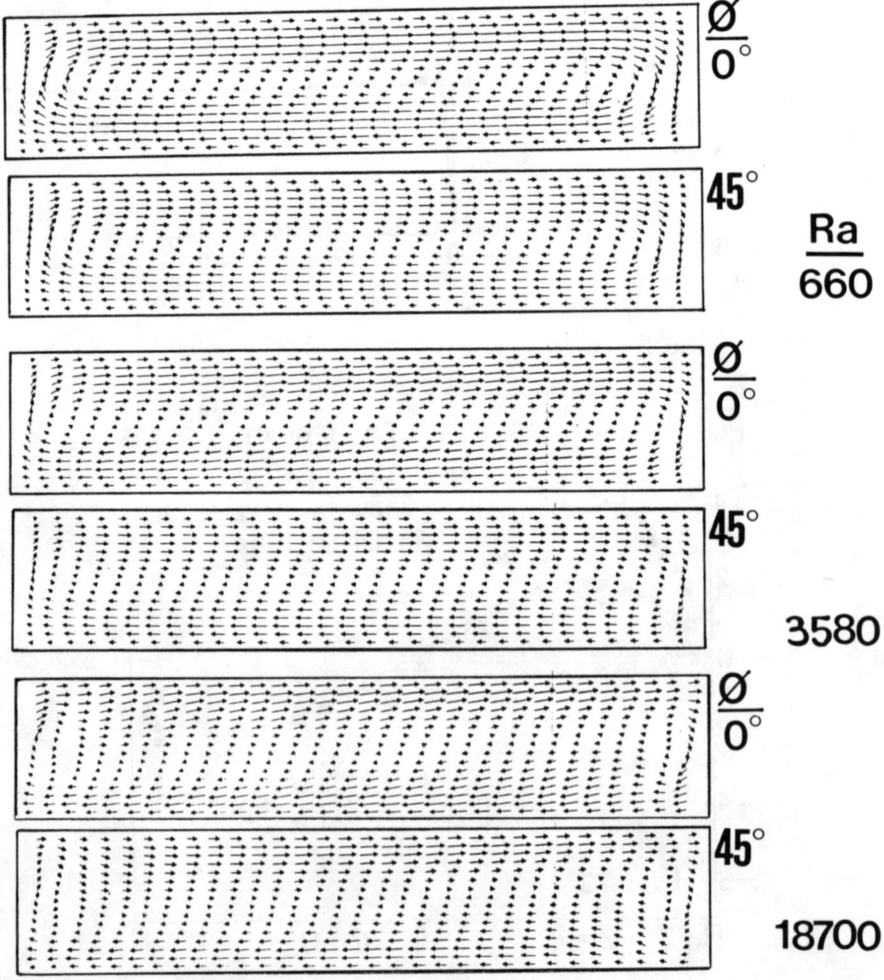
Ø
0°
45°
Ra
660
Ø
0°
45°
3580
Ø
0°
45°
18700

Fig. 7

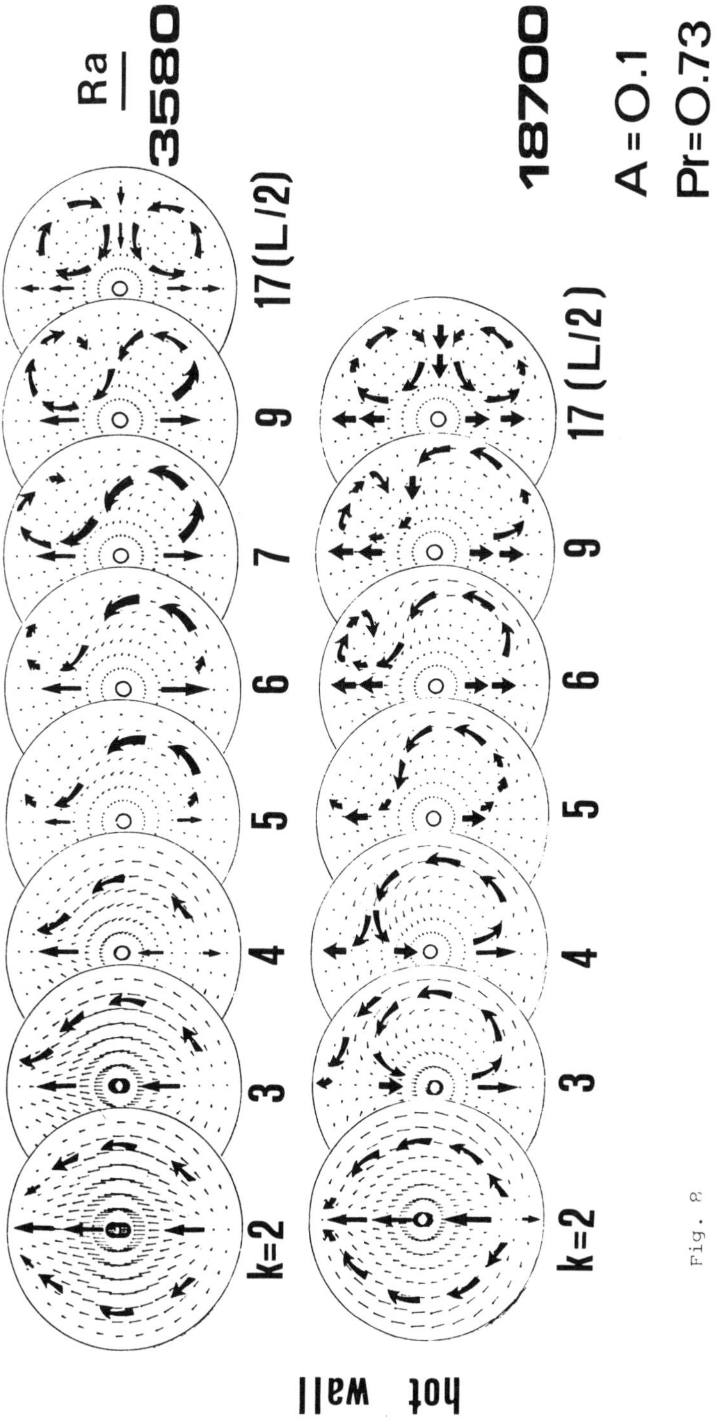

Fig. 8

MULTIPLE-GRID STRATEGIES FOR ACCELERATING THE CONVERGENCE OF THE EULER EQUATIONS

By

Robert M. Stubbs

National Aeronautics and Space Administration
Lewis Research Center
Cleveland, Ohio 44135

SUMMARY

A recently developed multiple-grid method has been applied to an implicit scheme for solving the unsteady Euler equations for quasi-one-dimensional transonic flows. Convergence acceleration is achieved by this technique for CFL numbers greater than one. It is shown that a small revision to the original multiple-grid method significantly improves the convergence rate. Several grid-cycling procedures are investigated to determine an optimum strategy for increasing the computational efficiency.

INTRODUCTION

A common method for finding the steady-state solution of the Euler equations is to march the unsteady equations of motion to their time-asymptotic limit. To accelerate this process Ni [1] developed one of the first multiple-grid methods applicable to hyperbolic systems. In the first part of his method fine-grid corrections are calculated by a one-step, second-order accurate, finite volume technique. In the second part these corrections are propagated rapidly throughout the computational domain by a coarse-grid procedure. Because Ni's method is explicit the size of the time step is limited by the CFL condition. The larger mesh spacings of the coarser grids allow larger time steps to be employed during the coarse-grid procedure. Since there is continuous cycling between the fine and coarse grids the accuracy associated with the fine grid is preserved.

Rather than use Ni's finite volume technique to calculate fine-grid corrections Johnson [2] found it to be more efficient to apply the coarse-grid procedure after using two-step Lax-Wendroff methods on the fine grid. The reason for the improved performance of the two-step methods is that the flux Jacobians need not be calculated as is the case in Ni's fine-grid procedure.

The coarse-grid procedure of Ni has proved to be a valuable addition to the tool box of the computational fluid dynamicist because of its easy adaptation to a number of Lax-Wendroff type solvers and its ability to accelerate the convergence of the Euler and Navier-Stokes equations [3]. Because of the relative newness of this procedure there are several questions regarding its use. For example, can it be used with implicit methods and is there any advantage to doing so? Are there modifications to the basic procedure which can improve its efficiency? Is the grid cycling procedure presented by Ni the optimum strategy to employ? There has been some recent progress in this area [4,5] and this paper continues to address these questions and presents the results of multiple-grid calculations of the quasi-one-dimensional Euler equations for transonic flow through nozzles. In the next section a relatively recent implicit method for solving the Euler equations will be presented. Then Ni's coarse-grid procedure will be outlined and applied to the implicit method in several variations. Finally, numerical results and the conclusions they imply will be given.

MACCORMACK'S IMPLICIT METHOD

The Euler equations for quasi-one-dimensional flow through a channel of varying cross section can be written as

$$\frac{\partial U}{\partial t} + \frac{\partial F}{\partial x} = G \tag{1}$$

where

$$U = \begin{bmatrix} \rho H \\ \rho u H \\ E H \end{bmatrix}, \qquad F = \begin{bmatrix} \rho u H \\ (P + \rho u^2) H \\ u(P + E) H \end{bmatrix}, \qquad G = \begin{bmatrix} 0 \\ P \dfrac{dH}{dx} \\ 0 \end{bmatrix} \tag{2}$$

Here ρ is the density, H the area, u the velocity, E the total energy per unit volume, P the pressure, and s and t are the spatial and temporal variables. An assumption of a perfect gas is made:

$$P = (\gamma - 1)\left(E - \frac{\rho u^2}{2}\right) \tag{3}$$

MacCormack [6] developed an implicit finite difference method for time-integrating the Navier-Stokes equations which is second order accurate in space and time, unconditionally stable, and very efficient in that no block or scaler tridiagonal inversions are required. This method is an extension to his well known explicit scheme [7] which is of the two-step Lax-Wendroff type. White and Anderson [8] applied this implicit scheme to Equation (1) in the following form. Predictor:

$$\left.\begin{aligned} &\Delta U_i^n = -\Delta t\left(\frac{F_{i+1}^n - F_i^n}{\Delta x}\right) + \Delta t G_i^n \\ &\left(I + \frac{\Delta t}{\Delta x}|A|_i^n\right)\delta U_i^{\overline{n+1}} = \Delta U_i^n + \frac{\Delta t}{\Delta x}|A|_{i+1}^n\,\delta U_{i+1}^{\overline{n+1}} \\ &U_i^{\overline{n+1}} = U_i^n + \delta U_i^{\overline{n+1}} \end{aligned}\right\} \tag{4}$$

Corrector:

$$\left.\begin{aligned}
\Delta U_i^{\overline{n+1}} &= -\Delta t\left(\frac{F_i^{\overline{n+1}} - F_{i-1}^{\overline{n+1}}}{\Delta x}\right) + \Delta t G_i^{\overline{n+1}} \\
\left(I + \frac{\Delta t}{\Delta x}\,|A|_i^{\overline{n+1}}\right)\delta U_i^{n+1} &= \Delta U_i^{\overline{n+1}} + \frac{\Delta t}{\Delta x}\,|A|_{i-1}^{\overline{n+1}}\,\delta U_{i-1}^{n+1} \\
U_i^{n+1} &= U_i^n + \frac{1}{2}\left(\delta U_i^{\overline{n+1}} + \delta U_i^{n+1}\right)
\end{aligned}\right\} \quad (5)$$

$|A|$ is a matrix with positive eigenvalues related to the Jacobian, $A = \partial F/\partial U$.

NI'S COARSE-GRID PROCEDURE

The basic philosophy of the coarse-grid procedure of Ni is to use the coarse grids to propagate the fine-grid corrections throughout the flow field more efficiently. To do this in one dimensional calculations the fine mesh, or h grid, is coarsened by removing every other point. The next coarser mesh, the 4h grid, results from removal of every other nodal point of the 2h grid, and so on. After fine grid corrections, δU_i^h, have been calculated by any of several time integration methods the first coarse-grid corrections, δU_i^{2h}, are calculated for locations i, i±2, i±4,... by

$$\delta U_i^{2h} = \frac{1}{2}\left[\delta U_{i-1}^h + \delta U_{i+1}^h + \frac{\Delta t}{\Delta x}\left(A_{i-1}\delta U_{i-1}^h - A_{i+1}\delta U_{i+1}^h\right) + \frac{1}{2}\,\Delta t\left(B_{i-1}\,\delta U_{i-1}^h + B_{i+1}\,\delta U_{i+1}^h\right)\right] \quad (6)$$

Here, B is the Jacobian $\partial G/\partial U$. After finding δU_i^{2h} at all points on the 2h grid, corrections to other points of the fine grid (i.e. points i±1, i±3,...) are calculated by linear interpolation of the corrections of Equation (6). Corrections on the next coarser, or 4h, grid are now calculated at points i, i±4, i±8,...

$$\delta U_i^{4h} = \frac{1}{2}\left[\delta U_{i-2}^{2h} + \delta U_{i+2}^{2h} + \frac{\Delta t}{\Delta x}\left(A_{i-2}\delta U_{i-2}^{2h} - A_{i+2}\delta U_{i+2}^{2h}\right) + \frac{1}{2}\,\Delta t\left(B_{i-2}\,\delta U_{i-2}^{2h} + B_{i+2}\,\delta U_{i+2}^{2h}\right)\right] \quad (7)$$

Again, the fine grid is updated by linearly interpolating these δU_i^{4h} onto every point of the fine grid. This process may be continued to progressively coarser grids. After computing corrections on the coarsest grid and interpolating these onto the fine grid the next iteration cycle of this multiple-grid scheme would begin by applying the time integration method on the fine grid.

In this paper a multiple-grid scheme for solving the Euler equations has been formulated by applying MacCormack's implicit method on the fine grid and coupling it with the coarse-grid procedure of Ni. Some motivation for doing this comes from the observation that, although the implicit MacCormack method is unconditionally stable in some situations, there are cases [9] where the size of the time step is stability limited and where accuracy deteriorates with increasing CFL number. Should the size of the time step be limited by either accuracy or stability considerations there might be an advantage to employing the multiple-grid scheme which could advance the solution further at every iteration with a minimum of work and without sacrificing the accuracy of the fine-grid method.

The way that the coarse-grid procedure described above is implemented with MacCormack's implicit method is to use the results of a predictor-corrector sequence as the fine-grid correction to be used in Equation (6). That is

$$\delta U_i^h = \frac{1}{2}\left(\delta U_i^{n+1} + \delta U_i^{\overline{n+1}}\right) \tag{8}$$

When the coarse-grid procedure was carried out with this substitution, effective convergence acceleration was achieved for CFL numbers greater than unity. Further improvement was achieved by modifying Equation (8) which will be discussed later.

GRID CYCLING STRATEGIES

The coarse-grid procedure of Ni involves cycling through a sequence of sucessively coarses grids. The number of cells at any grid level is 2^{-N} times the number of cells in the less-coarse grid, where N is the dimesionality of the problem. The question arises whether cycling in this sequence, i.e. h grid to 2h grid to 4h grid to 8h grid, etc., is the optimum sequence to pursue. Since there is less computational work involved on the coarser grids might there be an advantage to skipping some of the less coarse grids? To answer this type of question several grid skipping sequences were tested and compared. For example a cycling between the h and 4h grids, skipping the 2h grid, would involve ignoring Equation (6) and replacing Equation (7) with

$$\delta U_i^{4h} = \frac{1}{2}\left[\delta U_{i-2}^h + \delta U_{i+2}^h + \frac{\Delta t}{\Delta x}\left(A_{i-2}\delta U_{i-2}^h - A_{i+2}\delta U_{i+2}^h\right) + \frac{1}{2}\Delta t\left(B_{i-2}\delta U_{i-2}^h + B_{i+2}\delta U_{i+2}^h\right)\right] \tag{9}$$

applied at i, i±4, i±8,.... As will be shown, it was found that certain economies accrue from these grid skipping procedures.

RESULTS

Figure (1) shows the results of a multiple-grid, implicit scheme calculation of the quasi-one-dimensional, shocked flow through the converging-diverging nozzle whose cross-sectional area is also shown. The number of grid points is 81 and the flow has a maximum local CFL number of

1.4. The converged steady state Mach number distribution is plotted along with the exact analytic solution. These results were produced with one level of grid coarsening. The convergence history of this calculation is shown in Figure (2) where it can be compared to the performance of the basic implicit method of MacCormack. It can be seen that the effect of one grid coarsening is to cut in half the number of iterations required to reach any level of convergence. Typically, in runs of similar nozzle flows, the additional computational work required to apply multiple-gridding with one grid coarsening was approximately one eighth of the work done in the fine grid calculation. At a given time step, then, any level of convergence can be achieved significantly faster when the multiple-grid procedure is employed.

It is found that the unconditional stability enjoyed by MacCormack's implicit method for some flow configurations is removed when the coarse-grid procedure is added. For the shocked flow case treated here the maximum CFL number which can be used is approximately 1.8, and for unshocked nozzle flows this number is about four [4]. Employing the coarse-grid procedure with this implicit procedure, then, would be useful only in cases where either the accuracy deteriorated to an unacceptable level or where instabilities developed near these values of the CFL number.

A small modification to Ni's scheme will improve its performance significantly when used with either explicit or implicit methods. Ni's procedure takes advantage of the fact that a second-order-accurate temporal change in U can be written in terms of first-order changes. For example, if, in the homogeneous version of Equation (1), first order changes at cell centers were labeled ΔU,

$$\Delta U_{i+1/2} = \Delta t \left(\frac{\partial U}{\partial t}\right)_{i+1/2} = -\Delta t \left(\frac{F_{i+1}^{n} - F_{i}^{n}}{\Delta x}\right), \tag{10}$$

then second-order temporal changes, δU, at grid points can be written as

$$\delta U_i = U_i^{n+1} - U_i^n = \Delta t \left(\frac{\partial U}{\partial t}\right)_i + \frac{(\Delta t)^2}{2}\left(\frac{\partial^2 U}{\partial t^2}\right)_i$$

$$= \frac{1}{2}\left(\Delta U_{i-1/2} + \Delta U_{i+1/2}\right) + \frac{\Delta t}{2\Delta x}\left(A_{i-1/2}\,\Delta U_{i-1/2} - A_{i+1/2}\,\Delta U_{i+1/2}\right) \tag{11}$$

In Equation (6), then, δU_{i-1}^{h} is used as the first-order temporal change in the 2h cell with boundaries at grid points (i-2) and i. One efficiency of the Ni method is that, rather than compute the first order change by flux differencing as in Equation (10), the first order change is approximated by "injecting" the change freshly calculated on the less coarse grid at the grid point which is now at the cell center. That is, the second order change at a point at one grid level becomes the first order approximation of the cell change at the next coarser grid:

$$\Delta U_{i-1}^{2h} = \delta U_{i-1}^{h} \tag{12}$$

But such a substitution does not preserve the CFL number used on the fine grid. Since ΔU_{i-1}^{2h} represents the first order temporal change for a cell of twice the fine grid size one would expect twice the temporal change for the same CFL number. Since it would seem more efficient to operate at this higher CFL number the following modification in scaling was tested:

$$\Delta U_{i-1}^{2h} = 2\ \delta U_{i-1}^{h} \tag{13}$$

That is, all occurances of δU^h in Equation (6) were replaced by $2\delta U^h$. The result of this modification was a 50 percent improvement in convergence rate over the original multiple grid method with no increase in computational work. That is, this modification produces a three-fold decrease in the number of iterations for convergence as shown in Figure (2).

Figures (3),(4) and (5) compare the rates of convergence for various grid-cycling procedures involving grids of 80 intervals (h grid), 40 intervals (2h grid), 20 intervals (4h grid), and 10 intervals (8h grid). Figure (3) indicates that the usual procedure of cycling through successively coarser grids can be improved upon by skipping some levels. For example, cycling between grids h and 4h produces a steeper convergence curve than the more costly cycle of grids h, 2h, 4h. When only two grid levels are employed, skipping between grids h and 4h produces faster convergence than either the h-2h sequence or the h-8h sequence. This is indicated in Figures (3) and (4) for both the standard and the modified multiple-grid schemes, respectively. The computations of Figure (4) were all carried out after multiplying the first-order approximations recommended by Ni, δU^h, by a factor of two. When larger factors were tried on the coarser 4h and 8h grids, to scale with the grid size, the calculations became unstable. Proper scaling of these first order approximations for maximum convergence rate is not fully understood and remains a subject of study.

When three grid levels are used in these calculations the aforementioned scaling modification cannot be applied at all coarse grid levels without an excessive amount of artificial viscosity having to be invoked. Figure (5) shows that, although there is improvement when used on either coarse grid, the scaling modification is put to greater advantage when used on the finer of the coarse grids.

CONCLUSIONS

The adaptability of Ni's coarse-grid procedure to several time integration methods is extended to include an implicit scheme to which it can be appended with ease. The explicit coarse-grid procedure introduces stability limitations to the size of the time step which can be used. Improvements in the coarse-grid procedure's convergence properties, when used with either explicit or implicit methods, are shown to result from a modification to the scaling of first-order approximations. Further improvements in computational efficiency occured when a grid skipping strategy was employed. There are indications that some of these conclusions are valid at higher dimensions and such a study is being pursued.

Although much has been learned of the multiple-grid method through computational experimentation there is clearly a need for a stronger theoretical understanding of these findings.

REFERENCES

[1] Ni,R.H., "A Multiple-Grid Scheme for Solving the Euler Equations," AIAA Paper 81-1025, June 1981

[2] Johnson,Gary M., "Multiple-Grid Acceleration of Lax-Wendroff Algorithms," NASA TM 82843, March 1982

[3] Johnson,Gary M., "Multiple-Grid Convergence Acceleration of Viscous and Inviscid Flow Computations," NASA TM 83361, April 1983

[4] Stubbs,Robert M., "Multiple-Gridding of the Euler Equations With an Implicit Scheme," AIAA Paper 83-1945, July 1983

[5] Johnson,Gary M., "Flux-Based Acceleration of the Euler Equations," NASA TM 83453, July 1983

[6] MacCormack,R.W., "A Numerical Method for Solving the Equations of Compressible Viscous Flow," AIAA Paper 81-0110, January 1981

[7] MacCormack,R.W., "The Effect of Viscosity in Hypervelocity Impact Cratering," AIAA Paper 69-354, April 1969

[8] White,M.E., and Anderson,J.D., "Application of MacCormack's Implicit Method to Quasi-One-Dimensional Nozzle Flows," AIAA Paper 82-0992, June 1982

[9] Shang,J.S., and MacCormack,R.W., "Flow Over a Biconic Configuration with an Afterbody Flap - A Comparative Numerical Study," AIAA Paper 83-1668, July 1983

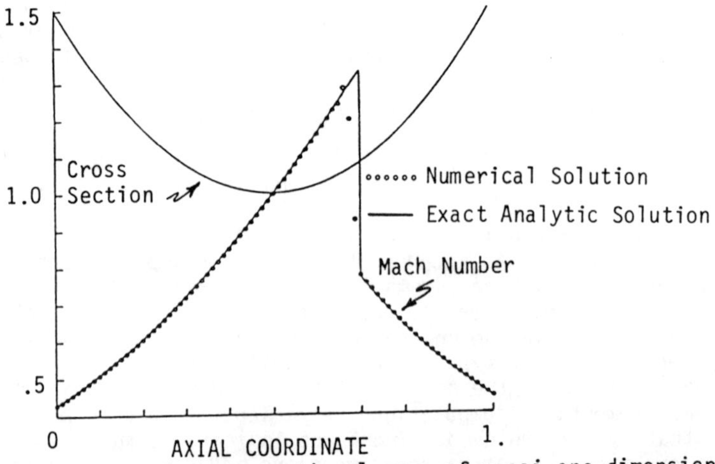

Figure (1) Mach number and cross-sectional area of quasi-one-dimensional nozzle.

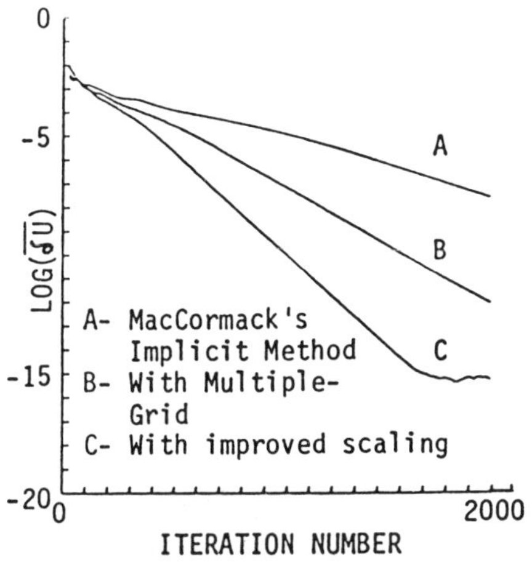

Figure (2) Convergence history of three implicit methods for solving the unsteady Euler equations for the model problem of Figure (1). The ordinate is the logarithm of the average change in the three components of the vector of conserved quantities, U.

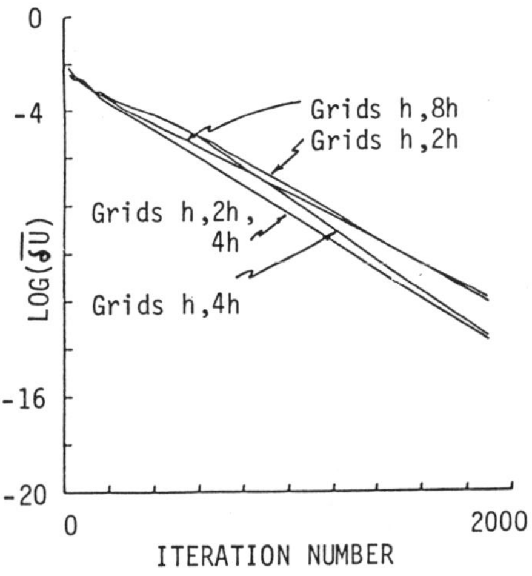

Figure (3) Convergence history of four grid-cycling procedures.

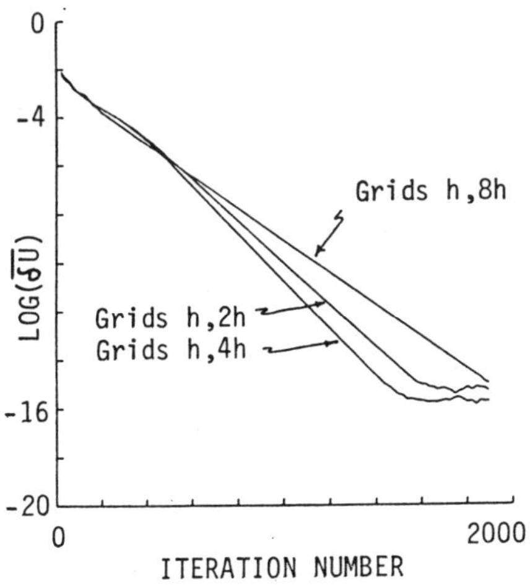

Figure (4) Convergence history of grid-cycling procedures using the improved multiple-grid method.

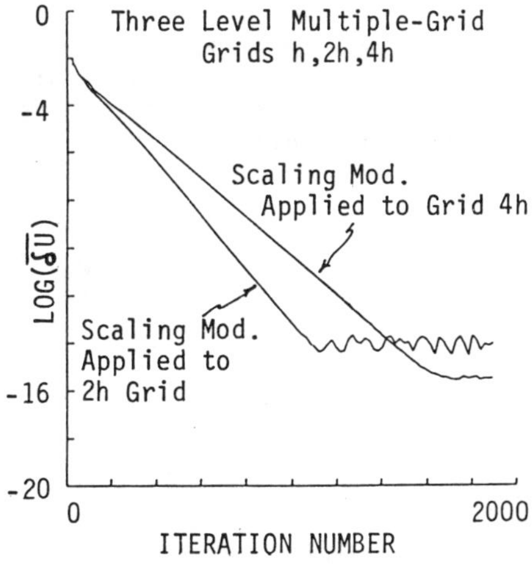

Figure (5) Convergence history showing effects of applying scaling modification at various grid levels.

A FAST, WELL POSED NUMERICAL METHOD FOR THE INVERSE DESIGN OF TRANSONIC AIRFOILS

G. Volpe
Research & Development Center
Grumman Aerospace Corporation
Bethpage, New York 11714

SUMMARY

A numerical method for designing closed airfoils that correspond to given surface speed distributions is described in this paper. Since this problem has a solution only if the target speed distribution satisfies three constraints, the speed can be prescribed only to within three adjustable parameters whose values are determined numerically. The computational scheme solves the full potential equation in the circle plane and in full conservation form by an approximate factorization-multigrid method. Several strategies for introducing the necessary freedom in the prescribed speed are described and airfoil solutions are presented.

INTRODUCTION

The design of airfoil profiles for transonic applications has followed one of three different approaches. One class of methods is based on the hodograph transformation. The mathematical elegance of these methods is, however, balanced by the difficulty of implementation. The inputs to be made to the methods are not easily translated into physical characteristics, and this requires the user to be familiar with the formulation of the method. A second approach is to use a sequence of Neumann-type "direct" solutions whereby the speed distribution over an airfoil surface is computed and then compared to a desired target speed. The difference between computed and target speed is then used in some rational way to modify the airfoil contour, and this process is repeated as long as desired. Control over the airfoil geometry is retained at all times with this type of approach but there is no guarantee that the iteration converges and that the differences between computed and target speeds can be reduced to arbitrarily small levels.

A third class of methods follows the classical approach described by Lighthill [1], in which the airfoil profile corresponding to a desired surface speed distribution is obtained as the solution to a Dirichlet problem. This is the approach that will be followed in this paper. One advantage of the inverse approach is that one can at all times maintain control over the flow characteristics and in particular over the boundary layer development on the airfoil surface. An important feature of design methods of this type is that the airfoil shape can always be found if the target speed distribution satisfies certain constraints as was first pointed out by Lighthill. Two constraints arise from the requirement that the airfoil profile have a specific trailing edge gap. An additional constraint, more subtle, requires that the specified surface speed distribution be compatible with the specified free stream speed. The existence of these three constraints implies that, in general, the target speed distribution must contain three free parameters to guarantee that the constraints are satisfied through proper adjustment of the parameters.

For incompressible flow, these constraints on the speed can be expressed in closed form. If the target speed distribution, q_o, is expressed as a function of the polar angle (ω) in the circle plane into which any airfoil can be mapped, one can expand q_o into a trigonometric series. Since the velocity q is analytic everywhere in the flow field, it can be expressed as a series in negative powers of r (measured from the center of the circle) and periodic in ω. At infinity, obviously q must be equal to the free stream speed q_∞. Hence, the zero order term for the series of q must be equal to q_∞. As a result, the zero order term of the series for the surface speed q_o must be q_∞. This constraint has been recognized and faithfully accounted for in incompressible design methods. However, it has been ignored in transonic flow applications until recently. It can also be shown that the requirement that the airfoil be closed precludes the series for q_o from containing terms of order one resulting in two additional constraints on q_o. For incompressible flow, having specified the target speed q_o as a function of ω on the unit circle enables one to immediately construct the airfoil profile since the ratio of q_o to the uniform flow over the circle (2 sin ω) is essentially the mapping modulus. If the latter two constraints are not enforced, the airfoil profile will not have the desired gap size. If the first constraint is not satisfied by q_o no airfoil profile exists because q_o on the circle does not represent a physical flow in which the circle itself is a streamline. Leaving three free parameters in q_o makes it possible to satisfy the three constraints at all times since the constraints themselves provide expressions for determining the proper values of the parameters. If q_o is prescribed as a function of the arc length s, the constraints are unchanged, but the evaluation of the parameters is more difficult because $q_o(\omega)$ will be known only after $s(\omega)$ is found. In practice the parameters and the airfoil shape (and thus $s(\omega)$) are found simultaneously.

Woods [2] extended Lighthill's analysis to compressible flows using a Karman-Tsien type gas, and derived explicit constraints on q_o which reduce to Lighthill's expressions in the incompressible limit. Explicit expressions for the constraints have not been derived for a perfect gas in the compressible, supercritical regime, but they must still exist since the incompressible limit is only a special case of the more general compressible flow. Alternate means of determining the proper values of the parameters must be formulated in order to guarantee an airfoil solution for every q_o.

TRANSONIC AIRFOIL DESIGN

The problem that is addressed is that of finding an airfoil profile of a specified trailing edge thickness (Δx, Δy) that corresponds to the speed distribution

$$\frac{q_o}{q_\infty} = F(s/s_{max};\, a_1,\, a_2,\, a_3) \qquad (1)$$

The arc length s is measured clockwise around the airfoil starting at the lower surface trailing edge point. Without loss of generality s_{max} can be set to unity. a_1, a_2, and a_3 are the three parameters that are to be found as part of the solution. The particular functional forms chosen to introduce the necessary freedom in F(s) affect the class of airfoils that can be obtained. The following form will be assumed for the remainder of this paper:

$$\frac{q_o}{q_\infty} = f_1(\omega;a_1)[f_o(s) + a_2 f_2(\omega) + a_3 f_3(\omega)] \qquad (2)$$

$f_o(s)$ specifies the target speed distribution and it is usually a tabulated function. f_1, f_2, and f_3 are specified functions that modify the initially prescribed distribution f_o by adjusting the values of a_1, a_2, and a_3 to assure the existence of the proper airfoil solution. Ideally a_2 and a_3 should be zero and $a_1 f_1(\omega)$ should be unity. It is therefore desirable to make $f_2(\omega)$ and $f_3(\omega)$ highly localized. Since the computational scheme to be described is formulated in the circle plane, making f_1, f_2, and f_3 functions of ω rather than s simplifies the formulation without loss of generality. The strategy followed is to iteratively modify some initial contour until the desired speed distribution, q_o, is achieved. This initial contour can be mapped into the unit circle by the transformation

$$\frac{dz}{d\zeta} = \left(1 - \frac{1}{\zeta}\right)^{(1-\varepsilon)} e^{P+iQ} \tag{3}$$

where $z = x + iy$ and $\zeta = re^{i\omega}$ are the coordinates in the physical and mapped planes, respectively, and $\varepsilon\pi$ is the included trailing edge angle. Equation (3) can be separated into its real and imaginary parts. Thus

$$\frac{ds}{d\omega} = \left[2 \sin \frac{\omega}{2}\right]^{(1-\varepsilon)} e^{P} \tag{4}$$

$$\theta = \frac{1}{2}(1 + \varepsilon)(\pi - \omega) - \frac{\pi}{2} + Q \tag{5}$$

where θ is the slope of the airfoil. P is the Fourier series

$$P = \sum_{n=0}^{N} (A_n \cos n\omega + B_n \sin n\omega) \tag{6}$$

and Q is its conjugate series.
The coefficients of the series are found by standard Fourier analysis. With this mapping procedure, the leading terms of the series are related to the trailing edge gap $(\Delta x, \Delta y)$ by

$$A_1 = g_1(\Delta x, \Delta y; B_o, \varepsilon) \quad , \quad A_2 = g_2(\Delta x, \Delta y; B_o) \tag{7}$$

where g_1 and g_2 are bilinear functions in Δx and Δy. Assuming irrotational flow a potential function ϕ can be defined from which the physical velocity components can be computed as

$$u = \frac{r}{h} \phi_\omega \quad , \quad v = \frac{r^2}{h} \phi_r \tag{8}$$

where h is the modulus of the transformation to the inside of the circle. Assuming now that the target speed q_o is equal to the tangential velocity u at the airfoil surface, the potential distribution on the unit circle can then be found by integration. The flow within the circle, corresponding to the flow over the airfoil, can then be computed numerically. In this Dirichlet-type problem, if the unit circle were a true streamline of the flow, the normal velocity component, v, would be zero on it and the airfoil would be the profile corresponding to the specified q_o. In general v is not zero. This means that the actual streamline is (to first order) rotated from the assumed boundary by an angle of magnitude

$$\delta\theta = \tan^{-1}\left(\frac{v}{u}\right) \tag{9}$$

This is used to modify the initial slope distribution θ in Eq (5). New series for Q and P can then be computed, as well as a new $\frac{ds}{d\omega}$ and

$$\frac{dx}{d\omega} = -\frac{ds}{d\omega}\cos\theta \quad , \quad \frac{dy}{d\omega} = -\frac{ds}{d\omega}\sin\theta \tag{10}$$

s, x, and y are then obtained by integration. Using the new airfoil shape a new Dirichlet problem can be set up, and the process is continued until a desired tolerance for $|v/u|$ is achieved. However, a look at Eq (9) will show that the procedure will fail if v is not zero at points where u is zero. As pointed out by Volpe and Melnik [3], v must be forced to be zero at points where the prescribed speed is zero. Physically, this means that, if the curve on which q_0 is prescribed is to be an airfoil, branching of the streamlines must be permitted, and this can happen only if the zeroes of q_0 are stagnation points of the flow. Also if the flow is to be a physically realistic flow there cannot be any mass flow at infinity. However, there is no guarantee that the integrated mass flow through the boundary is zero at any intermediate step of the design procedure. Thus a mass flow term $[\sigma \log r]$ must be introduced in the field. The value of σ is computed by making $\phi_r = 0$ at the trailing edge. This condition is analogous to the condition that ϕ_ω vanish at this point in Neumann problems providing an equation to evaluate the circulation. This condition guarantees that the trailing edge is a branch point in the Dirichlet problem also and that (v/u) remains finite there. σ goes to zero as the design problem approaches its solution. The requirement that v be zero at the leading edge stagnation point then leads to a regularity condition that replaces the first constraint. One of the parameters in Eq (2) can be adjusted to force v to be zero where u is zero at all times. The other two parameters are adjusted to drive the first order terms of the series of the mapping (Eq (6)) to its desired values which can be computed from Eq (7). Details of this can be found in Ref. 4.

The solution to each Dirichlet problem is obtained numerically after discretizing the full potential equation in full conservation form on a polar coordinate grid. Obviously several Dirichlet problems must be solved in succession, and during each, a_1, a_2, a_3, and σ are continuously adjusted causing a shifting of the boundary conditions. The procedure would be hopelessly lengthy without a fast algorithm for the solution of the flow field. The algorithm used is basically the multigrid-ADI scheme developed by Jameson [5] modified to accept Dirichlet boundary conditions. The search for the parameters is facilitated by choosing f_1, f_2, and f_3 in Eq (2) in such a way that each constraint affects only one of the three parameters strongly.

Satisfaction of the first constraint is accomplished by adjusting a_1. From Eq (2) it can be seen that this causes a scaling of (q_0/q_∞). It could be interpreted as a scaling of the prescribed surface speed q_0 or of q_∞. Regardless, the effect on the resulting pressure coefficient is the same. By making $f_1 = a_1$, the scaling can be made uniform along the airfoil. It can be concentrated in the front half by choosing

$$f_1(\omega;a_1) = \sqrt{1 + a_1 \sin^2(\tfrac{\omega}{2})} \tag{11}$$

Control over Δy can be exercised by defining

$$f_2(\omega) = \sin\,(\tfrac{8}{3}\,\omega) \quad , \quad \omega \leqslant \tfrac{3}{4}\,\pi \tag{12}$$

Outside this range f_2 is zero. This function alters the target speed only on the lower surface of the airfoil and would thus be unsatisfactory if a symmetric design was to be accomplished. In the computational plane the velocity on the lower surface is negative. A form of f_2 that would change the target speed and the pressure coefficient in the physical plane symmetrically is

$$f_2(\omega) = 1 - \frac{\omega}{\omega_1} \quad , \quad \omega \leqslant \omega_1$$

$$= 1 - \left(\frac{2\pi - \omega}{\omega_1}\right), \quad \omega \geqslant 2\pi - \omega_1 \qquad (13)$$

Δx is primarily affected by the location of the leading edge stagnation point. However, a change in the gradient $(dq_o/d\omega)$ at the stagnation point would significantly affect Δy and the first constraint. Thus, to maintain a loose coupling among a_1, a_2, and a_3, the shift in the stagnation point must be accomplished without noticeably altering the local velocity gradient. A function that manages this has been found to be

$$f_3(\omega) = \max(1,\mu) \ , \quad \frac{3\pi}{4} \leqslant \omega \leqslant \frac{5\pi}{4} \qquad (14)$$

where

$$\mu = \max\left[1,\ 0.4\left(\frac{dq_o}{d\omega} - 1\right)\right]$$

Outside the specified range f_3 is taken as zero. Occasionally f_3 does not vanish smoothly near the limits of this range and introduces some local overshoots in the target speed distribution.

The strategy for solving each Dirichlet problem is then as follows. The flow field is swept once on each grid down to the most coarse achievable grid and up again. At this point a_1 and σ are determined by making ϕ_ω zero at the leading edge stagnation point and at the trailing edge. The sequencing through the grids is then resumed. ϕ_ω at the leading edge stagnation point goes to zero quite fast, and when it is below a given tolerance (typically 10^{-5} - 10^{-6}) estimates of what the values of A_1 and B_1 would be if the airfoil was to be constructed at that stage are made. These values are compared with the values they should have based on the desired trailing edge gap (Eq (7)) and the differences, δA_1 and δB_1, used to change a_2 and a_3, respectively. The change in a_2 is made directly proportional to δA_1 and that in a_3 proportional to $-\delta B_1$. a_1 must be updated continuously to keep the surface speed and the free stream from becoming incompatible. This would become apparent in the algorithm becoming quickly divergent. a_2 and a_3 have to be updated only infrequently because the convergence qualities of the algorithm are not affected by the gap size, but each update alters the boundary conditions raising the residuals everywhere in the flow field.

RESULTS

Several schemes for modifying the speed distribution to satisfy the constraints have been investigated and results will be presented for each. In every instance f_3 has the form defined by Eq (14). With a free stream Mach number of 0.800, the speed distribution depicted by the symbols in Fig. 1 does not lead to a solution even if trailing edge closure is not enforced. It does not satisfy the first constraint. With $f_1 = a_1$, $a_2 = a_3 = 0$, this speed can be scaled into one for which a solution exists (the dashed line), even though the airfoil is open in both x and y, as can be seen by the results in Fig. 2. A closed airfoil can be obtained as the solution by modifying the original target not only by a constant scaling but also by the addition of f_3 and of f_2 as defined by Eq (12) (this will be called scheme 1). Despite the fact that in this case f_3 introduces an overshoot near the leading edge, the solution is found.

The pressure distributions in Fig. 2 are the direct solutions computed for the corresponding airfoil contours. Since care has been taken to make the numerical schemes of the direct Nuemann problem and the Dirichlet problem identical, these pressures are undistinguishable from the pressure distributions computed from the modified target speeds which are shown in Fig. 1 and which were used to generate the contours. The trailing edge gap is less than 0.001%. The airfoil thickness, however, becomes slightly negative over a short distance upstream of the trailing edge. This result not ruled out by this formulation. Additional constraints on the speed distribution are needed to prevent crossovers such as this. Alternatively, the trailing edge thickness might be specified as a number greater than one to retain a realistic thickness throughout. A second example in Fig. 3 (also at M_∞ = 0.800) compares the solution obtained by defining f_2 by Eq (13) (scheme 2) with the solution obtained when f_2 is defined by Eq (12) (scheme 1). The trailing edge gap is again specified to be zero. In both cases a constant scaling ($f_1 = a_1$) is used. It can be seen that the changes introduced by scheme 1 are highly unsymmetrical. At times this may be a very desirable feature. Scheme 2 changes the pressure locally near the trailing edge and the sum total of the modifications, for this particular case, is less than that due to scheme 1. In Fig. 4 the modified target speed distributions (corresponding to the computed distributions in Fig. 3) are compared with the original target. A third example (Fig. 5) compares the solution obtained by scheme 2 with that obtained by defining f_1 by Eq (11), which biases the scaling in the forward portion of the airfoil (scheme 3). For this case, the solutions are very close and both exhibit the f_3 overshoot near the leading edge. A final example, at M_∞ = 0.683, shows (Fig. 6) that the method can be used even in cases where the target distribution has a shock. In this case there is a crossover of the surfaces near the trailing edge.

The above results were obtained by designing the airfoils first on a coarse 48 x 8 grid and then moving through a medium to a fine 192 x 32 grid. In the early design cycles on the coarse grid typically 100 multigrid sweeps were required to converge the Dirichlet problem. This means reduction of the maximum residuals in the flow field to a tolerance of 10^{-6} - 10^{-7}, and reducing the value of ϕ_r at the leading edge stagnation point to 10^{-5}. On the fine grid the number of multigrid sweeps typically reduces to 35-50. Conver-gence of the airfoil shape, and thus of the design problem, on a particular grid can be assumed when the maximum $|v/u|$ is reduced to a value of 10^{-2}, a tolerance that can be achieved in 8 to 10 design cycles (airfoil updates).

CONCLUSIONS

The method described in this paper is formulated to take into account the constraints that exist in the inverse airfoil design problem. The formulation is applicable at transonic speeds, even with shock waves. The method is quite general in the sense that $f_0(s)$ is general and an airfoil solution will always be found by modifying the target speed in order to satisfy the constraints. Also, the initial airfoil contour needed to start the procedure need not be close to the final contour to achieve convergence. The overshoots in the obtained pressure distribution near leading edges that arise in some of the results are due to the particular choices employed for the pressure adjustment functions f_1, f_2, and f_3. The particular forms proposed for these functions are by no means exhaustive or even necessarily best. They do, however, provide the freedom needed to satisfy the constraints automatically, without user intervention, and introduce only a weak coupling among their respective multipliers making their evaluation simpler. Other choices to eliminate overshoots are currently being studied. Considering the variety of cases done, the numerical method has proven to be an efficient and reliable scheme for the design of airfoil profiles of given trailing edge thicknesses at transonic speeds.

REFERENCES

1. Lighthill, M.J., "A New Method of Two-Dimensional Aerodynamic Design," R&M 2112, April 1945, Aeronautical Research Council, London, England.

2. Woods, L.C., "Aerofoil Design in Two-Dimensional Subsonic Compressible Flow," R&M 2845, March 1952, Aeronautical Research Council, London, England.

3. Volpe, G. and Melnik, R.E., "The Role of Constraints in the Inverse Design Problem for Transonic Airfoils," AIAA Paper 81-1233, 1981.

4. Volpe, G., "The Inverse Design of Closed Airfoils in Transonic Flow," AIAA Paper 83-504, 1983.

5. Jameson, A., "Acceleration of Transonic Potential Flow Calculations on Arbitrary Meshes by the Multiple Grid Method," AIAA Paper 79-1458, 1979.

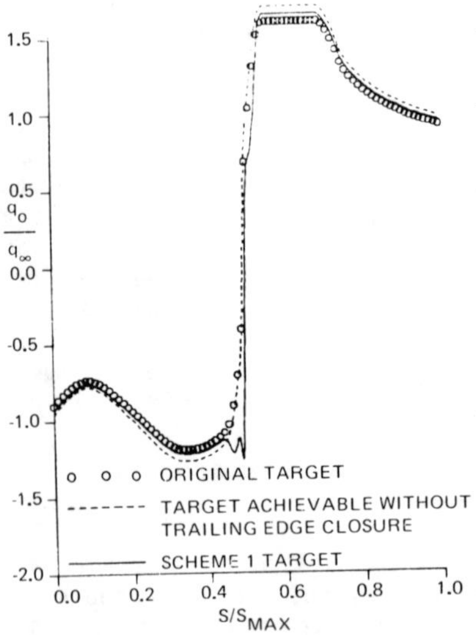

Fig. 1 Original and modified speed distributions; case A1, $M_\infty = 0.800$

1606-003D

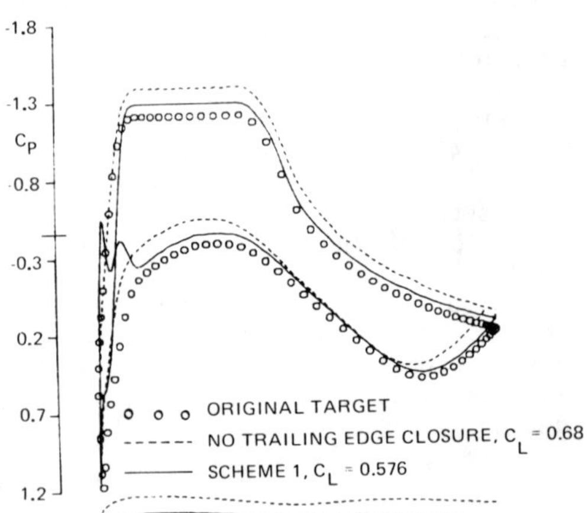

Fig. 2 Airfoil solutions with and without trailing edge closure; computed pressures and original target; case A1, $M_\infty = 0.800$

1606-004D

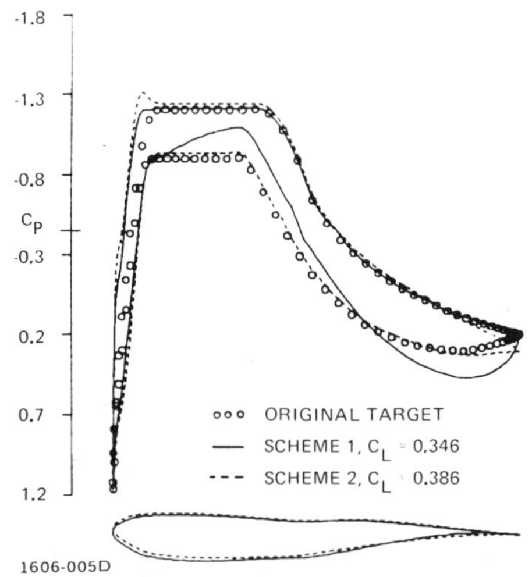

Fig. 3 Effect of closure scheme on airfoil solutions; computed pressures and original target; case A2, M_∞ = 0.800

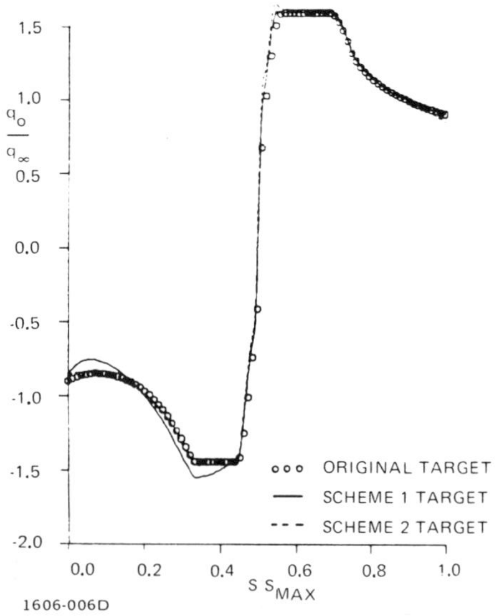

Fig. 4 Original and modified speed distributions; case A2, M_∞ = 0.800

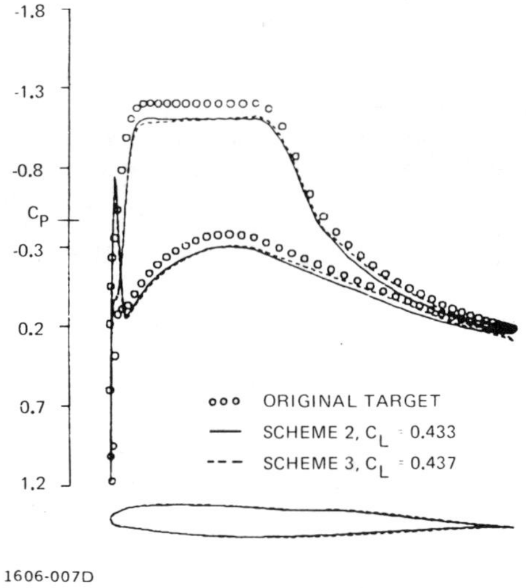

Fig. 5 Effect of closure scheme on airfoil solutions; computed pressures and original target; case A3, M_∞ = 0.800

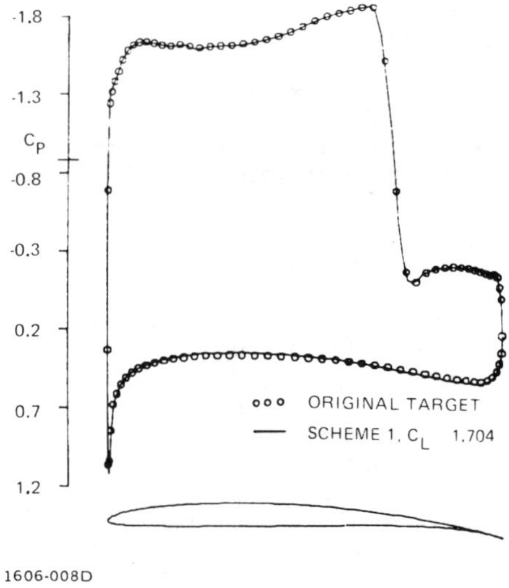

Fig. 6 Designed contour, computed pressures and original target; case E1, M_∞ = 0.683

A COMPARISON OF POTENTIAL- AND EULER-METHODS FOR THE CALCULATION OF 3-D SUPERSONIC FLOWS PAST WINGS

by
C. Weiland
Messerschmitt-Bölkow Blohm GmbH
Ottobrunn, FRG

Summary

This work deals with the calculation of three-dimensional flow fields past simple delta-wing configurations by integrating the Euler equations, and the full Potential equation. The freestream Mach number is always larger than unity. While the Euler equations are solved by using a semi-implicit space-marching procedure including an algorithm for bow shock fitting, the full potential equation is integrated applying a fully implicit, factored finite difference space-marching procedure, where all the arising shocks are captured. The "Butler"-wing at a freestream Mach number 2.5 and two angles of attack acts as test case for both methods.

1. INTRODUCTION

A task of modern aerodynamics is to give the magnitude of the flow variables at every point of a flow field in particular on body surfaces. From the theoretical point of view this problem may be treated with the numerical integration of a suitable set of governing equations. For the supersonic inviscid flow fields considered here, the problem is fully described by solutions of the Euler equations. Numerical approximations of the conservatively formulated Euler equations are able to deal with flow fields where the entropy function increases along streamlines due to shocks and where, owing to the curvature of shocks, a gradient of the entropy function normal to the streamlines occurs. The latter means that vorticity is produced and then transported by convection.

For three-dimensional flows the Euler equations are a set of five partial differential equations. The numerical solution of such a system may be very time consuming, if solutions are sought for a large parameter field consisting of several Mach numbers, angles of attack, and body contours.

Therefore the question is posed: To what extent can a, presumably cheaper, solution of the full potential equation replace the solution of the Euler equations for inviscid supersonic flows with shocks? The full potential equation, conservatively formulated, together with the isentropic relation for the density is capable to calculate strong discontinuities where, however, only the massflux is conserved.

The entropy-function of course is constant due to the assumption of irrotationality necessary for the introduction of a potential. The potential equation is only one scalar partial differential equation and therefore it can be expected that the computer time necessary is only a fourth of that for the Euler equations.

The flow fields treated here allow the application of space-marching methods for both the integration of the potential equation and the Euler equations. The procedure for the integration of the potential equation is explained here in some more detail, while the Euler method is described in earlier publications of the author [1-3].

2. THE EULER METHOD

The Euler equations are quasi-conservatively formulated using cylindrical coordinates as frame of reference. The bow shock is treated by a fitting method, which means that it coincides with a coordinate surface where the Hugoniot-relations are satisfied. The integration of the Euler equations is carried out with a semi-implicit space-marching procedure [1,4]. A set of analytical functions is employed for the generation of the computational grid [2,3], where the outer boundary(bow shock) is unknown and part of the solution.

3. THE POTENTIAL METHOD

3.1 The Potential Equation

The conservative formulated Potential equation in cylindrical coordinates (z,r,φ) reads

$$\frac{\partial \rho u}{\partial z} + \frac{\partial \rho v}{\partial r} + \frac{1}{r}\frac{\partial \rho w}{\partial \varphi} + \frac{\rho v}{r} = 0 \tag{1}$$

$$u = \phi_z \; ; \quad v = \phi_r \; ; \quad w = \frac{1}{r}\phi_\varphi$$

where the density is given by the isentropic relation

$$\rho = \{1 - \frac{\kappa-1}{2} M_\infty^2 \left[V^2 - 1 \right] \}^{\frac{1}{\kappa-1}} \quad . \tag{2}$$

The speed of sound and the static pressure are defined by

$$a^2 = \rho^{\kappa-1}/M_\infty^2 \quad ; \qquad p = \rho^\kappa/\kappa M_\infty^2 \quad .$$

The physical space is mapped into the computational space $\alpha_1,\alpha_2,\alpha_3$.

One obtains

$$\frac{\partial}{\partial \alpha_1}\left(\frac{\rho U^1}{J}\right) + \frac{\partial}{\partial \alpha_2}\left(\frac{\rho U^2}{J}\right) + \frac{\partial}{\partial \alpha_3}\left(\frac{\rho U^3}{J}\right) + \frac{1}{J}\frac{\rho v}{r} = 0 \tag{3}$$

where J is the Jacobian of the transformation

$$J = \frac{\partial(\alpha_1,\alpha_2,\alpha_3)}{\partial(z,r,\varphi)}$$

The U^j are the contravariant velocity components with

$$\phi_{\alpha_i} \equiv \frac{\partial \phi}{\partial \alpha^i} \quad ,$$

$$U^j = g^{ji} \phi_{\alpha_i} \quad ,$$

where the g^{ji} are the contravariant metric coefficients.
In this notation the square of absolut value of the velocity vector $\vec{v}$ is given by

$$v^2 = U^1 \phi_{\alpha_1} + U^2 \phi_{\alpha_2} + U^3 \phi_{\alpha_3} \quad . \tag{4}$$

Equation (3) is hyperbolic, if the velocity of the flow is larger than the speed of sound, and a space-marching procedure can be applied, if the component of the velocity vector $\vec{v}$ in the direction of the contravariant velocity component U^1 is larger than the local speed of sound.

3.2 The Numerical Approximation

The potential equation (3) with $\alpha_1=\bar{z}$, $\alpha_2=\xi$, $\alpha_3=\vartheta$ is implicit discretised and solved by applying a factorisation method [5]. Following the ideas given in [6,7,8,9,10] regarding the introduction of the artificial viscosity by a special treatment of the density and by an upwind-differencing of the derivatives of the contravariant velocity components in the cross-flow plane, one obtains a stable numerical procedure for supersonic flow fields.

The density in the first term of equation (3) is expanded with respect to the absolute value of the velocity vector $|\vec{v}|$ *) in terms of a known state denoted by the subcript o.

$$\rho = \rho_o + \left(\frac{\partial \rho}{\partial |\vec{v}|}\right)_o \Delta|\vec{v}| + \frac{1}{2!}\left(\frac{\partial^2 \rho}{\partial |\vec{v}|}\right)_o (\Delta|\vec{v}|)^2 + \ldots \tag{5}$$

Taking into account only the first order term one obtains

$$\frac{\partial}{\partial \bar{z}}\left(\frac{\rho\, U^1}{J}\right) = \frac{\partial}{\partial \bar{z}}\left\{ \rho_o \left(1 - \frac{|\vec{v}|_o\, |\vec{v}| - |\vec{v}|_o^2}{a_o^2}\right) \cdot \frac{1}{J} \cdot \right. \tag{6}$$

$$\left. \cdot \left[\phi_{\bar{z}}\, g^{11} + \phi_{\xi}\, g^{12} + \phi_{\vartheta}\, g^{13} \right] \right\} \; .$$

*) Other authors [6,9] expand the density with respect to the potential ϕ and find the derivative $\partial\rho / \partial\phi$ to be a differential operator. However, this is valid only for Cartesian coordinates [11].

The density of the second and the third term of eq. (3) is treated by iteration and is replaced by a symmetric upwind biased density [6,9,11]. For instance in the ϑ-direction one gets

$$(\bar{\rho}^q)_{j,k+1/2} = (1 - \nu_{j,k+s})\, \rho^q_{j,k+1/2} + \tag{7}$$

$$+ \frac{1}{2} \nu_{j,k+s} \left(3\, \rho^q_{j,k+s} - \rho^q_{j,k-1+3s} \right)$$

with
$$\nu_{j,k+s} = \left(1 - \frac{a_0^2}{|v|_0^2}\right)_{j,k+s}$$

and
$$U^3 > 0 \qquad s = 0 \qquad \xi_j = \Delta\xi(j-1)$$
$$U^3 < 0 \qquad s = 1 \qquad \vartheta_k = \Delta\vartheta(k-1)$$

The superscript q denotes the iteration number. The curvature term is approximated by

$$\frac{1}{J}\frac{\rho v_r}{r} = \frac{1}{J}\frac{\rho}{r}\left\{ \bar{z}_r\, \phi_{\bar{z}} + \xi_r\, \phi_\xi + \vartheta_r\, \phi_\vartheta \right\}$$

$$= \frac{1}{J}\frac{1}{r} \quad \rho^s\left(\bar{z}_r \frac{(\Delta\phi)^q}{\Delta\bar{z}} + \xi_r \left| \frac{\partial(\Delta\phi)^q}{\partial\xi} + \frac{\partial\phi_0}{\partial\xi} \right| \right. \tag{8}$$

$$\left. + \vartheta_r \frac{\partial(\Delta\phi)^q}{\partial\vartheta} + \vartheta_r \frac{\partial\phi_0}{\partial\vartheta} \right) \equiv K$$

$$(\Delta\phi)^q = \phi^q - \phi_0$$

The boundary condition for surface-oriented coordinates at the body is given by

$$U^2 = \phi_{\bar{z}}\, g^{21} + \phi_\xi\, g^{22} + \phi_\vartheta\, g^{23} = 0 \tag{9}$$

From this equation a formula for the potential at a dummy point inside the body is derived in order to preserve the order of the approximation in ξ-direction. The outer boundary is chosen such, that it lies completely in the undisturbed outer flow field. There the potential ϕ is defined in cylindrical coordinates. One obtains

$$\phi = u_\infty \cdot z - v_\infty \cdot r \cos\varphi$$

with $u_\infty = \cos\alpha$, $v_\infty = \sin\alpha$; α is the angle of attack. Thus the bow shock is captured. For further details see Ref. [11].

4. RESULTS

Fig. 1a shows the pressure distribution - calculated with the potential code - on a cone surface (half angle 10°) with an angle of attack $\alpha=20°$ at $M_\infty = 2.0$. This result agrees well with the one given in Ref. [12], based on the integration of the non-conservative full potential equation. Isobars in a crossflow plane (here defined by z=const) are plotted in Fig. 1b.

A simple delta-wing configuration developed by Butler [13] is used in order to compare the solutions of the Euler equations and the Potential equation. In addition the influence of the mesh discretization is studied for both methods. Fig. 2 shows in planes z=const (see Fig. 4) the isobars and for the potential method (at z=0.8) the grid used, which does not resolve sufficiently the leading-edge region, when the wing gets thinner. Calculations with grid II shown in Figs. 3a and 3d at z=0.8 and z=0.9 gives a much better resolution of the flow field near the leading edge. The agreement of both methods is satisfactory. The pressure distribution at the body (Fig. 3g) found with the Potential method shows near the trailing edge (z=0.9) between the plane of symmetry and the leading edge some deviations compared with the one predicted by the Euler method. Isobars at the body surface are plotted in Fig. 4 from the Euler results. The pressure distribution along the line of symmetry at the body surface is plotted in Fig. 5. In the conical part of the wing the static pressure predicted by the potential method is somewhat larger compared with the one found with the Euler method. This is due to the fact that the total pressure remains constant across the bow shock in the potential method. The Figs.6a-d show results for the case $M_\infty = 2.5$ and $\alpha=5°$ in the plane z=0.7. Again we have a satisfactory agreement between both methods.

In both methods the number of grid points in the circumferential direction φ (half space) is 37. in ξ-direction 17 points are used in the Euler code and 31 points in the potential code. The latter was necessary in order to capture the bow shock.

The Euler method uses with the given grid about four times more computation time than the potential method. The good mutual agreement of the results of both methods allows the conclusion that the full potential equation, although inherently limited because of the assumption of irrotationality, can substitute the Euler equations at least for parametrical studies. Thus the computational costs can be reduced without much sacrifice of accuracy.

REFERENCES

[1] Weiland C.: J. of Comp. Phys., 29, No. 2, pp. 173-198 (1978)
[2] Weiland C.: Lecture Notes in Mathem., Vol.953, pp.172-187 (1982)
[3] Weiland C.: AGARD-CP-342, Paper No. 19 (1983)
[4] Babenko K.I., Voskresenskii G.P., Lyubimow A.N., Rusanow V.V.: NASA TT F-380 (1966)
[5] Beam R.M., Warming R.F.: J.of Comp.Phys., 22,pp.87-110 (1976)
[6] Steger J.L., Caradonna F.X.: NASA-TM 81211, Oct. 1980
[7] Jameson A., Caughey D.A.: AIAA-Paper, No. 77-635, 1977
[8] Holst T.: AIAA-Paper, No. 79-1459, 1979
[9] Shankar V., Chakravarthy S.: AIAA-Paper, No. 81-1004, 1981
[10] Shankar V., Szema K., Osher S.: AIAA-Paper, No. 83-1887, 1983
[11] Weiland C.: Iterative Lösung der 3-D nichtlinearen Potentialgleichung der Gasdynamik für Überschallströmung MBB-UFE122-AERO-MT-665 (1983)
[12] Grossman B.: AIAA-Journal, Vol. 17, No. 8, 1979
[13] Butler D.S.: Proc. Roy.Soc. London, Vol. 255, pp.232-252 (1960)

Butler-Wing (Grid I) $M_{\infty} = 2.5$ $\alpha = 0$

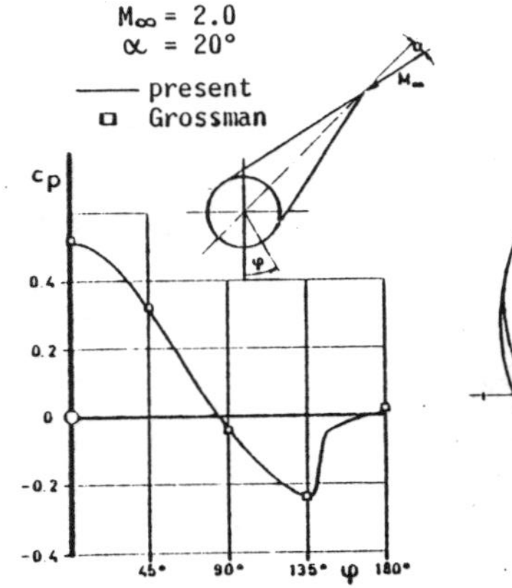

Fig. 1a Pressure distribution along the contour of a cone

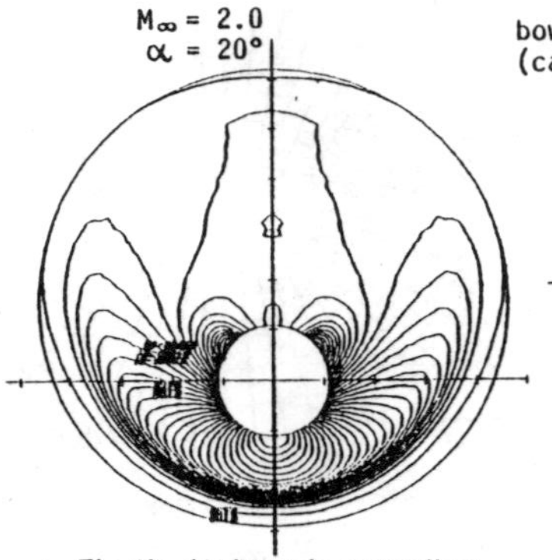

Fig. 1b Isobars in crossflow plane of cone

Potential-solution

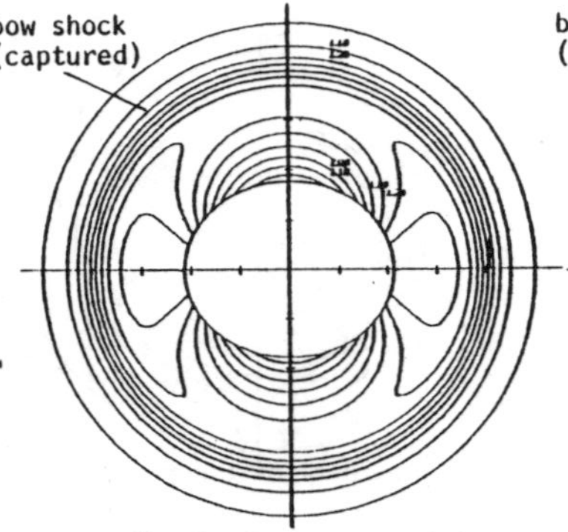

Fig. 2a Isobars z = 0.4

Euler-solution

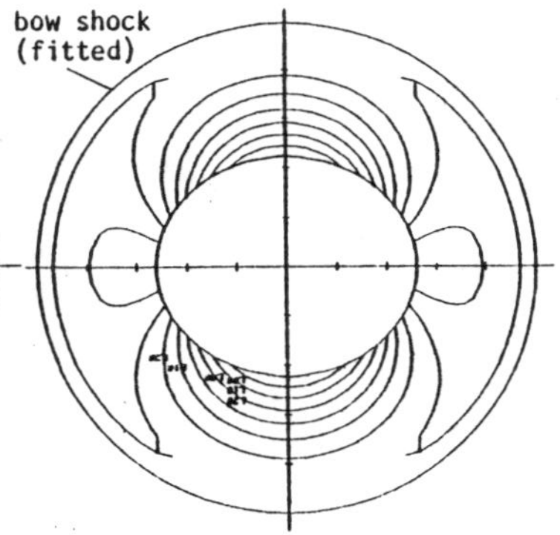

Fig. 2b Isobars z = 0.4

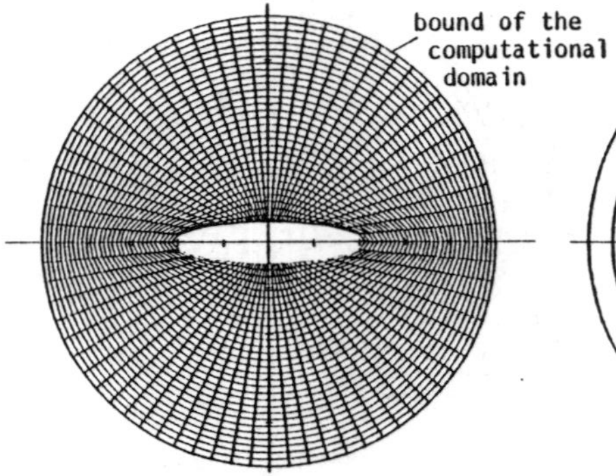

Fig. 2c Computational mesh Potential-solution, z = 0.8

Potential-solution

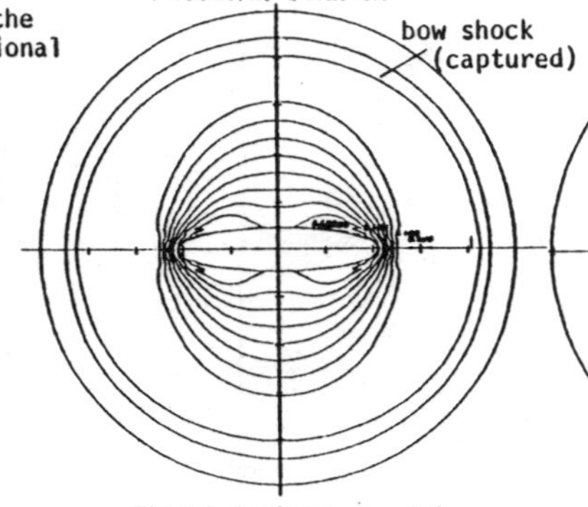

Fig. 2d Isobars z = 0.8

Euler-solution

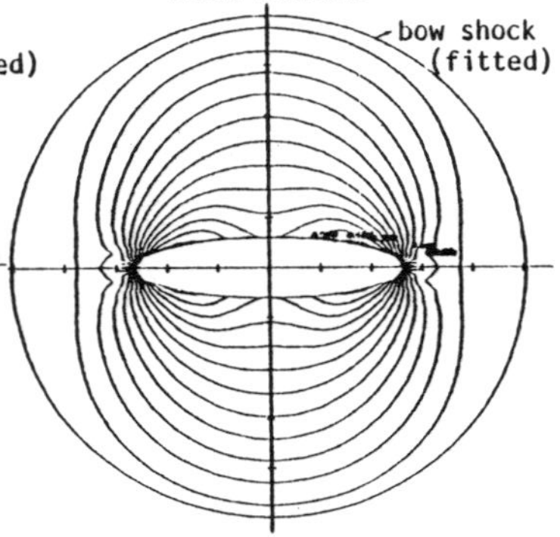

Fig. 2e Isobars z = 0.8

Butler-Wing (Grid II) Moo = 2.5 α = 0

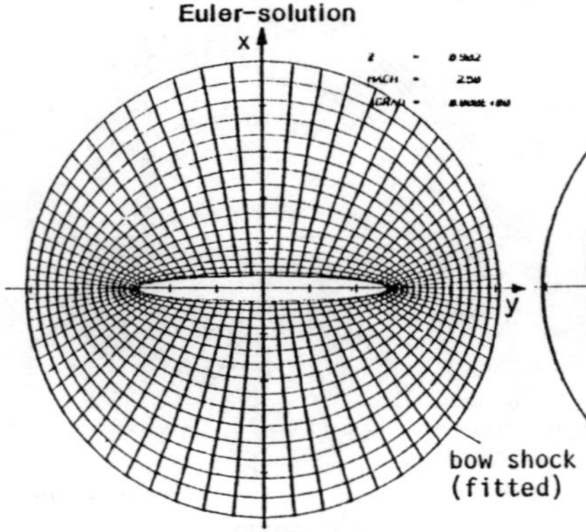

Fig.3a Computational mesh at z = 0.9

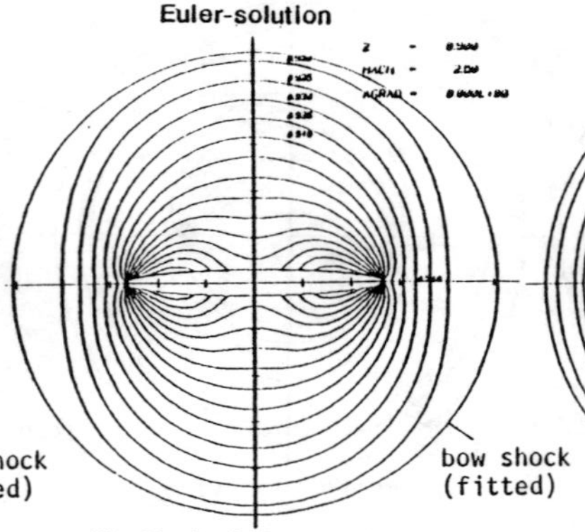

Fig.3b Isobars z = 0.9

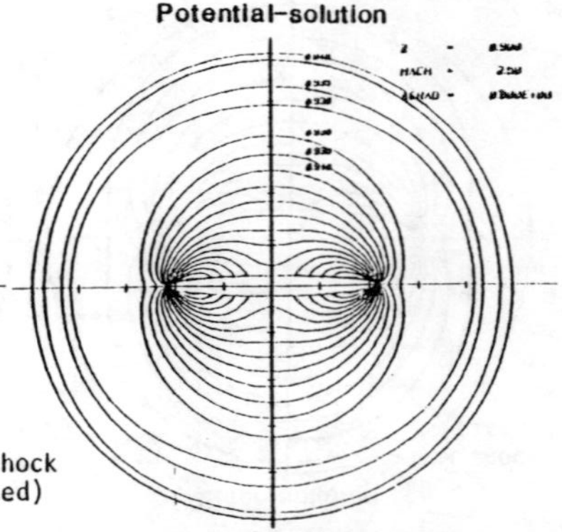

Fig.3c Isobars z = 0.9

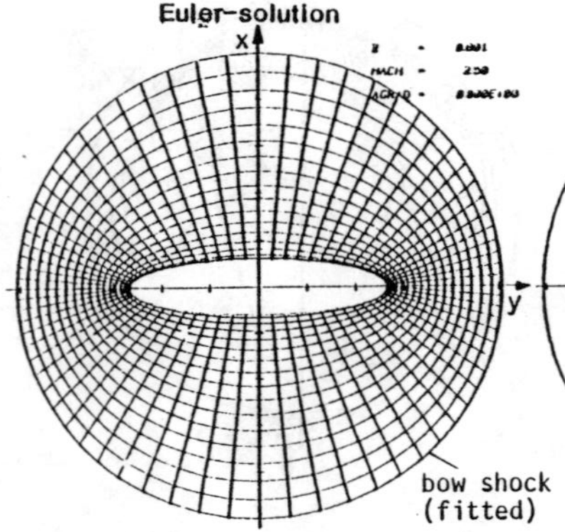

Fig.3d Computational mesh at z = 0.8

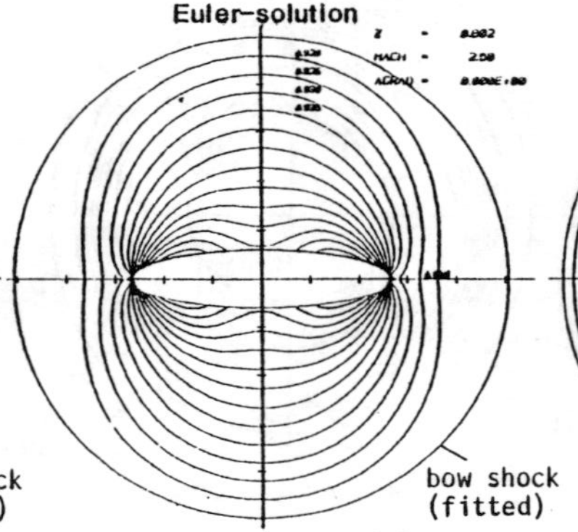

Fig.3e Isobars z = 0.8

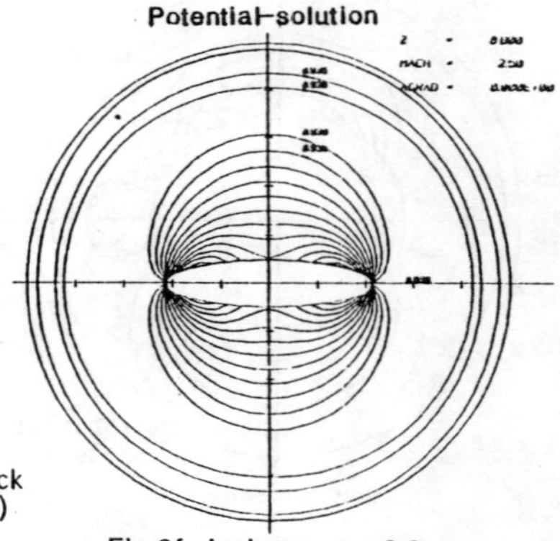

Fig.3f Isobars z = 0.8

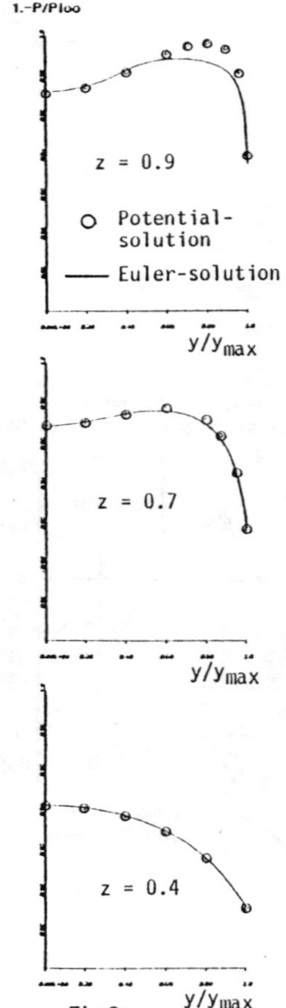

Fig.3g

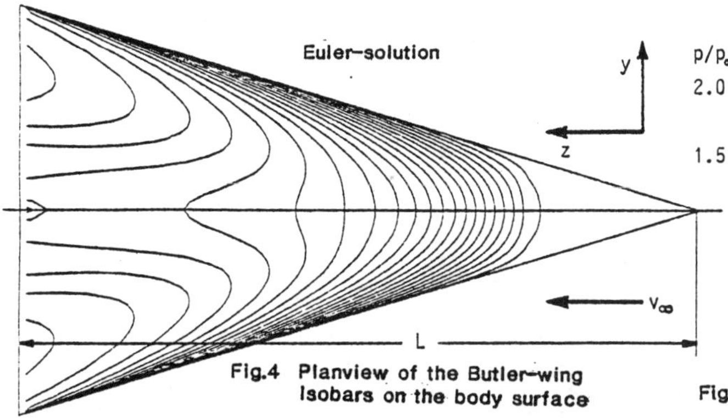

Fig.4 Planview of the Butler-wing
Isobars on the body surface

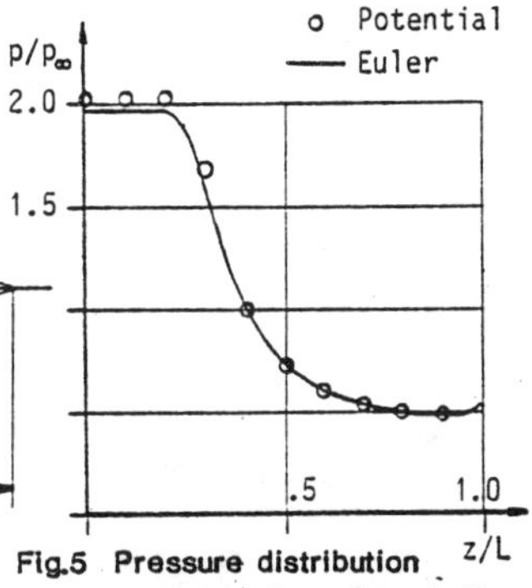

Fig.5 Pressure distribution on the body surface in the plane of symmetry

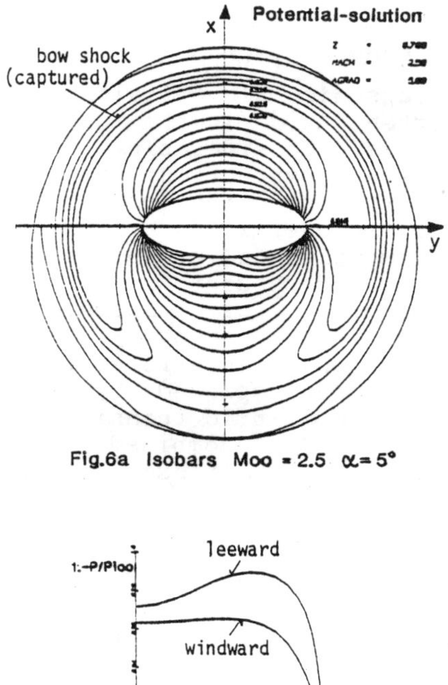

Fig.6a Isobars Moo = 2.5 $\alpha = 5°$

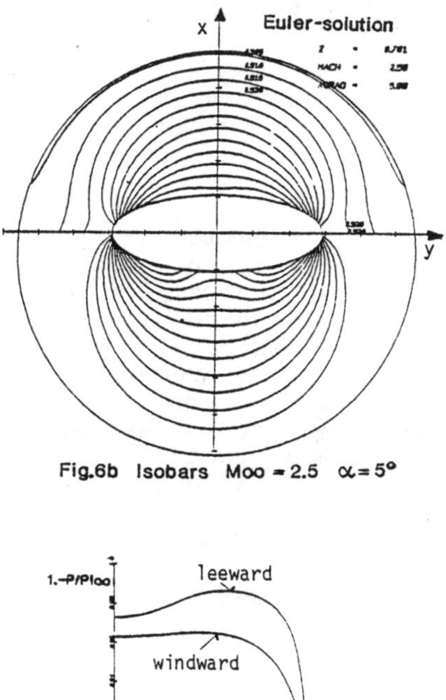

Fig.6b Isobars Moo = 2.5 $\alpha = 5°$

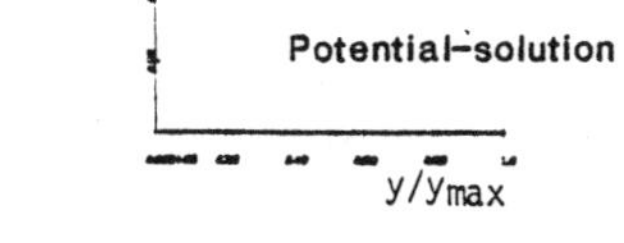

Fig.6c Pressure distribution along the body contour

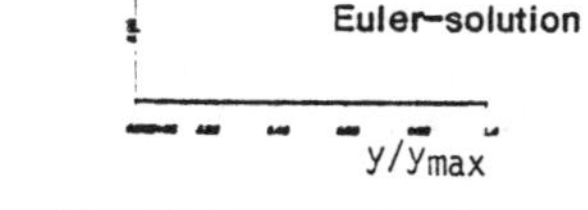

Fig.6d Pressure distribution along the body contour

AN INEXACT NEWTON-LIKE ITERATIVE PROCEDURE FOR THE FULL POTENTIAL EQUATION IN TRANSONIC FLOWS**

Yau Shu Wong

Department of Mathematics and Statistics
McGill University, Montreal, Quebec, CANADA.

SUMMARY

A fast iterative procedure for the solution of full potential equation in transonic flows is presented. The procedure consists of outer and inner iterations. The outer iterate is based on a Newton-like algorithm, and the minimal residual method is used to seek an approximation solution for each inner iterate. Comparison of the present method and the approximate factorization scheme for potential flows around NACA-0012 airfoil at different Mach numbers and different angles of attack are given.

INTRODUCTION

The ability to compute transonic flow fields around airfoils or wings is an important aid in the design of efficient modern transport aircrafts. Considerable effort has been spent, in recent years, on the construction of fast and accurate numerical methods for the solution of the full potential equation. However, to be useful as a design and analysis tool, the success of a computational procedure should neither be problem-dependent nor user-dependent.

The standard iterative procedure for transonic flow equation was based on the successive line over-relaxation (SLOR) method. Because of its slow convergence rates for many practical problems, the method has been replaced by many new iterative procedures: such as the multigrid technique [1], approximate factorization scheme [2], combined SLOR + Conjugate Gradient [3],...,etc. Although these procedures provide substandtial improvement in rates of convergence compared to the SLOR method, they all require one or more iteration parameters in order to accelerate the convergences, and for cases of approximate factorizate (AF) schemes, intermediate variables are also introduced into the iterative process. Consequently, the uncertainty as to what values should be used for the iteration parameters, and how to select the boundary conditions for the intermediate variables, may affect the convergence rates as well as the stability of the iterative process. To some extent a good knowledge of the physical problem and enough computational experience could usually help to resolve these questions. In this paper, we shall present an efficient iterative procedure which not only yields a rapid rate of convergence, but also eliminates some

**This work was supported by the National Aeronautics and Space Administrations under Contracts NAS1-15810 and NAS1-16394 while the author was in residence at ICASE, NASA Langley Research Center, Hampton, Virginia, U.S.A.

of the difficulties mentioned above. The method is based on the Newton-like iterative process, and the idea was first proposed by the author in [4]. Although our early computational results indicated that the method was not competitive with the AF scheme implemented by Holst [5], substantial progress has been made in our method.

For a two-dimensional problem in Cartesian coordinates, the governing partial differential equation for an inviscid isentropic fluid flow expressed in the conservation form is

$$(\rho\phi_x)_x + (\rho\phi_y)_y = 0 , \tag{1}$$

where

$$\rho = [1 - \frac{r-1}{r+1} (\phi_x^2 + \phi_y^2)]^{1/(r-1)}$$

Eq.(1) is known as the full potential equation, where ϕ is the velocity potential, ρ, the density of the fluid flow, and r, the ratio of specific heats. Eq. (1) is a nonlinear equation, since ρ is a function of ϕ_x and ϕ_y. For flow problems in transonic ranges, Eq.(1) changes its type from elliptic in subsonic regions to hyperbolic type in supersonic regions, but the boundary between these regions is unknown. Moreover, the equation also admits discontinuous solution, such as shocks may exist in the flow fields.

For a general flow problem with complex geometries, it is necessary to transform Eq.(1) from a physical domain in the Cartesian coordinates into the computational domain in a rectangle. Eq.(1) written in the computational coordinates ξ and η is given by

$$(\rho U/J)_\xi + (\rho V/J)_\eta = 0 , \tag{2}$$

where

$$\rho = [1 - \frac{r-1}{r+1} (U\phi_\xi + V\phi_\eta)]^{1/(r-1)}$$

Here U and V are the contravariant velocity components along the ξ and η directions, J is the Jacobian of the grid transformation. In order to eliminate the expansion shocks, which are physically meaningless, from the flow fields, an artificial viscosity term with an upwind bias is introduced into Eq.(2). Using the method of artificial density [6] the fluid density is modified in such a way:

$$\rho \leftarrow (\rho - \mu\rho_\xi\Delta\xi) , \tag{3}$$

where

$$\mu = \max [0, 1 - (1/M^2)]$$

Here $s \leftarrow (t)$ indicates that s is replaced by t. In the above expression μ is a switching function which is zero in subsonic flow fields and non-zero in supersonic flow fields, M is the local Mach number, and ρ_ξ is the desity gradient in the upwing direction.

SOLUTION PROCEDURE

Using the standard finite difference approximation, the problem of the full potential equation in transonic flows can be expressed as

$$L(\phi) = 0 \tag{4}$$

where L is the nonlinear operator. Now consider the following iterative procedure:

let ϕ^0 be an initial guess of the potential vector, and define the residual vector, $r^0 = L(\phi^0)$, then for n=0,1,2,..., until $||r^n||_2 < \varepsilon$

Solve
$$M_n(\phi^{n+1} - \phi^n) = -\tau_n r^n \tag{5}$$

where n is a iterarion number, τ is a parameter, M is a linear operator, and the subscript n denotes that M_n and τ_n may vary from iterations to iterations. If $M = L'(\phi)$, the Jacobian of L, then (5) becomes the classical Newton iteration process. Although Newton method yields a fast rate of convergence, the initial guess ϕ^0 must be inside a domain of attraction in order to ensure for convergence. Furthermore, for a large system of equations it is often difficult and expensive to compute L' for each iterate. However, if M is a linear operator and in some sence makes $||\ L(\phi^n) - M\phi^n\ ||$ almost insensitive to ϕ^n, then the iterative process (5) converges [7]. In this paper, instead of the exact Jacobian matrix, M is chosen so that it is an approximation to L. The construction for M can be described as follows:

Consider at the (n+1)th iterate, the fluid density has been calculated from the values of velocity potential at the nth iterate. The result of the application of a central difference approximation to $L(\phi^n)$ then leads to a nine-point formula

$$\begin{aligned}(L\phi)_{i,j} = {} & C(i,j)\phi(i,j) + W(i,j)\phi(i-1,j) + E(i,j)\phi(i+1,j) \\ & + N(i,j)\phi(i,j+1) + S(i,j)\phi(i,j-1) \\ & + NW(i,j)\phi(i-1,j+1) + NE(i,j)\phi(i+1,j+1) \\ & + SW(i,j)\phi(i-1,j-1) + SE(i,j)\phi(i+1,j-1)\end{aligned} \tag{6}$$

For the full potential equation in conservation form the nonzero values at the NW, NE, SW and SE position are due to the skewnees effect of the grid transformation, and they are usually much smaller than those at N, W, C, E and S positions. Now by ignoring the skewness effect M will have five non-zero diagonals only. A simple scaling is used so that the main diagonal elements of M is unity, and this is not done in our early paper[4]. Note that, setting the values at NW, NE, SW and SE to zero will no longer provide a good approximation for M if the equation is in a non-conservation form.

The procedure (5) works well only for purely subsonic flow calculations, and for mixed subsonic - supersonic flow problems it is necessary to introduce an upwind directional bias in the supersonic flow fields. This can be achieved by modifying the operator M so that a $\phi_{\xi t}$-type term is explicitly included. For stability reason a small negative value, α, is added to the main diagonal elements of M. The convergence rate, however, is not sensititive to α in the range from 0.025 to 0.1, and $\alpha = 0.05$ is used for all test problems. The operator M can thus be expressed as

$$M = L + E + \alpha I \pm \mu \overset{\rightarrow}{\overset{\leftarrow}{\partial}}_{\xi} \tag{7}$$

where the matrix E is due to setting the values at NW, NE, SW and SE to zero, and $||E|| << ||L||$ for the equation in conservation form. The procedure (5) can now be rewritten as:

$$(L + E + \alpha I \pm \mu \overset{\leftrightarrow}{\partial}_{\xi}) (\phi^{n+1} - \phi^{n}) = -\tau_n L(\phi^{n}) \tag{8}$$

Since $(\phi^{n+1} - \phi^{n})/\tau_n$ simulates a ϕ_t-type term, let $\tau_n = 1$ for all n and ignoring the error matrix E, it is not hard to observe that Eq.(8) simulates a type-depandent problem:

$$\phi_t \pm \mu\, \phi_{\xi t} + L(\phi^{n+1}) = 0 \tag{9}$$

Now define $d^n = \phi^{n+1} - \phi^n$ be the correction vector, then for each outer iteration we need to solve a linear system of equations

$$Md = -r \tag{10}$$

where M is a large sparse matrix operator. A Minimal Residual (MR) algorithm, which can be applied to symmetric or non-symmetric problems as long as all eigenvalues of M have positive real part, is used for the solution of Eq.(10). The numbers of iteration ,NI, requires to obtain a given accuracy ε is given by [8]

$$NI = 0.5\ K \ln (1/\varepsilon) \tag{11}$$

where K is the condition number of the matrix operator M, and $K(M) = ||M||\ ||M^{-1}||$. Thus the larger the value for K, the slower is the convergence rate that can be expected. In order to accelerate the iterative process, a preconditioning matrix C is introduced, and the preconditioned system is written as

$$MC^{-1}d^* = -r \tag{12}$$

where $d^* = Cd$. C is a non- singular matrix, and it is chosen so that C^{-1} is a good approximation to M^{-1}. Consequently, $K(MC^{-1}) << K(M)$, and thus the application of the MR algorithm to the preconditioned system (12) yields a faster convergence rate than that to the original system(10). The preconditioned MR algorithm is as follows:

Let d^0 be an initial guess, compute the residual vector $p^0 = -r^0 - Md^0$, and solve $Cz^0 = p^0$, then for $k=0,1,\ldots,k^*$, do:

$$d^{k+1} = d^{k} + \alpha_k z^{k}, \qquad p^{k+1} = p^{k} - \alpha_k Mz^{k}, \tag{13}$$

$$Cz^{k+1} = p^{k}, \qquad \text{where } \alpha_k = (p^{k}, Mz^{k}) / (Mz^{k}, Mz^{k}).$$

Since we are interested in the overall convergence for the nonlinear problem, it is not necessary to solve the linear system of equations exactly at each iterate. In our implementation the MR algorithm is used only to seek an approximate solution, and a small number of iterations (typically $k^* = 4$) is sufficient for this purpose. The iterative method described can thus be regarded as an Inexact Newton-like (IN) iterative procedure.

For a good preconditioned algorithm such as that defined in (13), not only the preconditioning matrix C should be chosen so that $K(MC^{-1})$ is much smaller than $K(M)$, but C^{-1}

should also be easily computed, otherwise the iterative procedures will not be efficient. Recall that the main computational work for each inner iterate in the preconditioned algorithm is in solving the linear system Cz = p. To satisfy the two criteria, C is taken to be an approximation M, and C can be factorizted into sparse triangular matrices:

$$C = LU = M + R \tag{14}$$

Here L and U are the sparse lower and upper triangular matrices, in which the three non-zero diagonals are in the same positions to those in the lower and upper triangular part of M, where

$$\begin{aligned}(L\phi)_{i,j} &= v(i,j)\phi(i,j) + t(i,j)\phi(i-1,j) + g(i,j)\phi(i,j-1) \\ &\quad + x(i,j)\phi(i+1,j-1)\,, \\ \text{and } (U\phi)_{i,j} &= \phi(i,j) + e(i,j)\phi(i+1,j) + f(i,j)\phi(i,j+1) \\ &\quad + y(i,j)\phi(i-1,j+1)\end{aligned} \tag{15}$$

In Eq.(14) R is the error matrix which measures how good is the approximation between C and M. The preconditioning matrix C is based on an incomplete factorization technique, L and U are computed from the coefficients of M, so that the row-sums equality condition [9] is satisfied. The algorithm for determining the elements of L and U can be found in [10].

The solution of Cp = z can now be obtained very efficiently: Since C = LU, the linear system can be rewritten as Ls = z, and Up = s, where s is a dummy vector. The solution of Ls = z is obtained through a forward substitution and Up = s by a backward substitution. Note that, unlike the usual AF scheme, no boundary conditions are required.

The preconditioning matrices L and U presented in this paper provide a better approximation to M compared to that being used in reference [4], in which L and U has three non-zero diagonals only (i.e. x(i,j)=0 in L, and y(i,j)= 0 in U for all i,j). Consequently, faster rates of convergence are obtained in our numerical results.

NUMERICAL RESULTS

In this section results of numerical experiments using the IN iterative scheme is presented. The computer program is based on the TAIR code, and they all have been carried out on the CDC CYBER 203 computer of NASA Langley Research Center. The author wishes to thank Dr. Terry Holst of NASA Ames Research Center for supplying the TAIR code. The problems to be considered are transonic potential flow fields around NACA 0012 airfoil, and the mesh system used in all examples is 149 * 30 = 4470 grid points.

The surface pressure coefficient distributions, and the lift coefficients are identical to those reported by Holst's experiments [5], and hence they will not be reported here. In this paper, we shall mainly focus on the performance of the IN scheme and the comparison of convergence rates with the AF scheme used in the original TAIR code. Figures 1-3 compare the rates of convergence of the two methods for the following cases: (1) $M_\infty = 0.85$, $\alpha = 0°$; (2)$M_\infty = 0.8$, $\alpha = 0.5°$; and (3) $M_\infty =$

0.75, $\alpha = 2°$. The first one is for zero angle of attack, i.e., a non-lifting condition, (2) and (3) correspond to lifting airfoil calculations. In Figures 1-3, the solid lines are the results obtained based on the AF scheme, and the dotted lines are those for the IN scheme. In Table 1 we give the total CPU time in seconds required to reduce an average residual (i.e., $||r||_2$) to less than 10^{-7}. The CPU time per iterate is 0.235 seconds for the AF scheme, and is 0.524 seconds for the IN scheme.

	$M_\infty = 0.85$, $\alpha = 0°$	$M_\infty = 0.8$, $\alpha = 0.5°$	$M_\infty = 0.75$, $\alpha = 2°$
AF	32.4	35.5	28.2
IN	41.8	44.5	31.3

TABLE 1. Comparison of CPU time for the AF and IN schemes

It should be noted that a considerable amount of computational work is needed for evaluating the residual vector at each iterate: since it requires to update the fluid density at each grid point, modifying the densities which are in the supersonic regions, and calculating the velocity potentials,...,etc. In fact, evaluating these residuals takes more computaing time than in solving the discrete potential equation using the AF scheme for each iterate. Consequently, although the work per iterate for the IN scheme takes twice the CPU time than that for the AF scheme, scince the total numbers of iteration is reduced, the overall computing time needed to attain the same accuracy for the IN scheme is not significantly larger than that based on the AF scheme. Our computational results presented in Figures 1-3, and in Table 1 indicate that the IN iterative method could be competitive to AF scheme. The IN algorithm produce generally smoother reductions in the residual compared with the AF algorithm (especially for the lifting airfoil calculations). Further improvement in computing time for the IN scheme may be possible: such as using a faster iterative algorithm to replace the MR algorithm, and including the skewness effect in the M operator,...,etc. These possibilities are under inve stigation, and experiments for the full potential equation in a non- conservation form are also being studied.

CONCLUSION

An Inexact Newton-like iterative scheme is presented for the solution of the full potential equation in transonic ranges. The method described here yields a comparable convergence rates as the AF scheme, However, it should be mentioned that in assessing these numerical results, one should keep in mind that although the AF scheme yields a rapid convergence rate, its formulation and application require specialized knowledge and experience of an individual user. In the sense that the performance of the AF scheme may be greatly affected by the choice of the acceleration parameters and also the treament of the boundary conditions for the intermediate variable. The IN iterative scheme, on the other hand, is easy to program and does not introduce an intermadiate variable.

Numerical results for transonic airfoil calculations are promising, and there is still room for improvement for the present method. Finally, two potential areas of application are as follows:

(1) Transonic Wing Calculations

It is straightforward to extend the IN method for a 3-D calculation. Moreover, the extra work needed over a 2-D problem is smaller for the present method compared to that for the AF scheme. For a 3-D equation in a conservation form, M will be a seven-point formula instead of five-point. However, the main computational work for each inner iterate is comparable to that required for a 2-D calculation, since a sparse LU factorization can be obtained with no difficulty. The AF scheme will now consist of three-step calculations [11], and it also require two boundary conditions for the two intermediate variables.

(2) Finite Element Method in Fluid Dynamics Problems:

Since the AF scheme is essentially based on the alternating direction splitting methods, they will not be applicable in the finite element formations, since it is no longer possible to partition the matrix operator in terms of the usual directional derivatives. The present method, however, does not suffer from this restriction.

REFERENCES

[1] Jameson, A., A Multigrid Scheme for Transonic Potential Calculations on Arbitrary Grids, Proc. of AIAA 4th Computational Fluid Dynamics Conference, Williamsbury, Virginia, 1979.

[2] Holst, T.L. and Ballhaus, W.F., Conservative Implicit Schemes for the Full Potential Equation Applied to Transonic Flows, NASA TM 78469, 1978.

[3] Wong, Y.S. and Hafez, M., Conjugate Gradient Methods Applied to Transonic Finite Difference and Finite Element Calculations, AIAA Journal, vol.20, pp.1526-1533, 1982.

[4] Wong, Y.S. and Hafez, M., A Minimal Residual Method for Transonic Potential Flows, Lecture Notes in Physics, 170, pp.526-532, Springer-Verlag, 1982.

[5] Dougherty, F.C., Holst, T.L., Gund, K.L. and Thomas. S.D., TAIR - A Transonic Airfoil Analysis Computer Code, NASA TM 81296, 1981.

[6] Hafez, M.M., Sonth, J.C. and Murman, E.M., Artificial Compressibility Methods for Numerical Solution of Transonic Full Potential Equation, AIAA Journal, vol. 17, pp.838-844, 1979.

[7] Axelsson, O., On Global Convergence of Iterative Methods, Lecture Notes in Mathematics, 953, pp.1-19, Springer-Verlag, 1982.

[8] Samarskii, A.A., Theory of Difference Schemes, Nauka, Moscow, 1977.

[9] Wong, Y.S., Conjugate Gradient Type Methods for Unsymmetric Matrix Problems, Inst. of Appl. Math. & Stat., University of British Columbia, Vancouver, Canada, TR 79-36, 1979.

[10] Wong, Y.S., Calculations of Transonic Potential Flows by a Parameter Free Procedure, Proc. of AIAA 6th Computational Fluid Dynamics Conference, Danvers, Massachusetts, 1983.

[11] Holst, T.L., and Thomas, S.D., Numerical Solution of Transonic Wing Flow Fields, AIAA paper #82-105, 1982.

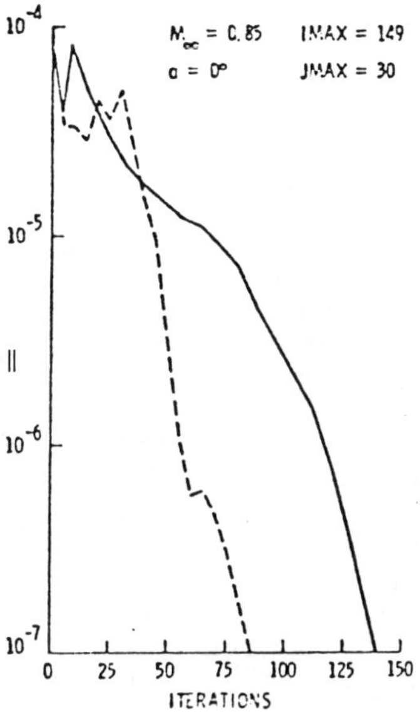

Figure 1.

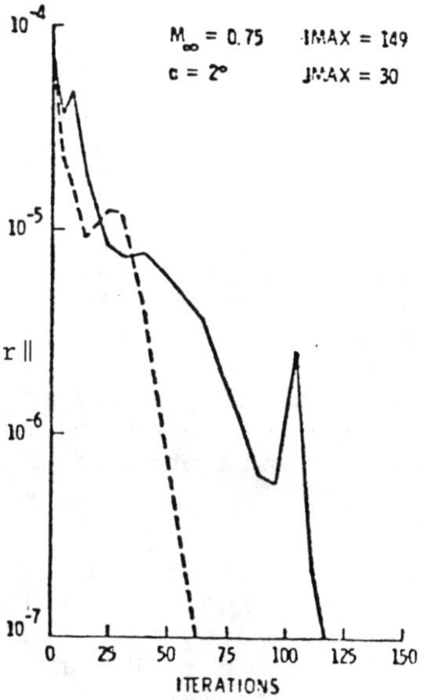

Figure 2.

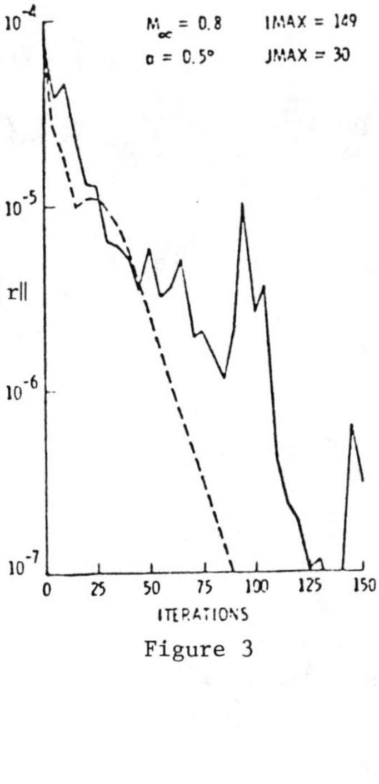

Figure 3

In Figures 1-3:

——————— AF scheme

- - - - - IN Scheme

UNSTEADY PERIODIC MOTION OF A FLEXIBLE THIN PROPULSOR USING THE BOUNDARY ELEMENT METHOD

A. ZERVOS,
National Technical University of Athens,Greece
G. COULMY,
Laboratoire d'Informatique pour la Mécanique et les Sciences de l'Ingénieur,Orsay,France

SUMMARY

A two-dimensional propulsor having infinitely thin and flexible walls is studied. The periodic motion of the propulsor should assure optimal propulsion, when a certain type of body deformation is imposed. The boundary element method is particularly well adapted to resolve the problem, since at each instant it takes into account the exact form of the walls and of the vortex wake.This type of body deformation brings three independant non-dimensional parameters into play. The systematic study of the influence of these parameters on the propulsive efficiency and the thrust allows us to define systems with optimized performances.

INTRODUCTION

The modes of propulsion of aquatic animals have, always, raised a great interest, due to their performances: relatively high hydrodynamic efficiency, low-noise characteristics, small drag etc. By analogy with the propulsion principles of highly efficient aquatic animals, motions were adopted which were characterized by a transverse wave with an amplitude increasing downstream.

The object of this study is the hydrodynamic design of a two-dimensional propulsor having flexible and infinitely thin walls, which are deformed periodically. Various mathematical models were constructed in the past in order to analyse the oscillating motion of a thin flexible propulsor |1,2|. All were based on small amplitude potential theory. In order, though, to achieve practical thrust levels, the oscillating amplitude of the propulsor must be large compared with its chord. The boundary element method, which is applied in this work, solves the problem using large oscillating amplitudes. This is, mainly, due to the fact that, at any moment, it can take into account the exact form of the walls and of the vortex wake.

FORMULATION OF THE PROBLEM

We consider a flexible infinitely thin two-dimensional propulsor S (Fig.1) of length l, which is advancing with constant velocity V_o in an incompressible and inviscid fluid, while at the same time its shape is deformed periodically. We define an absolute OXY and a relative O_1xy coordinate system as shown in Figure 1. The leading edge of the propulsor stays always on the OX axis.

The motion of such a propulsor in an unlimited fluid,being

time-dependent, necessarily generates a wake Σ of free vortices which, as long as the system is not heavily loaded, can be confused with the surfaces generated by the trailing edge of the body |3|.

We assume, also, that the fluid is irrotational in the region outside of the propulsor and its wake. This assumption permit us to introduce the potential function ϕ:

$$\vec{V}_{abs}=\nabla\phi=\vec{V}_{rel}+\vec{V}_{ent} \qquad (1)$$

where $\vec{V}_{abs}$ and $\vec{V}_{rel}$ are the velocities of the flow with respect to the absolute and the relative system, and $\vec{V}_{ent}$ the velocity of the relative system with respect to the absolute system. The fluid being also incompressible, the potential ϕ verifies the Laplace equation:

$$\nabla^2\phi=0 \qquad (2)$$

with the following boundary conditions:

a) The fluid is at rest at infinity: $(\nabla\phi)_\infty=0$

b) There is no flow through the surface S: $\vec{V}_{rel}\cdot\vec{n}=0$.

Upon using the relation (1), the above leads to a Newmann condition:

$$\frac{\partial\phi}{\partial n}=\phi'_n=\vec{V}_{ent}\cdot\vec{n} \qquad (3)$$

c) The Kutta-Joukowsky condition, which requires that the velocity at the trailing edge should be finite.

d) On the wake Σ, the value of the circulation Γ at each point P, corresponds to the difference $\delta\phi(P)$ between the values of the potential ϕ^+ on the upper surface et ϕ^- on the lower surface of the wake, at this point: $\Gamma(P)=\delta\phi(P)=\phi^+(P)-\phi^-(P)$.

Changes in circulation around the body show up as vorticity in the wake in such a manner that the total circulation, in circuit about both the body and the wake, remains constant (Kelvin's theorem).

The problem, thus, consists in determining ϕ from eq.(2) using the above boundary conditions. It involves finding the shed vorticity strength and the body circulation.

METHOD OF RESOLUTION

We know |4| that an harmonic field can be created by a distribution of singularities on the surface of the body and on the vortex wake. The boundary conditions are, then, expressed by integral equations, which are functions of the distributed singularities.

The method consists of substituting the continuous distribution of singularities by a discretised distribution. The body is divided into m elements and the wake into k elements and on each element the distribution of the singularity is considered uniform and equal to its mean value. A control point i is chosen at the center of each surface element. In satisfying the boundary conditions at each control point, we transform the integral equation into a system of linear equations. For this system to be well conditioned, the choice of the nature of the singularities is determinant |5|.

For our problem, in which the body walls are considered infinitely thin, the best choice of singularities consists in a doublet distribution of strength $\delta\phi_j$ on each element j of the surface of the propulsor and a point vortex of strength Γ_F at its trailing edge |6|.

The unsteady phenomenon is also discretised into elementary intervals of time Δt and the field is resolved at each of these intervals. Between instants $t-\Delta t$ and t a vortex filament will be created, leaving the trailing edge and schematised by a point vortex of strength Γ. Each period of motion is, thus, divided into N sequences ($T_o=N\cdot\Delta t$). Between time $t-T_o$ and t, N vortices will be created (fixed on the surfaces generated by the trailing edge of the body).

With this discretisation model, the boundary condition (3), at each control point i, can, then, be written as

$$\phi'_{n_i} = \sum_{j=i}^{m} \delta\phi_j W_{ind_{i,j}} + \Gamma_F V_{ind_{iF}} + \sum_{k=1}^{\infty} \Gamma_k V_{ind_{i,k}} = \vec{V}_{ent_i} \cdot \vec{n}_i \quad (4)$$

where the expressions $W_{ind_{i,j}}$, $V_{ind_{iF}}$ and $V_{ind_{i,k}}$ represent the projections on the normal $\vec{n}_i$ of the induced velocity, at the control point i, created respectively from the jth surface element j, the point vortex of the trailing edge and the kth point vortex of the wake.

The effective strengths of the singularities $\delta\phi_j$, Γ_F and Γ_k are the unknowns of the problem. The number of the unknowns can be diminished by using the other boundary conditions:

The Kutta-Joukowsky condition, which assumes a stagnation point at the trailing edge, requires that

$$\Gamma_F = \delta\phi_m \quad (5)$$

where $\delta\phi_m$ is the doublet strength on the last (closest to the trailing edge) element.

Kelvin's theorem permits us to create a mechanism of vortex shedding, where the strength Γ_1 of the vortex shed between two sequences, n-1 and n, is given by the relation $\Gamma_1=\delta\phi_m^{n-1}-\delta\phi_m^{n}$. The remaining unknowns, then, are the $\delta\phi_j$ and equation (3) represents a system of m equations with m unknowns.

Once the $\delta\phi$'s are determined the force $\vec{dF}$ on an element ds of the propulsor is calculated, using the lifting surface theory.

The mean force $\bar{F}$ applied to the propulsor during a period T_o of motion is, then, given by

$$\bar{F} = \frac{1}{T_o} \int_t^{t+T_o} \int_o^1 \vec{dF} dt \quad (6)$$

We define the propulsive efficiency η and the mean thrust coefficient $\bar{C}_T$ by the relations

$$\eta = \frac{\frac{1}{T_o} \int_t^{t+T_o} \int_o^1 (-V_o\vec{i}) \cdot \vec{dF} dt}{\frac{1}{T_o} \int_t^{t+T_o} \int_o^1 \vec{V}_{rel} \cdot \vec{dF}} \quad , \quad \bar{C}_T = \frac{-\bar{F}_x}{\frac{1}{2}\rho V_o^2 l} \quad (7)$$

where $\bar{F}_x$ is the projection of the mean force $\bar{F}$ on the axis OX.

DEFORMATION LAW

As it was mentioned in the Introduction, the type of body deformation used, consists of a transverse wave, the amplitude of which increases as the trailing edge of the body is approached.

A sinusoidal form with linearly varying amplitude has been chosen in order to modelise this kind of motion. The epxression used, in non-dimensional form, is the following

$$\tilde{y} = \alpha \frac{V_o}{C} \tilde{x} \sin 2\pi \left(\frac{V_o}{C} \tilde{x} - \tilde{t}\right) \qquad (8)$$

where $\tilde{y}=y/H, \tilde{\alpha}=\alpha/H, \tilde{x}=x/H, \tilde{t}=t/T_o=V_o \cdot t/H$

and H is the distance covered in one period T_o, l the length of the body, λ the wavelength of propagation, $C=\lambda/T_o$ the velocity of wave propagation.

The independant non-dimensional parameters are three and can be chosen by any combination of the dimensional parameters. We chose those which have more physical significance: a) V_o/C, the relative celerity, b) l/λ, the number of waves which constitute the body and c) A/H, which characterises the form of the wake (A is the total amplitude of motion).

RESULTS AND DISCUSSION

We have discretised the body in 15 elements and the period of motion in 20 sequences. Figure 2 shows the form of the propulsor and the distribution of pressure difference at each boundary element of the body, on different sequences, for a half-period of motion. We note that the pressures are distributed in such a manner that, at each instant, a propulsive force is created.

The object of the study being, also, the optimization of the propulsion, the influence of each parameter on the thrust coefficient $\bar{C}_T$ and on the propulsive efficiency η is investigated. Two different types of figures are produced. The first (Fig.3) which gives η and $\bar{C}_T$ versus V_o/C when one of the two remaining parameters varies (while the third remains constant) makes more explicit the influence of V_o/C. The second type (Fig.4) which gives η versus $\bar{C}_T$ for different values of V_o/C and of one of the other two parameters, help us to understand better the influence of A/H and l/λ.

We can summarize the information about the parameters, gathered from the above mentioned figures as follows:

V_o/C: In order to have a propulsion ($\bar{C}_T$ positive), V_o/C must be less than 1. The increase of V_o/C produces a decrease in the value of $\bar{C}_T$. Its optimal value depends on the values of the other parameters and more precisely on l/λ; it passes from 0.6 for $l/\lambda=0.4$, to 0.9 for $l/\lambda=2.0$ (Fig.4). On the other hand the influence of A/H on the optimal value of V_o/C is almost negligible.

l/λ: The influence of l/λ on η and $\bar{C}_T$ is very important. Its augmentation ameliorates the efficiency, under the condition that does not get bigger than 2.0, but diminishes, at the same time, the corresponding values of the $\bar{C}_T$.

A/H. As it is shown in Figure 5, the increase of the value of A/H produces a slight decrease of the efficiency, but it is accompanied with an important increase of the $\bar{C}_T$. On the other hand, this augmentation makes the results interesting for a bigger range of values of $\bar{C}_T$ (for example, for A/H=0.5, η stays above 0.7 when $\bar{C}_T$ is varying from 0.13 to 0.36).

From the combination of the above figures we can conclude that different sets of parameters allow us to obtain the same optimal value of efficiency (Figure 5).

It is also interesting to examine the structure of the wake for different sets of parameters. Figure 6 shows the positions of the wake vortices and their respective intensities, during a period of motion.

We note that, for the case where the hydrodynamic efficiency is high (η=0,83), the vortex intensity attains its minimum (negative) value at the highest position of the wake and its maximum (positive) value at the lowest position.(The same is not true in the case where η=0,53). This is something which could expect physically, since in this case we have a more pronounced ejection of the fluid opposite to the direction of the motion |7| and as a consequence a greater propulsive force. We see then that the knowledge of the wake structure can give us precise information on the quality of the motion.

The problem of the thin flexible propulsor was treated by Wu |1|, using the Fourier series method for the acceleration potential. His analysis is valid, though, only when the deformation amplitude of the propulsor stays very small with respect to its length. Under these conditions the amplitude does not constitute a parameter of the problem. We showed, though, that the influence of this parameter is very important. On the other hand Kelly |8| made an experiment with a mechanical model and found results close to the theoretical.

In order to compare these results with the results of our method, we adopted the same conditions used by the other authors. We present this comparison in Figure 7. We note that the two theoretical curves are almost identical for values of V_o/C between 0.65 and 1 but diverge notably for $V_o/C<0.65$. In this region the results of the linearized theory of Wu are less valid since the decrease of V_o/C increases implicitly the relative amplitude A/H of the motion.

The systematic difference between the theoretical curves on one hand and the experimental results on the other, is certainly due to the body friction. The better validity of our results is confirmed by the fact that our curve stays almost parallel to the experimental curve.

CONCLUSIONS

The above analysis shows that a flexible thin propulsor can attain excellent performances when a waving deformation of the body is used. The facility of adaptation of the boundary element method to any form of body deformation or motion permitted us to define optimized performances of the propulsor through a parametric analysis and to show the relative importance of each of the parameters of the problem.

REFERENCES

|1| Wu,T.Y., "Swimming of a waving plate", J.Fluid Mech.,Vol. 10, pp. 321-344, 1961

|2| Lighthill, M.J., "Note on the swimming of a slender fish", J.Fluid Mech., Vol.9, pp.307-317, 1960

|3| Luu,T.S. and Coulmy, G., "Some linear and non-linear problems in aero and hydrodynamics", Developments in Boundary Element Methods, Vol.3, 1983 (to be published)

|4| Hess,J.L. et Smith,A.M.O., "Calculation of potential flow about arbitrary bodies", Progress in Aeronautical Sciences, Vol.8, Pergamon Press, 1967

|5| Coulmy,G., "Calculation of flow field by means of panel methods", Euromech Colloquium 75, Braunschweig-Rhode,1976

|6| Luu, T.S., Coulmy,G.et Corniglion,J.,"Technique des effets élémentaires de singularités dans la résolution des problèmes d'hydro et d'aérodynamique", ATMA, 1969

|7| Zervos,A., "Systèmes propulsifs à paroi souple en mouvement périodique", Thèse de Docteur-Ingénieur,Paris 6,1981

|8| Kelly,H.R., "Fish propulsion hydrodynamics", dans Developments in Mechanics, Layland Malvern ed., North Holl. Publ.Comp., 1, pp.442-450, 1961

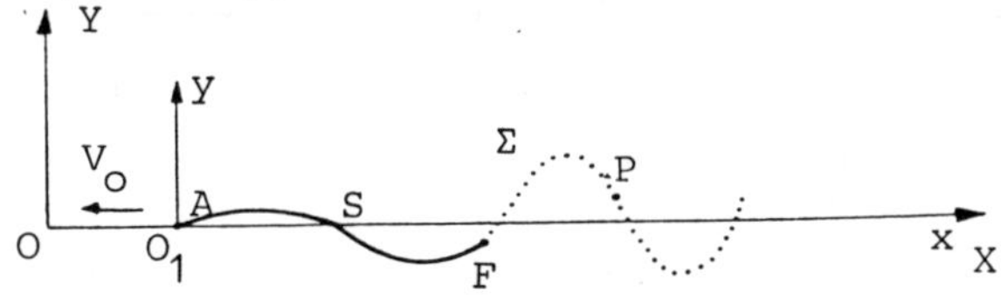

Fig. 1. Sketch of the propulsor.

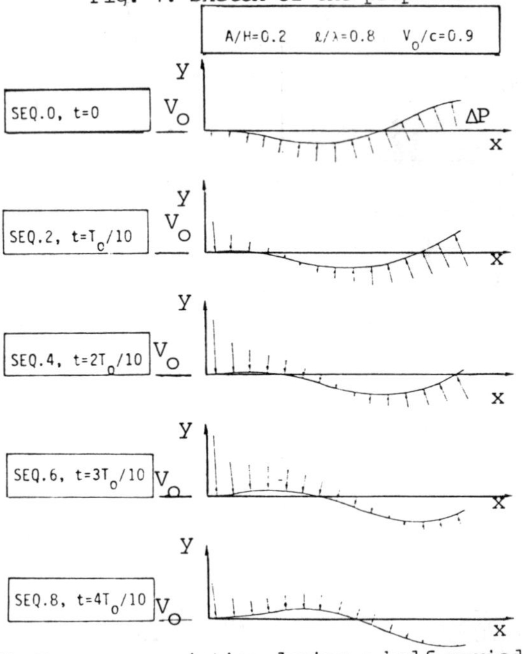

Fig. 2. Pressure variation during a half period of motion.

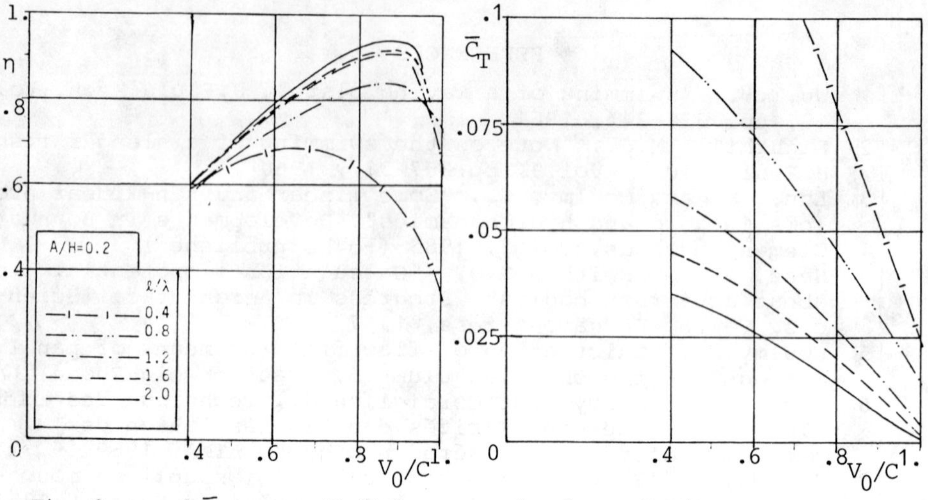

Fig. 3. η and $\overline{C}_T$ versus V_0/C for several values of A/H.

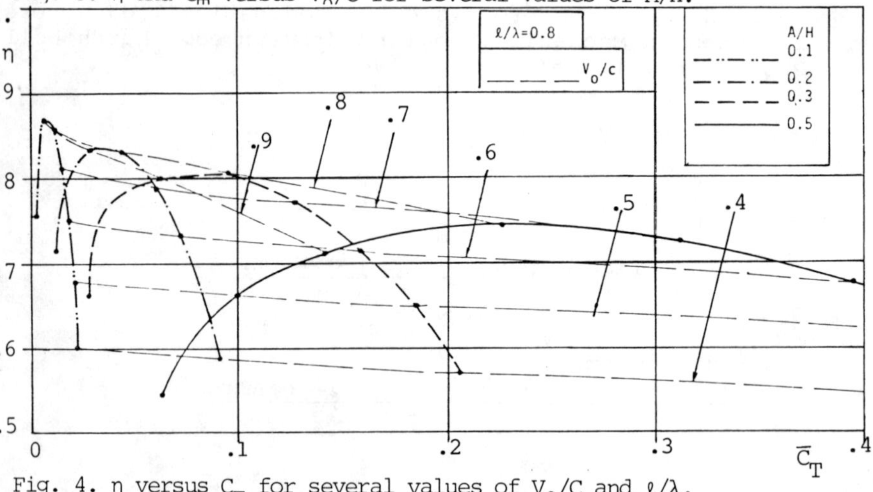

Fig. 4. η versus C_T for several values of V_0/C and ℓ/λ.

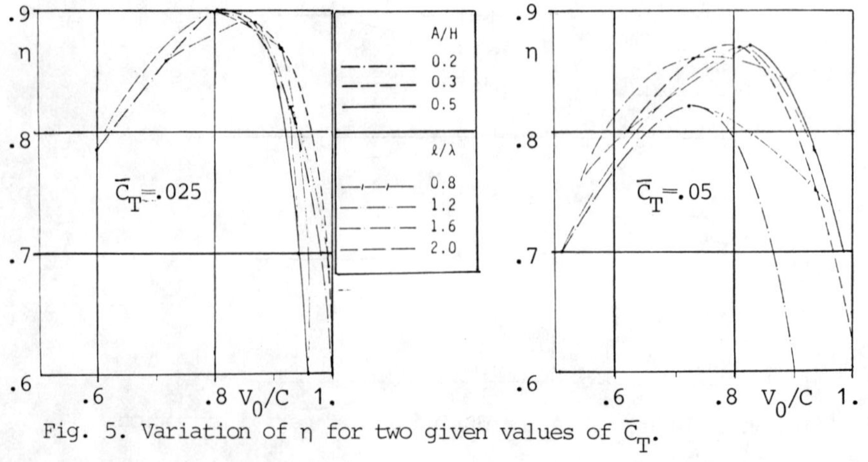

Fig. 5. Variation of η for two given values of $\overline{C}_T$.

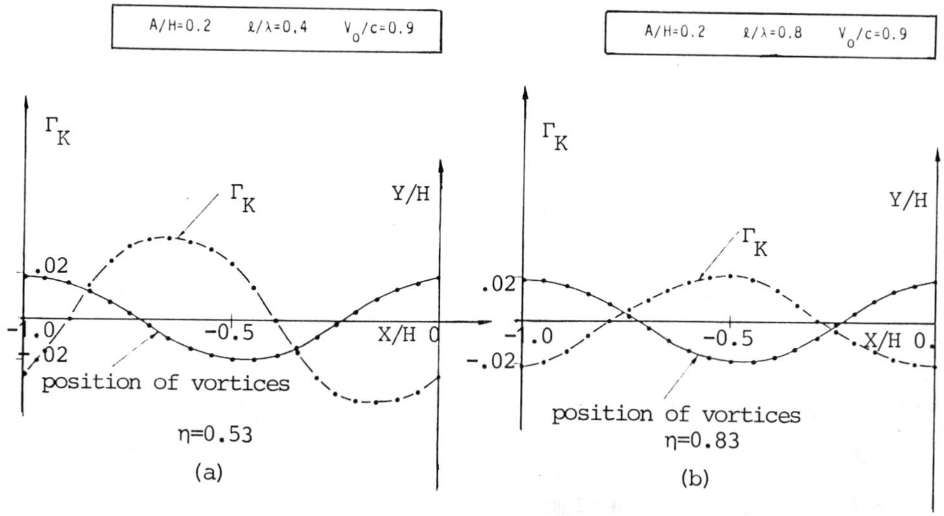

Fig. 6. η as a function of the wake structure.

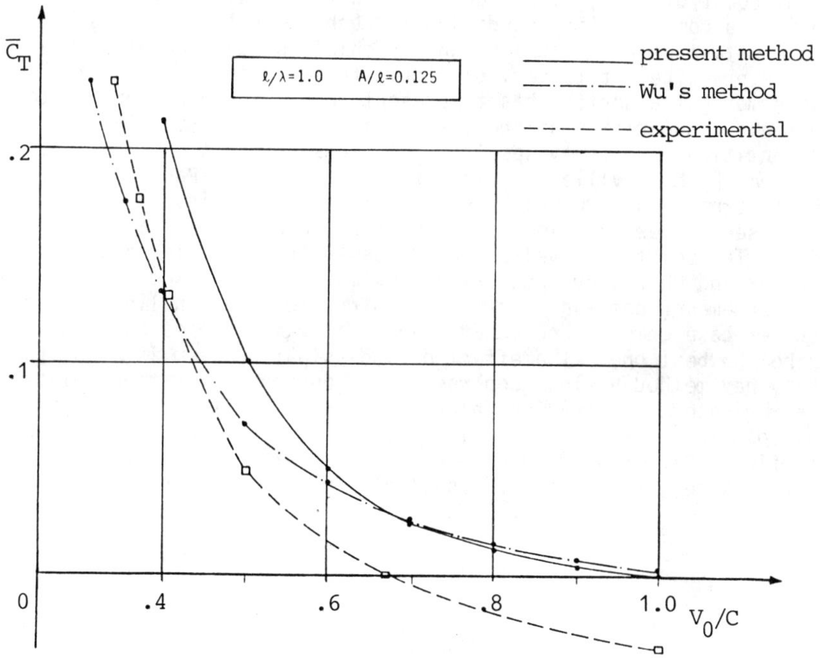

Fig. 7. Comparison of the results.

Short Report on the GAMM Workshop
"SPECTRAL METHODS"

M. Deville
Unité de Mécanique Appliquée - Université Catholique de Louvain
LOUVAIN-LA-NEUVE - Belgique

This workshop was held in Louvain-la-Neuve, Belgium, in October 1980 and about 20 participants coming from France, Germany and Belgium attended the meeting.

Two test problems were proposed. The first one is a Stokes problem for incompressible fluid flows. Several algorithms were examined : splitting methods, influence matrix technique, penalty formulation and pseudospectral space-time approximation. The main conclusions from the comparison of various schemes are that the influence matrix method, the penalty and pseudospectral space-time algorithms yield the best results. Particularly, the full pseudospectral technique reaches machine accuracy. From the mathematical point of view, it was detected that the initial velocity field must satisfy a compatibility condition in such a way that the resulting pressure field satisfies both Neumann and Dirichlet boundary conditions at every time. As, in practice, it is very difficult to build initial compatible fields, the numerical algorithm has to present good damping properties to cope with initial singularities at the very beginning of the time integration. A paper entitled "Chebyshev spectral solutions of the two-dimensional Stokes problem" by M. Deville, L. Kleiser and F. Montigny-Rannou is submitted to the International Journal Numerical Methods in Fluids.

The second test problem is the Burgers equation with a very small viscosity. The solution develops into a sawtooth wave after a short time. Fourier and Chebyshev spectral approximations were used with second-order time schemes (Adams-Bashforth or leapfrog for the non-linear terms). The Fourier case compares collocation and Galerkin schemes and finds the latter to be the best one. The effect of a dissipation term is also studied. The Chebyshev method yields good results if enough collocation points lie inside the region of high gradient near the origin. It turns out that the numerical solution of this problem is more easily computed than the analytical solution. Participants are C. Badesvant (Paris); Lacroix, Ouazzani and Peyret (Nice); Haldenwang (Marseille); Deville (Belgium).

Short Report on the GAMM Workshop "NUMERICAL METHODS IN LAMINAR FLAME PROPAGATION"

N. Peters
Institut für Allgemeine Mechanik, RWTH Aachen

and

J. Warnatz
Institut für Physikalische Chemie, TH-Darmstadt

West-Germany

The workshop was held at Aachen, West-Germany, during Oct. 12-14, 1981. The aim of the workshop was to

1. Establish the difficulties that result from non-equal diffusities of heat and matter and the consequences that these have upon the accuracy of the solution.
2. Compare different numerical schemes for two test problems:
   A) the unsteady propagating flame with one-step chemistry and Lewis number different from unity
   B) the steady, stoichiometric hydrogen-air flame with complex chemistry.
3. Exchange about work in progress to be presented by the participants.

The results of the numerical calculations of test problem A are challenging just as much for scientists employing numerical methods as for those devoted to large activation-energy asymptotics: Satisfactory agreement between the five different groups were obtained only for two out of six cases, those with Lewis number Le equal to one. The very strong oscillations that occur at Le=2 and and a nondimensional activation energy of 20 were accurately resolved only by one group. This case is particular interesting because the asymptotic theory so far predicts instability but not oscillations.

On the other hand, test problem B confirmed the expectation that the existing codes for one-dimensional flat flames with complex chemistry yield satisfactory results if a non-uniform grid is used with enough (about 25) grid points within the flame front. With the kinetics prescribed, the fairly large deviations in calculated flame speeds are due to the different transport models used. In spite of the computational effort involved it seems to be neccessary to use rather complex transport models.

The workshop was sponsored by the Deutsche Forschungsgemeinschaft, to which the organiser and the participants owe a special debt.

The results were reported in the book: Peters, N. and Warnatz, J.: Numerical Methods in Laminar Flame Propagation, Notes on Numerical Fluid Mechanics, Vol.6, Braunschweig, Vieweg-Verlag, 1982.

Short Report on the GAMM Workshop "FLOW OVER BACKWARDS FACING STEP"

J. Periaux (AMD-BA)
O. Pironneau (University of Paris VI)
F. Thomasset (INRIA)

INTRODUCTION

In 1981 it appeared to the organizers that a workshop to compare codes to solve the Navier Stokes equations would be of great interests to the community of numerical analists and users working on CFD (Computer Fluid Dynamics).

The backward step problem was selected for its simplicity although from the point of view of smoothness of boundary it is not so good ; on the other hand all codes would produce vorticity at the same detachment point. The challenge is on the position of the reattachment point.

1. THE PROBLEM FOR ANALYSIS

The geometries of the test cases are described of figure 1.

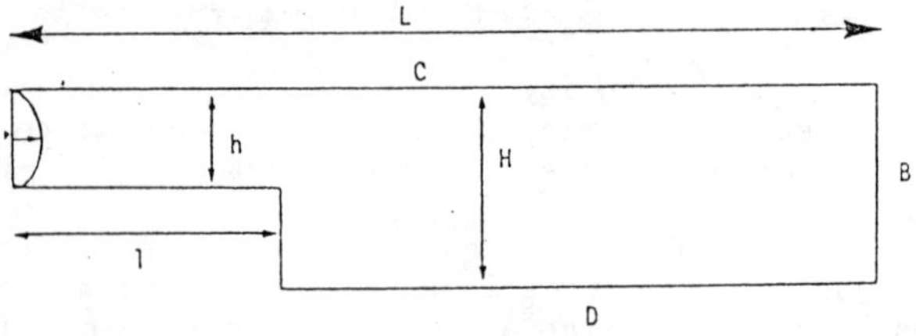

Figure 1

Four geometrical parameters define the domain of the flow : two of them are fixed :

(1) L = 22 and h = 3
(2) H = 1.5 and h = 1
(3) H = 1 and h = .5

Such that (1)-(2) and (1)-(3) generate two domains of computation.

The given boundary conditions are the standard adherence of the flow on the sides C and D of the channel while at the entrance (A) a fully developped laminar flow profile is prescribed and the exit (B) is left to the contributor choice.
Four laminar test cases had to be computed

(i) H = 1.5 , h = 1 , Re = 50
(ii) H = 1 , h = .5, Re = 50
(iii) H = 1.5 , h = 1 , Re = 150
(iv) H = 1 , h = .5, Re = 150

where the Reynolds number is defined by

$$Re = U_{max} (H-h)/ \gamma$$

U_{max} : the maximum value of the velocity profile at (A)
γ : the standard viscosity

The required numerical outputs of the steady solutions of (i)-(iv) consisted of

1) streamlines plots on whole domain
2) streamlines plots in recirculation zone
3) pressure level plots on whole domain

4) pressure level plots in recirculation zone
5) velocity profiles
6) wall shear stress values
7) length of recirculation zone

All the lengths in (1)-(7) had to be nondimensionalized with respect to (H-h)

Details of the computation such as the computer used, the storage of the code, the CPU and real time used and the number of unknowns were enclosed in the requirements.

2. RESULTS

Participation to the workshop was based on invitation to peoples who were known to have softwares for solving the Navier Stokes equations.

Surprisingly most of the results appeared quite good and similar : thus we must make some comments to help the reader on what to look for in the comparisons.

The computer results can be matched with the measurements of KUENY and BINDER.

To compare the plottings of streamlines one should watch for the three critical features :

- Position of the reattachment point
- Position of the center of the vortex
- Concavity or convexity of the streamline at the detachment point.

In the plottings or tables of velocity profiles the interesting features are

- Maximum and minimum values
- Slopes at the boundary
- General smoothness of the curves.

It is in the plotting of wall shear-stresses that the greatest variety appeared.

Comparison of pressure plots is the most important for pratical purposes but, unfortunatly, it was not possible to obtain measurements for the pressures.

3. CONCLUSION

At first sight the results appeared quite similar especially on the velocity profiles and streamline plots. However in the pressure contours and in the computed shear stresses significant differences appear. Perhaps it is due to the fact that these data involve first (or second) derivatives of the computed quantities, velocities or stream function.

At higher Reynolds number (500 or more) the results vary even more perhaps because the codes run beyond their domain of validity and also perhaps due to the presence of multiple steady solutions ; there the choice of algorithms for solving the nonlinearities may also play a role (review of methods by O. PIRONNEAU).

It appeared quite important to participants to evaluate the assets and drawbacks of each formulation in these simple cases of incompressible viscous flows at moderate Reynolds number because of the extensions to turbulent flows and to compressible flows for which much less is known at present. A similar workshop for compressible flows will be held in SOPHIA-ANTIPOLIS-FRANCE in December 1985.

The results of the workshop will appear in the Vieweg Series "Notes on Numerical Fluid Mechanics".

Short Report on the GAMM Workshop "LECTURES ON NUMERICAL METHODS IN FLUID MECHANICS"

B. Gampert
Fachgebiet Strömungslehre, Universität Essen-GHS - Essen 1, FRG

The workshop "Lectures on Numerical Methods in Fluid Dynamics" took place at the University of Essen from February 25th to 26th, 1982. As a first step it was intended to discuss the situation at German universities thus the workshop was confined to German participants. The following papers were presented.
K.G. Roesner (TH Darmstadt): "Classification and properties of partial differential equations".
B. Gampert (University of Essen): "Finite difference discretization methods for partial differential equations and methods for the solution of systems of algebraic equations".
W. Schönauer (University of Karlsruhe): "The numerical integration of parabolic differential equations".
K. Förster (University of Stuttgart): "The numerical integration of hyperbolic differential equations".
U. Schumann (Kernforschungszentrum Karlsruhe): "The numerical integration of elliptic differential equations and advanced methods for the solution of systems of algebraic equations".
K. Förster (University of Stuttgart): "The numerical integration of ordinary differential equations".

Lectures in the traditional fields of fluid mechanics given at various German universities generally contain approximately the same material. In the young discipline of numerical fluid mechanics in contrast the topics presented up to now often strongly depend on the special research experience of the lecturer while other subjects of equal importance are neglected. This was one of the motivations for holding a workshop not dealing with research problems but with lecturing.

It was intended to obtain a survey of the material presented altogether at various German universities. Thus the workshop should help to put lectures in numerical fluid mechanics on a broader basis by teaching the teachers those subjects wherein they themself were not experts. Another task was to discus and to exchange problems including computer programs appropriate for introducing the students into the practical application of numerical methods.

Although motivation and encouragement of the students and vividness in presenting the material were accepted principles of teaching, practical conclusions were difficult to harmonize as they very much depend on the different situations at the various universities.

Important subjects in connection with the topic of our workshop are the school education in mathematics and the contents of university courses on mathematics for engineering students.

The participants intend to publish the papers of this workshop in a monograph.